Lecture Notes in Electri

Volume 640

Series Editors

Leopoldo Angrisani, Department of Electrical and Information Technologies Engineering, University of Napoli Federico II, Naples, Italy

Marco Arteaga, Departament de Control y Robótica, Universidad Nacional Autónoma de México, Coyoacán, Mexico

Bijaya Ketan Panigrahi, Electrical Engineering, Indian Institute of Technology Delhi, New Delhi, Delhi, India

Samarjit Chakraborty, Fakultät für Elektrotechnik und Informationstechnik, TU München, Munich, Germany

Jiming Chen, Zhejiang University, Hangzhou, Zhejiang, China

Shanben Chen, Materials Science and Engineering, Shanghai Jiao Tong University, Shanghai, China

Tan Kay Chen, Department of Electrical and Computer Engineering, National University of Singapore, Singapore, Singapore

Rüdiger Dillmann, Humanoids and Intelligent Systems Laboratory, Karlsruhe Institute for Technology, Karlsruhe, Germany

Haibin Duan, Beijing University of Aeronautics and Astronautics, Beijing, China

Gianluigi Ferrari, Università di Parma, Parma, Italy

Manuel Ferre, Centre for Automation and Robotics CAR (UPM-CSIC), Universidad Politécnica de Madrid, Madrid, Spain

Sandra Hirche, Department of Electrical Engineering and Information Science, Technische Universität München, Munich, Germany

Faryar Jabbari, Department of Mechanical and Aerospace Engineering, University of California, Irvine, CA, USA

Limin Jia, State Key Laboratory of Rail Traffic Control and Safety, Beijing Jiaotong University, Beijing, China

Janusz Kacprzyk, Systems Research Institute, Polish Academy of Sciences, Warsaw, Poland

Alaa Khamis, German University in Egypt El Tagamoa El Khames, New Cairo City, Egypt

Torsten Kroeger, Stanford University, Stanford, CA, USA

Qilian Liang, Department of Electrical Engineering, University of Texas at Arlington, Arlington, TX, USA

Ferran Martín, Departament d'Enginyeria Electrònica, Universitat Autònoma de Barcelona, Bellaterra, Barcelona, Spain

Tan Cher Ming, College of Engineering, Nanyang Technological University, Singapore, Singapore

Wolfgang Minker, Institute of Information Technology, University of Ulm, Ulm, Germany

Pradeep Misra, Department of Electrical Engineering, Wright State University, Dayton, OH, USA

Sebastian Möller, Quality and Usability Laboratory, TU Berlin, Berlin, Germany

Subhas Mukhopadhyay, School of Engineering & Advanced Technology, Massey University, Palmerston North, Manawatu-Wanganui, New Zealand

Cun-Zheng Ning, Electrical Engineering, Arizona State University, Tempe, AZ, USA

Toyoaki Nishida, Graduate School of Informatics, Kyoto University, Kyoto, Japan

Federica Pascucci, Dipartimento di Ingegneria, Università degli Studi "Roma Tre", Rome, Italy

Yong Qin, State Key Laboratory of Rail Traffic Control and Safety, Beijing Jiaotong University, Beijing, China

Gan Woon Seng, School of Electrical & Electronic Engineering, Nanyang Technological University, Singapore, Singapore

Joachim Speidel, Institute of Telecommunications, Universität Stuttgart, Stuttgart, Germany

Germano Veiga, Campus da FEUP, INESC Porto, Porto, Portugal

Haitao Wu, Academy of Opto-electronics, Chinese Academy of Sciences, Beijing, China

Junjie James Zhang, Charlotte, NC, USA

The book series *Lecture Notes in Electrical Engineering* (LNEE) publishes the latest developments in Electrical Engineering—quickly, informally and in high quality. While original research reported in proceedings and monographs has traditionally formed the core of LNEE, we also encourage authors to submit books devoted to supporting student education and professional training in the various fields and applications areas of electrical engineering. The series cover classical and emerging topics concerning:

- Communication Engineering, Information Theory and Networks
- Electronics Engineering and Microelectronics
- Signal, Image and Speech Processing
- Wireless and Mobile Communication
- Circuits and Systems
- Energy Systems, Power Electronics and Electrical Machines
- Electro-optical Engineering
- Instrumentation Engineering
- Avionics Engineering
- Control Systems
- Internet-of-Things and Cybersecurity
- Biomedical Devices, MEMS and NEMS

For general information about this book series, comments or suggestions, please contact leontina.dicecco@springer.com.

To submit a proposal or request further information, please contact the Publishing Editor in your country:

China

Jasmine Dou, Associate Editor (jasmine.dou@springer.com)

India, Japan, Rest of Asia

Swati Meherishi, Executive Editor (Swati.Meherishi@springer.com)

Southeast Asia, Australia, New Zealand

Ramesh Nath Premnath, Editor (ramesh.premnath@springernature.com)

USA, Canada:

Michael Luby, Senior Editor (michael.luby@springer.com)

All other Countries:

Leontina Di Cecco, Senior Editor (leontina.dicecco@springer.com)

** **Indexing: The books of this series are submitted to ISI Proceedings, EI-Compendex, SCOPUS, MetaPress, Web of Science and Springerlink** **

More information about this series at http://www.springer.com/series/7818

Baoming Liu · Limin Jia · Yong Qin ·
Zhigang Liu · Lijun Diao · Min An
Editors

Proceedings of the 4th International Conference on Electrical and Information Technologies for Rail Transportation (EITRT) 2019

Rail Transportation Information Processing and Operational Management Technologies

Editors
Baoming Liu
National Innovation Center
of High Speed Train
Qingdao, China

Yong Qin
State Key Laboratory of Rail
Traffic Control and Safety
Beijing Jiaotong University
Beijing, China

Lijun Diao
Beijing Jiaotong University
Beijing, China

Limin Jia
State Key Laboratory of Rail
Traffic Control and Safety
Beijing Jiaotong University
Beijing, China

Zhigang Liu
Beijing Jiaotong University
Beijing, China

Min An
School of Science, Engineering
and Environment
University of Salford
Salford, UK

ISSN 1876-1100   ISSN 1876-1119 (electronic)
Lecture Notes in Electrical Engineering
ISBN 978-981-15-2916-0   ISBN 978-981-15-2914-6 (eBook)
https://doi.org/10.1007/978-981-15-2914-6

© Springer Nature Singapore Pte Ltd. 2020
This work is subject to copyright. All rights are reserved by the Publisher, whether the whole or part of the material is concerned, specifically the rights of translation, reprinting, reuse of illustrations, recitation, broadcasting, reproduction on microfilms or in any other physical way, and transmission or information storage and retrieval, electronic adaptation, computer software, or by similar or dissimilar methodology now known or hereafter developed.
The use of general descriptive names, registered names, trademarks, service marks, etc. in this publication does not imply, even in the absence of a specific statement, that such names are exempt from the relevant protective laws and regulations and therefore free for general use.
The publisher, the authors and the editors are safe to assume that the advice and information in this book are believed to be true and accurate at the date of publication. Neither the publisher nor the authors or the editors give a warranty, expressed or implied, with respect to the material contained herein or for any errors or omissions that may have been made. The publisher remains neutral with regard to jurisdictional claims in published maps and institutional affiliations.

This Springer imprint is published by the registered company Springer Nature Singapore Pte Ltd.
The registered company address is: 152 Beach Road, #21-01/04 Gateway East, Singapore 189721, Singapore

Contents

Improvement of Key Management Mechanism for RSSP-II
and Its Formal Modeling and Verification 1
Huozhen Lian, Lianchuan Ma and Yuan Cao

Charging Strategy of Temporal-SoC for Plug-in Electric
Vehicles on Coupled Networks 11
Xian Yang and Zhijian Jia

The Forecast of the Subway Passenger Flow Based on Smooth
Relevance Vector Machine 23
Ni Luo, Meng Hui, Lin Bai, Rui Yao and Qisheng Wu

Dynamic Debugging System of EMU Based on Cloud Platform 31
Yu Qiu, Lide Wang, Jie Jian, Chaoyi Li and Yu Bai

Health Assessment Method for Controller Area Network
in Braking Control System of Metro Train 41
Yueyi Yang, Xingwen Guan, Yinhu Liu and Bin Xue

High-Temperature Zone Localization Image Processing Algorithm
for Pantograph Infrared Image 51
Wen Xu, Zhaoyi Su, Guangtao Cong and Zongyi Xing

Design of PCI Bus High-Speed Data Acquisition System Based
on FPGA 61
Xiaoming Zhu, Xiaodong Wang, Jie Chen, Ruichang Qiu and Zhigang Liu

Design of Control Unit of Traction System Based on CPCI Bus 71
Decheng Tang, Cheng Yang, Jie Chen, Ruichang Qiu and Zhigang Liu

Design of Route Search Algorithm Based on Station Map Information
and Depth-First-Search 79
Yahan Yang, Shaobin Li, Kai Sun and Xiaobin Di

Adaptive Global Particle Swarm Algorithm-Based Train Recommend Speed Curve Optimization Study in Urban Rail Transit 87
Zhen Hu, Xiao Xiao and Feng Bao

An Improved SSD and Its Application in Train Bolt Detection 97
Jiabing Zhang, Zhouyi Su and Zongyi Xing

Research on the Wheelset Life Optimization of Urban Rail Transit Trains 105
Xiaoxiao Xu, Zhengjun Ye, Jianyu Zhang, Zongyi Xing and Yingshun Liu

Design of a 3D Traction Substation Based on Real-Time Infrared Simulation System 119
Liping Zhao, Yirui Yu and Shengyang Zheng

Multi-channel Man-in-the-Middle Attack Against Communication-Based Train Control Systems: Attack Implementation and Impact 129
Mengchao Chi, Bing Bu, Hongwei Wang, Yisheng Lv, Shengwei Yi, Xuetao Yang and Jie Li

Optimization on Metro Timetable Considering Train Capacity and Passenger Demand from Intercity Railways 141
Haiyang Guo, Yun Bai, Qianyun Hu, Huangrui Zhuang and Xujie Feng

Hybrid Intrusion Detection with Decision Tree and Critical State Analysis for CBTC 153
Yajie Song, Bing Bu and Xuetao Yang

Research on Pedestrian Traversal Time in T-Shaped Passage 163
Mengdi Liang, Jie Xu, Yiwen Chen and Siyao Li

Research on FN-Based MCVideo Service for Railway Communication System 175
Qingqing Wang, Mingchun Li, Bin Sun, Jianwen Ding and Zhangdui Zhong

End-to-End Communication Delay Analysis of Next-Generation Train Control System Based on Cloud Computing 187
Jiakun Wen, Lianchuan Ma and Yuan Cao

A New Method of High-Speed Railway Station Planning Based on Network Evolution 197
Chengchen Liu, Yanhui Wang, Limin Jia and Shuai Lin

A Rail Train Number Identification Algorithm Based on Image Processing 207
Shuangyan Yang, Ruifeng Xu, Zhihui Zhou, Zongyi Xing and Yong Zhang

Multi-stage Attack and Proactive Defense for CBTC 215
Xiang Li, Bing Bu and Li Zhu

**Design and Development of a General Equilibrium Calculation
System for Rail Vehicles**.. 225
Yuwen Liu, Fengping Yang, Lin Jin, Feng Liu and Jian Wu

**Research on Vehicle Number Localization of Urban Rail Vehicle
Based on Edge-Enhanced MSER** 239
Ruifeng Xu, Jiandong Bao, Shuangyan Yang, Chen Zhang
and Zongyi Xing

**Urban Rail Transit Power Monitoring System Techniques
Based on Synchronous Phasor Measurement Unit** 251
Shize Huang, Lingyu Yang, Ji Xue and Kai Yu

**Railway Capacity Calculation in Emergency Using Modified
Fuzzy Random Optimization Methodology**......................... 269
Li Wang, Min An, Limin Jia and Yong Qin

**Research on FlexRay Bus Communication Protocol Stack of Rail
Vehicle Electronic Control System Based on AUTOSAR Standard** 283
Zhi yuan Wang, Yong Liu, Yong liang Ni and Yan ming Li

Research on Indoor Location Technology in Metro Station 297
Zongyi Xing, Hang Yang and Yuan Liu

**Arrival Train Delays Prediction Based on Gradient Boosting
Regression Tress** 307
Rui Shi, Jing Wang, Xinyue Xu, Mingming Wang and Jianmin Li

The Subway Passenger Flow Macroscopic State Analysis 317
Yanhui Wang, Hao Shi and Suizheng Zhang

**Real-Time Monitoring System for Passenger Flow Information
of Metro Stations Based on Intelligent Video Surveillance** 329
Xiang ling Yan, Zheng yu Xie and Ai li Wang

Application of Target Detection Algorithms in Railway Intrusion 337
Xingwei Jia, Yumeng Sun and Zhengyu Xie

**A Method for Pedestrian Intrusion Detection of Railway Perimeter
Based on HOG and SVM**.. 347
Yumeng Sun and Zhengyu Xie

**Timetabling, Platforming, and Routing Cooperative Adjustment
Method Based on Train Delay** 357
Yinggui Zhang, Zengru Chen, Min An and Aliyu Mani Umar

Research on Service Ability Evaluation of Automatic Fare
Collection System Based on DEMATEL and VIKOR-Gray
Relational Analysis 369
Yuanyuan Zhou, Jiabing Zhang, Zhihui Zhou and Zongyi Xing

Cloud Computing System of Rail Transit Traction System
and Data Collection 379
Yijun Liang, Ruijun Chen, Jie Chen, Ruichang Qiu and Zhigang Liu

Exploration on Innovation Education of Higher Vocational
Automobile Specialty Based on Internet Thinking 389
Lei Yang, Fanling Zeng and Daoru Yao

A Study on the Redundancy of Ethernet Train Backbone
Based on Dual-Link Topology 397
Xiaodong Sun, Jianyu Jin, Jianbo Zhao and Chenyang Bing

System Implementation of a Processing Subcontracting Business
in SAP ERP Non-standard Subcontracting Scenarios 411
Chenyang Bing, Guohui Yang and Jinlong Kang

Research on Passengers' Choice of Subway Travel Intention
Based on Structural Equation Model 427
Yu ping Xu, Siwei Chen and Wei Xu

Adaptive Radio Resource Management Based on Soft Frequency
Reuse (SFR) in Urban Rail Transit Systems 437
Ye Zhu and Hailin Jiang

Research on Passenger Flow Distribution in Urban Rail
Transit Hub Platform 447
Long Gao, Lixin Miao, Zhongping Xu, Liang Chen, Zhichao Guan,
Hualian Zhai, Limin Jia, Yizhou Chen and Jie Zhuang

Simulation Analysis of Urban Rail Transit Hub Based
on Complex Network 457
Long Gao, Lixin Miao, Zhongping Xu, Liang Chen, Zhichao Guan,
Hualian Zhai, Limin Jia, Yizhou Chen and Jie Zhuang

Regret Minimization and Utility Maximization Mixed Route
Choice Model 467
Chenran Sun, Shiwei He and Yubin Li

Urban Rail Transit Passenger Flow Monitoring Method Based
on Call Detail Record Data 479
Lejing Zheng, Liguang Su and Honghui Dong

Research on Dynamic Route Generation Algorithms
for Passenger Evacuation in Metro Station 493
Weiming Dong and Zhengyu Xie

Single Image Dehazing of Railway Images via Multi-scale Residual Networks 503
Zhi wei Cao, Yong Qin and Zheng yu Xie

System Dynamic Model and Algorithm of Railway Station Passengers Distribution 513
Zhe Zhang, Limin Jia and Sai Li

Estimation of Line Capacity for Railway 525
Xu Wei and Jie Xu

Research on the Process Control for the Acceptance of Urban Rail Transit Equipment and Facilities 537
Cheng xin Du, Jun hua Zhao, Ming Zhang, Zhi fei Wang and Chao Zhou

A Model-Based Quantitative Analysis Methodology for Automatic Train Operation System via Conformance Relation 547
Ruijun You, Kaicheng Li, Yu Liu and Zhen Xu

Research on the Framework of New Generation National Traffic Control Network System and Its Application 557
Qian Li, Rusi Chu and Haiyang Wang

A Fuzzy and Bayesian Network CREAM Model for Human Error Probability Quantification of the ATO System 567
Jianqiang Jin, Kaicheng Li and Lei Yuan

Operation Organization Methods for Automated People Mover Systems at Airports 577
Wenqian Liu, Xiaoning Zhu and Li Wang

Real-Time Train Detection Based on Improved Single Shot MultiBox Detector 585
Yuchuan Xu, Chunhai Gao, Lei Yuan and Boyang Li

Application of Intelligent Video Surveillance Technology for Passenger Flow Detection in Urban Rail 595
Shi yao Yu, Zi teng Wang, Yuan Zhao, Xi li Sun, Ai li Wang, Xiang ling Yan and Zheng yu Xie

Planning Research for Automated People Mover System at Airports 605
Shuai Wang, Xiaoning Zhu and Li Wang

Design of Vibration Test Bench for Electric Locomotive Bogie 611
Yunxiao Fu and Boyang Li

Evaluation of Energy Consumption Level of Urban Rail Train Operation Plans: A Case Study 625
Jiao Zhang, Yujie Li and Hao Xie

Study on Dynamic Energy-Saving Adjustment Strategy of Metro Vehicle Air-Conditioning and Ventilation System 637
Yujie Li, Xuyang Wang, Jiao Zhang, Lu Wang and Liang Ma

Study on Energy Efficiency Performance of Rail Timetable Considering Track Condition 643
Jiao Zhang, Yujie Li, Lu Wang and Liang Ma

Coordinated Control Method of Passenger Flow Based on Anylogic 653
Xuyang Wang, Huijuan Zhou, Yu Zhao and Ying Liu

Train Operational Plan Optimization of Rail Transit Connecting to Airport Ground Transport Center 663
Yamin Li, Xiaohan Xu, Rui Xu and Ailing Huang

Defect Detection for Catenary Sling Based on Image Processing and Deep Learning Method 675
Jing Cui, Yunpeng Wu, Yong Qin and Rigen Hou

Rail Fastener Defect Inspection Based on UAV Images: A Comparative Study 685
Ping Chen, Yunpeng Wu, Yong Qin, Huaizhi Yang and Yonghui Huang

Defect Detection for Bird-Preventing and Fasteners on the Catenary Support Device Using Improved Faster R-CNN 695
Jiahao Liu, Yunpeng Wu, Yong Qin, Hong Xu and Zhenglu Zhao

Research on the Flexible Operation and Maintenance Management of Urban Rail Vehicle 705
Xiaoqing Cheng, Ruohong Lan, Zhiwen Liao and Yong Qin

Structural Health Monitoring of Cracks on Bogie Frame Using Lamb Waves 719
Kexin Liang, Guoqiang Cai and Ye Zhang

Research on Emergency Capacity Training System for Rail Transit Dispatchers 727
Qingmei Hu, Nan Lv, Lan Zhang and Yukun Gu

A Study on Detection Technology of Rail Transit Vehicle Wheel Web Based on Lamb Wave 735
Ge Wang, Guoqiang Cai and Xunhe Yin

Improvement of Key Management Mechanism for RSSP-II and Its Formal Modeling and Verification

Huozhen Lian, Lianchuan Ma and Yuan Cao

Abstract As the RSSP-II protocol has hidden dangers in the management of transport keys and authentication keys, in order to strengthen the safety of the key management of RSSP-II and make the communication between the safety-related entities of the train control system safer and more reliable, an improved scheme is presented in this paper. This scheme adopted the Raft algorithm combined with elliptic curve cryptography and time-triggered mechanism to get all safety-related devices in a certain area of the system to update and consistently share an authentication key in a way that works without key management center and reduces human intervention. Then, the specification language TLA+ is used to model the consensus process, and the TLC model checker is used to verify the properties of the model. The results show that the scheme is feasible, safe and can simultaneously avoid the deadlock problem. At last, the safety analysis shows that the scheme is safe and meets EN50159 standard.

Keywords RSSP-II key management · Raft · TLA+ formal verification · TLC model checking

1 Introduction

In recent years, China's 300–350 km/h passenger dedicated line and high-speed railway adopt CTCS-3 level train control system as a unified technology platform, according to the EN50159 safety communication standard [1], the railway safety communication protocol (RSSP-II) [2] is developed to ensure the integrity, authenticity, safety and confidentiality of the transmission of safety-related data.

H. Lian (✉) · L. Ma · Y. Cao
School of Electronic and Information Engineering, Beijing Jiaotong University, Beijing, China
e-mail: 17120237@bjtu.edu.cn

L. Ma · Y. Cao
National Engineering Research Center of Rail Transportation Operation and Control System, Beijing Jiaotong University, Beijing, China

The RSSP-II protocol has set three levels of keys: transport keys (KTRANS), authentication keys (KMAC) and session keys (KSMAC), and these keys are used in 3DES symmetric encryption algorithm to ensure the safe interaction of data. However, there are some hidden dangers in the key management mechanism of these keys. First, KTRANS and KMAC are almost completely controlled by the key management center (KMC), once the KMC fails, is threatened by an attack or even replaced by a maliciously disguised center, and it will endanger the communication of the entire system. Second, the distribution of KTRANS is carried out through the offline physical channel, which is interfered by the operator, so the key is unchanged for long term in the practical applications, and one key management area uniformly uses the same transport key, so the key might be cracked or compromised; and at the same time, because the KMAC is encrypted and distributed to safety entities through KTRANS, the safety risks of KTRANS will endanger the safety of KMAC and KSMAC.

To solve the above problems, this paper proposes an improved scheme for the key management mechanism of RSSP-II protocol, which removes the key management center and replaces the transport key with asymmetric keys to reduce human intervention, and on this basis, the Raft algorithm [3–5] is appropriately modified to be combined with the elliptic curve cryptosystem (ECC) [6, 7] to improve the management of KMAC, so that a certain range of safety-related devices can update and share the same KMAC. Finally, the TLA+ [8] language is used to formally model the management scheme of KMAC based on the Raft algorithm, and then, the TLC model detector is run to verify the safety and feasibility of the program.

2 Improvement of RSSP-II Key Management Mechanism

For open transmission systems, the premise of safety-related communication is the identity authentication between the two parties, and then, safety-related information can be transmitted through encryption technology.

2.1 Verification of Authenticity and Legality of Identities

This paper proposes to replace the transport key with asymmetric key pairs, so the primary task of the safety-related devices is to authenticate each other and obtain the public keys of the authenticated devices. The execution is as follows:

(1) Pre-acquisition of relevant knowledge for identity authentication: Secret information is generated offline for each safety-related device (device A has its secret information as SI_A) and stored on all safety-related devices in the area.
(2) Identity authentication and acquisition of public keys by using a hash function: Record that the private key of device A is sk and the public key is pk, and the

hash function is $H(x)$, where x is the content that needs to be hashed. Device A calculates the hash value $h = H(SI_A + pk)$ and then announces (pk/h) to other devices in the area. Device B receives pk and h, then finds the SI_A stored by itself and calculates $h' = H(SI_A + pk)$. If $h' = h$, it means that pk's owner (A) belongs to the local area and is authentic and legal, so device B saves pk for later use. In this way, each device in the cluster stores the public keys of other devices with which it needs to establish safety communication.

2.2 Generation, Update and Consensus of KMAC

To ensure the legality of the KMAC shared by the cluster, a certain number of legal keys with size of 192-bit are generated offline in advance, and these keys are randomly sorted so that each key obtains a random and unique sequence number r, then the hash value $H\ (key/r)$ of each key is calculated, and thereby, we obtain a key book including sequence numbers, keys and hash values. The key book is stored in all safety-related devices of the cluster, and therefore, the cluster can obtain the legal authentication key that is to be agreed upon from the key book, and only the hash value, not the key itself is treated as data to be transmitted, making the transmission more confidential and safe.

In order to reduce workloads of the safety-related devices, the Raft process is performed only when the KMAC used currently needs to be replaced with a new one, so the scheme introduces a time-triggered mechanism to start the process, where each KMAC is attached with a random update time as the trigger condition. And when the cluster implements consensus on a new key, the cluster ends the process. Figure 1 shows the state transition diagram of a node in the Raft process.

In this process, each node is in one of three states: Follower, Leader or Candidate. When the cluster is triggered to start the Raft process, the cluster enters the initial state of the process, that is to say, every node is in the Follower state with an election term that is initially numbered with 1, and each Follower sets a random timer, which is called the election timeout. A Follower node can just respond to requests from Leaders and Candidates, and if it does not receive requests, it becomes a Candidate to start an election, and when a Candidate gets the majority of the votes, it becomes a Leader. A Leader node is responsible for sending information that needs to be

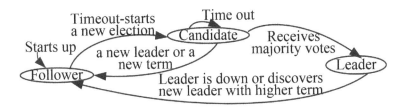

Fig. 1 Node's state transition diagram

agreed upon or sending heartbeat RPC to certify that it is working. Besides, there are some changes for the Raft algorithm used in this paper: The log entries are generated autonomously and independently by the Leaders, and a Leader generates only one log entry in its term. And this paper calls the changed Raft process as the consensus process.

The consensus process can be divided into two stages, leader election, and new authentication key's consensus:

(1) Leader Election: when a node does not receive a heartbeat RPC from a Leader or a request from a Candidate until its timeout elapses, it becomes a Candidate as soon, then it increases the number of its election term by 1, resets its election timeout, sends vote requests to other nodes and votes for itself; when receiving a request, a node gives its response, updates its term and resets its election timeout. Figure 2 shows the workflow of a Candidate node in the election process.

(2) The consensus of a new KMAC, in order to ensure that the cluster can reach consensus on the key information as quickly as possible, the key information is included in the first log entry of the Leader, and only one log entry can be generated by a Leader during its term. Thus, as long as all nodes agree on the first entry, the cluster gets consensus on the key. When a Leader is elected in a new term T, the first action of the Leader is to create a new entry whose term equals T, which contains two conditions. If the Leader's log is empty, it selects a new key from the key book, then uses public-key encryption algorithm (see Fig. 3) to encrypt the key-related information, which generates Kn, and the first

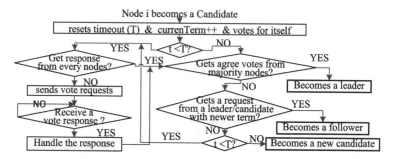

Fig. 2 Workflow of a candidate

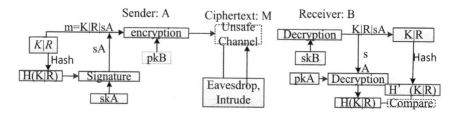

Fig. 3 Schematic of the asymmetric encryption algorithm

log entry (<T, Kn>, its index is 1) of the Leader is created; if the log part is not empty, the entry <T, 0> is directly generated and then attached to the tail of the Leader's log. When T in the latest entry of the Leader's log coincides with the current term of the node, the Leader starts to periodically send RPC requests to other nodes, receives replies and determines the content of the next RPC requests according to the replies.

3 TLA-Based Formal Modeling and Verification

Leslie Lamport's Temporal Logic of Action [9] (TLA) is a combination of temporal logic and behavioral logic which can use one language to simultaneously express system models and system properties. This paper adopts TLA+ language to describe the system as a state machine, and the syntax and semantics of TLA+ language see reference [8]. And after completing the description of the model, it is time to run the TLC detector for the model detection. The TLC first generates a set of initial states that satisfy the specification, then it performs a breadth-first search for all defined state transitions until all states are traversed, in the meantime, if TLC finds a state that violates the specification, it stops running and provides a state tracking path to this violation, which is called the counterexample.

3.1 TLA+ Description of the State Machine and Specification

TLA+ constructs a system in the form of a module, and the module of the key's consensus mechanism mainly includes the following parts:

(1) Defining the module name: *MODULE kmacraftnew*;
(2) Introducing external modules that contain definitions of basic definition: *EXTEND Naturals, FiniteSets, Sequences, TLC, Integers*
(3) Declaring constants and variables: *CONSTANTS SRE, Value, F, C, L, Nil, RVReq, RVRes, AEReq, AERes*; *VARIABLES M, Kn, CTerm, state, votedFor, CL, GetKmac, VR, VG, NI, MI, CI, log*; *CVars* $==\ll VR, VG\gg$; *LVars* $== \ll NI, MI\gg$; *logVars* $== \ll log, CI\gg$; *SREVars* $== \ll CTerm, state, votedFor, CL, GetKmac\gg$; *vars* $== \ll M, SREVars, CVars, LVars, logVars, Kn\gg$;

For the constants, SRE is a collection of all the nodes of a cluster; value is a collection of the keys from which a Leader can choose one to use; and besides, the other constants are module values that are not strings, numbers or Booleans, but just equal to themselves. For the variables, *GetKmac[i]* is the set of the nodes that are known by node i to have obtained *Kn*; *CL[i]* means the Leader that the node i is following during its current term period; *VR[i]* is a collection of the nodes who have given replies to the Candidate i; *VG[i]* is a collection of the nodes who have already

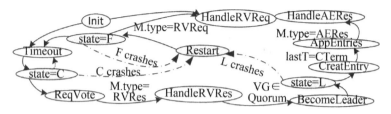

Fig. 4 Specific state transition of next action

voted for the Candidate *i*; a log is composed of several entries; *CI* is the index of the latest entry that most nodes have obtained from Leader *L*; *NI[L][j]* is the index of the next entry that *L* wants to send to *j*; *MI[L][k]* is the index of the latest entry that node *k* has successfully got from *L*; and besides, all requests and responses transmitted between nodes are recorded in *M*.

(4) Defining the initial state and the "*Next*" action:

Next == ∨ \E i \in SRE: Restart(i) ∨ \E i \in SRE: Timeout(i) ∨ \E i, j \in SRE: ReqVote(i, j) ∨\E i \in SRE: BecomeLeader(i) ∨\E i \in SRE: CreatEntry(i) ∨\E i \in SRE: AdvanceCI(i)∨\E i \in SRE: AdvanceGetKmac(i)∨\E i, j \in SRE: AppEntries(i, j) ∨\E m \in ValidM(M): Receive(m) ∨\E m \in SingleM(M): DuplicateM(m) ∨\E m \in ValidM(M): DropM(m) ∨ Endmodule;

(5) A node decides what sub-action of "*Next*" to do according to its current states (see Fig. 4), and at first, all nodes are in the initial state.

When a node crashes and then restarts, the "*Restart*" action is executed to initialize the state; when a node's election timeout elapses, the "*Timeout*" action is executed, and then, the node becomes a candidate with a new timeout to start a new election; a candidate executes the "*ReqVote*" action to send requests to other nodes, and then, it receives responses from other nodes; when a candidate obtains votes from most nodes of the cluster, it becomes a leader, and it first executes the "*CreatEntry*" action to create a new entry <*CTerm[i], v*> , then performs the "*AppEntries*" action to periodically send RPC requests to its followers and handles responses from the follower.

(6) Defining the specification "$KmacSpec == Init \wedge \Box[Next]_{vars}$" to present the safety property which asserts that the system cannot take a step that does not satisfy the next-state action.

TLC mode:	Breadth-first search
Last checkpoint time:	Thu Mar 28 22:14:09 CST 2019
Current status:	Computing reachable states
Errors detected:	No errors

Fig. 5 Detection result

3.2 TLC Model Checking of the Module

TLC model checking is performed by the following steps:

(1) Set parameters. The model uses device's ID to represent the device, and for the rationality and simplicity of the model, the number of the nodes is set to 5, and their IDs are ri $(i = 1,\ldots,5)$, then $SRE = \{r1, r2, r3, r4, r5\}$. Besides, the model properties such as *KmacSpec* and *Deadlock* are selected to be checked.
(2) Run the TLC detector to perform model detection, and then obtain the result from Fig. 5, where "no error" shows two conclusions. First, the state machine conforms to the safety property, which indicates that all actions do not perform steps that are not allowed by the model. Second, there is no deadlock problem, which means that the cluster can always carry out the consensus process properly.
(3) Run the TLC to check the formula $\exists i \in SRE$: $GetKmac[i] \neq SRE$, then TLC gives a counterexample ending in a state that makes the formula false, and this state is called the terminate state. Figure 6 is the concrete transformation of variables' states of the counterexample.

The state path of the counterexample can be summarized into 12 states: (a) at initial time, the cluster enters the Raft stage; (b)–(d) $r2$ becomes a Candidate, it works as Fig. 2 shows, and when it gets votes from the majority of nodes, it becomes a Leader; (e) $r2$ creates an entry, then sends requests with the entry to others; (f) $r1$ successfully adds the entry to its own log, update "*logvars*" and "*SREvars*," then replies to $r2$; (g) $r2$ receives $r1$'s reply and updates the "*Lvars*" and "*logvars*;" (h)–(k) $r1/r3/r4/r5$ does the same thing as $r1$, and then $r2$ updates relevant variables based on the replies from others, then sends requests with the updated parameters to others; (l) finally, after handling enough messages on the same principle, the formula $\forall i \in SRE$: $GetKmac[i] = SRE$ becomes true, it indicates that the state machine has reached the terminate state, which means that the process can realize the function of updating and sharing KMAC in a cluster.

3.3 Proof of Safety for the Consensus Process

The result of the TLC model checking shows that the consensus process is safe, which means that the model has to satisfy the following 3 conditions:

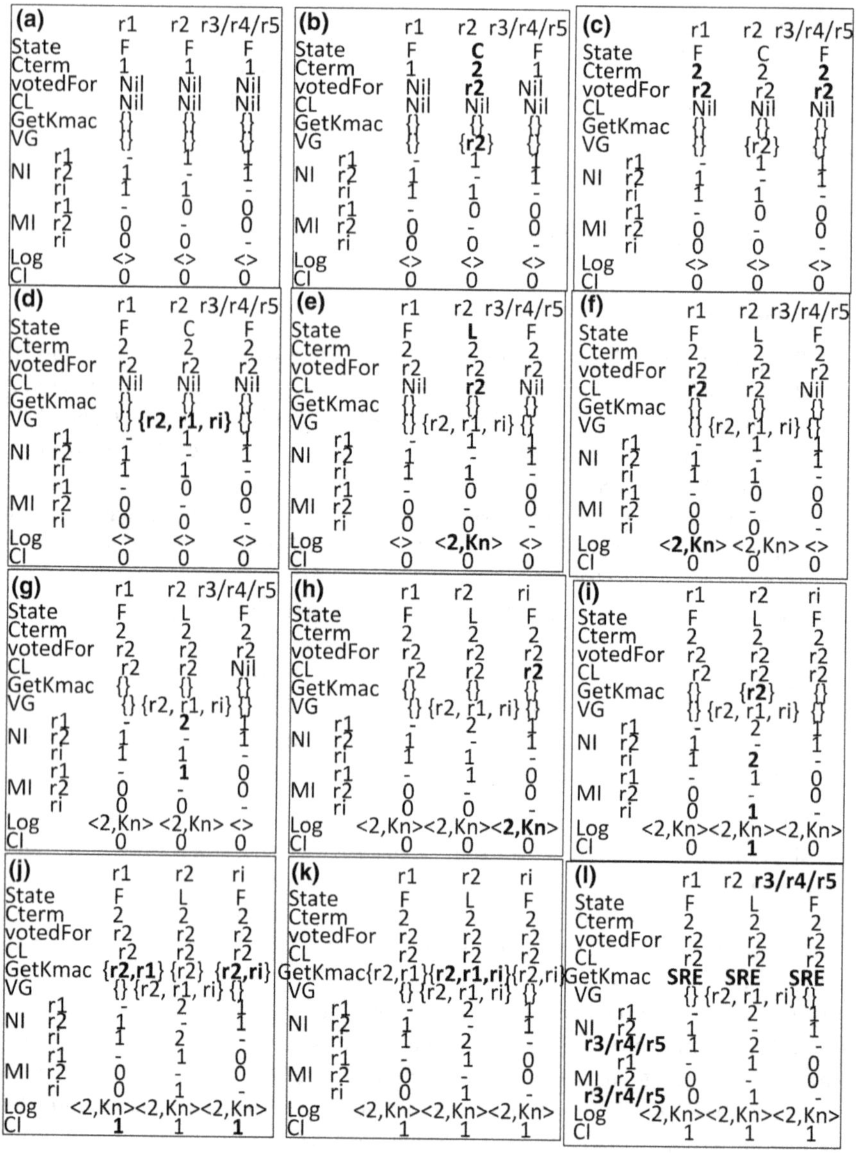

Fig. 6 Concrete transformation of variables' states

(1) The term of each device in cluster *SRE* is monotonically increasing; (2) there is only one Leader at most in one term; and (3) the node whose log is not the latest cannot be elected as a Leader. Here are logical proofs of these conditions:

Firstly, based on the definition, the following two formulas are true.

$$\forall i \in SRE : CTerm[i] \leq CTerm'[i] \qquad (1)$$

$$state[i] = L \Rightarrow VG[i] \in Quorum \qquad (2)$$

where *Quorum* is the set of the subsets of *SRE* who have more than half elements of *SRE*.

Secondly, each node can only vote once at most in the same term, so, for every node i in *SRE*, the number of *votedFor[i]* can only be 0 or 1, so,

$$CTerm[i] = CTerm[j] \Rightarrow VG[i] \cap VG[j] = \emptyset \qquad (3)$$

$$\forall i, j \in SRE : state[i] = state[j] = L \wedge CTerm[i] = CTerm[j] \Rightarrow i = j \qquad (4)$$

Thirdly, only the Leaders can create and send entries, which means Followers' logs are the part of the Leader's log, so,

$$\forall i \in SRE : state[i] = L \Rightarrow \forall j \in SRE : \wedge len(log[i]) \geq len(log[j]) \qquad (5)$$

Formula (1), (4) and (5) show that the above three conditions are valid, which also proves that the model is safe.

3.4 Safety Analysis

This paper treats the KMAC that needs to be shared in a cluster as the safety data to be transmitted, so the improved scheme has to follow EN50159 standard.

EN50159 defines seven threats to the transmission system, and they are repetition, deletion, re-sequencing, delay, insertion, corruption and masquerade, at meanwhile, it also provides eight defensive measures as well, and the RSSP-II protocol uses these measures to defend against the corresponding threats. And because the scheme proposed in this paper is an improvement of the key management in the RSSP-II, its implementation still needs to be based on the protocol itself, which means that, when transmitting information, the first four threats are still countered through a sequence number, time stamp and timeout. And for the insertion, there is the identification procedure, the feedback message and source and destination identifiers as the defenses, these measures can be found intuitively from the improved scheme. In addition, since the scheme performs identity authentication first and adopts the asymmetric encryption algorithm (ECC) which has commendable encryption strength [10], the scheme can also avoid the last two threats and ensure that the devices share a KMAC legally and secretly, and besides, the private key signature and the hash function can keep the KMAC authentic and integral.

4 Conclusion

Based on the safety risk of the key management mechanism of RSSP-II protocol of the train control system, this paper proposes a new scheme for the key management, where the Raft algorithm is used to make the key management work without the key management center and reduce human interventions, the time-triggered mechanism is introduced to make the Raft process be triggered only when the KMAC needs to be updated, and at the same time, the elliptic curve cryptography algorithm makes the KMAC more confidential and safer. Then, in order to verify the properties of the scheme, the Raft consensus process is described as a state machine through TLA+ language, and the TLC model checker tells that the state machine can realize the function of a new key's update and consistent sharing, and what's more, the consensus process is safe, which is also proved through logic formulas in Sect. 3.3. Finally, combined with the analysis in Sect. 3.4, it can be seen that the scheme meets the EN50159 standard, and the KMAC-related information can be transmitted with integrity, error-free, authenticity and confidentiality among the authentic and legal devices.

Acknowledgements This work is supported by National Key R&D Program of China under Grant (No. 2016YFB1200602) and Beijing Laboratory of Urban Rail Transit.

References

1. CENELEC EN.50159 (2010) Railway applications-communication, signaling and processing systems-safety-related communication in transmission system
2. Railway Signal Safety Communication Protocol II (V1.0) (2010) Transport Bureau of the Ministry of Railways (in Chinese)
3. Ongaro D, Ousterhout J (2014) In search of an understandable consensus algorithm. USENIX, pp 305–320
4. Raft consensus algorithm, website: https://raft.github.io
5. Huang D, Ma X, Zhang S (2018) Performance analysis of the Raft consensus algorithm for private blockchains. IEEE Trans Syst Man Cybern: Syst 99:1–10
6. Lozupone V (2018) Analyze encryption and public key infrastructure (PKI). Int J Inf Manage 38(1):42–44
7. Loi KCC, An S, Ko SB (2014) FPGA implementation of low latency scalable Elliptic Curve Cryptosystem processor in GF(2 m). In: IEEE international symposium on circuits and systems. IEEE
8. Lamport L (2002) Specifying systems, the TLA+ language and tools for hardware and software engineers. Addison-Wesley Longman Publishing Co. Inc
9. Lamport L (1994) The temporal logic of actions. ACM
10. Ahuja P, Soni H, Bhavsar K (2018) High performance vedic approach for data security using elliptic curve cryptography on FPGA. In: 2018 2nd International conference on trends in electronics and informatics (ICOEI)

Charging Strategy of Temporal-SoC for Plug-in Electric Vehicles on Coupled Networks

Xian Yang and Zhijian Jia

Abstract The integration of a massive number of plug-in electric vehicles (PEVs) into infrastructure systems accelerates the complex interactions between the distribution network and the road network and brings the direct challenges to the coupled traffic and energy transportation networks, which have stunted the further development of PEVs. To solve this problem, the dynamical PEVs behaviors, including PEV travel behaviors and charging behaviors, need to be better understood. This paper mainly proposed a charging strategy for modeling PEV travel behaviors and charging behaviors. The charging strategy with objective functions considering minimization of the travel route distance on the road network, the voltage fluctuation on the distribution network, and the number of failure PEVs, respectively, was developed to explore the correlation between the PEV behaviors and state of charge (SoC) range on the coupled networks. Furthermore, the variation coefficient method was used to calculate the weights of indexes in the objective functions to optimize the threshold of SoC. Besides, the improved IEEE-33 bus system with the geographic information was taken as an example to verify the effectiveness of the proposed method.

Keywords Dynamical PEVs behaviors · Coupled networks · Plug-in electric vehicles (PEVs) · State of charge (SoC) · Variation coefficient method

1 Introduction

The plug-in electric vehicles (PEVs), as the coupling points of the distribution network and road network, are being integrated into the infrastructure systems and playing a momentous role in the electrified transportation. On the one hand, by

X. Yang (✉)
School of Mechanical and Electrical Engineering, Hunan City University,
Yiyang 413002, China
e-mail: xianyang922@hnu.edu.cn

Z. Jia
State Grid Yiyang Power Supply Company, State Grid, Yiyang 413000, China
e-mail: zhijianjia@outlook.com

minimizing road network's reliance on oil-based fuels, the PEVs provide a cleaner environment with low carbon emission [1]. On the other hand, combined with renewable energy [2], the PEVs, as controllable loads and storage devices, provide more flexibility to increase the system efficiency in demand-side management [3, 4]. In general, the PEVs are considered to be an important part of the next-generation smart grid system [5], and their market penetration remains to be raised, which faces a number of challenges.

The temporal and spatial distribution characteristics of PEV charging demand imposes a significant impact on the load profiles of distribution networks, resulting in unexpected voltage fluctuation and increased peak-valley gap [6–8]. Most of the existing solutions assumed that the PEV charging demand merely depends on the arrival rate at charging spots [6] and daily travel route distance [7], ignoring the temporal and spatial characteristics of PEV travel, such as the next arrival location [8]. Furthermore, PEVs should be routed to the convenient charging stations when necessary, which is influenced by the travel route from the origin to the destination. Reference [9] proposed a travel strategy to optimize the route choice based on the lowest total fuel consumption. Reference [10] presented a methodology of modeling the PEV charging load considering the travel pattern and willingness of customs. Thus, the PEV travel behaviors, combined with the charging behaviors, need to be better understood to ensure the reliable and economical network operation. From this point of view, the vehicle travel route is crucial for routing of PEVs considering position of the charging stations and limited driving range. Based on systematic integration of power system analysis and transportation analysis [11], a spatial-temporal model with origin–destination (OD) analysis was used to model the characteristics of PEV charging load.

Generally speaking, the optimized PEV charging strategies can benefit the PEV owners and the power grid operators by decreasing the charging cost and enhancing the reliability and the flexibility of distribution network [12, 13]. Thus, most of the researchers make efforts to formulate the charging behaviors, e.g., charging start time, charging duration, and state of charge (SoC). The charging start time was usually considered based on survey statistics [14], or assumed to follow a given normal distribution [15]. Reference [16] defined daily driving distance to calculate the energy depletion of battery, which represents the charging duration. And reference [6] presented a smart charging strategy by offering multiple charging options, with multi-objective to minimum charging time, travel time, and charging cost. However, previous studies usually only took consideration of single network, rather than considering the distribution network, the road network, and user experience comprehensively. Furthermore, it should be noticed that SoC state is divided into initial SoC and finished SoC. Commonly, the initial SoC could be from historical statistics [17], and the PEVs are not necessarily to be fully charged. Thus, the finished SoC is meaningful to analyze the PEVs charging behaviors, which are not considered in the previous study.

In this paper, we presented a charging strategy for modeling PEV travel behaviors and charging behaviors. Firstly, we built the system model based on the complex network theory and formed the PEVs characteristics, including the origin location, the destination location, the battery capacity, and the initial SoC. Due to the limited battery capacity, the PEV was needed to take a detour to the fast-charging stations (FCSs) for the emergency power supply when necessary. Besides, the shortest way algorithm (Floyd, Bellman-Ford, and Dijkstra) was used to make the shortest way analysis with graph theory [18] for minimizing the cost and energy. Considering the temporal and spatial distribution characteristics of PEV, the objective of the proposed charging strategy is to find the thresholds of SoC to minimize the travel route distance, the voltage fluctuation, and the number of failure PEVs within a set period of time. Furthermore, the variation coefficient method was used to calculate the weight coefficients of the objective. Besides, the improved IEEE-33 bus system with the geographic information was taken as an example to verify the effectiveness of the proposed method.

2 Charging Strategy

2.1 System Model

The networks can be represented by the complex network G which is composed of nodes and edges. The distribution network can be represented by a graph $G_E = [E(N), E(L)]$, where $E(N)$ represents the set of buses, and $E(L)$ represents the set of branches. The road network can be abstractly represented by a graph $G_R = [R(N), R(L), W]$, where $R(N)$ represents the set of crossroads, $R(L)$ represents the set of roads, and W represents the road weight matrix. The road weigh is described by the road length in this paper. It should be noticed that the FCSs are the coupling points between the distribution network and the road network. Thus, the energy can be exchanged between the PEVs and the distribution network when the PEVs are charging at the FCSs.

In our system model, each PEV indexed by i can be attributed by its current state of charge $SoC_i(t)$, current location $S_i(t)$, and intended destination $D_i(t)$ at a time instant t. Besides, all FCSs and PEVs are connected to the distribution network and can exchange information in real time.

Firstly, the PEVs formed the OD pairs [19]. OD pair means that a PEV_i starts from the origin node and plans to go to the destination node. If its current $SoC_i(t)$ suggests that it remains little stored energy on the way, it will have to recharge at the nearest FCS to make urgent electricity supplement. Notably, if the PEV_i cannot reach any FCS within the current $SoC_i(t)$, the PEV_i, equivalent to stop on the roadside due to the vehicle failure, is regarded as the failure PEV. Otherwise, the PEV_i, will be charging at the nearest FCS until the SoC_i showing enough stored energy. How to measure the stored energy? We define the thresholds, the low limit of state of charge (SoC_{min}) and the upper limit of state of charge (SoC_{max}), to represent the little stored energy and enough stored energy, respectively. The research question then translates into finding the optimum thresholds that do not significantly increase the travel route distance in the road network and voltage fluctuation in the distribution network and offer better user experience, which described by the number of failure PEVs.

2.2 Proposed Model

The objective function, F, is given as

$$F = \omega_1 \cdot x(D) + \omega_2 \cdot y(v) + \omega_3 \cdot z(N_{\text{failure}}) \tag{1}$$

$$x(D) = \frac{D(o,d) - D_0}{D_0} \times 100\% \tag{2}$$

$$D(o,d) = \frac{1}{N_{\text{PEVs}}} \sum_{i=1}^{N_{\text{PEVs}}} D_i(o,d) \tag{3}$$

$$D_i(o,d) = \frac{1}{N_{ts}} \sum_{t \in N_{ts}} D_i(t) + D_i(s(t), d) \tag{4}$$

$$D_i(t) = \frac{E^{\text{ca}}(\text{SoC}_i - \text{SoC}_i(t))}{E^{\text{travel}}} \tag{5}$$

$$y(v) = \frac{1}{N_{ts}} \sum_{t \in N_{ts}} v(t) \tag{6}$$

$$v(t) = \sqrt{\frac{1}{N_E} \sum_{j=1}^{N_E} (v_j - \bar{v})} \tag{7}$$

$$z(N_{\text{failure}}) = \frac{N_{\text{failure}}}{N_{\text{PEVs}}} \times 100\% \tag{8}$$

where ω_1, ω_2, and ω_3 are weight coefficients of the objective function. $x(D)$ is the increasing rate of the travel route distance. D_0 is the average original travel route based on $\text{SoC}_{\min} = +\infty$, which means the PEVs are with full capacity without consideration of emergency power supply. Taking the charging behaviors into consideration, Eq. (4) represents the average travel route distance of PEV_i from the origin node to the destination node. $D_i(t)$ represents the travel route distance from origin node to the current location node, and $D_i(s(t), d)$ represents the travel route distance from current location node to destination node. As shown in Eq. (5), E^{ca} and E^{travel} represent the battery capacity of PEVs and energy consumption per unit driving distance, respectively, and SoC_i and $\text{SoC}_i(t)$ represent the initial and current state of charge, respectively. $y(v)$ is used to describe the voltage fluctuation, and N_{ts} is the number of time steps. $v(t)$ is the standard deviation of node voltage in the distribution network, and N_E is the number of buses. $z(N_{\text{failure}})$ is the failure rate of PEVs in the simulation time to describe the user experience, and N_{PEVs} is the number of PEVs.

$$\text{s.t. } D_i^{\max}(t) \leq (E^{\text{ca}}\text{SoC}(t)/E^{\text{travel}}) \tag{9}$$

$$\text{SoC}_{\min} \leq \text{SoC}(t) \leq \text{SoC}_{\max} \tag{10}$$

$$v_{\min} \leq v_j(t) \leq v_{\max} \tag{11}$$

$$p_{\min} \leq p_j(t) \leq p_{\max} \tag{12}$$

In the optimization problem, the summation of increasing rate of the travel route distance, the voltage fluctuation, and the failure rate of PEVs should be minimized. Constraint (9) shows the maximum distance that the PEV can travel based on its SoC at a current time t. Constraint (10) indicates that the amount of SoC for a PEV at every time t should be at the range from $\text{SoC}_{\min}$ to $\text{SoC}_{\max}$. Besides, the constraints of the voltage and power with power flow calculation are described in (11) and (12), respectively.

2.3 Variation Coefficient Method

The variation coefficient method is used to directly obtain the weight coefficients of the objective by calculating the indexes data, avoiding the subjective factor.

Phase 1: Evaluation Matrix

There are m evaluation indexes in the cases, and the evaluation index vector of Case j is denoted by $Y_j = \begin{bmatrix} y_{1j}, y_{2j}, \ldots, y_{mj} \end{bmatrix}^T$ ($m = 3$ in this paper. y_{1j} represents the index of the increasing rate of the travel route distance, y_{2j} represents the index of voltage fluctuation, and y_{3j} represents the index of user experience, respectively). Then, the evaluation matrix can be denoted by $Y = [Y_1, Y_2, \ldots, Y_n]$ with n cases, and weight vector is denoted by $\omega = [\omega_1, \omega_2, \ldots, \omega_m]$.

Phase 2: Non-dimensional Processing

The evaluation matrix $Y = (y_{ij})_{m \times n}$ can be translated into standard matrix Y' by non-dimensional processing, as follows.

$$y'_{ij} = \frac{y_{ij}}{\sqrt{\sum_{j=1}^{n} y_{ij}^2}} \quad i = 1, 2, \ldots, m \tag{13}$$

Phase 3: The coefficient of variation index

$$V_i = \frac{\sigma_i}{\overline{y_i}} \tag{14}$$

$$\bar{y}'_i = \frac{1}{n}\sum_{j=1}^{n} y'_{ij} \quad i = 1, 2, \ldots, m \tag{15}$$

$$\sigma_i = \sqrt{\frac{1}{n}\sum_{j=1}^{n}(y'_{ij} - \bar{y}'_i)^2} \quad i = 1, 2, \ldots, m \tag{16}$$

where V_i denotes the variation coefficient of index i. $\bar{y}'_i$ is the arithmetic mean of index i, and σ_i represents the standard deviation of index i.

Phase 4: The weight of variation coefficient

$$\omega_i = \frac{V_i}{\sum_{i=1}^{m} V_i} \tag{17}$$

where ω_i denotes the weight of variation coefficient.

Figure 1 is the flowchart of the proposed charging strategy model, and the detail process is described as follows:

Step1: Set up the initial operation condition. Input the system power flow calculation files, the topology statistics of coupled networks and PEV parameters such as battery capacity, energy consumption per unit driving distance, initial SoC; set time $t = 0$;
Step2: Set $i = 0$;

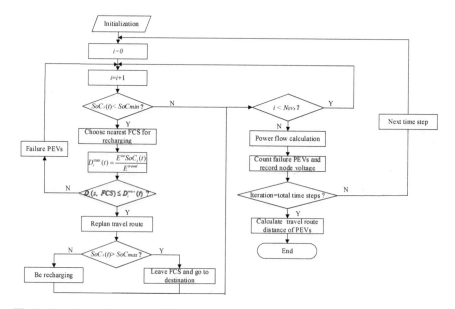

Fig. 1 Flowchart of the proposed charging strategy

Step3: $i = i+1$; check whether the current SoC(t) is lower than SoC$_{min}$ for PEV_i. If so, choose the reachable nearest FCS for recharging, then replan the travel route, and go to the next step; otherwise, the PEV$_i$ is regarded as a failure PEV, and go to Step 5;

Step4: Check whether the current SoC$_i$(t) is over SoC$_{max}$ for PEV$_i$. If so, leave the FCS, and go to the destination; if not, continue being recharging;

Step5: If all PEVs have been considered, calculate the power flow, then count the failure PEVs and record the node voltage; otherwise, go back to Step 3.

Step6: If total time steps have been iterated, calculate the travel route distance; otherwise, go back to Step 2.

3 Simulation and Results

3.1 Simulation Model

The distribution transformer of rating 100 MVA was considered to be supplying the load. Based on the improved IEEE 33-bus system, the road network was built with 33 crossroads and 53 roads [20]. The topology of the coupled networks was shown in Fig. 2. As shown in road network, the numerical value on the line, between two corresponding nodes, represented the route distance. For example, the route distance between the #1 node and #2 node was 4.2 km. Besides, the average path length of road was 4.12 km. Furthermore, the FCSs should be selected in the position where traffic is convenient for PEVs. Taking eight nodes as the FCSs denoted by $\Omega_T = \{\#3, \#5, \#6, \#10, \#20, \#21, \#30, \#31\}$ [20].

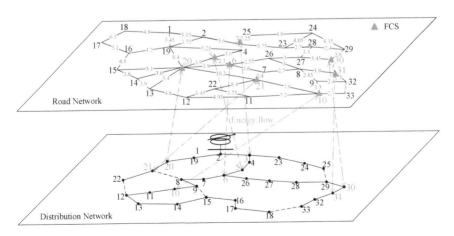

Fig. 2 Topology of coupled networks

Table 1 Possible optimal schemes

	SoC$_{min}$	SoC$_{max}$
Case 1	0.1	0.5
Case 2	0.2	0.7
Case 3	0.3	0.9

In the simulation, 1000 PEVs with battery capacity of 30 kWh were considered to pour into the road network, with random OD pair, and their SoC is modeled as a uniform distribution in the 10~90% range. PEVs were assumed to travel at the speed of 30 km/h, consuming 0.12 kWh/km [21]. Once the SoC$_i(t)$ beyond SoC$_{min}$, the PEV$_i$ would choose the nearest FCS to make urgent electricity supplement, and PEV$_i$ would be supplying 20 kW charging power until the SoC$_i(t)$ up to SoC$_{max}$. It is assumed that there are enough charging points in each charging station. Besides, it is also assumed that the possible values of SoC$_{min}$ are set from 0.1 to 0.3, and the possible values of SoC$_{max}$ are set from 0.5 to 0.9. The simulation time is from 8 a.m. to 10 a.m., taking 5 min as a time step, totaling 24 steps.

3.2 OD Pair Strategy of EVs

Firstly, determine the possible optimal schemes primarily. Based on the range of SoC$_{min}$ and SoC$_{max}$, considering the objective to minimize the increasing rate of route distance, the node voltage fluctuation, and the number of failure PEVs, we choose three possible optimal schemes as shown in Table 1.

As shown in Table 1, there are three possible optimal schemes in this paper. Case 1: when SoC beyond 0.1, PEVs need to be recharging until SoC up to 0.5; Case 2: when SoC beyond 0.2, PEVs need to be recharging until SoC up to 0.7; and Case 3: when SoC beyond 0.3, PEVs need to be recharging until SoC up to 0.9.

The average travel route distance based on SoC$_{min}$ = +∞ is 14.11 km, and the average travel route distances under Cases were 14.34 km, 14.76 km, and 15.23 km, respectively. Figure 3 shows power tail of the travel route distance of PEVs in the road network. Obviously, PEVs made the detour to find a FCS when SoC beyond SoC$_{min}$. Furthermore, with the larger SoC$_{min}$, more PEVs, planned to go to destination directly, may change to be recharging at FCS. Thus, compared with SoC$_{min}$ = +∞, the cumulative probability of travel route distance of PEVs is lowest in Case 1, which means the increasing rate of the travel route distance is minimum in Case 1.

As shown in Fig. 4a, we compared the number of charging PEVs at every time interval with Case 1, Case 2, and Case 3, respectively, and the result showed that there were more charging PEVs in Case 3. Due to the larger SoC$_{min}$ in Case 3, more PEVs' current SoC exceeded the minimum threshold. For example, the EV$_i$ with current SoC$_i(t)$ of 0.25 did not be recharging at FCS in Case 2, but it did in Case 3. Combined with the larger gap between SoC$_{min}$ and SoC$_{max}$, resulting in charging time was longer and the charging load was heavier in Case 3, as shown in Fig. 5a.

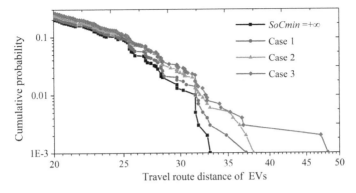

Fig. 3 Cumulative distribution function of travel route distance of PEVs

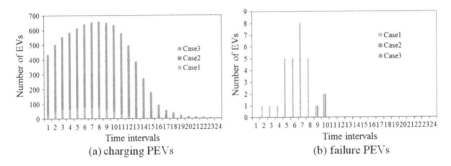

Fig. 4 Number of PEVs at every time interval

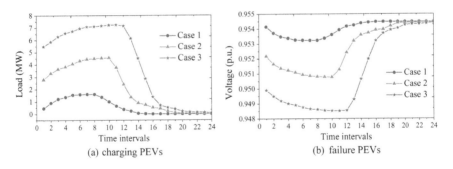

Fig. 5 PEVs charging load in distribution network

Generally speaking, far from the transformer, the terminal node voltage decreased severely. Thus, comparing the voltage of several terminal nodes in the whole distribution network, we recorded the minimum node voltage under different cases. As shown in Fig. 5b, it is obviously that the node voltage is lowest in Case 3, and it

Table 2 Indexes of possible optimal schemes

	Case 1	Case 2
Increasing rate of route distance	1.63%	4.63%
Voltage fluctuation	0.0122	0.0125
Number of failure PEVs	0.029	0.003

dropped to nearly 0.948 p.u., which may have a bad effect on the operation of distribution network. On the contrary, the node voltage is nearest to 1 p.u. in Case 1, which means the standard deviation of node voltage is minimum.

From the above analysis alone, it seemed that Case 1 was the best choice of optimal scheme in terms of coupled networks. However, taking the user experiment into consideration, the result changed. From Fig. 4b, it showed that there were eight failure PEVs in the 7th time interval, and 29 failure PEVs totally under Case 1. Compared to the three failure PEVs under Case 2 and 0 failure EV under Case 3, the user experience was worst in Case 1.

According to Eq. (2)–(8), the increasing rate of route distance, the voltage fluctuation, and the failure rate of PEVs were calculated as shown in Table 2.

Then, the evaluation matrix was translated to standard matrix by non-dimensional processing according to Eq. (13), and the standard matrix was described as follows:

$$Y' = \begin{bmatrix} 0.17414 & 0.49464 & 0.85147 \\ 0.56185 & 0.57566 & 0.59409 \\ 0.99469 & 0.10290 & 0 \end{bmatrix}$$

According to Eqs. (14)–(17), the arithmetic mean $\bar{y}'_i$, the standard deviation σ_i, the variation coefficient V_i, and the weight of variation coefficient ω_i were calculated, as shown in Table 3.

The results showed that there were a large weight of number of failure PEVs, middle weight of increasing rate of route distance, and the small weight of voltage fluctuation. Thus, according to Eq. (1), the objective function was as follows:

$$F = 0.3050x(D) + 0.0128y(v) + 0.6822z(N_{\text{failure}}) \quad (18)$$

Combing the indexes in Table 2, the results were $F1 = 0.02491$, $F2 = 0.01633$, and $F3 = 0.02433$. Obviously, $F2 < F3 < F1$, it demonstrated that Case 2 was the optimal scheme in this paper.

Table 3 Results of index parameter calculation

Indexes	$\bar{y}'_i$	σi	Vi	ωi
Increasing rate of route distance	0.5068	0.2767	0.5459	0.3050
Voltage fluctuation	0.5772	0.0132	0.0229	0.0128
Number of failure PEVs	0.3659	0.4466	1.2208	0.6822

4 Conclusion

Based on the temporal-SoC analysis, a charging strategy with dynamic PEV behaviors was proposed in this paper. For PEVs, whether to be recharging at FCSs was determined by the current SoC. Under different scenarios of SoC range, the travel behaviors and charging behaviors are different, which could be reflected by the travel route and charging load. Taking the coupled networks and user experiment into consideration, the objective functions were to minimize the three indexes including the increasing rate of travel route distance, the average standard deviation of node voltage, and the failure rate of PEVs. From the objective point of view, the weight of three indexes was determined by the variation coefficient method based on the simulation results. Then, by multiplying the simulation results and the corresponding weights, the scenario with minimum value of objective function was regarded as the optimal scheme of SoC range. Based on the improved IEEE-33 bus system with topology of the road network, the effectiveness and validity of the proposed method have been verified.

References

1. Masayoshi W (2009) Research and development of electric vehicles for clean transportation. Environ Sci 21:745–749
2. Gao S, Chau KT, Liu C et al (2014) Integrated energy management of plug-in electric vehicles in power grid with renewables. IEEE Trans Veh Technol 63:3019–3027
3. Falahati B, Fu Y, Darabi Z (2011) Reliability assessment of power systems considering the large-scale PHEV integration. In: Proceedings of vehicle power and propulsion conference (VPPC), Chicago, IL, USA, pp 6–9
4. Monteiro V, Carmo JP, Pinto JG et al (2016) A flexible infrastructure for dynamic power control of electric vehicle battery chargers. IEEE Trans Veh Technol 65:4535–4547
5. Kennel F, Görges D, Liu S (2013) Energy management for smart grids with electric vehicles based on hierarchical MPC. IEEE Trans Industr Inf 9:1528–1537
6. Moghaddam Z, Ahmad I, Habibi D et al (2018) Smart charging strategy for electric vehicle charging stations. IEEE Trans Transp Electrification 4(1):76–88
7. Yang X, Li Y, Cai Y et al (2018) Impact of road-block on peak-load of coupled traffic and energy transportation networks. Energies 11(7):1776
8. Sun SS, Yang Q, Yan WJ (2018) A novel Markov-based temporal-SoC analysis for characterizing PEV charging demand. IEEE Trans Industr Inf 14(1):156–166
9. Ericsson E, Larsson H, Brundell-Freij K (2006) Optimizing route choice for lowest fuel consumption: potential effects of a new driver support tool. Transp Res Part C: Emerg Technol 14(6):369–383
10. Zhang P, Qian K, Zhou C et al (2012) A methodology for optimization of power systems demand due to electric vehicle charging load. IEEE Trans Power Syst 27(3):1628–1636
11. Mu Y, Wu J, Jenkinsa N et al (2014) A spatial–temporal model for grid impact analysis of plug-in electric vehicles. Appl Energy 114:456–565
12. Faddel S, Al-Awami AT, Mohammed OA (2018) Charge control and operation of electric vehicles in power grids: a review. Energies 11:701
13. Monteiro V, Carmo JP, Pinto JG et al (2016) A flexible infrastructure for dynamic power control of electric vehicle battery chargers. IEEE Trans Veh Technol 65(6):4535–4547

14. Qian K, Zhou C, Allan M et al (2011) Modelling of load demand due to ev battery charging in distribution systems. IEEE Trans Power Syst 26(2):802–810
15. Agiiero JR, Chongfuangprinya P, Shao S et al (2012) Integration of plug-in electric vehicles and distributed energy resources on power distribution systems. In: Proceedings of IEEE international electric vehicle conference, Greenville, SC, pp 1–7
16. Li G, Zhang XP (2012) Modeling of plug-in hybrid electric vehicle charging demand in probabilistic power flow calculations. IEEE Trans Smart Grid 3(1):492–499
17. Kong C, Bayram IS, Devetsikiotis M (2015) Revenue optimization frameworks for multi-class PEV charging stations. IEEE Access 3:2140–2150
18. Goldberg AV, Harrelson C (2005) Computing the shortest path: A* search meets graph theory. In: Proceedings of SODA. Society for Industrial and Applied Mathematics
19. Jacopo I, Ginestra B (2016) Extracting information from multiple networks. Chaos: Interdisc J Nonlinear Sci 26(6):065306
20. Liu BL, Huang XL, Li J et al (2015) Multi-objective planning of distribution network containing distributed generation and electric vehicle charging station. Power Syst Technol 39(2)
21. Badin F, Berr FL, Brike H et al (2014) Evaluation of EVs energy consumption influencing factors, driving conditions, auxiliaries use, driver's aggressiveness. In: Proceedings of IEEE electric vehicle symposium & exhibition, 1–12

The Forecast of the Subway Passenger Flow Based on Smooth Relevance Vector Machine

Ni Luo, Meng Hui, Lin Bai, Rui Yao and Qisheng Wu

Abstract Metro passenger flow could be taken as time series. The short-term passenger flow forecast is significant for the metro management. Many experts have researched this issue, but there also have some problems such as how to forecast the real-time passenger flow for holidays effectively. Compare with the support vector machine (SVM), relevance vector machine (RVM) takes model sparsity criterion as the priority of model weights. Smooth relevance vector machine (SRVM) is an effective method to control sparsity in Bayesian regression. It introduces an elastic noise-dependent smoothness prior into the RVM. In this paper, we use SRVM to forecast the passenger flow, and the simulation results indicate that it is effective.

Keywords Subway · Passenger flow forecast · SRVM

1 Introduction

Due to the characteristics such as high speed, punctuality and large traffic volume and so on, the metro system is considered as the most effective way to alleviate the adverse effect of fast urbanization and traffic congestion. Since 1990s, the metro system construction has entered a blowout period in China. Till now, there are dozens of large and medium-sized cities are building subway systems [1]. The expanding subway system will stimulate sharp rise of ridership. In this paper, we take Xi'an subway system as an example. The subway system of Xi'an is comprised of four lines and 95 stations with 126.35 km till 2019. The passenger flow has reached 3.3 million per day, and the passenger transport intensity ranks the first in China. The blowout growth of ridership has triggered a series of problems, for instance, crowdedness in carriages and the insufficient transport capacity of subway facilities [2]. These phenomena are common in modern cities [3]. The public transport managers face more severe challenges because of the explosive growth of the passenger flow. One possible solution for these challenges is the proactive forecast of passenger flow.

N. Luo · M. Hui (✉) · L. Bai · R. Yao · Q. Wu
School of Electronic and Control, Chang'an University, 710064 Xi'an, ShaanXi, China
e-mail: ximeng@chd.edu.cn

© Springer Nature Singapore Pte Ltd. 2020
B. Liu et al. (eds.), *Proceedings of the 4th International Conference on Electrical and Information Technologies for Rail Transportation (EITRT) 2019*, Lecture Notes in Electrical Engineering 640, https://doi.org/10.1007/978-981-15-2914-6_3

This information could help the manager to optimize the service schedules. Also, it could help the passenger to adjust their plan to avoid the uncomfortable ride.

Many experts have been devoted to short-term passenger flow forecasting model research in order to improve the accuracy of the forecasting results. The existing approaches for short-term passenger flow predict could be categorized into linear and nonlinear methods [4, 5]. Based on the assumption that the passenger flow prediction trends are linear and stationary, the existing research works mainly studied historical average model, smoothing technique and so on [6–9]. Later, time-series analysis models have been applied to forecast the passenger flow. Autoregressive integrated moving average (ARIMA) [10] and seasonal autoregressive integrated moving average (SARIMA) [11] have been proposed. The studies verified that the time-series analysis methods have better performance than other linear methods.

Unlike linear methods, the nonlinear short-term passenger flow prediction models could reflect the nonlinear characteristics and achieve more accurate prediction results in transportation systems. Along with the development of artificial intelligence, many new intelligent models are extensively studied. Some algorithms, such as artificial neural networks [12, 13] and SVM [14–16], have been applied in modern transportation systems. The SVM is a supervised learning models with associated learning algorithms. In addition, SVM has better performance than other nonlinear techniques. Regression models need to establish a relationship between metro ridership and lots of influential factors, such as demographic, economic and so on [17–20].

Although many passenger flow forecast models have been proposed, but there also have some problems to be resolved. The passenger flow in metro system is regularity and randomness, and most of the early short-term prediction methods are not accuracy enough. In this paper, we proposed a new short-term passenger flow forecast model based on SRVM.

2 Smooth Relevance Vector Machine

A function of interest y can be approximated by a linear combination of the input vector x in nonlinear regression. Projected these linear combinations onto a set of nonlinear basis functions, $\{\phi_m\}_{m=1}^{M}$.

$$y(x) = \omega_0 + \sum_{m=1}^{M} \omega_m \phi_m(x) \tag{1}$$

Therefore, given a set of N training input vectors $\{x_n\}_{n=1}^{N}$ and corresponding targets t_n, the work is to find the M + 1 weights ω_m that will give the most credible approximation to y. Choosing $\phi(x) \equiv x$ regains linear regression. The latest research results of SVM in various fields show that a sparse representation is needed to adjust the bias tradeoffs in regression and classification problems in an appropriate space.

In SVM regression, an ideal level of sparsity is brought about indirectly by determining an error or margin parameter through a cross-validation plan, and the Bayesian formula RVM for regression problems allows for a prior structure that explicitly coded sparse representation.

Complementing the standard likelihood function as:

$$p(t|w, \sigma^2) = (2\pi\sigma^2)^{-\frac{N}{2}} \exp\left(-\frac{\|t - \Phi w\|^2}{2\sigma^2}\right) \quad (2)$$

where Φ is a $N \times M$ matrix, w is N-vector of ω_m. With an "automatic relevance determination" takes precedence over the weights:

$$p(t|\alpha) = (2\pi)^{-\frac{M}{2}} \prod_{m=1}^{M} \alpha_m^{\frac{1}{2}} \exp\left(-\frac{1}{2}\alpha_m \omega_m^2\right) \quad (3)$$

that has the effect of "switching off" basis functions that are not supported in the data.

While $p(\alpha)$ is uniform, a standard inverse gamma prior is placed over the noise variance σ^2:

$$p(\sigma^2) = \frac{h^g}{\Gamma(g)} \sigma^{-2(g+1)} e^{-\frac{h}{\sigma^2}} \quad (4)$$

where g and h are fixed parameters.

Although RVM has the advantages of a probabilistic formulas, unfortunately, it still does not go far enough in terms of Bayesian encoding of the sparsity constraint. Fortunately, an advantage of Bayesian models is its inherent extensibility through additional prior structure. Equation (3) does not seem to strongly favor sparsity, but the overall influence on the weights depends on prior assigned to α. Here, we only consider the priors of the form like $p(\alpha, \sigma^2) = \prod_{m=1}^{M} p(\alpha_m|\sigma^2)p(\sigma^2)$. Next, the effective prior on ω_m is

$$p(\omega_m|\sigma^2) = \int p(\omega_m|\alpha_m)p(\alpha_m|\sigma^2)d\alpha_m \quad (5)$$

Equation (5) shows that $p(\omega_m)$ is a scale mixture of Gaussians, and in most conditions, it has positive kurtosis. Because almost any prior on $\alpha_m|\sigma^2$ will favor sparsity to a certain extent, there is considerable freedom in its choice. Empirically, the $p(\omega_m)$ resulting from a uniform $p(\alpha_m|\sigma^2)$ does not enforce sparsity for elastic kernel types, and sparser prior over $\alpha_m|\sigma^2$ is desirable.

Based on the introduction of SRVM, the algorithm flowchart is shown in Fig. 1.

Algorithm 1 The sRVM algorithm.

```
 1:  σ² ← 0.1 × var(t)                              initialization for σ²
 2:  α ← [∞ ··· ∞]ᵀ                                 start with the empty model
 3:  i ← argmax (‖φₘᵀt‖/‖φₘ‖)                       pick an i that stands a good chance of being relevant
 4:  S ← {i}                                        include it in the set of included components
 5:  update(α_i, Σ_S, μ_S, s, q)                    compute initial values for all model paramters
 6:  R ← 10                                         reestimate noise every R steps
 7:  step ← 1                                       already made the first step
 8:  until converged()
 9:      i ← randint(M)                             pick a random component i
10:      DID-NOTHING ← False
11:      if ( q_i² − s_i > 0 and ...
              has-real-positive-root(numerator(ê'(α_i)))) if component i is relevant
12:          unless α_i < ∞                         unless it is already included
13:              S ← S∪{i}                          add it
14:      else
15:          if α_i < ∞                             the component is irrelevant but currently included
16:              S ← S \{i}                         delete it
17:          else
18:              DID-NOTHING ← True                 component is and was irrelevant
                                                    no need for action
19:      unless DID-NOTHING                         otherwise update everything
20:          step ← step + 1
21:          update(α_i, Σ_S, μ_S, s, q)            update the model parameters
22:          if step mod R = 0
23:              reestimate(σ²)
```

Fig. 1 The algorithm of SRVM

3 Passenger Flow of Xi'an Subway System

In this paper, we choose the subway line 2 in Xi'an as an example. The 5 min scale passenger flow in September 2017 is shown in Fig. 2.

4 Forecast Result of Short-Term Passenger Flow

We take five days passenger flow data of September 2017 as training data, and according to the SRVM algorithm, the passenger flow data of September 19, 2017, were predicted. The simulation results are shown in Fig. 3. Mean absolute percentage error (MAPE) was calculated based on the simulation, MAPE = 6.85%.

5 Conclusion

Passenger flow prediction is significant for public traffic management. In this paper, the influence of choosing prior structure on a short term passenger flow data sets is presented. The SRVM algorithm is used to predict the short-term passenger flow. The

The Forecast of the Subway Passenger Flow Based on Smooth ...   27

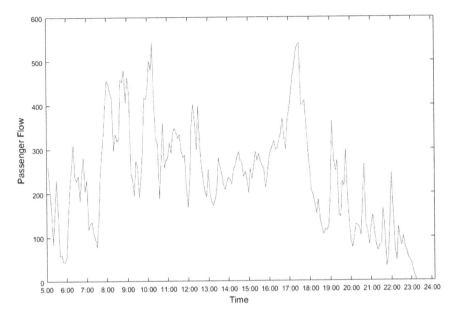

Fig. 2 Passenger flow with 5 min of Beikezhan station

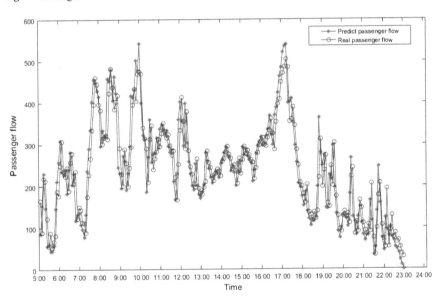

Fig. 3 The prediction results of passenger flow

forecast results reveal that SRVM algorithm could forecast the short-term passenger flow effectively.

Acknowledgements This work was supported in part by the National Natural Science Foundation of China (No. 51407012) and in part by the Fundamental Research Funds for the Central Universities, CHD (No. 300102329102, 300102328201).

References

1. Xianjinet (2018) Opinions on further strengthening the planning and construction management of urban rail transit. Available via DIALOG. https://www.xianjichina.com/news/details_78814.html (in Chinese)
2. Zhong C, Batty M, Manley E, Wang J, Wang Z, Chen F et al (2016) Variability in regularity: mining temporal mobility patterns in London, Singapore and Beijing using smart-card data. PLoS ONE 11(2):e0149222
3. Ma XL, Wang YH, Chen F, Liu JF (2012) Transit smart card data mining for passenger origin information extraction. J Zhejiang Univ Sci C 13(10):750–760
4. Leng B, Zeng J, Xiong Z, Weifeng LV, Wan Y (2013) Probability tree based passenger flow prediction and its application to the beijing subway system. Front Comput Sci 7(2):195–203
5. Sun Y, Leng B, Guan W (2015) A novel wavelet-SVM short-time passenger flow prediction in Beijing subway system. Neurocomputing 166:109–121
6. Bin YU, Shan-Hua WU, Wang MH, Zhao ZH (2012) K-nearest neighbor model of short-term traffic flow forecast. J Traffic Transp Eng
7. Chan KY, Dillon TS, Singh J, Chang, E (2011) Traffic flow forecasting neural networks based on exponential smoothing method. Ind Electron Appl
8. Frejinger E, Bierlaire M (2007) Capturing correlation with subnetworks in route choice models. Transp Res Part B (Methodol) 41(3):363–378
9. Dou F, Jia L, Qin Y, Jie XU, Wang L (2014) Fuzzy k-nearest neighbor passenger flow forecasting model of passenger dedicated line. J Central South Univ 45(12):4422–4430
10. Xu C, Li Z, Wei W (2016) Short-term traffic flow prediction using a methodology based on autoregressive integrated moving average and genetic programming. Transport 31(3):343–358
11. Li Z, Bi J, Li Z (2017) Passenger flow forecasting research for airport terminal based on SARIMA time series model. In: IOP conference series: earth and environmental science, vol 100, p 012146
12. Zheng D, Wang Y (2011) Application of an artificial neural network on railway passenger flow prediction. In: International conference on electronic & mechanical engineering and information technology
13. Xie MQ, Li XM, Zhou WL, Fu YB (2014) Forecasting the short-term passenger flow on high-speed railway with neural networks. Comput Intell Neurosci 2014(9):23
14. Sun Y, Leng B, Wei G (2015) A novel wavelet-SVM short-time passenger flow prediction in beijing subway system. Neurocomputing 166(C):109–121
15. Yang J, Hou Z (2013) A wavelet analysis based LS-SVM rail transit passenger flow prediction method. China Railway Sci 34(3):122–127
16. Qian C, Li W, Zhao J (2011) The use of LS-SVM for short-term passenger flow prediction/maziausiu kvadratu atraminiu vektoriu metodo taikymas trumpalaikiam keleiviu srautui prognozuoti. Transport 26(1):5–10
17. Hu Y, Chong W, Liu H (2011) Prediction of passenger flow on the highway based on the least square support vector machine/maziausiu kvadratu atraminiu vektoriu metodo taikymas keleiviu srautui greitkelyje prognozuoti. Transport 26(2):197–203

18. Bhadra D, Gentry J, Hogan B, Wells M (1971) Future air traffic timetable estimator. J Aircr 42(42):320–328
19. Zhang Z, Xu X, Wang Z (2017) Application of grey prediction model to short-time passenger flow forecast. In: American institute of physics conference series
20. Kong C, Xu K, Wu C (2006) Classification and extraction of urban land-use information from high-resolution image based on object multi-features. J China Univ Geosci 17(2):151–157

Dynamic Debugging System of EMU Based on Cloud Platform

Yu Qiu, Lide Wang, Jie Jian, Chaoyi Li and Yu Bai

Abstract At present, China's EMU debugging system is still in the "paper-based stage," realizing the automation of EMU debugging is the future development trend. Cloud platform has the characteristics of resource integration and parallel processing of large amounts of data, and it can comprehensively process and store a large amount of data. Thus, it can process and store large amounts of data in an all-round way to meet the real-time processing and storage requirements of dynamic debugging system of EMU. This paper proposes a cloud-based dynamic debugging system of EMU, puts forward the overall structure of the system, and designs the cloud platform of dynamic debugging system in detail. Finally, tacking the debugging of CRH380 EMU as an example, the system function is explained and verified. The example shows that compared with the traditional dynamic debugging system of EMU, this system integrates the decentralized debugging process into a whole, realizes the collaborative work among departments, and improves the debugging efficiency, which is a feasible scheme.

Keywords Rail transit · EMU · Dynamic debugging · Cloud platform

1 Introduction

Dynamic debugging of electric motor train unit (EMU) is an important means to verify whether its functions reach its standard before leaving the factory. Therefore, the dynamic debugging technology of EMU plays a decisive role in the operation quality of rail transit. The traditional dynamic debugging of EMU is still in the "manual recording stage," and there are four main problems [1]:

1. Dynamic debugging data cannot be stored in real time. If a transient fault is caused by some coincidence during the debugging process, it is difficult to reproduce and analysis in the subsequent debugging process.

Y. Qiu (✉) · L. Wang · J. Jian · C. Li · Y. Bai
School of Electrical Engineering, Beijing Jiaotong University, Beijing 100044, China
e-mail: 17121484@bjtu.edu.cn

2. The debugging of each system is relatively independent, there is no unified data processing platform, and the debugging work is scattered.
3. Manually record and summarize the debugging results, and manually store the debugging results to the database. There is no redundancy in storage devices, and the data stored is relatively single. Once the storage device fails, it can only be used after the device is replaced or repaired.
4. Lack of a unified historical database.

In view of the above problems, many scholars have studied the EMU debugging system, and the research mainly focusing on data acquisition and data processing [1, 2]. EMU debugging system includes data processing, data storage, and data query functions. The system structure is scattered and there is no unified management platform integration system function. For the current EMU debugging system, (1) the debugging management personnel cannot monitor the debugging progress in real time, nor can they adjust the debugging results after debugging. (2) The debugging departments are relatively independent, unable to coordinate and adjust the debugging progress in real time, affecting the real time and reliability of debugging.

The advent of 5G era has promoted the development of cloud platform. The rail transit industry has begun to build its own private cloud to serve its industry. For example, China has established a railway information cloud platform [3]. At present, the research of cloud platform in the field of rail transit focuses on train ticketing system, information system, and other fields [4–8], mainly for system integration and big data processing.

Based on the above analysis, it is an inevitable trend to develop a dynamic debugging system for EMU, that can integrate decentralized debugging resources, comprehensively analyze and store a large amount of data, and has high reliability and efficiency. This paper presents a dynamic debugging strategy for EMU based on cloud platform. This strategy takes advantages of the heterogeneity and the resource sharing of the cloud platform, improves the efficiency and reliability of dynamic debugging, and realizes intelligent dynamic debugging of EMU.

2 System Design

2.1 Overall Architecture of Dynamic Debugging System for EMU

The cloud-based dynamic debugging system for EMU is composed of data acquisition device and dynamic debugging cloud platform. The structure is shown in Fig. 1.

The data acquisition device (DAD) is mainly responsible for collecting train operation status data and transferring the collected data to the cloud platform. In the process of data transmission, firstly, data is packaged into stream data model, and then

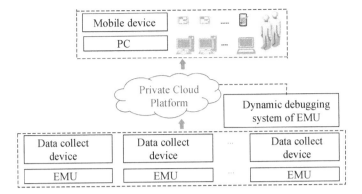

Fig. 1 Dynamic debugging system architecture based on cloud platform

Fig. 2 Stream data structure model

Type of information (2Byte)	Information length (4Byte)	Data content (n Byte)

data transmission between DAD and cloud platform is completed by using multi-threaded socket technology. The data information structure is shown in Fig. 2. For the boundary problem of data transmission, length information is added in the data content part to solve the boundary problem.

2.2 Design of Cloud Platform Architecture for Dynamic Debugging of EMU

The dynamic debug cloud platform uses a distributed architecture for data processing, data storage, and data management. The cloud platform designed by this system is mainly composed of four parts: basic resource layer, basic service layer, service support layer, and application layer. Its structure is shown in Fig. 3.

The basic resource layer is the foundation of the cloud platform, which mainly includes hardware devices such as servers, storage, and networks. The virtual resource pool is the data center of cloud platform, which integrates these hardware resources into virtual resource pool by virtualization technology. The virtual resource pool can provide basic resource services such as computing, storage, and network for applications according to the needs of the upper layer.

The basic service layer consists of the basic components of the software. It calls the resources of the basic resource layer through the program interface and opens the basic resources to the service support layer. The basic service layer is divided into three sub-platforms: development layer, scheduling layer, and management layer. The development sub-platform provides users with tools and language execution environments for software development and debugging. The scheduling sub-platform

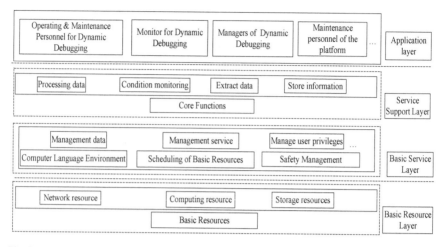

Fig. 3 Structural model of dynamic debugging cloud platform for EMU

is used to monitor resource usage and instance running status, deploy, and allocate basic resources according to user requests, which can meet the requirements of multi-user and multi-task concurrent requests. The management sub-platform realizes data management, service management, authority management, and security management of the cloud platform.

Service support layer provides data resource interface, hardware and software resource interface, and communication interface for dynamic debugging platform, which is used to process and store data.

The application layer is a human–computer interaction interface. Staff can log into the cloud platform through the application layer for debugging work. The cloud platform provides the user with debugging services within its authority according to the user login information.

As the data processing center of the dynamic debugging system, the cloud platform is the core component of the intelligent dynamic debugging of the EMU. The dynamic debugging model based on the cloud platform is shown in Fig. 4.

2.3 Monitoring Object of Dynamic Debugging System for EMU

Dynamic debugging of emus is conducted for all kinds of vehicle systems, such as traction operation test, and so on. Different systems have their own characteristics. For example, the traction operation test is divided into start-up test, traction performance test, constant speed operation test, etc.

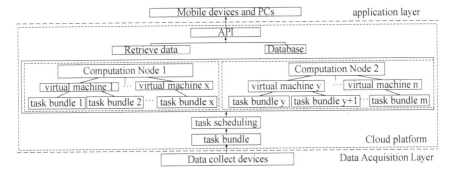

Fig. 4 Cloud-based dynamic debugging model

Dynamic debugging data was obtained by various types of onboard sensors, and the data package of train operation is 616 bytes in total, including train number, running speed, traction voltage value, motor current feedback, BS pressure, and so on.

2.4 Realization of Automatic Dynamic Debugging for EMU

The system is divided into interface layer, logic layer, debugging service layer, data access layer, database layer, and system layer. The interface layer displays the status monitoring data, debugging task form, fault data, and database management of EMU by calling business logic layer and data access layer. The business logic layer is used to parse and process the debug data. The data access layer implements the object-relation mapping (ORM) function [9], maps the debug data to the database table, and provides an interface for the upper layer component to access the database, including functions such as adding, modifying, and querying data. Software development components all depend on the basic components of the underlying service layer. In addition, the virtual network technology is used in the dynamic debugging cloud platform of EMU to realize the communication between systems and the transmission and reception of background data. The service layer encapsulates the data buffer for data transmission, and the background service thread fetches the data in the buffer through the business logic layer, the data access layer, and the database layer to persist it in the database. The software design realizes intelligent collaborative processing and real-time storage of dynamic debugging data. The software flow is shown in Fig. 5.

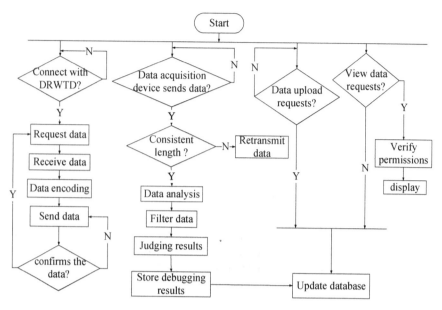

Fig. 5 Intelligent dynamic debugging software flow of EMU

3 Example Analysis

Taking the dynamic debugging of CRH380 EMU as an example, the workflow of the dynamic debugging system based on cloud platform is illustrated.

1. Data acquisition. At the beginning of debugging, data acquisition devices communicate with vehicle-mounted wireless transmission device (DRWTD) via Ethernet. The data acquisition device requests to establish a connection with DRWTD. After the connection is established, the device is turned on, the software starts automatically, and the registration frame and data request frame are started to be sent. The real-time data of the train running status received is displayed in TextBox. The collected data is transmitted to the cloud platform server through protocol conversion. The data acquisition interface is shown in Fig. 6. As shown. When the data is displayed in TextBox, the communication between the vehicle data acquisition device and DRWTD has been established.
2. Data processing. Dynamic debugging software is used to process the received data. Firstly, the data is parsed according to the specific decoding method, and then the data is brought into the set discriminant function processing model. Finally, the debugging results are generated according to the model. The data processing flow is shown in Fig. 7. The cloud platform parses and displays the data. Take the system status monitoring interface as an example, as shown in Fig. 8.

Debugging managers can check the running status of trains in real time by landing on the cloud platform. From the figure, we can see that the number of

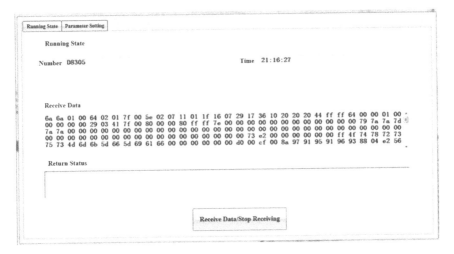

Fig. 6 Data acquisition interface

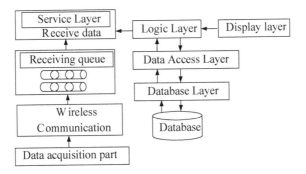

Fig. 7 Data processing flow

trains debugged is D8305, the group number is 2400A, the accumulated mileage of the train is 208760 km, and the current running gear is P4, the speed is 20 km/h, and the acceleration is 1.3 km/h/s. The curve shows the change of train speed.

The generated debugging results and train data are automatically stored in the data management center. The debugging results are shown in Fig. 9.

The debugging results contain debugger information, vehicle information, and debugging project information. The debugging results show the status of the train in the current debugging project. The staff can develop a follow-up debugging plan based on the debugging results and historical debugging data.

3. Equipment maintenance and commissioning. Workers log on to the dynamic debugging cloud platform to view the data processing results. The debugging manager can adjust and assign the debugging plan according to the debugging information, and the debugging staff can view and download the debugging task table for a new round of debugging work.

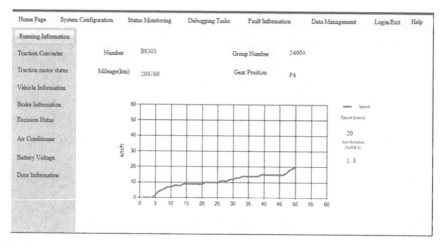

Fig. 8 Train operation information

Quality Inspector	Vehicle Type	Experiments	Project Page	Edition	Vehicle Type	T1	M1	M2	T2	T3	M3	M4	T4
	CRH2	Dynamic Debugging		A									
Project	Operations/Steps		Confirmation/Criteria			Judgment/Data							
	1	Vehicle Inspection Bottom of the train	The equipment cover of the train bottom has been fixed			good	good	good	good	good	good	good	good
			The connection between the two carriages is in good condition			good	good	good	good	good	good	good	good
			No abnormal connection state between bogies			good	good	good	good	good	good	good	good
			Confirm that the wire of the bottom plate mounting bolt removed in the static adjustment is in good condition and the			good	good	good	good	good	good	good	good

Fig. 9 Partial debugging results

4 Conclusion

Aiming at the decentralization of dynamic debugging system of EMU, this paper designs a debugging system of EMU based on cloud platform. The dynamic debugging system of EMU is designed and implemented and the cloud platform is applied to the dynamic debugging system of EMU. The functions of data processing, data storage, data access, debugging, and monitoring of the dynamic debugging system are integrated into a whole. What is more, relevant staff work cooperatively through landing on the cloud platform. As an information management center, cloud platform enables debuggers of various departments to view the debugging progress of different departments in real time, which is conducive to coordinated debugging operations. It also enables debugging managers to monitor debugging work in real time, meanwhile timely adjusts the debugging plan and allocation debugging tasks according to the debugging conditions. Finally, the dynamic debugging of CRH380 EMU is taken as an example to illustrate and verify the system.

Acknowledgements This work is supported by the Beijing Municipal Natural Science Foundation under Grant L171009.

References

1. Zhao J, Shen H, Li C (2016) Design and realization of the automatic test system for EMU train based on MVB dataflow. Comput Meas Control 24(6) (in Chinese)
2. Chang J (2015) Application of data analysis technology in the application of EMU digital debugging line. Comput Meas Control 23(11):3759–3761+3765 (in Chinese)
3. China Railway Information Department (2017) Introduction to the achievements of china railway information technology center. J China Railway Soc 39(03):129 (in Chinese)
4. Ning B, Mo Z, Li K (2019) Application and development of intelligent technologies for high-speed railway signaling system. J China Railway Soc 41(03):1–9 (in Chinese)
5. Du S, Ma M, Wang P (2015) A Cloud-based high-speed rail information service platform architecture. In: 2015 Ninth international conference on frontier of computer science and technology. IEEE
6. Zhu L, Yu FR, Fellow et al (2018) Big data analytics in intelligent transportation systems: a survey. IEEE Trans Intell Transp Syst 20(1):383–398
7. Zhao S, Zhang N, Wang J (2018) System architecture and key technologies of rail transit AFC system based on private cloud platform. Urban Mass Transit 189(06):99–102 (in Chinese)
8. Ruscelli AL, Cecchetti G, Castoldi P (2016) Cloud networks for ERTMS railways systems (short paper). In: IEEE international conference on cloud networking. IEEE
9. Guo Y, Ma X, Zhang Li (2016) Design and implementation of the.Net ORM Middleware based on database control. Electron Des Eng 24(3):29–31 (in Chinese)

Health Assessment Method for Controller Area Network in Braking Control System of Metro Train

Yueyi Yang, Xingwen Guan, Yinhu Liu and Bin Xue

Abstract Controller area network (CAN) has been widely used in many control subsystems of metro trains because of its high real-time performance and low cost. In the braking control system, CAN bus is used to connect multiple electronic brake control units (EBCU) and transmit braking instructions. Once the network fails, it will endanger the safety of driving. In order to warn early faults and detect network performance degradation in time, the common faults of CAN bus are carefully investigated, and a health assessment method based on self-organizing map network (SOM) is proposed to monitor network performance and detect early network anomaly. The characteristic parameters of the physical layer are extracted, and the SOM model representing the normal patterns of the CAN devices is trained by the normal dataset. Performance degradation is indicated by the distance between the test sample and the SOM best matching unit (BMU). Experimental results show the validity of the proposed method for anomaly detection of CAN bus.

Keywords Self-organizing map · Health assessment · Anomaly detection · Controller area network

1 Introduction

Under the rough environment of metro train, CAN bus is easier to suffer from interference. Accumulative degradation of network performance will lead to network failure, which will cause unnecessary economic losses and even security risks. Therefore, reasonable network health monitoring and condition assessment are very important.

In recent years, some studies focused on the probability distribution of the worst response time of messages [1–3], but the researchers took no account of many random factors, the variation of message response time could not be reflected effectively. Lei

Y. Yang (✉) · X. Guan
School of Electrical Engineering, Beijing Jiaotong University, 100044 Beijing, China
e-mail: 17117405@bjtu.edu.cn

Y. Liu · B. Xue
Nanjing CRRC Puzhen Haitai Brake Equipments Co., Ltd., 211800 Nanjing, Jiangsu, China

et al. [4] extracted the features from the physical-layer information and fused the timestamp information of data link layer, and multi-dimensional clustering algorithm was used to detect faults and evaluate performance degradation. This method plays a forward-looking role in the diagnosis and prediction of fault. Suwatthikul et al. [5] constructed an adaptive-network-based fuzzy inference system (ANFIS), using the fault signals of the existing CAN bus such as error frames, the CAN network-level fault pre-diagnosis based on ANFIS was adopted, and the health status of the CAN bus was divided into different levels, but the equivalent resistance value is unknown in practice which cause low classification accuracy. Some scholars have proposed that the network should be regarded as a discrete event system, and the data flow path of the network can be constructed into a Markov model. In view of the sudden failure, the alarm information can be sent out through state prediction [6]. Literature [7] presented an evaluation method based on the reliability of node disconnection time, which predicts the off-line time of nodes by monitoring the change of transmit error counter (TEC) of any node.

In view of the existing problems, a SOM-based health assessment method is proposed, which uses the physical-layer information of CAN bus to monitor the network status, detect the performance degradation in time, and provide a scientific theoretical basis for condition-based maintenance.

2 Overview of Health Assessment Method for CAN Bus

2.1 Common Faults of CAN Bus

In this paper, a metro train with 6 compartments is taken as an example, each train carriage is equipped with an electronic brake control unit (EBCU), the braking instructions are received by EBCU in trailer car and transmitted among six EBCUS through CAN bus, the schematic diagram of braking control network for metro train is shown in Fig. 1.

The fault types of CAN bus include degradation of cable performance, mismatch of network impedance, intermittent connection fault, and grounding problem [4]. The degradation of cable performance and mismatch of network impedance are common faults in braking control system in practice and considered in this paper. These faults can result in physical waveform distortion, and the fault symptoms are persistent, so we can use the characteristic parameters of the physical layer for health assessment.

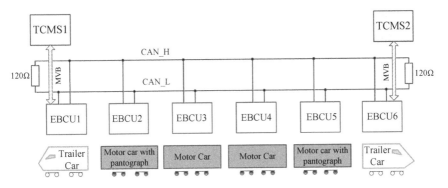

Fig. 1 Schematic diagram of braking control network for metro train

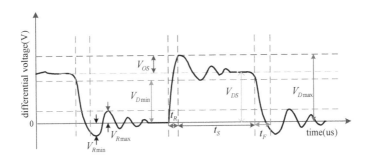

Fig. 2 Characteristic extraction of the differential signal physical waveform

2.2 Feature Extraction

According to the requirement of ISO11898 for conformance testing of CAN bus physical layer and related enterprise standards, the conformity characteristic parameters and their ranges of CAN bus differential voltage physical waveform in normal operation can be obtained. The principle diagram of CAN bus differential voltage features is shown in Fig. 2.

The physical waveform characteristic parameters are extracted and are presented in Table 1:

2.3 Health Assessment Method Based on SOM

SOM, also known as Kohonen maps, is a 2D projection of a multi-dimensional feature space, which has the capability to maintain the relevant information contained in the raw data. SOM has the advantages of automatic feature mapping, simple network

Table 1 Characteristic parameters of differential voltage

Feature name	Minimum	Maximum	Average	Standard deviation
Amplitude of dominant bits	V_{Dmin}	V_{Dmax}	V_{DS_ave}	V_{DSD}
Amplitude of recessive bits	V_{Rmin}	V_{Rmax}	V_{RS_ave}	V_{RSD}
Rise time			t_{R_ave}	t_{RSD}
Fall time			t_{F_ave}	t_{FSD}
Overshoot			σ_{ave}	σ_{SD}
Bit time			t_{B_ave}	

construction and good stability, which can be used for fault detection [8, 9]. In this paper, a health assessment method for CAN bus based on SOM is presented.

Literature [10] summarizes the principle of SOM algorithm and gives suggestions on the selection parameters. Each neuron i in the SOM is represented by an m-dimension weight vector $w_i = [w_{i1}, w_{i2},\ldots, w_{im}]$. After obtaining the optimal weight matrix of the corresponding training sample set, Euclidean distance MQE_T between each test sample and its best matching neurons can be obtained:

$$\mathrm{MQE\_T} = \|T - m_{\mathrm{BMU}}\| = \sqrt{(x_i - w_{\mathrm{BMU}})^2} \tag{1}$$

The standard deviation of MQE trained by normal samples is obtained:

$$\delta = \sqrt{\frac{1}{N_h - 1} \sum_{i=1}^{N_h} (\mathrm{MQE}_i - \mathrm{MQE\_ave})^2} \tag{2}$$

where N_h is the number of normal training samples and MQE_ave is the average value of MQE trained from normal samples, which can define the current network health index quantity H:

$$H = |\mathrm{MQE\_T} - \mathrm{MQE\_ave}| - 3\delta \tag{3}$$

when $H > 0$, the network is abnormal, otherwise, the network is normal.

On the basis of the formulas (2) and (3), the expressions representing the health degree D of the CAN nodes can be obtained:

$$D = \frac{c}{\sqrt{H} + c} \times 100\%, H > 0 \tag{4}$$

where c is a scale parameter, which can be determined according to the MQE value trained from normal samples. The c can be formulated as:

$$c = \sqrt{\frac{1}{N_h} \sum_{i=1}^{N_h} |\mathrm{MQE}_i|} \tag{5}$$

Therefore, when H is larger, the node health D is lower, and the abnormal degree is higher.

It is noteworthy that D is only for the performance quantization index of CAN node level. Because of CAN bus communicates with message as its basic unit, we can consider that the more the number of messages sent by a node in a time window, the more important the node is to the network. Assuming that node X sends P messages in a time window, and the total number of messages is Q, the proportion of messages sent by node X is:

$$\text{ratio} = P/Q \tag{6}$$

The proportion of messages sent by nodes in the observation time window is taken as an index of the importance of nodes to the whole network. Therefore, the relationship between the quantitative index D of node performance and the quantitative index D_C of network performance can be obtained:

$$D_C = \frac{\sum_{i=1}^{n} \text{ratio} \times D_i}{T} \tag{7}$$

where n is the total number of nodes participating in sending and receiving messages in CAN bus, and T is the number of message cycles contained in the observation time window. Besides, D_C is the quantitative result of the CAN bus performance that we need.

3 Experiment Analysis

3.1 Setup of Experiments

A CAN bus communication platform is built as shown in Fig. 3, it includes five STM32 microcontrollers and a USB-CAN Analyzer, and the communication speed is 250 Kbit/s. The fault injection device can be used to inject analog interference into CAN bus. The data acquisition device with sampling frequency of 20 MS/s is used to collect physical waveform data of CAN bus, and the data collected is transmitted to PC software through USB for physical waveform restoration and feature extraction.

Finally, in the normal state of the network, 700 training samples were collected, and 1168 test samples were collected in different network states. According to the actual distribution of 1168 test samples collected, we collect data in nine time windows for different network states and count the source nodes of messages according to the order of test samples saved, as shown in Table 2.

The principal component analysis (PCA) is used to reduce the feature dimension. The number of neurons in the competitive layer of SOM is 132, the initial learning rate is 0.5, the radius of the initial winning field is 2, and the maximum number of

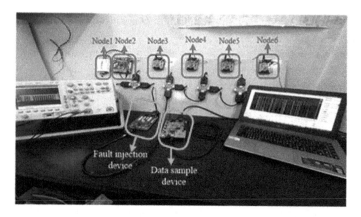

Fig. 3 CAN bus test-rig

training times is 10,000. The Euclidean distance MQE between each SOM training sample and the best matching neuron was analyzed, the results showed that the data presented a trend of approximate normal distribution, which accorded with the presupposition.

3.2 Experiment Results

According to the formula (3), corresponding network health index H is obtained, as shown in Fig. 4. When SOM_H < 0, the sample label is classified as 1 indicating the normal state, and when SOM_H > 0, the sample label is classified as-1 indicating the abnormal state. After the training phase, the average Euclidean distance between training samples and the best matching neurons was MQE_ave = 0.41589, and the standard deviation was $\delta = 0.16359$.

Three metrics including the precision, the recall, and F1-score are used to evaluate the performance of the SOM model. The test set is made up of 1168 samples from each node in different fault types, the precision is up to 0.99902, the recall is 0.94717, weighted harmonic mean F1-score is 0.97240, and test accuracy achieves 95%.

The master–slave structure is adopted in CAN bus communication platform, the percentage of periodic messages in the master node is 1/2, and that of each slave node is 1/8, which is equivalent to half of the periodic messages of master node and all slave nodes, respectively. In order to facilitate calculation, feature acquisition is carried out in a complete sequence of periodic messages, so the proportion of each node in the observation window at a certain time is equivalent to the proportion of periodic messages. According to the obtained H, combination (4) and (5), the quantization results of node state can be obtained. The final result is shown in Fig. 5.

SOM_D is the corresponding node health D. Therefore, according to the formula (7) and the periodic message proportion of each node, the node health D of different observation time windows can be mapped to the network health D_C.

Table 2 Test samples in different fault types

Message source node	Remove a terminal resistance (Ω)	Parallel a terminal resistance (Ω)				Connect a resistor in series from slave 1 (Ω)				Normal
		40	60	80	100	20	40	60		
Slave1~4	120	46	39	39	60	82	40	72		46
Master	198	34	39	37	56	75	39	50		42
Total	184	70	78	76	116	157	79	122		88
	382									

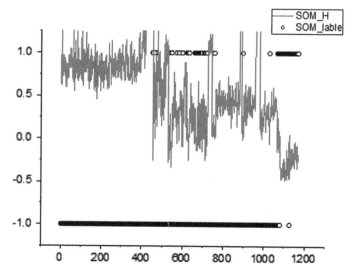

Fig. 4 Node health indicators

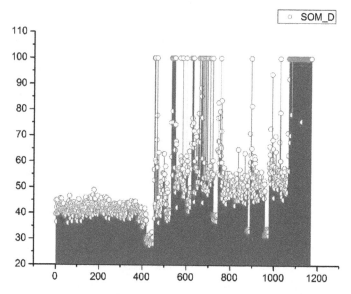

Fig. 5 Node state quantization based on SOM

Therefore, the network state quantization results of nine observation time windows are obtained as shown in Fig. 6, which conform to the actual situation of network state of different observation time windows.

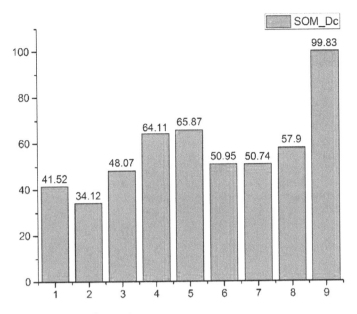

Fig. 6 Quantitative results of network state

4 Conclusion

In this paper, the features are extracted from the physical layer of CAN bus in the braking control system, and a health assessment approach based on SOM is proposed. The performance of node and network are quantified and anomaly threshold can be set, which can provide assistant decision-making for the condition-based maintenance of metro train. Finally, the validity of the proposed method is verified by the experiments of health assessment for different faults.

Acknowledgements This work was supported in part by the Fundamental Research Funds for the Central Universities under Grant 2018YJS148, and in part by the Beijing Municipal Natural Science Foundation under Grant L171009.

References

1. Zeng H, Natale MD, Giusto P (2009) Stochastic analysis of CAN-based real time automation systems. J IEEE Trans Industr Inf 388–401
2. Gao D, Wang Q, Gao D, Wang Q (2014) Health monitoring of controller area network in hybrid excavator based on the message response time. In: IEEE/ASME international conference on AIM, pp 1634–1639
3. Sun Y, Yang F, Lei Y (2015) Message response time distribution analysis for controller area network containing errors. In: 2015 Chinese automation congress (CAC), pp 1052–1057

4. Lei Y, Djurdjanovic D, Barajas L, Workman G, Biller S, Ni J (2011) Devicenet network health monitoring using physical layer parameters. J Intell Manuf 22(2):289–299
5. Suwatthikul J, Mcmurran R, Jones RP (2011) In-vehicle network level fault diagnostics using fuzzy inference systems. Appl Soft Comput 11(4):3709–3719
6. Short M, Sheikh I, Rizvi SAI (2011) Bandwidth-efficient burst error tolerance in TDMA-based CAN networks. Short M 19(6):1–8
7. Zhang LM, Tang LH, Lei Y (2017) Controller area network node reliability assessment based on observable node information. Front Inf Technol Electron Eng 18:615–626
8. Chong B, Yang SK et al (2019) Degradation state mining and identification for railway point machines. Reliab Eng Syst Saf 188:432–443
9. Lu WP, Yan XF (2019) Visual monitoring of industrial operation states based on kernel fisher vector and self-organizing map networks. Int J Control Autom Syst 17:1535–1546
10. Rigamonti M, Baraldi P, Zio E, Alessi A, Astigarraga D, Galarza A (2016) Identification of the degradation state for condition-based maintenance of insulated gate bipolar transistors: a self-organizing map approach. Microelectron Reliab 60:48–61

High-Temperature Zone Localization Image Processing Algorithm for Pantograph Infrared Image

Wen Xu, Zhaoyi Su, Guangtao Cong and Zongyi Xing

Abstract The infrared temperature of the pantograph is one of the important technical parameters in the train operation and it is important to monitor the infrared of the train in time to ensure the safety of the train. This paper proposes a high-temperature region localization image processing algorithm for pantograph infrared image. Smoothing the image Gaussian filtering should be performing on the image data collected in the field at first; After performing grayscale processing on the filtered image, image morphology is enhanced according to image pixel grayscale differences; The k-means algorithm is used to cluster the regions with higher temperatures to locate the target detection regions. Compared with manual observation, the system detection consistency is better than manual observation, and the calculation formula can meet the requirements of measurement accuracy and real time. Comparing with the effect photos in different values, the best parameters are obtained, which can help for further research.

Keywords Gaussian filtering · Grayscale processing · Image morphology · K-means algorithm

1 Introduction

During the running of the train, arc heat, frictional heat, and Joule heat are generated between the pantograph and the catenary. Thereby, the temperature of the pantograph is raised.

W. Xu · G. Cong · Z. Xing (✉)
School of Automation, Nanjing University of Science and Technology, Jiangsu, China
e-mail: xingzongyi@163.com

Z. Su
Guangzhou Metro Group Co., Ltd., Guangzhou, China

© Springer Nature Singapore Pte Ltd. 2020
B. Liu et al. (eds.), *Proceedings of the 4th International Conference on Electrical and Information Technologies for Rail Transportation (EITRT) 2019*, Lecture Notes in Electrical Engineering 640, https://doi.org/10.1007/978-981-15-2914-6_6

By detecting the temperature condition of the pantograph, the pantograph fault can be effectively detected and also can reduce the accident and ensure train safety [1].

Jinbiao Zhu and others from Southwest Jiao Tong University chose the full-radiation temperature measurement method as the way to realize the temperature measurement of the bow network, and proposed a temperature measurement system based on pyroelectric detector.

In this article, by taking an infrared image of the pantograph, the image is used to locate the temperature rise zone of the pantograph [2].

2 System Architecture

The pantograph temperature detection system is mainly responsible for real-time detection of the temperature condition of the pantograph during train operation.

By analyzing the temperature condition of the critical part of the pantograph, it is an important part to ensure the safety of the bow to judge whether the pantograph is faulty.

The hardware structure of the pantograph temperature detection system mainly includes a thermal imager triggering device and an infrared image capturing device.

The specific equipment layout is shown in Fig. 1.

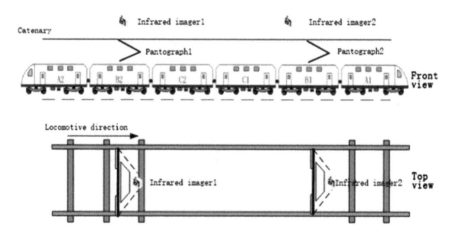

Fig. 1 Schematic diagram of equipment arrangement of pantograph temperature detection system

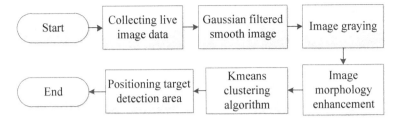

Fig. 2 Algorithm flow framework

3 System Data Processing Algorithm

3.1 Algorithm Structure

After obtaining the on-site pantograph infrared image, it needs to be processed to obtain the temperature rising area, thereby positioning the target area [3]. The flow chart is shown in Fig. 2.

3.2 Gaussian-Filtered Smooth Image

Gaussian filtering eliminates noise on infrared images. In order to generate a 3 × 3 Gaussian filter template, the center of the template is taken as the coordinate origin. The template coordinates for each position are shown in Fig. 3.

The coordinates of each position are input into the Gaussian function, and the obtained value is the coefficient of the template. For the size of 3 × 3 in the window template, the formula for calculating the value of each element in the template is as follows:

$$H = \frac{1}{2\pi\sigma^2} e^{-\frac{(i-k-1)^2+(j-k-1)^2}{2\sigma^2}}$$

The original image is discretized by Gaussian filtering, and the Gaussian distribution parameter sigma determines the width of the Gaussian function. A

Fig. 3 3 × 3 Gaussian filter template

(-1,1)	(0,1)	(1,1)
(-1,0)	(0,0)	(1,0)
(-1,-1)	(0,-1)	(1,-1)

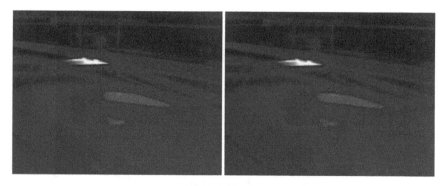

Fig. 4 Infrared detection original image and infrared detection Gaussian filter

two-dimensional Gaussian function is as follows:

$$H = \frac{1}{2\pi\sigma^2} e^{-\frac{(i-k-1)^2+(j-k-1)^2}{2\sigma^2}}$$

(x, y) is the point coordinate, which can be considered as an integer in image processing; σ is the standard deviation.

The original image data and the processed data are shown in Fig. 4.

3.3 Grayscale Processing

It is necessary to convert colorful images into grayscale images through image processing technology during the arcing process of the pantograph [4]. This process is called image grayscale processing.

Grayscale the image as follows:

The collected original image data is subjected to maximum-scale gradation processing, and each picture composed of one pixel is present, and the expression is as follows:

$$f(x, y) = \begin{bmatrix} f(0,0) & f(0,1) & \cdots & f(0,0) \\ f(1,0) & f(1,1) & \cdots & f(0,0) \\ \vdots & \vdots & & \vdots \\ f(i,0) & f(i,1) & \cdots & f(i,j) \end{bmatrix}, i \in (0, x-1), j \in (0, y-1) \quad (1)$$

It will present a dot matrix which each dot is a pixel after enlarging the image. Different colors can be displayed by matching the colors of RGB, wherein each pixel is composed of three channels [5].

In this paper, the maximum value of the three-channel luminance in the colorful image is processed as the grayscale value:

Fig. 5 Infrared detection grayscale map

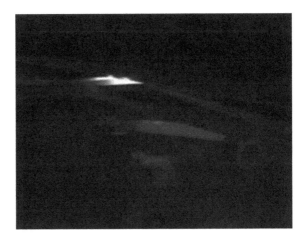

$$\text{Gray}(i, j) = \max(R(i, j), G(i, j), B(i, j))$$

In the field image data obtained, the pantograph and the background are relatively large, it is better to process the image by brightness maximum grayscale processing.

Figure 5 shows the grayscale image.

3.4 Morphological Enhancement (Expansion Corrosion)

An important research direction of image processing theory is mathematical morphology. It extracts useful image components in the image by expressing and characterizing the shape of the region by using specific structural elements.

Among them, the white top cap transformation refers to subtracting the image that has been first etched and then swelled from the original image, and the calculation formula is:

$$\text{WTH}_{\text{Gray},b}(x, y) = \text{Gray}(x, y) - ((\text{Gray} \otimes b) \oplus b)(x, y) \tag{2}$$

The black top-hat transformation refers to the image after the first expansion and corrosion, minus the original image, and the calculation formula is:

$$\text{BTH}_{\text{Gray},b}(x, y) = ((\text{Gray} \otimes b) \oplus b)(x, y) - \text{Gray}(x, y) \tag{3}$$

The contrast can be enhanced by the pass-through method:

Fig. 6 Infrared detection morphology enhancement map

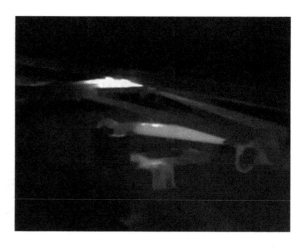

$$\text{TH}_{\text{Gray},b}(x, y) = \text{WTH}_{\text{Gray},b}(x, y) + f(x, y) - \text{CTH}_{\text{Gray},b}(x, y) \qquad (4)$$

where (x, y) is the position of the picture after the histogram enhancement, and $f(x, y)$ is the gray value of the place (Fig. 6).

3.5 K-means Clustering Algorithm

The image processed through the above steps can be taken as an input, and at the same time the pixels of each point can be taken as a sample set.

Since the temperature of the contact area between the pantograph and the contact net is much higher, and the temperature of the outer area is lower, the infrared image after the previous steps is displayed as different grayscale values [6]. The k-means clustering algorithm is used to aggregate regions with large grayscale values.

Introduction to the principle: For a given sample set, the k-means clustering method can automatically assign samples to different classes, but it is not possible to decide which classes to divide.

At each iteration, the k-means clustering algorithm assigns observations to the class closest to them and then moves the center of gravity to the average of all member positions in the class.

Determine the k value: Firstly, observing whether the dataset is suitable for more than one class. Since the light and dark distribution in the picture is obvious and the demand data exceeds one type, the k value can be determined by different methods.

By calculating the average contour coefficient of the current k, the k corresponding to the value with the largest contour coefficient is finally selected as the final number

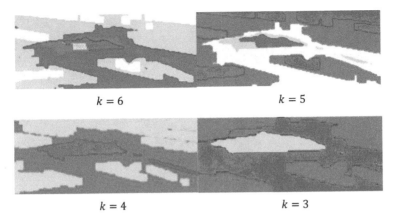

Fig. 7 k = 6 friction part map (partial); k = 5 friction part map (partial); k = 4 friction part map (partial); k = 3 friction part map (partial)

of clusters [7]. The final clustering effect can be observed by selecting different k values. The result can be shown in Fig. 7.

Initial centroid selection: Choosing the appropriate initial centroid is a key step in the basic k-means algorithm.

The method of choosing the initial centroid: since there are few outliers (more than outliers), most of them will not appear in random samples, and the amount of calculation will be greatly reduced.

By this method, it is ensured that the initial centroid of the selection is not only random but also scattered.

Distance metric: The distance metric measures the absolute distance between points in the space, directly related to the position coordinates of each point (the value of the individual feature dimension) [8]. Since the infrared detection image mainly detects the area block at a higher temperature, the positional feature is larger than the difference in the direction. So the Euclidean distance is selected as the calculation method.

For $n = 1, 2 \ldots N$

(1) Initialize cluster partitioning to $C_t = \emptyset, t = 1, 2 \ldots k$.
(2) For $i = 1, 2 \ldots m$, calculate the distance between the sample and each centroid vector: the smallest xi mark is the category λ_i corresponding to d_{ij}. Update now.

After the standardization is completed, the Euclidean distance is obtained. The formula is as follows:

$$D = \sqrt{(x_1 - x_2)^2 + (y_1 - y_2)^2} \qquad (5)$$

In summary, the first step is to find the clustering center for the points to be clustered. The second step is to calculate the distance from each point to the cluster center, and cluster each point into the cluster closest to the point. The third step is

Fig. 8 Infrared detection k-means diagram

to calculate the coordinate average of all points in each cluster, and use this mean as the new cluster center. It should be executed repeatedly until the cluster center no longer moves over a large area or the number of clusters reaches the required level (Fig. 8).

4 Conclusion

In this paper, an on-line detection method based on k-means clustering infrared image is proposed. By clustering the friction temperature increasing regions together, the on-line non-contact measurement of the friction part of the pantograph and the contact net is realized [9]. Through the image processing of the infrared image data collected in the field, the friction heating zone with the contact net is located.

By performing image processing on the infrared image data collected on-site, the temperature rising zone rubbed with the contact net is located. The next step is to classify the area by machine learning [10]. Therefore, the temperature rise zone is used to distinguish whether the pantograph is in contact with the contact net in normal. Improve train speed and pantograph system safety during system inspection, and further promote it to trams and other railway industries. By improving the speed of the train and the safety of the pantograph system during system inspection, it is further extended to trams and other railway industries.

Acknowledgements This work is supported by National Key R&D Program of China (2017YFB1201201).

References

1. Zhu X, Gao X, Wang Z, Wang L, Yang K (2010) Study on the edge detection and extraction algorithm in the pantographslipper's abrasion. In: 2010 International conference on computational and information sciences (ICCIS) (in Chinese)
2. Ma L, Wang Z, Gao X, Wang L, Yang K (2009) Edge detection on pantograph slide image. In: 2nd International congress on image and signal processing (in Chinese)
3. Hulin B, Schussler S (2007) Concepts for day-night stereo obstacle detection in the pantograph gauge. In: 2007 5th IEEE international conference on industrial informatics
4. Bouras C, Tsogkas V (2011) Clustering user preferences using W-kmeans. In: 2011 Seventh international conference on signal-image technology and internet-based systems (SITIS)
5. Tan M, Zhou N, Cheng Y, Wang J, Zhang W, Zou D (2019) A temperature-compensated fiber Bragg grating sensor system based on digital filtering for monitoring the pantograph–catenary contact force. In: Proceedings of the institution of mechanical engineers, 233(2) (in Chinese)
6. Wei W, Liang C, Yang Z, Xu P, Yan X, Gao G, Wu G (2019) A novel method for detecting the pantograph–catenary arc based on the arc sound characteristics. In: Proceedings of the institution of mechanical engineers, 233(5) (in Chinese)
7. Ren Y, Wang K, Yang H (2019) Stability analysis of stochastic pantograph multi-group models with dispersal driven by G -Brownian motion. Appl Math Comput 355 (in Chinese)
8. Guo J, Peng J, Li J, Gao X, Yuan M (2018) 3-Dimensional surface inspection system for pantograph in railway nondestructive testing based on laser line-scanning. In: Other Conferences (in Chinese)
9. Yaman O, Karakose M, Aydin I, Akin E (2014) Image processing and model based arc detection in pantograph catenary systems. In: 2014 22nd Signal processing and communications applications conference (SIU)
10. Tang P, Jin W, Chen L (2014) Visual abnormality detection framework for train-mounted pantograph headline surveillance. In: 2014 IEEE 17th international conference on intelligent transportation systems (ITSC) (in Chinese)

Design of PCI Bus High-Speed Data Acquisition System Based on FPGA

Xiaoming Zhu, Xiaodong Wang, Jie Chen, Ruichang Qiu and Zhigang Liu

Abstract PCI bus is a standard bus of the computer. It adopts European standard and has the advantages of plug and play, interrupt sharing, high-speed transmission, and safety fastening. However, the PCI communication protocol is complex and difficult to develop. This article uses the PCI9054 protocol conversion chip to build a bridge between the upper computer and the lower computer. The data acquisition card uses AD7606 to complete the conversion between analog signal and digital signal. The FPGA controls the acquisition of AD data and the data transfer between the AD7606 and PCI9054. PLX provides API functions for the PCI protocol on the computer side. Based on these functions, the data acquisition and display program are written in the VC++ 6.0 environment, and then the program is packaged as a dynamic link library. The dynamic link library is called by LabVIEW, and the data display and storage are done in LabVIEW. The experimental results show that the scheme works stably, the transmission speed is fast, and the data is accurate.

Keywords PCI9054 · FPGA · PCI bus · LabVIEW

X. Zhu (✉) · J. Chen · R. Qiu · Z. Liu
School of Electrical Engineering, Beijing Engineering Research Center of Electric Rail Transportation, Beijing Jiaotong University, 100044 Beijing, China
e-mail: 18121557@bjtu.edu.cn

J. Chen
e-mail: jiechen@bjtu.edu.cn

R. Qiu
e-mail: rchqiu@bjtu.edu.cn

Z. Liu
e-mail: zhgliu@bjtu.edu.cn

X. Wang
Hohhot Urban Rail Transit Construction Management Co., Ltd., Beijing, China
e-mail: wxd721111@126.com

© Springer Nature Singapore Pte Ltd. 2020
B. Liu et al. (eds.), *Proceedings of the 4th International Conference on Electrical and Information Technologies for Rail Transportation (EITRT) 2019*, Lecture Notes in Electrical Engineering 640, https://doi.org/10.1007/978-981-15-2914-6_7

1 Introduction

Data acquisition technology is a very important application of signal processing, and it is widely used in communications, radar, image processing, and other fields. With the rapid development of railways, rail transit has put forward new requirements for communication technologies, such as intelligent railway technology, rail transit big data analysis technology, and rail transit Internet of things technology. The data acquisition card in this system mainly uses the PCI9054 bridge chip provided by PLX and the Cyclone IV EP4CE10F17C8N chip of Intel Corporation. There are four main functional modules: AD7606 data acquisition module, FPGA control module, PCI9054 bridge module, and LabVIEW display module [1]. The system finally realizes the display and storage of the data collected by the lower computer in the upper computer.

The PCI9054 bridge chip integrates the PCI bus protocol, which greatly reduces the user's workload. FPGA programming is flexible and can be reprogrammed. Therefore, FPGA and PCI9054 are the main architectures in the design. Figure 1 is the overall design of the system.

The subject uses AD7606 as the AD acquisition chip to achieve eight channels of high-accuracy data acquisition. One end of the FPGA is connected to the AD7606 to control the AD7606 to collect data and receive the collected data [2]. The other end of the FPGA is connected to the local bus of the PCI9054. The connected clock signal uses the 33 MHz clock pulse inside the FPGA. The collected data can be cached in FLASH. The PCI9054 bridge chip is used to connect the underlying design to the PC. The PCI bus part of the other end of the PCI9054 bridge chip is connected to the

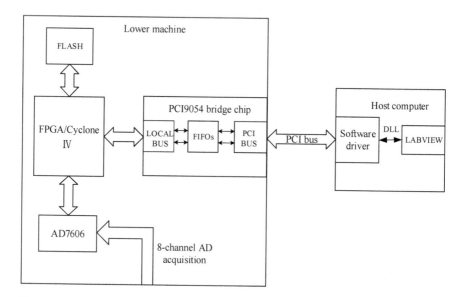

Fig. 1 Overall design

host computer through the PCI bus, and the driver is a function interface of the bus and the PC operating system [3]. The host computer mainly completes four tasks: developing a dynamic link library (DLL) in the VC++ 6.0 environment, designing the display interface and calling the DLL in LabVIEW, and finally realizing data communication between the data acquisition card and the PC.

2 FPGA Communication Interface Design

Since its invention in the 1980s, FPGA has been widely used in many fields such as smart home appliances, industrial control, communication engineering, and aerospace. This article uses Intel's research and development of Cyclone IV series chips. Cyclone IV devices integrate a low-cost transceiver that can be selected by users, which greatly saves the power consumption and cost of the chip without affecting performance. Such chips are ideal for applications in small and low-cost fields such as wireless communications, wired communications, and other fields. FPGA mainly completes the design of AD7606 and PCI9054 communication ports. At the same time, the FPGA also needs to cache the collected data into the FLASH.

2.1 PCI Local Bus Mode Selection

FPGA should design the interface with AD7606 on the one hand and design the communication interface with PCI9054 on the other hand. The control of the AD7606 is relatively simple, and the difficulty lies in the design of the FPGA and PCI9054 communication interface. PCI9054 was originally produced by the US PLX company's general interface chip; it can achieve 32-bit data bits and 33-MHz clock frequency data transmission. The data transfer rate between the local bus and the PCI bus can even reach 133 MB/s. PCI9054 works reliably, and users can apply it without too complicated configuration. Its series of bridge chips have been widely used both at home and abroad [4].

PLX has set up three different working methods for PCI9054 for different application development environments. Table 1 shows the different modes of PCI9054.

Table 1 Three modes of local bus

Mode	M mode	C mode	J mode
Application	The model developed by PLX for Motorola	In this mode, address lines and data lines cannot be multiplexed, and the setting is relatively simple	In this mode, the address and data bus can be multi-multiplexed, and the setting is relatively complicated

Among the three modes, the M mode was originally developed specifically for Motorola. The address and data in J mode can be reused to reduce the resource occupancy, but its running timing and setting method are too complicated. Therefore, in most cases, users use C mode. This design uses the C mode [5].

The PCI9054 has already packaged the PCI bus protocol. The user only needs to set the corresponding pin of PCI9054 according to a certain timing to control the PCI9054 for data transmission. Figure 2 shows the data transmission timing diagram of PCI9054 in C mode.

LCLK is the PCI9054 local clock. In this design, 40 M crystal oscillator is used as the input clock. LHOLD is the local bus use request signal, and LHOLDA is the local bus arbitration signal. PCI sends the LHOLD signal and sets high to indicate that the local bus is applied. If the FPAG agrees to the bus request, it will set the LHOLDA signal high. The ADS, BLAST, LW/R, and READY signals all indicate a valid signal when they are low. The PCI9054 will send an ADS signal before transmitting the address and data. After receiving the ADS signal, the FPGA will send a READY signal indicating that it will receive the data and address. PCI9054 will send a BLAST signal to indicate the end of the data transfer, when the FPGA will convert the READY signal to a high level. LA is the address bus, and LD is the data bus. They are both input signals and output signals. The LW/R signal indicates the flow of data. A high level indicates that PCI9054 writes data to the FPGA, and a low level indicates that PCI9054 reads data from the FPGA. The LBE signal indicates

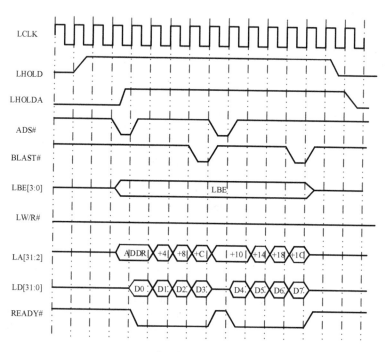

Fig. 2 Data transmission timing diagram of PCI9054 in C mode

the effective number of bits of the local data. When it is equal to zero, it indicates that the local bus 32-bit data is valid.

The FPGA designs the timing of transmitting data according to Fig. 2. The operating state of the PCI9054 is controlled by controlling arbitration signals such as LHOLD, LHOLDA, and ADS.

2.2 PCI Data Transmission Mode Selection

As a bridge chip connecting the PCI bus and the local bus, the PCI9054 is divided into different modes according to the data transmission mode and the bus that initiates access: DMA burst mode and direct data transfer modes [6]. Among them, the direct data transmission mode is divided into two modes: target mode and initiator mode.

This design uses target mode. The PC is the initiator of the access and applies for the arbitration signal of the PCI bus. The host bridge agrees to connect. Through the set address mapping rule, the computer can directly access the PCI9054 local bus terminal connected logic module after connecting the PCI bus port of the PCI9054. In this process, the local bus acts as a slave device in response to the local logic module connected to the local bus, and the PCI bus acts as the master device to initiate a data connection to the computer; that is, the computer is the master and the local module acts as the slave.

2.3 FPGA Local Bus State Machine Design

It not only need to master the timing of PCI9054 data transmission, but also need to set the state of each signal in the data communication process when writing the communication program of FPGA and PCI9054 local bus. Figure 3 shows the different state settings in the FPGA.

State0 is the initial state. When $LHOLD = 1$, the system enters state1; otherwise, the state does not change. State1 is the bus hold state. When $LHOLD = 0$, the system returns to state0. When $ADS = 0$, $LWR = 0$, the system goes to state3; if $ADS = 0$, $LWR = 1$, the system goes to state4. State2 indicates that the local signal can be written. If $BLAST = 0$, it indicates that the write operation is completed and goes to state4; if $BLAST = 1$, it remains in this state. State3 indicates that the local signal can be read. If $BLAST = 0$, it indicates that the write operation is complete and goes to state4; if $BLAST = 1$, it remains in this state. State4 indicates that a cycle of information operations has been completed. If there is no arbitration signal request for the next cycle, it returns to the initial state; if $LHOLD = 1$, it indicates that the local bus is requested, then it goes to state1 to perform the next read and write operation.

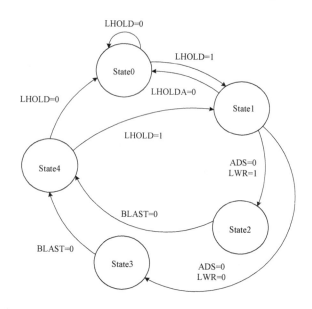

Fig. 3 Timing control of PCI9054 in FPGA

3 Dynamic Link Library Programming

The dynamic link library implements the sharing of library files, which is an indispensable file for computer software integration development. The programming of DLL has four major elements that cannot be ignored:

1. There are two main ways to create project files: Win32 Dynamic Link Library and MFC AppWizard. This topic uses the former method to build a dynamic link library [7].
2. Use the output description. The output program should be marked in the main program for other software to recognize.
3. Initialize the function. The DllMain function provides an environment for all initialization and exit handlers. If initialization is not required, the program will automatically generate a DllMain function that will not be processed.
4. The main function, which is the main API function interface mentioned above. Among them, four API functions are mainly used: PlxPciDeviceOpen, PlxPciDeviceClose, PlxBusIopRead, PlxBusIopWrite. The host function control program of PCI9054 is designed by API function interface which is packaged as dynamic link library.

4 LabVIEW Communication Test

LabVIEW is a graphical programming software that can display and store data. It can also be connected with programs in other development environments using dynamic link libraries. Therefore, LabVIEW is used to complete DA conversion and graphical display programming [8].

4.1 Calling the Dynamic Link Library

LabVIEW's background optional program has been designed to connect to DLL interface functions in other software environments. Need to set the call library function node: the library name/path is the location of the compiled DLL file; because a DLL sometimes has more than one function to be exported, users need to select the function name to be selected by the user. Generally, the UI thread is selected. The calling specification selects C, and the parameter variable in the parameter setting is set to a pointer.

There is a module in LabVIEW that calls the library function node. The difficulty in calling the library function node is not in LabVIEW, but in how to write the host program in the VC++ 6.0 environment. It should be noted that the main function is required when writing a host program, and the main function is not required for writing a dynamic link library. The main work done in LabVIEW is to reverse the DA conversion according to the formula of AD conversion. In addition, the display and storage of data are done in LabVIEW.

4.2 Communication Test

Because FPGAs and AD7606 are actually hardware circuits, they run very fast. In order to ensure the data transmission rate, the data acquisition program of the lower computer is actually running all the time.

The following are the results of the acquisition of sine, square and triangle waves. The sine wave frequency of the three kinds of waves is 1 kHz, and the peak-to-peak value of them is 6 V. The square wave duty ratio is 50% (Fig. 4).

According to the collection result, the value of the acquired analog quantity is within the reasonable error range and the waveform display meets the requirements.

5 Conclusion

In view of the demand for data acquisition by rail transit information technology, the design of 8-channel high-precision data acquisition system is realized by using AD7606, FPGA, PCI9054, and other chips.

The lower computer takes FPGA as the core and PCI9054 as the bridge between data acquisition card and computer. The FPGA controls the acquisition of data and passes this data to the PCI9054. The host computer uses the dynamic link library

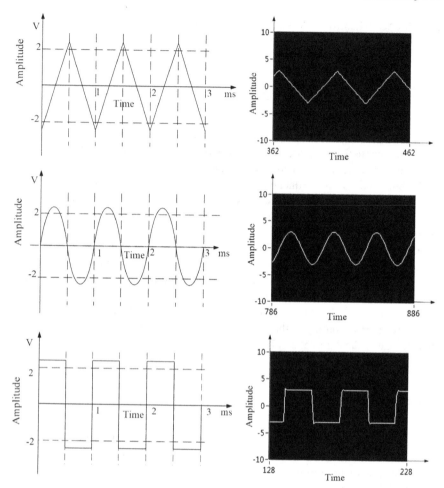

Fig. 4 Waveform test: (top left) triangle wave reference waveform, (top right) triangle wave test waveform, (middle left) sine wave reference waveform, (middle right) sine wave test waveform, (bottom left) square wave reference waveform, (bottom right) square wave test waveform

as a bridge between LabVIEW and PCI9054. LabVIEW controls the data transfer between the host computer and the lower computer and is responsible for displaying and storing the received data.

The experimental results show that the scheme works stably, the transmission speed is fast, and the data is accurate. This design has certain application value for PCI board design, PCI protocol transmission, and data acquisition and analysis.

Acknowledgements This work is supported by the National Key Research and Development Program (2017YFB1200802-01).

References

1. Zhao Z, Zhao X (2018) Design of high speed sampling and processing device based on FPGA. Mar Power Technol 38(1):40–42 (in Chinese)
2. Cai X, Che Y, Wu B (2018) Design of multi-channel analog acquisition circuit based on FPGA. Aeronaut Comput Technol 48(2):101–104,108 (in Chinese)
3. Li M, Huang Y (2013) Liu Y (2013) Design and implementation of high speed data acquisition card based on PCIe bus. Meas Control Technol 32(7):19–22 (in Chinese)
4. Zhang D, Zhang X, Liu W (2013) Based on PCI9054 data communication interface card design. J Wuhan Univ (Inf Manage Eng Ed) 35(3):305–308 (in Chinese)
5. Wang LN (2008) The design of FPGA-based PCI bus data acquisition. North China University of Technology, pp 13–25 (in Chinese)
6. Kavianipour H, Muschter S, Bohm C (2014) High performance FPGA-based DMA interface for PCIe. IEEE Trans Nucl Sci 61(2):745–749
7. Shuang W (2017) Design and implementation of video surveillance area intrusion detection system based on MFC + OpenCV. Shandong Normal University (in Chinese)
8. Dai J, Dai S (2017) Glass defect detection based on LabVIEW programming platform. Meas Control Technol 36(07):9–12 (in Chinese)

Design of Control Unit of Traction System Based on CPCI Bus

Decheng Tang, Cheng Yang, Jie Chen, Ruichang Qiu and Zhigang Liu

Abstract With the development of high-speed railway, it is necessary to design a traction control system with high performance, high stability, and high system integration. The CPCI bus is widely used in industrial automation because of its system integration and good stability. The CPCI bus is used as the connection mode between the board and the host computer, which enables the system to be compatible with a variety of operating systems and processors. The design of the CPCI bus enables the board to transmit data at high speed and on the host computer. In addition, the CPCI bus's unique European pin standard has improved the reliability and stability of the board. The main content of this paper is the development of traction master control unit based on CPCI bus. The design of the main control system includes AD acquisition module, digital signal processing module, and CPCI bus module. Based on the high-speed data transmission of CPCI bus and the coordinated control of DSP and FPGA, a high-efficiency and high-performance traction control system is designed.

Keywords Traction control unit · CPCI bus · FPGA

D. Tang (✉) · J. Chen · R. Qiu · Z. Liu
School of Electrical Engineering, Beijing Engineering Research Center of Electric Rail Transportation, Beijing Jiaotong University, Beijing 100044, China
e-mail: 18121494@bjtu.edu.cn

J. Chen
e-mail: jiechen@bjtu.edu.cn

R. Qiu
e-mail: rchqiu@bjtu.edu.cn

Z. Liu
e-mail: zhgliu@bjtu.edu.cn

C. Yang
Taiyuan Railway Transit Development Co., Ltd., Taiyuan 030000, Shanxi, China
e-mail: 21341240@qq.com

1 Introduction

The national high-speed rail operating mileage exceeded 29,000 km by 2018, accounting for 67% of the world's total high-speed rail. Traveling by rail vehicle has become a very important way of travel for our people. At the same time, with the improvement of China's rail transit system, people's demand for rail vehicles is also increasing [1]. All of this requires us to design a safe, stable, and fast rail traffic traction control system. Nowadays, the mainstream electric locomotives in China are basically equipped with foreign train traction control systems [2]. For a traction control system, the core component is the main control unit of the traction system. The main control unit of the rail transit traction control system is the core control component of the electric locomotive AC drive. Its hardware performance directly affects a series of operating characteristics such as start, speed regulation, energy flow feedback, and braking of the electric locomotive. The main control unit of the traction system plays the role of monitoring, protection, and control in the entire traction drive system, which is a necessary condition for the safe and stable operation of electric locomotives [3]. Therefore, the development of a main control unit of the traction control system is of great significance to the development of China's rail transit industry.

The traction drive control system is an important part of the electric locomotive. Its running performance is closely related to the safety and stability of the train. The traction control unit is the core of the traction drive control system, with functions of control, monitoring, communication, etc. Its stability to traction transmission control system performance plays a decisive role.

2 System Design

2.1 Design Theory

A complete control system can be divided into two important parts—one is the basic hardware circuit part, and the other is the software program part for control and signal conversion [4]. The hardware part plays a crucial role in the stability of the system. Whether the hardware design conforms to the relevant specifications and whether the layout is reasonable is directly related to the performance of the entire system. The development of science and technology has made the design software function of integrated circuits more and more complete, and the manufacturing technology of integrated circuits has been gradually developed. The functions of circuit systems, which are an important part of the system, are also more powerful, and the stability and efficiency of the system are also getting higher and higher.

2.2 CPCI Bus

The CPCI bus is an open, international technical standard developed and supported by the PCI Industrial Computer Manufacturer Group. The CPCI bus has strict standards and specifications to ensure compatibility and supports multiple processors and operating systems, and CPCI-compliant expansion cards can be plugged into any CPCI system and work reliably [5]. The CPCI standard combines the Peripheral Component Interconnect (PCI) standard features with the rugged mechanical form factor that supports embedded applications, and its performance characteristics are tailored to the industrial environment.

CPCI is based on PICMG2.0 specification. Its electrical characteristics are the same as that of PCI bus. Therefore, the user's software is compatible with ordinary PC. The existing PCI peripheral card can also be easily transplanted to the CPCI platform. CPCI uses high-density pinhole bus connector. Compared with PCI card with golden finger connector, it has the characteristics of more reliable connection, complete air tightness, high seismic and corrosion resistance of module. In today's industrial control, communications, military, and many other areas requiring high reliability, CPCI bus has replaced ISA, STD, and other buses and become the mainstream bus form.

2.3 System Architecture Design

According to the design requirements, the traction control unit can be divided into five functional modules: logic operation module, monitoring module, CPCI network communication module, traction control module, and remote program loading module.

These modules take the CPU as the core and output control signals in multiple stages according to the input turntable signal. The overall architecture of the system is shown in Fig. 1.

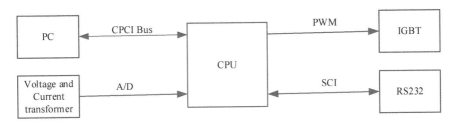

Fig. 1 System framework

2.4 Main Control Board Device Selection

The CPU design of this design adopts DSP + FPGA architecture. The DSP chip adopts TI's floating-point DSP-TMS320F28335. Compared with other digital control chips, the chip has high logic block integration, high control precision, and low cost, and its power consumption is small, performance balance, and other advantages. The chip can well meet the computing power of the traction control unit. The FPGA chip uses Altera's Cyclone 10CL080YF484I7G chip. This series of chips is one of Intel's latest FPGAs, with up to 423 users free to use IO ports to fully meet the needs of this design. At the same time, the chip has 8000 logic units, which can meet the logic operation record of the control unit. Compared with other series of FPGAs, the chip consumes less power and has higher data bandwidth, which is suitable for the design of intelligent control systems.

The monitoring module uses the AD7656 chip from analog devices. The chip can use the parallel port to transfer data directly to the FPGA chip through the data bus parallel bus, or multi-chip daisy-chain serial transmission, and the data is directly transmitted to the serial port of the DSP chip through serial communication.

The CPCI network communication module uses the PCI9054 chip of PLX company. The chip can convert the data collected by the AD into the data stream on the PCI bus, so that the upper PC can directly read the corresponding data on the bus.

3 System Hardware Design

3.1 System Hardware Circuit Design

DSP chip refers to a chip with digital signal processing capability. DSP chip adopts the Harvard structure commonly used by the processor [6]. The program and data under this structure are separated from each other, and the chip has hardware multiplication dedicated for multiplication operation. Using the appropriate programming instructions and logic, the DSP can quickly implement various digital signal processing algorithms. Therefore, DSP chips have a very wide range of applications in the field of hardware control.

FPGA is called field-programmable gate array, and its predecessor is CPLD, GAL, PAL, and other programmable devices. In general, an FPGA chip is equivalent to a collection of a large number of logic gates, and combining these logic gate arrays is to achieve the circuit functions required by the user. Compared to general-purpose CPUs and GPUs, FPGAs can run programs in parallel, so they run more efficiently and run faster. Table 1 shows several FPGA chips in the same series from Altera. According to the table parameters, we can see that the logic unit of the 10CL080 series chip and the user's maximum I/O pin are more than the other two series of chips, meeting the design requirements.

Table 1 Cyclone 10 LP chip parameter comparison

Parameter	10CL025	10CL040	10CL080
Logical unit	25,000	40,000	80,000
PLL	4	4	4
Global clock network	20	20	20
Maximum user I/O pin	325	325	423

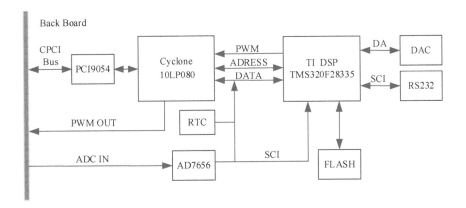

Fig. 2 Traction control unit system

Figure 2 shows the system block diagram of the main control unit of the traction system designed for this project. From the perspective of the entire system framework, the control core of the entire system is composed of a combination of DSP and FPGA. From the function points, the FPGA chip is mainly responsible for the input and output of various digital signals.

The CPCI bus is connected to the PCI9054 chip by the FPGA to realize communication between the lower-layer hardware and the upper-layer PC. The RTC chip is used to supply the local hardware with the same time as the real time. It transmits the real-time clock data to the DSP and FPGA chips through the data bus. DSP is mainly responsible for the logic operation of the whole system. When the amount of data is too large, the DSP can store the processed data into the external FLASH to prevent data loss. In addition, the system inputs 16 high-precision ADC data from the outside, and the data stream is transmitted to the FPGA through the data bus by the AD7656 chip and then transmitted to the upper CPCI bus. The data stream can also be transferred to the DSP chip in serial mode and converted to the FPGA by the DSP. The alternating use of these two methods allows the data bus to be fully utilized [7].

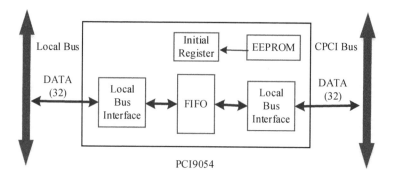

Fig. 3 PCI9054-chip architecture

3.2 CPCI Bus and PCI9054

Referring to the circuit scheme of the CPCI interface, directly implementing the CPCI interface with the FPGA is higher in time and technical difficulty than using the mature PCI protocol chip, so the subject uses the PCI9054 chip of the PLX company as the interface chip [8].

As shown in Fig. 3, in the whole system, the PCI9054 chip is equivalent to a bridge. Its function is mainly to complete the bridge between the local bus and the CPCI bus and complete the conversion of commands between two different interfaces. The information requested by the user is pre-stored in an EEPROM. When the chip is powered on, the PCI9054 automatically loads the information and completes the initialization.

3.3 System Power Supply Design

The entire traction control system unit uses multiple-echelon, multi-voltage supply voltages, so in the power supply design, the power supply of each module must be strictly allocated. The system's power input is 5 V, and all power supplies to the entire board are converted from 5 V. The input voltage mainly has two input paths—one is input voltage through a 5 V power interface, and the other is to provide 5 V voltage to the circuit board through the J1 port of the CPCI bus.

The 3.3 V voltage of the entire board is obtained by the TPS75733KTT voltage conversion chip. The FPGA has strict requirements for voltage ripple. Intel company's EN5339QI chip is used here. The N5339QI is a pin-compatible small, high-efficiency synchronous DC-DC step-down. The chip is designed to provide scalable power solutions for embedded computing, FPGA, ASIC, and embedded processors.

The power requirements of the A/D and D/A chips used in the design are quite special. These two modules require higher than the initial power supply voltage. Therefore, in the design process, the EC2SA-05S15D power conversion module is

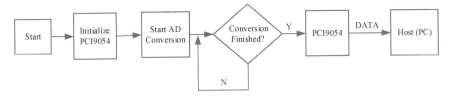

Fig. 4 Flowchart of system software

used. This module is A module based on a DC-DC boost circuit that converts the input voltage of 3–5 V into a voltage of ±15 V.

4 System Software Design

This design uses the Verilog HDL language to write FPGA programs. The FPGA program design mainly includes the initial configuration of PCI9054, the driver of AD7656 chip, and the data transmission function. The flow of the whole program is as follows. After the hardware circuit is powered on, the FPGA resets and initializes the entire system according to the designed program. Then control the AD chip for AD data acquisition, and transfer the data to the local bus of PCI9054. The host computer receives the data through the CPCI bus and stores the data in the upper computer memory. The specific process is shown in Fig. 4.

After the entire system design is completed, the designed system is connected to the industrial computer for testing. Each module of the system can work normally, and the test conditions are good and meet the expected requirements.

5 Conclusion

This paper introduces a traction system based on CPCI bus. According to the requirements of this project design, after consulting a lot of domestic and foreign literature and component manuals, this paper proposes a hardware solution that meets the design requirements and designs according to the scheme. Verification has resulted in a traction control unit that meets the requirements.

In order to reduce the power consumption of the system and improve the operating efficiency of the design board of this project, the design uses Altera's Cyclone 10 LP 080 chip. The low power consumption and high efficiency of this chip are very suitable for this design. Another advantage of this design is the replacement of the conventional 64-pin board with the CPCI bus interface, which enables the board to transmit data at a high data rate, so that real-time monitoring of the rail train can be achieved. Upload data to a PC for more sophisticated analysis and processing operations.

Acknowledgements This work is supported by the National Key Research and Development Program (2016YFB1200504-C-02).

References

1. Xu S (2017) Analysis of the development status and trend of traction control for rail transit vehicles. Sci Technol Inf 15(32):84–85 (in Chinese)
2. Wang Y (2015) Design and development of software and hardware for train traction control unit. Southwest Jiaotong University (in Chinese)
3. Guo Q, Kang J, Feng J (2017) Development status and trend of traction control for rail transit vehicles. J Power Supply 15(02):40–45 (in Chinese)
4. Yang Z (2013) Design of high-speed data acquisition and processing module based on CPCI bus. University of Electronic Science and Technology (in Chinese)
5. Yuan J (2005) Design and implementation of A/D module based on CPCI bus. Comput Digital Eng 129–133 (in Chinese)
6. Wei J (2018) Design and implementation of CPCI system based on FPGA. Appl Electron Technol 44(11):50–52 (in Chinese)
7. Wu R, Chou J, Liu T (2012) Development and application of loongson 2GJ-based CPCI-bus board. In: Second international conference on instrumentation, measurement, computer, communication and control (IMCCC)
8. Wang R, Youfang Z, Yanyun C (2010) Design of CPCI interface based on data acquisition card of FPGA. Electron Technol 37(02):35–37 (in Chinese)

Design of Route Search Algorithm Based on Station Map Information and Depth-First-Search

Yahan Yang, Shaobin Li, Kai Sun and Xiaobin Di

Abstract Route search is an important part in the interlocking system. The result can affect the work in the station. Through the study of information such as turnouts, track number and the coordinates of equipment, this article shows a new method to search the route. By adding a variety of conditions, the Depth-First-Search algorithm is modified and applied for the route searching process. The existing route search methods usually select multiple paths and compare them to select the best one. The result usually is uncertain, and searching progress has low efficiency. The improved algorithm in this article imitates how the people search routes. Thus, only one path will be selected, which is the best one. Also, this algorithm can improve the efficiency in the searching process, and the storage of data is reduced. By simulating the interlocking system, the correctness of the algorithm is verified. And the performance comparison between different methods is showed in this article.

Keywords Route search · Depth-First-Search · Interlocking

1 Introduction

Traditional way of route search in computer interlocking system is mostly based on the interlocking table. Because the system searches for the routes only by the data already in the interlocking table, the correctness of the routes cannot be guaranteed. Also, the path which is not written in the table cannot be selected. If the signal equipment is replaced, the interlocking table needs to be rewritten. This brings a lot of work.

    In recent years, many scholars have studied data structures or algorithms to find ways to search routes. For example, using A* algorithm in route search can reduce the problem of excessive data caused by traversal. But the heuristic functions of each node need to be recorded [1]. Using data structures of the form of binary trees and searching data onto the opposite direction can improve efficiency. But the structure

---

Y. Yang (✉) · S. Li · K. Sun · X. Di
School of Electronic and Information Engineering, Beijing Jiaotong University, Beijing, China
e-mail: 18120280@bjtu.edu.cn

of the binary tree will change when the route is chosen in the opposite direction [2]. Other search algorithms like ant colony optimization can find better results. But the algorithm is complex [3]. In this paper, a new route search algorithm is designed which is relatively simple and can avoid circuitous problems. Also, it can reduce memory occupancy and improve search efficiency.

2 Search Conditions Analysis

2.1 Search Direction Analysis

Station map is a two-dimensional plane coordinate map, and each point has its corresponding coordinates as shown in Fig. 1 [4]. From the map, we can get much information. It can be defined that horizontal coordinate increases from left to right and vertical coordinate increases from top to bottom. The direction of route search can be determined by comparing coordinates to avoid the reverse direction search.

2.2 Route Analysis

The selected routes should have the least impact on other work in the station and circuitous routes are not allowed [5]. When the turnout is searched, the reverse position should be preferred. This can avoid the situation that the route cannot be selected because of the direction which is not changed in time. When we press the button, we can know the beginning and the end nodes of the route. By calculating the number of tracks, we can get the approximate number of reversed position turnouts that the train needs to go through. Limited by this number, the search can be stopped in time when an invalid route is found.

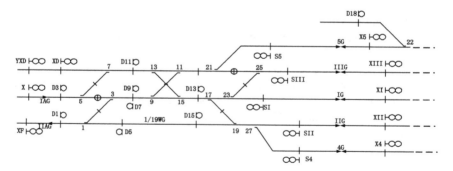

Fig. 1 Station map

3 Algorithm Design

3.1 Direction Algorithm Design

Set (x_i, y_i) and (x_k, y_k) as the coordinate of the starting signal and terminal signal. If $x_i > x_k$, the train goes to the left. If $x_i < x_k$, the train goes to the right. As for the vertical direction, if $y_i > y_k$ the train goes to the above direction. Else if $y_i < y_k$ the train goes to the below direction. Else the train goes straight. Compare the coordinates of the current node and neighbor node. If the search direction is correct, the neighbor node will be recorded.

3.2 Turnouts Algorithm Design

The method of calculating the number of reversed turnouts by the number of tracks is as follows. Because of the difference between single-track and double-track railway, there are two kinds of numbering methods for tracks, and the calculation method is also based on the actual station map.

As shown in Fig. 2, the numbering method of single-track railway is the number increases from near to far based on the signal building.

Set N as the number of reversed turnouts, S_{i1} and S_{i2} as the number of the tracks where the beginning and the end nodes are located. So, the calculation equation can be described as follows:

$$N = 2|S_{i2} - S_{i1}| \qquad (1)$$

The numbering method of double-track railway is the number increases from main tracks to sidetracks. The main tracks are numbered in Roman numerals while the sidetracks are numbered in Arabic numerals. Tracks of up-side direction are numbered in even numbers, and tracks of down-side direction are numbered in odd numbers which are as shown in Fig. 3.

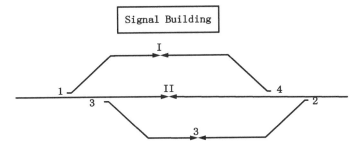

Fig. 2 Single-track railway

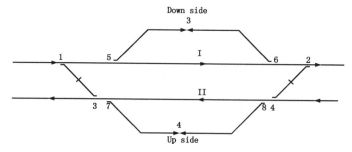

Fig. 3 Double-track railway

Because of different situations, there are three ways to calculate.

The first situation is that both the start button and the end button are on the tracks of odd number and the calculation formula can be described as follows:

$$N = 2\left|\frac{S_{i1}-1}{2} - \frac{S_{i2}-1}{2}\right| \quad (2)$$

The second situation is that both the start button and the end button are on the tracks of even number and the calculation formula can be described as follows:

$$N = 2\left|\frac{S_{i1}}{2} - \frac{S_{i2}}{2}\right| \quad (3)$$

The third situation is that the start button and the end button are on the tracks of even and odd numbers, respectively. The calculation formula can be described as follows:

$$N = 2\left(\frac{S_{i1}-1}{2} + \frac{S_{i2}}{2}\right) \quad (4)$$

$$N = 2\left(\frac{S_{i1}}{2} + \frac{S_{i2}-1}{2}\right) \quad (5)$$

In the formula (4), S_{i1} is odd number and S_{i2} is even number. And in the formula (5), S_{i1} is even number and S_{i2} is odd number.

3.3 Depth-First-Search

Depth-First-Search algorithm is often used to search graphs. The specific steps are as follows. Choosing one of the nodes in the graph and then searching its neighbor nodes and recording them. Repeat these steps to deeply search until find the route [6].

This algorithm belongs to blind searching, therefore has low efficiency [7]. However, it is flexible and easily to be modified. In this paper, the improved Depth-First-Search algorithm will be applied in route search. According to the above analysis and design, the route search process can be designed as follows, and the flowchart is shown in Fig. 4.

- Read the information of the start button and the terminal button, judge the type of route and record it.
- Read coordinates to determine the direction.
- Read the number of the track where the button is located and calculate the number of reversed turnouts.
- Read the coordinates of the next node to determine whether it is the end point or whether it conforms to the route search direction. The eligible nodes will be recorded.
- If the next node is a turnout, determine the current number of reversed turnouts that whether it has reached the maximum value. If so, this turnout will be in the normal position. If not, proceed to the next step.
- Determine whether the turnout conforms to the route search direction. If so, this turnout will be in the reversed position and the number of reversed turnouts increases. If not, this turnout will be in the normal position.

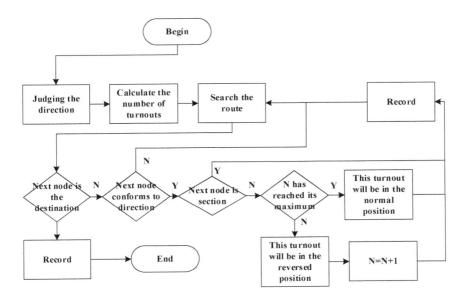

Fig. 4 Program flow chart

4 Simulation

To verify the correctness of the algorithm, the simulated interlocking system programmed with C# is shown in Figs. 5 and 6. Figure 5 is the result of using Deep-First-Search algorithm, and Fig. 6 is the result of using an improved algorithm. The visited nodes and time cost in the searching process are showed in the system.

From the figures, we can see both the results are correct. The improved algorithm takes less time to search the route than the non-improved search algorithm. Also, the improved algorithm visits fewer nodes, which is more efficient.

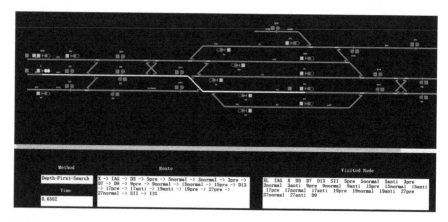

Fig. 5 Result of Deep-First-Search algorithm

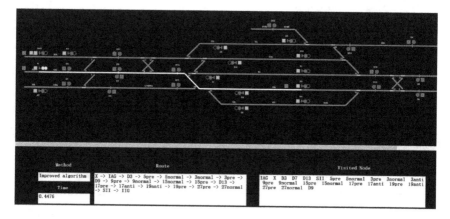

Fig. 6 Result of improved algorithm

5 Performance Analysis of Algorithm

5.1 Algorithm Structure

The route search using the heuristic algorithm needs to evaluate the location of each node and decide whether to select the node according to the evaluation results. There are many formulas in the algorithm and is not easy to understand. Also, the programming will be more complex. For example, the heuristic function needs to be calculated in the route search algorithm of A*. The formula of bionics algorithm is more complex and has more parameters. Some of the parameters are uncertain, and the values of parameters have great influence on the results [8]. Moreover, heuristic search algorithms often need multiple searches to get ideal results.

The algorithm in this paper does not use complex formulas, only by comparing coordinates can get information. The search direction is relatively determined, which greatly improves the search speed.

5.2 Data Storage

The route search using the heuristic search algorithm needs multiple searches to get the results, and many invalid data will be generated in the search process. There is also more data generated by complex formulas [9].

Because of the decrease of invalid routes, and the formula for the algorithm is simple, the algorithm in this paper can reduce the data storage.

6 Conclusion

This paper analyzes the characteristics of the station map. By reading much useful information, a new Depth-First-Search algorithm applied in the route search process is designed. Due to various searching conditions, the shortcoming of blind search is solved. Therefore, the invalid search is reduced, and the search efficiency is greatly improved. Because the algorithm is not complex and will not produce too much invalid data, this method can reduce memory space. Routes searching of special station maps needs to be further explored.

References

1. Yifan L, Li T, Ting F (2013) Research and implementation of A* route search algorithms. Railway Stand Des 2:117–119 (in Chinese)

2. Wuchen W, Xiaoming W (2008) Data structure and route search algorithms of computer interlocking. J Chongqing Univ Technol (Nat Sci) 22(6):51–53 (in Chinese)
3. Dorigo M, Stützle T (2003) The ant colony optimization metaheuristic: algorithms, applications, and advances. New Ideas Optim 57(3):251–285
4. Fantechi A, Haxthausen AE, Macedo HD (2017) Compositional verification of interlocking systems for large stations. In: International conference on software engineering and formal methods
5. Cappart Q, Christophe L, Schaus P, Quilbeuf J, Traonouez LM, Legay A (2017) Verification of interlocking systems using statistical model checking. In: IEEE international symposium on high assurance systems engineering
6. Bazin A (2018) A depth-first search algorithm for computing pseudo-closed sets. Discrete Appl Math
7. Peng W, Weihua K, Munan X, Dapeng L (2017) Research on path search algorithms based on Dijkstra and depth-first-search. J Transp Eng Inf (4) (in Chinese)
8. Wenquan Z, Lijian Y (2015) A path search algorithm based on genetic algorithms. Railway Signal Commun 51(9):9–11 (in Chinese)
9. Kanza Y, Safra E, Sagiv Y, Doytsher Y (2008) Heuristic algorithms for route-search queries over geographical data. In: ACM SIGSPATIAL international conference on advances in geographic information systems

Adaptive Global Particle Swarm Algorithm-Based Train Recommend Speed Curve Optimization Study in Urban Rail Transit

Zhen Hu, Xiao Xiao and Feng Bao

Abstract For the sequence of five working conditions in urban rail transit lines, generating the train recommend speed curve is established into a multi-objective optimization problem including safety, punctuality, parking accuracy, comfort, and energy consumption. The working condition conversion point data is used as the target to be optimized. An adaptive global particle swarm optimization (AGPSO) algorithm is proposed. Furthermore, AGPSO and two variants of PSO are used to solve the multi-objective optimization problem. The results show that only the optimization results of AGPSO meet the requirements of various indicators of automatic train driving system (ATO) control strategy.

Keywords Dynamic model · Speed curve · Multi-objective · Particle swarm optimization

1 Introduction

At present, domestic and international research on the automatic driving control strategy for urban rail transit trains mainly uses numerical algorithms and evolutionary algorithms [1–3]. They have low operating efficiency and high energy consumption [4, 5]. The ATO control strategy based on the working condition sequence switching point can solve the problem of frequent speed regulation in train operation. Evolutionary algorithms such as genetic algorithm (GA) [6], ant colony optimization (ACO), and particle swarm optimization (PSO) have lower requirements for mathematical models and can optimize the train recommendation speed curve optimization model to generate better quality recommendations speed curve. Comparing with other intelligent algorithms, the PSO has fewer parameters, simple structure, and easy implementation. In the iterative process, relying on optimal particles for information transmission, the search speed is fast. Therefore, particle swarm optimization is widely used in data mining, power system economic dispatching, and

Z. Hu (✉) · X. Xiao · F. Bao
Traffic Control Technology, TCT, Beijing, China
e-mail: 15150092932@163.com

artificial intelligence [7]. Since the PSO lacks dynamic adjustment speed, it is easy to fall into local optimal solution, resulting in poor performance of different problems [8]. In order to overcome these shortcomings, scholars have made a lot of improvements: adjusting parameters, choosing different learning strategies, and combining with other algorithms.

This paper proposes an adaptive global particle swarm optimization algorithm (AGPSO): The cosine mapping is used to adjust the inertia weight; dynamically changing individual factors and social factors assist in speed updates; new global location update policy. Furthermore, for the sequence of five working conditions in urban rail transit lines, the sequence of working condition conversion points is taken as the target to be optimized. A model is established which meets train safety, parking accuracy, on-time, passenger comfort, and energy efficiency. AGPSO and the other two algorithms solve the optimization model. The simulation results show that only the AGPSO meets the requirements of each indicator.

2 The AGPSO

The PSO is a heuristic algorithm developed by Kennedy et al. [9]. It is inspired by the foraging behavior of animals. It updates particles as follows:

$$V_i^d = V_i^d + c_1 r_1 \left(\text{pbest}_i^d - X_i^d\right) + c_2 r_2 \left(\text{gbest}^d - X_i^d\right) \tag{1}$$

$$X_i^d = X_i^d + V_i^d \tag{2}$$

where V_i^d and X_i^d are the current velocity and position of the particle; d is search space dimension; $c_1 = 2$ is a self-knowledge factor; and $c_2 = 2$ is a social factor; r_1 and r_2 are random numbers uniformly distributed inside [0, 1]. pbest_i^d and gbest_i^d are the individual optimal values and global optimal values.

Shi and Eberhart (1998) added linear decreasing inertia weights to the speed update formula [10–12]:

$$V_i^d = \omega V_i^d + c_1 r_1 \left(\text{pbest}_i^d - X_i^d\right) + c_2 r_2 \left(\text{gbest}^d - X_i^d\right) \tag{3}$$

$$\omega = \omega_{\max} - \frac{\text{ni}}{\text{NI}} (\omega_{\max} - \omega_{\min}) \tag{4}$$

where ni and NI are the current and maximum number of iterations, $\omega_{\max}$ and $\omega_{\min}$ are the maximum and minimum values of the inertia weight [13].

After analyzing the standard PSO model, in order to improve the convergence of PSO, the AGPSO is proposed [14].

Adjusting the inertia weight by cosine mapping and random factors:

$$\omega = \cos(\pi \times \chi) \tag{5}$$

Introducing adaptive self-awareness factors and social factors to adjust effects of individual optimal particles and global optimal particles on particle flight trajectories:

$$c_1 = 2 - 1.5\frac{ni}{NI}, \quad c_2 = 0.5 + 1.5\frac{ni}{NI} \tag{6}$$

Inspired by the speed update strategy, a new location update strategy was designed:

$$X_i^d = \lambda\left(X_i^d + V_i^d + \rho \text{gbest}^d\right) \tag{7}$$

where λ has a value of 0.5 and ρ is a random number uniformly distributed within [0, 1]. The new location update strategy helps to increase the convergence speed of the algorithm and avoid falling into local optimum [15, 16].

Figure 1 shows the specific flow of the AGPSO. Initializing parameters, position and velocity of the AGPSO; calculating fitness value of each particle; the speed and

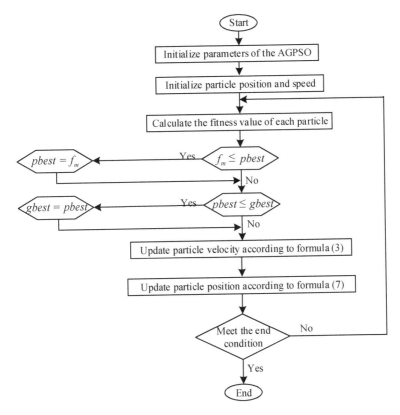

Fig. 1 AGPSO flowchart

position of the particles are continually updated according to Eqs. (3) and (7); pbest and gbest are updated according to the fitness value f_m; the AGPSO is terminated by satisfying the algorithm end condition.

3 The Model and Analysis

The ATO controls safe and efficient operation of the train. Therefore, advantages and disadvantages of the ATO control strategy are particularly important. Based on the previous studies, this paper is aimed at the sequence of five working conditions in urban rail transit lines. The working condition conversion point sequence is used as the mathematical model of the ATO target speed curve to be optimized. The AGPSO is applied to the model to obtain the best automatic driving scheme under the conditions of safe operation, punctuality, on-time, energy-saving, and passenger comfort. The simulation results prove that this method is effective.

Figure 2 is a graph showing the train running speed–distance curve in a sequence of five operating conditions in an ideal state. The abscissa indicates the train running distance; the ordinate indicates the train running speed; S_1, S_2, S_3, S_4, and S_5 represent distance of train under each working condition; v_0, v_1, v_2, v_3, v_4, and v_5 indicate speed of train at each working condition turning point; v_{max_1} and v_{max_2} are speed limit levels; L_1 and L_2 represent the distance range of two speed limit levels. The sum of L_1 and L_2 are equal to the distance between the two stations. Assume that the running time of the five working conditions is t_1, t_2, t_3, t_4, and t_5, respectively; The train running time is T_{max}. The train traction acceleration is $a_{jia} = 0.6\,\text{m/s}^2$, and the brake deceleration is $a_{jian} = -0.6\,\text{m/s}^2$. The following is an analysis of the train operation process:

$$S_1 + S_2 \le L_1 \tag{8}$$

First acceleration:

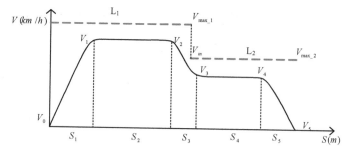

Fig. 2 Five working condition sequences of speed–distance curve

$$v_1^2 - v_0^2 = 2a_{jia}S_1 \qquad (9)$$

$$S_1 = \frac{v_1^2}{2a_{jia}} \qquad (10)$$

$$v_1 = v_2 \qquad (11)$$

Third stage deceleration:

$$v_m^2 - v_2^2 = 2a_{jian}(L_1 - S_1 - S_2) \qquad (12)$$

$$v_m = \sqrt{v_2^2 + 2a_{jian}(L_1 - S_1 - S_2)} \leq v_{max\_2} \qquad (13)$$

Fifth stage deceleration:

$$v_5^2 - v_4^2 = 2a_{jian}S_5 \qquad (14)$$

$$S_5 = -\frac{v_4^2}{2a_{jian}} \qquad (15)$$

Obtained by Formulas (8)–(15):

$$v_1 = \sqrt{2a_{jia}S_1} \qquad (16)$$

$$v_4 = \sqrt{-2a_{jian}S_5} \qquad (17)$$

$$v_3^2 - v_2^2 = v_4^2 - v_1^2 = 2a_{jian}S_3 \qquad (18)$$

$$S_2 \leq L_1 - \frac{v_{max\_2}}{2a_{jian}} + \left(\frac{a_{jia}}{a_{jian}} - 1\right)S_1 \qquad (19)$$

$$S_3 = \frac{v_4^2 - v_1^2}{2a_{jian}} = \frac{-2a_{jian}S_5 - 2a_{jia}S_1}{2a_{jian}} = -S_5 - S_1\frac{a_{jia}}{a_{jian}} \qquad (20)$$

$$S_4 = L_1 + L_2 - S_1 - S_2 - S_3 - S_5 \qquad (21)$$

Based on the above derivation, the upper limit of S_2 is determined by S_1; S_3 is determined by S_1 and S_5 together. Therefore, it is only to initialize S_1 and S_5 to establish an ATO recommended speed curve optimization model. In order to meet the safety index of the automatic operation of the train, the speed of the train running in the L_1 section is not greater than v_{max_1}. The speed of the train running in the L_2 section is not greater than v_{max_2} then:

$$\frac{v_{\max\_2}^2}{2a_{jia}} \leq S_1 \leq \frac{v_{\max\_1}^2}{2a_{jia}} \tag{22}$$

$$0 \leq S_5 \leq -\frac{v_{\max\_1}^2}{2a_{jian}} \tag{23}$$

In the process of optimizing the model, the sum of the distances of L_1 and L_2 is equal to the distance between the two stations. Therefore, the model satisfies the parking accuracy index. The punctuality indicator is set as a constraint, and the tolerance of the constraint is 0.01 s. The formula is as follows:

$$K_t = |T_{\max} - t_1 - t_2 - t_3 - t_4 - t_5| \tag{24}$$

Passenger comfort is related to the magnitude of the acceleration and deceleration and the duration. The cumulative sum of the running time and the absolute value of the acceleration product in different operating conditions are the objective function of the comfort index:

$$K_c = \sum_{i=1}^{i} |t_i a_i| \tag{25}$$

The energy consumption during train operation mainly includes the energy consumption caused by traction. This paper studies the impact of train control strategies on energy consumption. In the case of a given operating time of the ATS system, the auxiliary power consumption is constant. Therefore, the energy consumption objective function is simplified to:

$$K_e = \sum_{i}^{n-1} f_i(s_{i+1} - s_i) \tag{26}$$

where f_i represents the amount of traction or braking force within a time step, $f_i = M_i a_i$.

Through above analysis, the problem model becomes a two-objective constraint problem. The overall fitness formula is as follows:

$$f_{\text{total}} = \omega_1 K_c + \omega_2 K_e + \lambda \phi(K_t) \tag{27}$$

where $\omega_1 = 0.52 \times 10^6$ and $\omega_2 = 0.48$ are comfort objective function and energy objective function weight coefficient. $\lambda = 10^{20}$ is coefficient. $\phi(K_t) = \max(K_t - \text{Tolerance}, 0)$ is violation. Tolerance $= 0.01$ is tolerance for punctuality constraints.

4 MATLAB Simulation

Two stations of Guangzhou Metro Line 1 have an experimental background, and total length is 1228.40 m. Considering the safety margin of the two stations, the distance between the two stations is 969.22 m. The station has a speed limit of 75 km/h in 0–658.4 m and a speed limit of 55 km/h in 658.4–969.22 m. Two stations stipulate that the train running time is 90 s. Considering the ideal situation, the entire line ignores the curvature and the slope is zero. The train parameters are shown in Table 1.

By taking the MATLAB as the simulation platform, in order to demonstrate the effectiveness of AGPSO in urban rail transit, AGPSO, SPSO, and XRPSO [17] are used to solve the multi-objective optimization model of urban rail train automatic driving.

The parameter settings of SPSO and XRPSO are set according to their references. The simulation experiment runs for 30 rounds, and the number of iterations per round is 300. The optimization result data is shown in Table 2 The AGPSO achieved the best results in the optimization of comfort fitness value, energy fitness value, and overall fitness value. Only the AGPSO satisfies the constraint tolerance of the punctuality index of 0.01 s, and the other two algorithms do not meet the requirements.

The data results in Table 3 describe the time and location of the specific operating conditions obtained by the three algorithms for solving the recommended speed curve of the train.

Figure 3 is the convergence curve of the three algorithms optimization model to solve the overall fitness value. Compared with XRPSO, AGPSO and SPSO have better optimization effect on the overall fitness value. However, the results of SPSO cannot satisfy the constraint tolerance of the punctuality index. Therefore, only the AGPSO of the three algorithms achieves the expected results of the experiment and

Table 1 Train parameter captions

Parameter name	Numerical value	Unit
Static quality	144,000	kg
Dynamic quality	158,400	kg
Wheel diameter	770–840	mm
Traction cutoff delay	0.2	s
Brake establishment delay	1.5	s
Emergency braking deceleration	1.2	m/s^2

Table 2 Fitness optimization result

Algorithm	Comfort	Energy	Fitness	Punctual constraints
SPSO	3.4200E+01	4.6319E+07	4.0017E+07	6.2060E−02
XRPSO	3.4973E+01	4.8435E+07	3.9707E+07	1.0273E+01
AGPSO	3.0555E+01	3.8972E+07	3.4595E+07	6.7721E−04

Table 3 Time (s) and position (m) of working condition change

Algorithm	SPSO		XRPSO		AGPSO	
	Time	Position	Time	Position	Time	Position
1	28.50	243.68	29.14	254.81	25.46	194.50
2	3.94	73.30	4.87	85.21	25.93	396.19
3	4.78	74.89	5.88	92.53	2.12	31.10
4	28.70	408.54	36.65	374.37	13.14	184.01
5	23.71	168.78	23.25	162.28	23.33	163.40

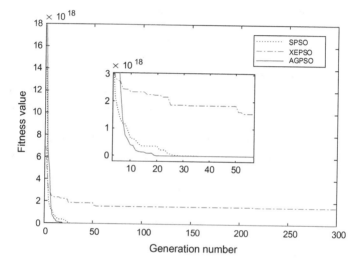

Fig. 3 Overall fitness convergence curve

verifies the effectiveness of the AGPSO in optimizing the recommended speed curve of urban rail transit trains.

Figure 4 shows the ATO recommended speed curve corresponding to the automatic driving time of the train for 90 s. Obviously, the ATO recommended speed curve solved by the AGPSO can better reflect the superiority of the ATO system automatic driving system. Meet the experimental expectations.

5 Conclusion

This paper proposes the AGPSO which can balance exploration and development to improve search capabilities. Moreover, it can effectively avoid falling into the local optimal solution. The ATO recommended speed curve problem is a multi-objective optimization problem. The mathematical model of the ATO recommended speed

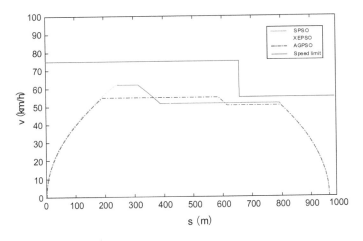

Fig. 4 ATO recommended speed curve

curve is established for the five working conditions. This paper takes two stations' data in the line of Guangzhou Metro Line 1 as the background. SPSO, XEPSO, and AGPSO are used to optimize the mathematical model. The simulation results show that only AGPSO can achieve the expected optimization effect and meet the model constraints. The valid of AGPSO is verified in the recommended speed curve of urban rail transit trains.

References

1. Calderaro V, Galdi V, Graber G et al (2014) An algorithm to optimize speed profiles of the metro vehicles for minimizing energy consumption. In: International symposium on power electronics, electrical drives, automation and motion. IEEE, pp 813–819
2. Corman F, Quaglietta E (2015) Closing the loop in real-time railway control: framework design and impacts on operations. Transp Res Part C 54:15–39
3. Han SH, Byen YS, Baek JH et al (1999) An optimal automatic train operation (ATO) control using genetic algorithms (GA). In: TENCON 99. Proceedings of the IEEE region 10 conference. IEEE Xplore, pp 360–362
4. Bocharnikov YV, Tobias AM, Roberts C (2010) Reduction of train and net energy consumption using genetic algorithms for trajectory optimisation. In: Railway traction systems. IET, pp 1–5
5. Carvajal-Carreño W, Cucala AP, Fernández-Cardador A (2014) Optimal design of energy-efficient ATO CBTC driving for metro lines based on NSGA-II with fuzzy parameters. Eng Appl Artif Intell 36:164–177
6. Mishra S, Mishra D, Satapathy SK (2012) Fuzzy frequent pattern mining from gene expression data using dynamic multi-swarm particle swarm optimization. Procedia Technol 4(5):797–801
7. Zou D, Li S, Li Z et al (2017) A new global particle swarm optimization for the economic emission dispatch with or without transmission losses. Energy Convers Manage 139:45–70
8. Zou D, Li S, Kong X et al (2018) Solving the dynamic economic dispatch by a memory-based global differential evolution and a repair technique of constraint handling. Energy 147:59–80
9. Kennedy J, Eberhart RC (1995) Particle swarm optimization. In: International conference on neural networks, pp 1942–1948

10. Shi Y, Eberhart RC (1998) Parameter selection in particle swarm optimization. In: International conference on evolutionary programming. Springer, Berlin, pp 591–600
11. Sun Y, Wang Z (2016) Improved particle swarm optimization based dynamic economic dispatch of power system. Procedia Manufact 7:297–302
12. Jensi R, Jiji GW (2016) An enhanced particle swarm optimization with levy flight for global optimization. Appl Soft Comput 43(C):248–261
13. Panigrahi BK, Pandi VR, Das S (2008) Adaptive particle swarm optimization approach for static and dynamic economic load dispatch. Energy Convers Manage 49(6):1407–1415
14. Tang B, Zhu Z, Shin HS et al (2017) A framework for multi-objective optimisation based on a new self-adaptive particle swarm optimisation algorithm. Inf Sci 420:364–385
15. Sun Y, Wang Z (2017) Improved particle swarm optimization based dynamic economic dispatch of power system. Procedia Manufact 7:297–302
16. Sancaktar I, Tuna B, Ulutas M (2018) Inverse kinematics application on medical robot using adapted PSO method. Eng Sci Technol Int J 21:1006–1010
17. Tang D, Dai M, Salido MA et al (2016) Energy-efficient dynamic scheduling for a flexible flow shop using an improved particle swarm optimization. Comput Ind 81(C):82–95

An Improved SSD and Its Application in Train Bolt Detection

Jiabing Zhang, Zhouyi Su and Zongyi Xing

Abstract A multi-scale fusion and multi-window target detection method is proposed. Firstly, the model and principle of the classical single shot multi box detector (SSD) method are expounded. From the principle analysis, explain the reason why the SSD method is not good in detecting small targets. On this basis, a multi-window and multi-scale fusion model is proposed and explains its model and working principle. This paper evaluates the detection accuracy of the original SSD model and the improved SSD model by using the same data set. The experimental results show that the accuracy of the original model is about 0.8, and the improvement is about 0.95. Therefore, the improved SSD model is more advantageous and accurate than the traditional SSD detection model in bolt detection.

Keywords Deep learning · Target detection · Multi-scale fusion · Small target

1 Introduction

Target detection is a hotspot and an important issue in computer vision research. Target detection technology is widely used and has important research value in the fields of pedestrian tracking, license plate recognition and driverless driving [1]. Before the emergence of deep learning, the target detection is mainly carried out in the traditional way, which is mainly divided into two directions: first, using the combination of features and models, such as Hough transform, using the characteristics of the data to establish the corresponding mathematical model, by solving the corresponding mathematical model to achieve the results of the target detection and second, through feature extraction and classifier to achieve the target detection effect. The feature extraction of the target object is performed by means of histogram of oriented gradients (Hog), shift, etc., and the classifiers such as support vector machine, short

---

J. Zhang · Z. Xing (✉)
Nanjing University of Science and Technology, Nanjing 210014, China
e-mail: xingzongyi@163.com

Z. Su
Guangzhou Metro Group Co., Ltd., Zhuhai District, China

for SVM [2] and AdaBoost [3] are trained to achieve the result of the classification target detection. In recent years, with the improvement of image classification accuracy by deep learning, the target detection method based on deep learning has gradually become the mainstream.

At present, the target detection methods based on deep learning are mainly divided into two categories: the first category is combined with the region proposal and convolutional neural networks (CNN) classification detection framework including region-CNN [1], fast R-CNN [4] and so on. The second is an end-to-end target detection framework represented by You Only Look Once (YOLO) [5] that transforms target detection into a regression problem, mainly including YOLO and SSD [6]. The region candidate uses information such as texture/edge in the image to find out in advance where the target may exist. Regression thinking does not require intermediate candidate regions to find the target but directly returns to the determination of the completed position and category. This paper uses SSD framework to study. SSD has the fastness of YOLO and the accuracy of faster R-CNN, but the SSD detection algorithm is not good for detecting small target objects. Because SD's feature maps for prediction are not reused, and there is not enough semantic information for the detection of small target objects, building more complex feature maps can further improve the detection accuracy of SSD2 SSD target detection.

2 SSD Introduction

2.1 SSD Detection Principle and Model

The SSD uses the feature pyramid structure for inspection. Through the convolutional layers conv4_3, conv_7, conv6_2, feature maps of different sizes are extracted, and classified regression and position regression are simultaneously performed on multiple features maps. The purpose is to be able to accurately detect objects of different scales, because the feature map in the lower layer has a smaller field of view, and the high-level perception field is larger, convolving in different feature maps, thereby achieving multi-scale purposes and improving detection. Robustness of targets is at different scales. At the same time, the default box in SSD is similar to the anchor mechanism of faster R-CNN. Some target preselection boxes are preset, and the final target position is obtained through soft-max classification and bounding box regression.

The architecture of the SSD is mainly divided into two layers: One is the basic network such as VGG-16, which is used for the initial extraction of target features. The other part is a multi-scale feature detection network located in the latter part, which is composed of a series of convolutional neural networks. In order to carry out the multi-scale extraction of the features extracted from the basic network as described above. The SDD framework is shown in Fig. 1 [7].

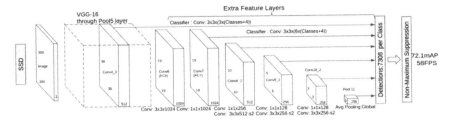

Fig. 1 SSD network structure

2.2 SSD Loss Function

The SSD training also returns the position and target categories. The target loss function is the sum of confidence loss and position loss [7]. The expression is

$$L(z, c, l, g) = \frac{1}{N}(l_{\text{conf}}(z, c) + \partial l_{\text{loc}}(z, l, g)) \quad (1)$$

In the above formula, N represents the number of default boxes matching the real box, l_{conf} is the confidence loss function, l_{loc} is the position loss function, z is the result of matching the real box of the different categories with the default box, and c is the confidence of the predicted object frame. Degree, l is the position information of the prediction frame, g is the real frame position information, and ∂ is the weight loss confidence and position loss function, generally 1:1 weight is selected.

By continuously reducing the value of the loss function, the SSD continuously optimizes the training model, so that the credibility of the prediction box category is continuously improved, and the credibility of the position of the prediction frame is continuously improved, thereby achieving an ideal target detection result.

2.3 SSD Feature Map Mapping

The SSD uses multi-scale methods to obtain feature maps of different sizes and obtains feature maps of different sizes from conv4_3, conv_7, conv6_2, conv7_2, etc. If the M-layer feature map is used for model detection, the default proportion of the k-th feature map is [7]

$$S_k = S_{\min} + \frac{S_{\max} - S_{\min}}{m - 1}(K - 1) \quad k = 1, 2, \ldots, m \quad (2)$$

$S_{\min}$ is generally taken as 0.2, $S_{\max}$ is taken as 0.95, the scale of the lowest layer is 0.2, and the scale of the highest layer is 0.95. There is a different size default box for each feature map unit, and the aspect ratio of each default section is different a_r = {1/3, 1/2, 1, 2, 3}. According to the above, the aspect ratio and feature map can calculate the length and width of the default model. The calculation formula is as follows:

$$W_k = S_k \sqrt{a_r} \tag{3}$$

$$H_k = S_k / \sqrt{a_r} \tag{4}$$

The prior box is matched with the real box according to the intersection over union (IoU). If the matching is successful, it is a positive sample, and vice versa is a negative sample.

Although SSD adopts multi-scale prediction, the different scale features are independent of each other, and the low-level position information is better, but the classification accuracy is poor [8]. In order to reduce the error caused by low level in SSD, SSD changed the pooling layer of VGG-16 into the convolution layer and constructed multi-scale information from the conv-4 convolutional layer in the network, but the shallow features lack deep feature information. Therefore, it is impossible to have a good detection effect in the small target detection process.

3 Depth Information Fusion

In view of the shortcomings of the traditional SSD detection algorithm above, based on the idea of feature fusion, this paper proposes an algorithm network structure that fuses shallow features with deep feature information and achieves features by merging shallow feature information with deep feature information. Enhance the effect and improve the accuracy of small target detection.

3.1 Model Introduction

In order to enhance the semantic information of the feature map, this paper proposes a MIF-SSD, as shown in Fig. 2. The low-level semantic information in the SSD is merged with the feature map of the adjacent high-level semantic information through a series of operations to enrich the semantic information of the low-level feature map. When the input image is 300 × 300, twice the up-sampling is required because the upper feature map is merged with the lower feature map. Since conv4_3 is the lowest level of feature map acquisition, conv4_3 is merged with conv5_3 from conv4_3 and so on.

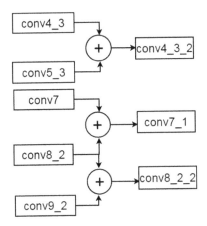

Fig. 2 Fusion between different feature maps

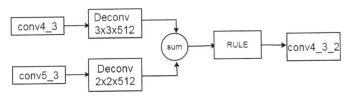

Fig. 3 Process diagram of feature map fusion

3.2 Fusion

For the merging of adjacent feature layers, taking conv4_3 and conv5_3 as an example, the conv5_3 is deconvolved before fusion, so that the width and height of the feature map are the same as conv4_3, and then the two layers are passed through the element-sum module. The feature maps are added directly, as shown in Fig. 3.

3.3 Training Methods

The training method in this paper is similar to the traditional SSD. The default box corresponds to the real box. The loss function is calculated end-to-end, and the back propagation is calculated. The training flowchart is shown in Fig. 4.

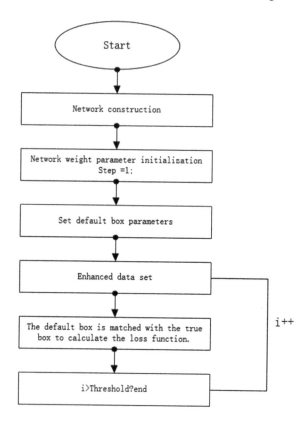

Fig. 4 Network training flowchart

3.4 Experiment and Analysis

In practical applications, the more accurate the positioning, the more accurate the classification is what we want. An excellent object detection method should detect the target with the highest possible confidence and should have a high recall rate and accuracy. In this paper, the improved SSD censor detection method is compared with the traditional SSD method to verify the effectiveness of the improved algorithm for small target detection.

Comparison of test results. In order to compare the performance of the improved SSD network with the traditional SSD network, the paper verifies by including the picture of the underbody bolt of the tram. A total of 120 images were selected, a total of 950 objects were recorded in the real frame, and the classic SSD target detection and improved SSD target detection were performed, respectively. For the same scene graph, the improved network has a significant improvement compared to the traditional SSD detection effect as shown in Fig. 5.

The above is the classic SSD test results, the following is the improved test results, and we found that the improved SSD is more accurate than traditional SSD detection.

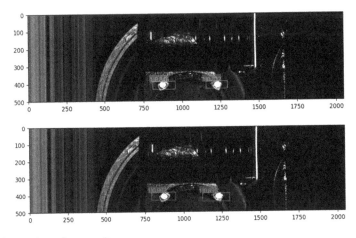

Fig. 5 Comparison of test results

Table 1 Bolt detection of VOC2007 data set

Method	mAp	Bolt
SSD	0.36	0.632
Improved SSD	*0.432*	0.578

Comparison of object detection accuracy. In object detection, the accuracy is generally expressed by mAP: for the above 120 images, the classic SSD and the improved SSD mAP are calculated for the target detection accuracy under the same conditions. The results are shown in Table 1.

The improved SSD of the modified SSD is 0.432, which is 0.07 higher than the traditional SSD detection. Therefore, the improved SSD in the bottom bolt detection is better than the traditional SSD.

4 Conclusion

The article first describes the SSD framework and its working principle. Then according to the traditional SSD's performance shortcomings and the idea of feature fusion based on practical experience, the feature information of the lower layer network and the adjacent network semantic information are merged into each other as a new feature extraction network to obtain more rich semantic information to improve the shortcomings of traditional SSDs that are not sensitive to small targets. Experiments show that the modified network has a greater improvement in the detection of small targets than the traditional SSD network.

Acknowledgements This work is supported by National Key R&D Program of China (2017YFB1201201).

References

1. Yongshun L, Cong T, Kedong Z (2008) Multi-window SSD target detection method based on deep learning. Infrared Laser Eng 1:302–310 (in Chinese)
2. Szegedy C, Reed S, Erhan D, Anguelov D (2015) Scalable, high-quality object detection. arXiv preprint v3
3. Wenguang W, Qiang L, Maosong L, Xianzhen H (2019) An efficient target detection method based on improved SSD. Comput Eng Appl 1–13 (in Chinese)
4. Girshick R (2015) Fast R-CNN. In: ICCV
5. Chuanhua C, Zhiqiang H, Wangsheng Y (2018) Improved target fine positioning detection algorithm based on SSD. J Air Force Eng Univ (Nat Sci Ed) (6):73–78 (in Chinese)
6. Yuming X, Wei Z, Yongying Z (2019) Improved SSD aerial target detection method. J Softw (3):738–758 (in Chinese)
7. Liu W (2016) Single shot multibox detector. Lect Notes Comput Sci 9905:1–13
8. Glorot X, Bengio Y (2010) Understanding the difficulty of training deep feedforward neural networks. In: AISTATS

Research on the Wheelset Life Optimization of Urban Rail Transit Trains

Xiaoxiao Xu, Zhengjun Ye, Jianyu Zhang, Zongyi Xing and Yingshun Liu

Abstract In the case of long-term operation, the wear level of the wheelset is intensified with the increase of the operating mileage. In addition, the unreasonable repair strategy will accelerate the consumption of the wheelset life and increase the operating cost. In order to extend the life of the wheelset, firstly, a large number of on-field measurement wheel size data are initially analyzed and relevant laws are got. Based on this, a data-driven wheelset life prediction model is established to predict the wheelset life. Finally, according to the statistical law of historical data, the wheelset reversal strategy and multi-template selection strategy are proposed. Based on the two strategies, a hybrid repair strategy is proposed. After comparison and analysis, it verifies the optimization effect of the proposed strategy. It can effectively extend the wheelset life.

Keywords Wheel wear · Repair strategy · Wheelset life

1 Introduction

During the operation of the urban rail transit train, continuous contact friction occurs between the train wheels and the rail, and the wheelset will also suffer from continuous wear, which is easy to cause the roundness of the wheel to be increased, and the tread damage is serious, which jeopardizes the safety of the urban rail transit train [1]. In response to this situation, the technical staff will periodically measure the wheel size parameters and repair or replace the wheelset in time whose wheel size parameters exceed the wear limit [2]. However, in the process of wheelset repair, it is prone to shorten the life of the wheelset due to the problems such as the improper repair. Therefore, it is important to study how to optimize the repair strategy to extend the wheelset life and reduce the operating cost of the wheelset whole life cycle [3].

X. Xu · Z. Ye · J. Zhang · Z. Xing (✉) · Y. Liu
School of Automation, Nanjing University of Science and Technology, Nanjing 210094, China
e-mail: xingzongyi@163.com

The research on extending the wheelset life began in the 1970s. In literature [4], the Talgo company in Spain analyzes the measured data of the wheelset wear and considers that the wheelset is repaired to restore the rim thickness parameter to 30.5 mm when it is reduced to 27.5 mm, which minimizes the cost of repairing the wheelset. In literature [5], a wheelset wear data prediction algorithm based on field data analysis is proposed. Based on the prediction results, a targeted repair strategy based on wear parameters is proposed. In literature [6], Zhao Wenjie based on the wear data of the wheel size parameters of the Guangzhou Metro train, the wear process of the rim thickness parameter and the diameter parameter is divided. According to Markov theory, the mathematical model of wear prediction is established based on the rim thickness parameter and the wheel diameter parameter. Finally, the control limit strategy is adopted, and the Monte Carlo simulation method is used to verify the proposed strategy. Based on the above, four optimization schemes of the wheelset repair strategy are proposed. In literature [7], Pang Songlin used the CRH2 EMU as the experimental model to establish the dynamic model of the CRH2 EMU and used the NURBS curve modeling principle to design the economical tread wheel tread shape. According to the actual measured data, the analysis results are compared with the existing repair strategies, and suggestions for improvement of wheelset repair are proposed. In literature [8], Wang Ling established a wheelset wear data driving model on the basis of analyzing the correlation between rim thickness and wheel diameter wear and repairing proportion coefficient. Based on the wear model, the wheelset repair strategy (S_{dh}, S_{df}) is proposed to optimize the wheelset repair strategy with the minimum expected cost rate, and the remaining service life forecast of the wheelset is given.

2 Establish the Wheelset Life Prediction Model

2.1 Overview of the Whole Life Cycle Status of the Wheelset

The Guangzhou Metro will periodically measure and record the wheel size parameters. The parameters include the rim height parameter, the rim thickness parameter, the wheel diameter parameter, etc. The standard of the wheel size parameters is shown in Table 1.

The wheelset life is ultimately limited by the wheel diameter parameter. The diameter of the new wheel is 840 mm. When the wheel is worn, repaired, etc. until the wheel diameter reaches 770 mm, the wheelset will be scrapped. During the operation of the train, if other parameters are found to exceed the standard, the wheelset will be repaired, and the parameters of the wheel will be repaired to the standard and continue to be used.

By analyzing a large number of wheel size history data, the following laws can be found:

Table 1 Standard table of wheel size parameters

Wheel parameters	Standard lower limit	Standard upper limit
Rim thickness Sd (mm)	26.5	32.5
Rim height Sh (mm)	28	34
Wheel diameter D (mm)	770	840

1. There is a big difference in wheel wear trends between the left and right wheels of the same wheelset;
2. The rim thickness changes regularly with the increase of the running mileage of the train and is usually the first parameter of exceeded the standard;
3. The wheel diameter parameter of the train will decrease with the running mileage;
4. There is no obvious regular change in the rim height parameter.

Based on the above laws, it is feasible to model the wear of the wheel based on data driving.

2.2 Wear Modeling Based on the Rim Thickness Parameter

Preliminary analysis shows that the rim thickness parameter is one of the most important parameters to guide the technical staff to repair the wheelset [9]. In this paper, the high-order polynomial fitting curve is used to model the wear of the rim thickness and to estimate the trend of the rim thickness parameter. The expression based on the high-order polynomial fit curve is:

$$f = \sum_{i=0}^{n} a_i x^i \qquad (1)$$

where each term consists of the product of the coefficient a_i and a rising integer power uncertain term x^i.

Considering that high-order polynomials are oscillating, while low-order polynomials more likely to fall between data and can show the trend of data points in a closed interval [10, 11], this paper chooses to use the fourth-order polynomial to fit the raw rim thickness data, and the fitting result is shown in Fig. 1.

As shown in Fig. 1, the data in the figure is the wear of the rim thickness of the 89A car after a repair and before the next repair. The data after the data preprocessing basically conforms to the linear trend. After the fourth-order polynomial fitting, the wear trend of the rim thickness parameter can be well reflected.

In a single repair cycle, the repair is usually carried out as long as the rim thickness parameter of either side of wheelset exceeds the limit. As can be seen in Fig. 1, the rim

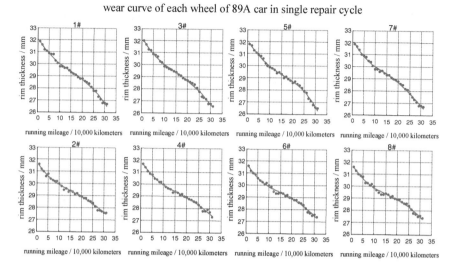

Fig. 1 Rim thickness wear curve of each wheel of 89A car in single repair cycle

thickness values of the odd wheels are worn faster than the even wheels. Therefore, the odd wheels of the wheelset first exceed the standard, and the wheelset needs to be repaired.

Based on the fitted polynomial expression and the length of the single repair cycle (running mileage between two repairs), the wear rate of the rim thickness of each wheel in the single repair cycle can be calculated by Eq. 2:

$$v = \frac{Sd_D - Sd_{D0}}{l} \quad (2)$$

where v represents the wear rate of the rim thickness in single repair cycle, Sd_{D0} represents the rim thickness value of a new wheelset or a just repaired wheelset, Sd_D represents the rim thickness value of a wheelset that needs repair, and l represents the length of the single repair cycle.

As shown in Table 2, the rim thickness wear rate of each wheel of 89A car is between 0.1 mm and 0.2 mm/million, which is similar to the experience of the technical staff. And, because the odd wheel of the 8990 train exceeds the standard first, the 8990 train wheel rim thickness wear rate takes an odd wheel average wear rate of 0.1529 mm/10,000 km.

Table 2 Wear rate of each wheel of 89A car in single repair cycle (unit: mm/10,000 km)

Wheel number	1#	2#	3#	4#	5#	6#	7#	8#
89A	−0.1605	−0.1217	−0.1613	−0.1273	−0.161	−0.1245	−0.1623	−0.1248

2.3 Wear Modeling Based on the Wheel Diameter Parameter

Wheel diameter parameter is one of the most important parameters of the whole life cycle of the wheelset [12]. And, the wear of the wheel diameter parameters comes from two aspects: one is that the wheel diameter decreases with the train operating; the other is that the wheelset is repaired to restore the rim thickness and the rim height parameters, which will sacrifice a certain wheel diameter. Therefore, it is not only necessary to get the wear rate of the wheel diameter in a single repair cycle but also to calculate the cutting amount of the wheel diameter in the whole life cycle.

Taking the 1# wheel diameter parameter of the 90A car as an example, the values are as shown in Fig. 2a after the data preprocessing. The figure can roughly see that the 1# wheel diameter values decrease in the whole life cycle. In the figure, the wear trend of the 1# wheel diameter in the whole life cycle can be roughly seen. And, the fourth-order polynomial fitting is performed in the single repair cycle in the same way. Similarly, based on the fourth-order polynomial, the wheel diameter wear rate and the average wear rate can be calculated. The average wear rate is 0.2045 mm/10,000 km.

The essence of wheel repair is to restore the tread of the wheel to the standard profile by cutting the wheel. This process is actually a process of sacrificing wheel diameter [13]. Therefore, in order to establish a wheelset life prediction model based on the wheel diameter parameter, it is necessary to count the amount of cutting for each wheelset. Take the 90A car wheels as an example. The effective range of the wheel diameter in the wheel whole life cycle is 770–840 mm, and the empirical value of the cutting amount in every repair is about 7 mm. Therefore, as long as the wheel diameter parameter changes more than 4 mm at a time, it is determined that the wheelset is repaired once. The difference between the wheel diameter values before and after in a repair is taken as the cutting amount. When the wheel diameter is lower than 770 or reaches 840 again, it indicates that the wheelset life cycle reaches the limit and the wheelset will be forcibly scrapped. The cutting amount got is shown in Table 3.

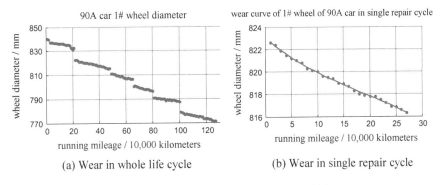

Fig. 2 Wear of the wheel diameter parameter of the 1# wheel of the 90A

Table 3 Cutting amount of each wheel of 90A car (unit: mm)

Number of repairs	1#	2#	3#	4#	5#	6#	7#	8#
1	9.89	8.8	8.43	7.21	8.9	9.5	9	10.5
2	4.71	6.48	7.4	5.9	5.65	6.45	4.68	6.72
3	5.96	6.76	6.57	5.7	8.92	9.6	9.71	9.52
4	5.27	6.96	5.86	6.07	5.3	5.12	4.54	4.12
5	8.83	9.86	10.17	10.8	7.92	9.18	8.11	8.56
Total amount of cutting	34.66	38.86	38.43	35.68	36.69	39.85	36.04	39.42

The effective range of the wheel diameter parameter is 770–840 mm, the effective diameter change is 70 mm, and the cutting amount is 34–40 mm. About half of the wheel diameter is consumed during the repair process. If this problem is solved, it can reduce the cutting amount to increase the wheelset life and reduce the operating cost of the wheelset.

2.4 Wheelset Life Prediction Model

The wheelset life prediction model is mainly composed of two parts. The first part is the rim thickness wear model, and the second part is the wheel diameter wear model. The length of the single repair cycle is usually determined by the rim thickness. The number of repair cycles in the whole life cycle of the wheelset is limited by the wheel diameter parameter. Based on these, the wheelset life can be predicted. The wear of the rim thickness and the wheel diameter in the whole life cycle of the wheelset is shown in Fig. 3.

From the before analysis, 0.1529 mm/10,000 km is used as the rim thickness wear rate in single repair cycle, and 0.2045 mm/10,000 km is used as the wheel diameter wear rate in single repair cycle, and the cut amount is 4–10 mm/time.

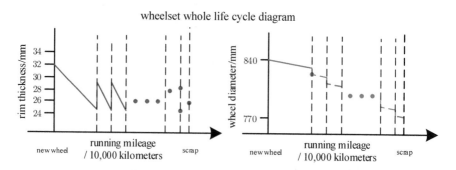

Fig. 3 Wheelset whole life cycle diagram

The average life of the wheelset is calculated as shown in Eq. 3:

$$E(T) = \sum_{1}^{N} \frac{Sd_N}{P_{(Sd,N)}} + T_{last}$$

$$\sum_{1}^{N} R_N + D_{last} + \sum_{1}^{N} \left(\frac{Sd_N}{P_{(Sd,N)}} P_{(D,N)} \right) \leq D - D_0 \quad (3)$$

where $E(T)$ represents the average life of the wheelset, N represents the number of repairs, Sd_N represents the effective value of the rim thickness parameter in the single repair cycle, $P_{(Sd,N)}$ represents the rim thickness wear rate, and T_{last} represents the mileage of the wheelset after the last repair, R_N represents the single cutting amount, D_{last} represents the wheel diameter wear after the last repair to the wheelset scrapped, and $P_{(D,N)}$ represents the wheel diameter wear rate, D represents the initial wheel diameter value of the new wheel, and D_0 represents the wheel diameter value when the wheelset is scrapped. Limit the length of the single repair cycle with the rim thickness parameter, and limit the length of the whole life cycle with the wheel diameter parameter. And, the wheelset life is predicted based on the above conditions.

3 Wheelset Life Optimization Strategy Analysis

The wheelset life optimization strategy mainly studies the wheelset repair strategy. The current strategy is mainly the fault repair strategy, that is, when the wheel size parameters exceed the standard or the wheel has serious wear, the wheelset is repaired. This kind of repairing method is very easy to cause the wheel cutting amount to increase and make the wheel diameter parameter consume too fast, and the wheelset life is greatly shortened. Therefore, this paper proposes the wheelset right and left wheels reversal strategy and the multi-template selection strategy for this problem. And based on this, the hybrid repair optimization strategy is proposed to extend the life of the wheelset [14].

3.1 Wheelset Left and Right Wheels Reversal Strategy

It can be seen from the before analysis that the wear trend of the left and right wheels is not the same. And, their rim thickness and wheel diameter parameter wear conditions are also different. Therefore, the left and right wheels of a certain wheelset are analyzed as follows.

As shown in Fig. 4, the thickness wear of the left wheel's rim of the wheelset is obviously larger than that of the right wheels, and the left wheels exceed the standard first, so the wheelset needs to be repaired, but the right wheels' rim thickness is still

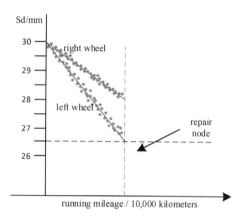

Fig. 4 Analysis of the difference between the left and right wheels

in normal range at this time. According to the above laws, this paper proposes a left and right wheels reversal strategy. As shown in Fig. 5, after adopting the wheelset reversal strategy, the left and right wheels can exceed the standard at the same time, which can delay the wear of the left wheels; thereby, the life of the wheelset in the single repair cycle can be extended.

Taking the 8990 train as an example, after the optimization of the reversal strategy, the average wear rate of the rim thickness is 0.1358 mm/10,000 km, and the average wear rate of the conventional strategy rim thickness is 0.1529 mm/10,000 km. Therefore, in a single repair cycle, the wear of the rim thickness can be effectively reduced, and the life of the wheelset in a single repair cycle is extended.

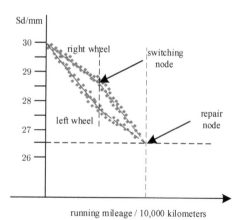

Fig. 5 Effect diagram after wheelset reversal

3.2 Multi-template Selection Strategy

Even with the left and right wheels reversal strategy, the wheelset wear will continue to occur until exceeding the standard. At this time, the tread of wheel must be repaired to the standard. Usually, there are multiple repair templates that can be selected. Take Guangzhou Metro Line 8 as an example. There are three repair templates to choose from. See Table 4 for specific information.

Taking Guangzhou Metro Line 8 as an example, the use of template C for repairs is rare. A large number of wheelset is repaired according to the 30 mm template combined with the 28 mm template. The rim thickness parameter values are analyzed before and after the repair with template A and template B and the result is shown in Fig. 6.

As shown in Fig. 6, if the wheelset is repaired with the template A, the rim thickness parameter of the wheelset will increase with the running mileage increasing, while repaired with template B, the rim thickness will decrease with the running mileage increasing. The reason why the template A rim thickness parameter is increasing continuously is that the measurement of the rim thickness is based on the reference position on the wheel tread. It will experience the wear of the rim and the tread, which will cause the rim thickness to no longer decrease or even increase when the tread wear exceeds the rim wear, so the measured rim thickness parameter can be increasing as wear occurs [15].

Table 4 Wheelset repair templates

Template model	Template A	Template B	Template C
Rim thickness (Sd/mm)	28	30	32

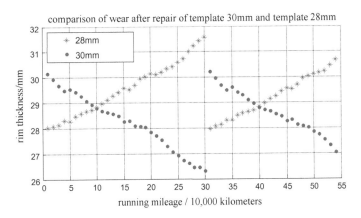

Fig. 6 Comparison of wear after repair of 30 and 28 mm template

Table 5 136A car whole life cycle repair cutting amount (unit: mm)

Number of repairs	1#	2#	3#	4#	5#	6#	7#	8#
1	8.33	9.25	8.27	9	8.23	8.84	8.91	8.46
2	5.19	4.86	6.04	5.17	5.06	5.64	5.72	6.12
3	5.01	4.69	6.17	5.77	5.17	5.82	5.98	5.33
4	6.01	5.99	5.75	5.33	5.64	6.13	5.47	5.24
5	5.75	5.26	4.7	5.56	5.17	5.1	5.65	5.26
Total amount of cutting	30.29	30.05	30.93	30.83	29.27	31.53	31.73	30.41

Based on the characteristics of the variation of the rim thickness parameter values of the template A and the template B, the multi-template selection strategy is proposed to reduce the wheel diameter cutting amount during the repair:

1. If the template A is used for the first repair of the wheelset, then with the increase of the rim thickness parameter, the template B is chosen when the repair needs to be performed again;
2. If the template B is used for the first repair of the wheelset, then with the decrease of the rim thickness parameter, the template A is chosen when the repair needs to be performed again.

The multi-template selection strategy has been experimented in the field. The experimental train 135/136 of Guangzhou Metro Line 8 adopts the multi-template selection strategy, while train 89/90 is selected as a reference which adopts the conventional strategy. In order to verify the optimization of the multi-template selection strategy, the wheelset of the same position of the two trains is selected to measure the wheel diameter repair cutting amount. The results are shown in Tables 3 and 5.

It can be seen from Tables 3 and 5 that the cars 90A and 136A both have 5 repairs according to their strategies, but the average cutting amount of 136A is much less than the 30.63 mm of the 90A. Therefore, the multi-template selection strategy can effectively reduce the cutting amount in the whole life cycle of the wheelset.

3.3 Hybrid Repair Optimization Strategy

The wheelset reversal strategy is combined with the multi-template selection strategy, and then a hybrid repair optimization strategy is proposed. The multi-template selection strategy is used to reduce the cutting amount of the wheel diameter value, and the wheelset reversal strategy extends the length of a single repair cycle. The strategy flowchart is shown in Fig. 7.

The specific steps are as follows:

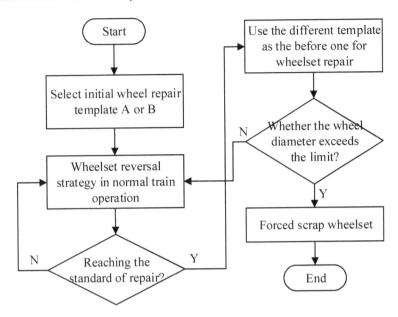

Fig. 7 Hybrid repair optimization strategy flowchart

- Step 1: Select template A or template B as the initial template;
- Step 2: When the train is running, the left and right wheels are worn. The reversal strategy is adopted to balance the wear of the left and right wheels, so that the wheels on both sides are as close as possible before reaching the next repair;
- Step 3: Measure the wheel parameter and judge whether the repair standard of the multi-template selection strategy is reached. If it is still not reached, proceed to Step 2. If the standard is reached, proceed to Step 4;
- Step 4: Select the template different from the last template to repair;
- Step 5: Determine whether the wheel diameter parameter exceeds the standard. If it is lower than 770 mm, force the wheelset to be forcedly scrapped. Otherwise, proceed to Step 2.

Based on Eq. 3, taking the 8990 car 1# wheels as an example, the new rim thickness wear rate with the wheelset reversal strategy replaces the rim thickness wear rate with the conventional strategy, and the new number of repairs with the multi-template selection strategy replaces the number of repairs with the conventional strategy. The average cutting amount of the wheel diameter with the multi-template selection strategy replaces the average cutting amount of the wheel diameter with the conventional strategy, and then to calculate respectively the 1/2# wheelset life with the conventional strategy and the hybrid optimization strategy. The results are shown in Table 6.

It can be seen from Table 6 that according to the conventional strategy, the life of the 1/2# wheelset is about 1.38 million kilometers. As a reference strategy, after the hybrid optimization strategy, the 1/2# wheelset life reaches. 2.28 million kilometers,

Table 6 Comparison of repair strategy results (unit: 10,000 km)

Optimization strategy	General strategy	Hybrid optimization strategy
Wheelset life	138	228

an extension of nearly 900,000 km, extended service life of 65.2%, also greatly reduced the cost of the wheelset.

4 Conclusion

The rim thickness wear model and the wheel diameter wear model are established based on the wheel size data measured in Guangzhou Metro, and then the wheelset life prediction model is established. On this basis, the wheelset reversal strategy is proposed, which can extend the length of a single repair cycle. The multi-template selection strategy can reduce the wheel diameter cutting amount during the repair; finally, combine two optimization strategies. The hybrid optimization strategy is proposed. Compared with the conventional repair strategy, it is proved that the hybrid optimization strategy can effectively extend the life of the wheelset.

Acknowledgements This work is supported by National Key R&D Program of China (2017YFB1201102).

References

1. Li J (2009) Introduction to urban rail transit system. Mechanical Industry Press (in Chinese)
2. Murali P, Ordóñez F, Dessouky MM (2016) Modeling strategies for effectively routing freight trains through complex networks. Transp Res Part C 70:197–213
3. Xu H, Yuan H, Wang L et al (2010) Wheelset wear modeling and optimization of metro vehicles based on Gaussian process. J Mech Eng 46(24):88–95 (in Chinese)
4. Pascual F, Marcos JA (2004) Wheel wear management on high-speed passenger rail: a common playground for design and maintenance engineering in the Talgo engineering cycle. In: Proceedings of the 2004 ASME/IEEE joint rail conference, 2004, pp 193–199
5. Ward A, Lewis R, Dwyer RS (2003) Incorporating a railway wheel wear model into multi-body simulations of wheelset dynamics. Tribology 41(3):367–376
6. Zhao W, Wang L, Yuan H et al (2014) Modeling and maintenance strategy optimization of metro vehicle wheelset wear based on Markov process. Sci Technol Eng 14(36):116–119 (in Chinese)
7. Pang S (2016) Research on the repair strategy of CRH2 high-speed EMU wheel. Southwest Jiaotong University (in Chinese)
8. Wang L, Yuan H, Nan W et al (2011) Optimization of wheelset repair strategy and residual life prediction based on wear data driven model. Syst Eng Theory Pract 31(6):1143–1152 (in Chinese)
9. Zeng Q (2005) Analysis of wheel life of metro vehicles. Electr Locomotive Urban Rail Veh (2):47–49 (in Chinese)

10. Li X, Qian X, Wang Z (2010) An efficient algorithm for high-dimensional anomaly data mining. Comput Eng 36(21):34–36 (in Chinese)
11. Dempster AP, Laird NM, Rubin DB (1977) Maximum likelihood from incomplete data via the EM algorithm. J R Stat Soc 39(1):1–38
12. Zhu W, Yang D, Guo Z et al (2015) Data-driven wheel wear modeling and reprofiling strategy optimization for metro systems. In: Transportation research board meeting
13. Jang F (2006) Discussion on the selection of cutting amount during wheel sweeping. Railway Veh (12):36–37 (in Chinese)
14. Wang W, Cheng M, Hua W et al (2010) Hybrid modeling and applications of virtual metro systems. In: Vehicle power & propulsion conference. IEEE
15. Niu M, Feng Q, Chen S (2006) A review of online dynamic measurement methods for column wheels. Railway Locomotive 26(2):32–35 (in Chinese)

Design of a 3D Traction Substation Based on Real-Time Infrared Simulation System

Liping Zhao, Yirui Yu and Shengyang Zheng

Abstract Considering the inefficiency and insecurity of traditional training methods which focused on outdoor work, one design of traction substation simulation system based on Unity3d framework is proposed by combining three-dimensional scene with real-time infrared simulation system. Also, we propose a more accurate computing model of declination angle δ' to improve the infrared simulation system. After that, our simulation result proves that the improved infrared simulation system can describe temperature distribution on the surface of objects accurately, which makes the training mode more concrete and vivid.

Keywords Substation simulation · Infrared simulation · Declination angle

1 Introduction

In recent years, the USA, Canada and other countries have developed three-dimensional simulation training system for substations based on virtual reality technology in succession [1]. For example, one power system simulation software based on MATLAB was connected with the virtual substation to simulate the operation state of the substation [2]. The roaming function provided by the platform VRP realizes the application function of the virtual substation training system by operating and controlling the devices in the scene; another case uses the 3ds Max modeling tool and the DynaVR-Net platform to design the virtual substation simulation training system [3].

L. Zhao (✉) · Y. Yu · S. Zheng
University of Southwest Jiaotong University (SWJTU), Chengdu, China
e-mail: lpzhao@home.swjtu.edu.cn

With the continuous improvement of computing hardware conditions, many works began to explore how to quickly establish infrared simulation models. Sheffer et al. [4] and Biesel et al. [5] established a heat balance equation based on the material of the object under external environmental conditions and then obtained a calculation model of the temperature field distribution of the objects' surface.

Owens et al. [6] improved the forum about how external environmental factors influence object's surface temperature and obtained a more accurate computing model compared to Sheffer's.

In response to the above, our group conduct 3D modeling of traction substation and then build a virtual three-dimensional substation scene by open-source engine (we choose Unity3d here). By improving the calculating formula of the declination angle δ', the infrared simulation has been proved effect is better, especially in rendering mode.

2 The Design of Framework

The daily work of the traction substation mainly includes electrical equipment inspection, switching, work ticket processing. So, there exist problems including large number of scene nodes and child-objects imported, also, the concentrated distribution of electrical equipment and the logic operations realized by codes are complex. In this project, the structure of the system is divided into three layers: the bottom resource layer, the data interface layer and the UI interaction layer. The framework is shown in Fig. 1.

Detailed models will increase the reality of the virtual environment, allowing users to be further immersed. This system is designed on basis of 220 kV/27.5 kV traction substation in northeastern China, including two V/v oil-immersed main transformers, pressure device, main transformer relay protection device, high and low voltage side line transformer, circuit breaker, isolation switches, HV surge arresters, etc., as well as the construction of peripheral environments, such as line towers, station wall and gravel floor. The ultimate effect of models based on LOD is shown in Figs. 2 and 3 (LOD is used to optimize the rendering speed).

Then we take SQL database for data storage. Considering the development and changes of the system in the future as well as the flexibility principle of system design, our group separate the definition layer and implementation layer. Also, the service interface layer is defined in the system. On this basis, the operational logic of the training process is constructed to make the 3D scenes relevant and logical, thus completing the design of UI layer.

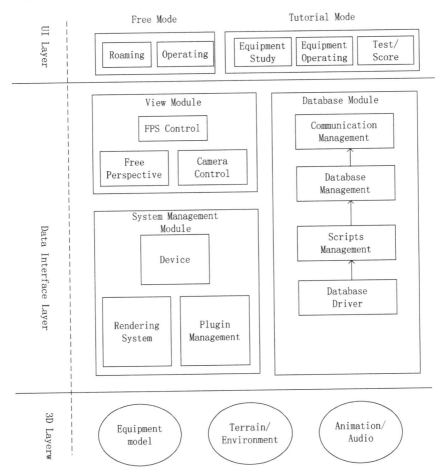

Fig. 1 Framework

Fig. 2 LOD model

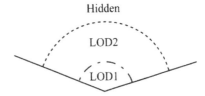

3 Real-Time Infrared Simulation Algorithm

Infrared simulation is an image of the actual infrared imaging device. We can calculate the value of infrared radiation on the object's surface. The computed radiation value will enter Unity3d and convert the radiation value into grayscale.

Fig. 3 Outdoor 3d models built by 3ds Max

3.1 Zero-Range-Distance Radiation Model

We can see from Fig. 4a that the sun moves around the earth at a uniform speed, the circumferential angle passing through the unit time is θ, with B as the reference point, and the sun moves to the point S after some time. Figure 4b is the sectional view of plane AOB. Assuming that $OA = R$, and the obliquity of the ecliptic $\theta_{tr} = 23.26°$, and we can get OA: $OA = \frac{AH_1}{\sin\theta_{tr}} = \frac{R\sin\delta}{\sin\theta_{tr}} = R\cos\theta$, thus we get: $\sin\delta = \cos\theta * \sin\theta_{tr}$.

Now, we take 2019.1.1 as t_0, spring equinox is 3.21, $\Delta t = 79$ days, we can know from Fig. 4 that δ is $0°$ on spring equinox. So, we can get δ' with:

$$\delta' = \arcsin\left(\sin\theta_{tr}\left[\frac{360}{365.2422}(N - 79)\right]\right) \tag{1}$$

And, the value of solar radiation received by the surface I'_d is:

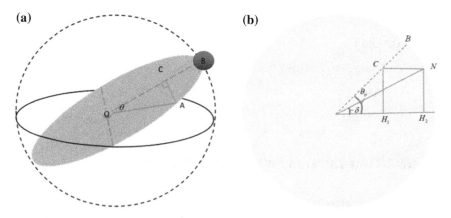

Fig. 4 Orbital diagram of sun

$$I'_d = \frac{1}{2} I_0 \sin\left(h \frac{1 - P^{\frac{1}{\sin(h)}}}{1 - 1.4 \ln P}\right) \quad (2)$$

where h is related to the value of δ' given by our formula.

Considering the atmospheric radiation, we finally get the algorithm model about radiation emitted by the sun and radiation received by objects in the scene with:

$$E_{\text{sun-object}} = \left(\varepsilon_1 C_b \left(\frac{T_1}{100}\right)^4 - \varepsilon_2 C_b \left(\frac{T_2}{100}\right)^4\right) \psi \quad (3)$$

Lastly, by calculating the discrete formula of the one-dimensional heat conduction differential equation, the temperature distribution on the surface and inside of the object can be obtained.

3.2 Infrared Heat Source Algorithm

We take camera component in Unity3d as radiation emitter; it has six cameras that are perpendicular to each other as shown in see Fig. 5.

We can place the cube camera in an appropriate position inside the 3D model, and the images in the six camera fields are rendered onto a texture image [7]. Then by camera render channel, computer calculates the radiation received by each texel by using the depth information in the camera.

Finally, the radiation information will be saved into the texture when the grayscale increment of the texture under the heat source is calculated. Compared to other algorithms, this method utilizes the parallel computing ability of the GPU, and the calculation speed is faster. The algorithm flowchart is shown in Fig. 6 as follows.

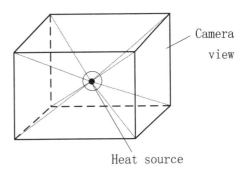

Fig. 5 Cube camera heat source

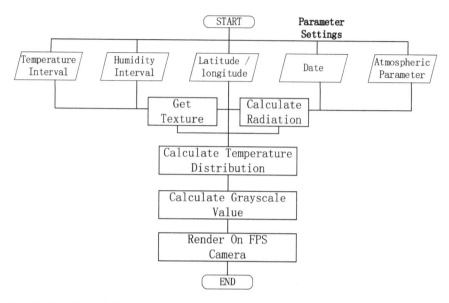

Fig. 6 Algorithm flowchart

4 Simulation

Figure 7a is the temperature distribution chart of main transformer. On this basis, we can select static points (S, S', S'') on the main transformer's surface texture (Fig. 7b) and simulate their changes of temperature under the condition of 300 K, 70% of humidity, at latitude of 114.26°, longitude of 38.03° in 24 h on the same day [8, 9]. Compared with the real measured data read by the main transformer's thermometer, we can see the accuracy of system easily in Fig. 8.

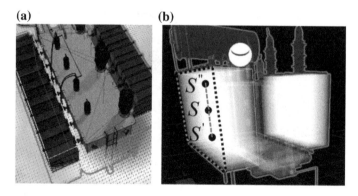

Fig. 7 Infrared imaging of the main transformer

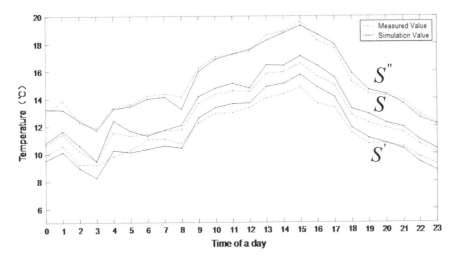

Fig. 8 The simulation of temperature distribution on the transformer (off state)'s surface

The simulation chart indicates that the temperature distribution differs from different zone of the same object. We can explain that it is because the radiation received from heat source differs in every part of the same texture. Also, as the radiation intensity decreases, the error of temperature distribution simulation has a tendency to increase, which will have a certain impact on the final infrared system. It is also because there are many factors affecting the temperature in the natural environment, and the system simulation process only considers the main influencing factors. So, when radiation is strong, the grayscale value is high and other errors can be ignored while with the decline of radiation, the grayscale value is more easily affected by environmental factors. In general, the system simulation results are more accurate.

Below are pictures about GW4 isolation switch in the scene simulated by Unity3d compared within the real scene. The normal scene is shown in Fig. 9a; the scene rendered by infrared heat source is shown in Fig. 9b.

The simulation image simultaneously shows the infrared effects of various objects under normal and radiation condition, for example, the distinction of texture between the insulator and the pillar of the isolation switch lead to different gradation of the radiation reaching the camera. Another detail, we can see that the gradation of the turf illuminated by the sunlight is higher, and the gradation of the turf at the back is smaller.

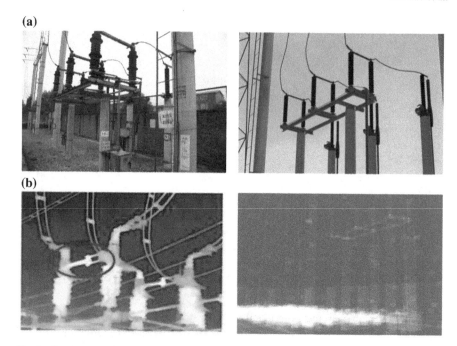

Fig. 9 Simulation of system

5 Conclusion

The contributions of this paper mainly include two parts:

(1) Based on the system, a more accurate theoretical derivation formula is applied, such as the calculation formula of the solar declination angle δ'.
(2) Considering the surface texture properties and internal heat source properties of the scene model, we calculate the surface temperature distribution decided by the heat source and convert the value into grayscale value.

With the development of virtual reality and augmented reality technology, the infrared simulation system will be extended to the virtual and real combination environment with hardware devices such as helmets and binoculars and may further improve the users' experience.

Acknowledgements This work is partially supported by the Undergraduate Education Research and Reform Key Project at Southwest Jiaotong University (1802046).

References

1. Gao M (2013) Discussion on the planning framework system of regional railway. J Railway Eng 6:10–14 (in Chinese)
2. Jiang C (2011) Research on the scheme of traction power supply safety monitoring system for high speed electrified railway. Power Autom Equip 9:18–21 (in Chinese)
3. Liao BB, Wang ZY, Ke XD et al (2017) IR scene image generation from visual image based on thermal database. In: Proceedings of SPIE, vol 6787. Society of Photo-Optical Instrumentation Engineers Press, Bellingham, pp 1–8
4. Sheffer AD, Cathcart JM (1988) Computer generated IR imagery: a first principles modeling approach. In: Proceedings of SPIE, vol 933. Society of Photo-Optical Instrumentation Engineers, Bellingham, pp 199–206
5. Biesel H, Rohlfing T (1986) Real-time simulated forward looking infrared (FLIR) imagery for training. In: Proceedings of SPIE, vol 781. Society of Photo-Optical Instrumentation Engineers, Bellingham, pp 71–80
6. Owens WR (1986) Data-based methodology for infrared signature projection. In: Proceedings of SPIE, vol 636. Society of Photo-Optical Instrumentation Engineers, Bellingham, pp 96–99
7. Wang Z (2012) Research on infrared image synthesis and multi-spectral information fusion for objects on the ground. Zhejiang University, Hangzhou (in Chinese)
8. Clough SA, Kneizys FX, Shettle EP et al (1986) Atmospheric radiance and transmittance-FASCOD2. In: Proceedings of conference on atmospheric radiation. American Meteorological Society Press, Boston, pp 141–144
9. Shi C (2016) Real-time simulation of infrared land and sea scene. Zhejiang University, Hangzhou (in Chinese)

Multi-channel Man-in-the-Middle Attack Against Communication-Based Train Control Systems: Attack Implementation and Impact

Mengchao Chi, Bing Bu, Hongwei Wang, Yisheng Lv, Shengwei Yi, Xuetao Yang and Jie Li

Abstract Communication-based train control (CBTC) systems utilize continuous, high-capacity, and bidirectional train-to-wayside wireless communications to ensure the safe and efficient operation, where wireless local area networks (WLANs) based on 802.11 protocols are adopted as the main method. WLANs are working at the public frequency, and malicious interference and attacks are inevitable, which can badly affect the performance of CBTC systems. Due to the fail-safe mechanisms of CBTC systems, malicious attacks can cause the emergency braking of trains, and the operation efficiency can be reduced. Based on the vulnerabilities of WLANs, a multi-channel man-in-the-middle (MitM) attack on WLAN-based train-to-wayside communications is considered. The proposed attack method can manipulate the 802.11 frames, and bring delays, packet losing, and even modification of messages transmitted between trains and wayside equipment. Based on the operation principles of CBTC systems, the impacts of MitM attack are quantified according to the comparison between the train trajectories under attacks and the preset optimal running profiles of trains. Simulation results demonstrate principles and consequences of MitM attack.

Keywords Communication-based train control · Multi-channel Man-in-the-Middle attack · WLAN · Cyber-physical system information security · Attack impact

---

M. Chi (✉) · B. Bu · H. Wang
Beijing Jiaotong University, Beijing, China
e-mail: 18120212@bjtu.edu.cn

Y. Lv
Institute of Automation, Chinese Academy of Sciences, Beijing, China

S. Yi
China Information Technology Security Evaluation Center, Beijing, China

X. Yang
Traffic Control Technology, Beijing, China

J. Li
Beijing Jiaotong University Microunion Technology Co., Ltd., Beijing, China

1 Introduction

CBTC systems utilize continuous, high-capacity, and bidirectional train-to-wayside wireless communications to improve the capacity of urban rail transit systems [1]. WLANs based on 802.11 protocol are commonly adopted as the main method of train-to-wayside communications due to the open standards and the available commercial-off-the-shelf equipment [2]. In the past several years, information security in rail transit has attracted more and more attention. In [3], by craftily disturbing wireless signals of balise telegrams, three types of attacks exploiting the train localization mechanism are proposed. In [4], impacts of signal jamming attacks against CBTC systems are present and analyzed where train journey time and passenger congestion are considered. More importantly, WLANs have been exposed more and more serious security problems. In [5], a novel channel-based man-in-the-middle attack is proposed, which enables reliable manipulation of encrypted traffic. In [6], the adversary performs key reinstallation attacks (KRACK) to trick a victim into reinstalling an already-in-use key by using a multi-channel technique. In [7], a downgrade attack is developed which forces usage of RC4 to encrypt the group key by performing a channel-based man-in-the-middle attack.

Considering the inherent characteristics of WLANs, a multi-channel MitM attack against the train-to-wayside wireless communications is considered in the paper. The adversary can manipulate specific 802.11 frames to perform operations such as delay, drop, and even modification, which may result in unnecessary traction, service braking, or even emergency braking of trains. Therefore, we analyze the impacts of MitM attack on CBTC systems and measure the reduction of system performance.

The rest of the paper is organized as follows. Section 2 presents an overview of CBTC systems. Multi-channel attack is presented in Sect. 3. Section 4 describes impacts of MitM attack on CBTC systems and quantifies the impacts by simulations. Finally, we conclude this paper in Sect. 5.

2 CBTC Systems

CBTC systems consist of five parts: control center equipment, station equipment, wayside equipment, vehicle on-board equipment, and data communications system (DCS), as shown in Fig. 1. The control center equipment is mainly the automatic train supervision (ATS) subsystem. The station equipment includes zone controller (ZC), computer interlocking (CI), data storage unit (DSU), and station ATS extension. The wayside equipment mainly includes balises and axle counting equipment. The vehicle on-board equipment mainly includes ATP, ATO, and man–machine interface (MMI), where ATP and ATO constitute the vehicle on-board controller (VOBC). DCS is composed of wired network and train-to-wayside wireless communications network.

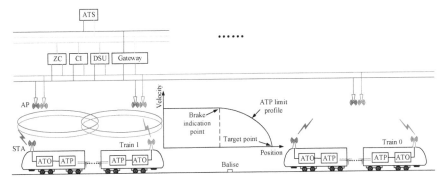

Fig. 1 A basic structure of CBTC systems

WLAN-based train-to-wayside wireless communication network is composed of wayside access point (AP) and mobile wireless station (STA). Based on redundancy design, two APs working at different channels are deployed at the same position along the tracks. Similarly, two STAs are installed at the front and rear of the train respectively [8]. The wireless links between AP and STA can provide train-to-wayside data channel between ZC and VOBC.

In CBTC systems, with train-to-wayside communications, trains can operate efficiently and safely based on moving-block mode. The train needs to change driving strategies according to the relevant data of the preceding train at all times so as to maintain a certain safe distance between two adjacent trains. VOBC sends the position, travel direction, and velocity of the train to ZC periodically. ZC calculates the movement authority (MA) of the train according to information from ATS, CI, and DSU, and also sends to the train periodically. Normally, MA can be the position of the preceding train's tail.

According to MA, ATP calculates a velocity/position profile to guarantee the safe operation of the train. When the velocity exceeds the limit velocity, ATP triggers the emergency braking until the train stops. When the communication time latency reaches the maximum allowable interruption time, the train will trigger the emergency braking in order to ensure the safe operation of the train. If the train doesn't receive the updated MA within a period, the old MA received in the previous period will be used as MA of the current period.

3 Multi-channel MitM Attack

Multi-channel MitM attack employs the multi-channel technique [9], which clones the real AP into the rouge AP on the rogue channel different from the real channel, and forces STA to connect with the rouge AP [5]. The adversary acts as the real AP by copying all frames sent by the real AP to the working channel of STA. At the same time, the adversary acts as STA by copying all frames sent by STA to the working

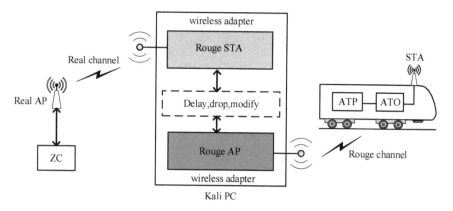

Fig. 2 Multi-channel MitM attack model

channel of the real AP. By forwarding frames between both channels, STA and the real AP can communicate.

To reliably intercept all traffic, the adversary forwards traffic between the real channel and the rogue channel, which requires two special wireless network adapters as shown in Fig. 2, where one operates on the real channel while the other one operates on the rogue channel [5]. Once the rouge AP is started, the adversary wants to force STA to connect to it. The adversary can forge channel switch announcement (CSA) element inside beacons or probe responses to force STAs to switch to the rogue channel [9]. The CSA is used by an AP to advertise the channel number of the new channel when it changes to a new channel [9]. After STA switches to the rouge channel, it will not disconnect from the real AP, so there is no need for authentication and association. Once STA has switched, the adversary starts forwarding frames between STA and the real AP.

The adversary can manipulate traffic and enable to delay, drop, or modify specific frames transmitted between STA and the real AP. The adversary can save the frame for a period of time after receiving it and then forward it or selectively drop it to increase the communication time latency between VOBC and ZC. Due to the safety requirements of CBTC systems, when the communication time latency reaches to a threshold value, it can be regarded as communication interruption. In the above attacks, the adversary manipulates the encrypted frames. In CBTC systems, the AP adopts the WPA2 protection mechanism based on CCMP. If the adversary wants to modify the encrypted frame, he must decrypt the frame into plaintext, modify it, and then encrypt and send it, as shown in Fig. 3. To decrypt the encrypted frame or encrypt the plaintext, the adversary must have the preshared key of AP, MAC addresses of AP

Fig. 3 A flowchart of modifying frames

and STA, and Anonce and Snonce generated by four-way handshake to derive TK for encryption and decryption [10]. MAC addresses of AP and STA can be obtained from the frame. In order to get Anonce and Snonce, the adversary needs to send a forged disassociation frame to STA after STA has switched, which causes four-way handshake to be restarted. We use pyDot11 library to decrypt and encrypt 802.11 frames, which can encrypt and decrypt 802.11 frames on-the-fly [11]. By modifying the frames, the train control data and train state data transmitted between VOBC and ZC can be modified.

As the train-to-wayside wireless communications network is redundant dual-network design, it is ineffective to attack only one network. It's very difficult to attack both networks at the same time, because the both adversaries must be strictly synchronized. Therefore, we consider the following scenario: An adversary keeps attacking a STA all the time through continuous disassociation attacks to disassociate it with AP, and the other adversary performs multi-channel MitM attack.

4 Impacts and Simulations

Through multi-channel MitM attack, the adversary can increase communication time latency, interrupt communication, and modify data between VOBC and ZC. In this section, we analyze impacts of these three types of attacks on CBTC systems. Based on operation principles of CBTC systems, the impacts of MitM attack are quantified according to the comparison between the train trajectories under attacks and the preset optimal running curves of trains. We compare the train journey time in different situations and use the area surrounded by two velocity/position profiles divided by the running distance to represent velocity difference per meter of the train. The larger the difference, the greater the impact of the attack on the train. Simulation parameters are shown in Table 1.

Table 1 Simulation parameters

Item	Value	Item	Value
Number of stations	5	Station spacing	2000 m
Balise spacing	500 m	Length of train	100 m
Maximum velocity	100 km/h	Traction acceleration	0.9 m/s^2
Resistance acceleration	−0.1 m/s^2	Service braking acceleration	−0.7 m/s^2
Emergency braking acceleration	−1.1 m/s^2	Communication period	200 ms

4.1 Communication Latency Attack

Through communication latency attack, MA update interval of the train is increased, but less than the maximum allowable interruption time. We define the MA update interval as $T_{interval}$. When there is no attack, the interval is approximately equal to the communication period T. The scenario we assume is as Fig. 4a, the distance between the train and target point is L, and the distance between the brake indication point and target point is X. If there are n consecutive communication periods when the train doesn't receive the updated MA, the time interval $T_{interval}$ between two adjacent different MAs is $(n + 1) \cdot T$. When the interval meets:

$$\int_0^{T_{interval}} v_f dt > L - X \tag{1}$$

where v_f is the velocity of the following train, and the following train will apply the service braking to maintain a safe distance from the leading train. When the following train finally receives the updated MA, it will stop the service braking and apply the traction to accelerate.

Our simulation scenario configurations are as follows. Two trains with the same performance start from Station 0 (0 m). The attack increases the communication latency by 3 s. As shown in Fig. 5, the train applies the traction and service braking frequently. When an attack occurs, the train decelerates from 600 to 680 m and

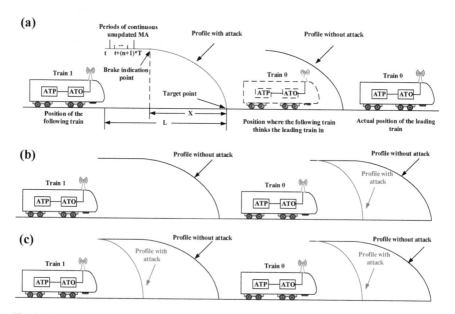

Fig. 4 Scenario of attacks

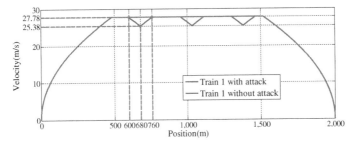

Fig. 5 Velocity/position profiles of Train 1 with or without communication latency attack

its velocity decreases from 27.78 to 25.38 m/s due to it travels close to the brake indication point. After the attack, the train receives the updated MA and accelerates from 680 to 760 m until the velocity returns to 27.78 m/s. Under this attack, the velocity difference per meter is 0.275 m/s between Station 0 and Station 1 (2000 m).

4.2 Communication Latency Attack

Through communication interruption attack, the time interval of train-to-wayside communications is greater than the maximum allowable interruption time, and the train will trigger the emergency braking, as shown in Fig. 4b. When the velocity drops to 0, the train can't re-operate in CBTC mode even if the communications is restored. The position of the train isn't accurate because of the slipping and locking of the wheels in the emergency braking process. In order to re-operate in CBTC mode, the train must firstly operate in restricted mode (RM) with velocity limit of 25 km/h to the next balise for position calibration.

In the simulation, the departure interval is 90 s. The attack begins when the leading train travels to 1000 m and finishes 90 s after the leading train stops. As shown in Fig. 6, the journey time of the leading train without and with attack is, respectively, 106 s and 232 s between Station 0 and Station 1. The journey time of the leading train without and with attack is, respectively, 455 s and 581 s between Station 0 and Station 4 (8000 m). Although the following train is not attacked, based on the operation principles of CBTC systems, its journey time increases accordingly. When the leading train is not or is attacked, the journey time of the following train is 106 and 176 s between Station 0 and Station 1. When the leading train is not or is attacked, the journey time of the following train are 455 and 530 s between Station 0 and Station 4.

Figure 7 demonstrates the velocity/position profiles of two trains with and without attacks. With the attack, the leading train triggers the emergency braking at 1000 m and then operates in RM until it restores to CBTC mode at 1500 m. Therefore, the following train applies the traction and service braking frequently between Station 0 and Station 1. The velocity difference per meter of the leading train is 4.446 m/s

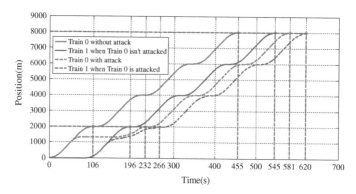

Fig. 6 Position/time profiles of two trains with or without communication interrupt attack

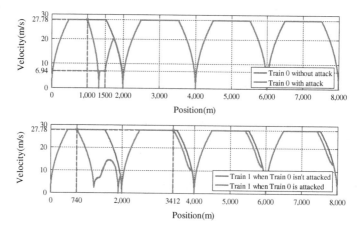

Fig. 7 Velocity/position profiles of two trains with or without communication interrupt attack

between Station 0 and Station 1. The velocity difference per meter of the following train is 7.022 m/s between Station 0 and Station 1. In addition, when the leading train is attacked, the distance between the leading train and the following train is too close. In order to maintain a certain safe distance, the following train applies the service braking at 3412 m earlier than that when the leading train is not attacked. And the velocity difference per meter of the following train is 0.772 m/s between Station 1 and Station 4.

4.3 Communication Latency Attack

Through data modification attack, the train control data and train state data transmitted between VOBC and ZC are tampered. This allows them to get the wrong data

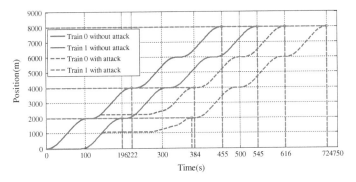

Fig. 8 Position/time profiles of two trains with or without modification data attack

about each other. When MA generated by ZC is modified, the adversary can make the train travel to the wrong position. When the position of the train sent by the train is modified, ZC will go down due to the large difference in the position of the train received twice in succession, which means all the trains in the ZC jurisdiction will trigger the emergency, braking shown as Fig. 4c. Restarting ZC requires manual confirmation and takes a long time.

Since modifying the train control data may only affect one train and modifying the train state data will cause the ZC to go down and affect multiple trains, we only consider the scenario of modifying the train state data. We assume that the recovery time of ZC is 120 s. Figure 8 shows both trains decelerate to a velocity of 0 and stop between stations for a long time, so the train journey time increases significantly compared without attack. The journey time of the leading train without and with attack is, respectively, 106 s and 268 s between Station 1 and Station 2 (4000 m). And the journey time of the leading train without and with attack is, respectively, 455 s and 616 s between Station 0 and Station 4. The journey time of the following train without and with attack is, respectively, 106 s and 286 s between Station 0 and Station 1. The journey time of the following train without and with attack is, respectively, 455 s and 634 s between Station 0 and Station 4.

As shown in Fig. 9, both trains trigger the emergency braking and then operate in RM until they restore to CBTC mode. With the attack, the velocity difference per meter of the leading train is 4.422 m/s between Station 1 and Station 2, and the velocity difference per meter of the following train is 6.701 m/s between Station 0 and Station 1.

5 Conclusion

The train-to-wayside wireless communications network in CBTC systems plays an important role in ensuring the safe and efficient operation of train. Based on the vulnerabilities of WLANs, we consider the adversary use multi-channel MitM attack

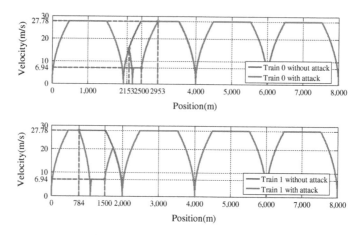

Fig. 9 Velocity/position profiles of two trains with or without modification data attack

against the train-to-wayside wireless communications network. With this attack, the adversary can obtain the MitM position and perform operations such as delay, drop, and even modification, which will result in unnecessary traction, service braking, and even emergency braking of the train. We classify these attacks into three categories: communication latency attack, communication interruption attack, and data modification attack. Based on the operation principles of CBTC systems, the impacts of MitM attack are quantified according to the comparison between the train trajectories under attacks and the preset optimal running profiles of trains. Our results show that the attack increases train travel time and velocity difference per meter compared with no attack.

Acknowledgements This paper was supported by grants from the National Natural Science Foundation of China (No. 61790575, 61603031), Beijing Natural Science Foundation (No. L181004) projects (No. RCS2018K008), and Beijing Laboratory for Urban Mass Transit.

References

1. Schifers C, Hans G (2000) IEEE standard for communications-based train control (CBTC) performance and functional requirements. In: Vehicular technology conference proceedings, VTC, pp 1581–1585
2. Aguado M, Jacob E, Saiz P, Unzilla JJ, Higuero MV, Matías J (2005). Railway signaling systems and new trends in wireless data communication. In VTC-2005-Fall. 2005 IEEE 62nd vehicular technology conference, vol 2. IEEE, pp 1333–1336
3. Weng J, Wu Y, Wei Z et al (2018) Position manipulation attacks to balise-based train automatic stop control. IEEE Trans Veh Technol 67:5287–5301
4. Lakshminarayana S, Karachiwala JS, Chang SY, Revadigar G, Kumar SLS, Yau DK, Hu YC (2018) Signal jamming attacks against communication-based train control: attack impact and

countermeasure. In: Proceedings of the 11th ACM conference on security & privacy in wireless and mobile networks. ACM, pp 160–171
5. Vanhoef M, Piessens F (2014) Advanced Wi-Fi attacks using commodity hardware. In: Proceedings of the 30th annual computer security applications conference. ACM, pp 256–265
6. Vanhoef M, Piessens F (2017) Key reinstallation attacks: forcing nonce reuse in WPA2. In: Proceedings of the 2017 ACM SIGSAC conference on computer and communications security. ACM, pp 1313–1328
7. Vanhoef M, Piessens F (2016) Predicting, decrypting, and abusing WPA2/802.11 group keys. In: 25th USENIX security symposium (USENIX security 16), pp 673–688
8. Zhu L, Ning B (2010) The design of the CBTC train-ground communication system based on IEEE 802.11g standard. China Railway Sci 31(5):119–124 (in Chinese)
9. Vanhoef M, Bhandaru N, Derham T, Ouzieli I, Piessens F (2018) Operating channel validation: preventing Multi-Channel Man-in-the-Middle attacks against protected Wi-Fi networks. In: Proceedings of the 11th ACM conference on security & privacy in wireless and mobile networks. ACM, pp 34–39
10. IEEE Std 802.11 (2012) Wireless LAN Medium Access Control (MAC) and Physical Layer (PHY) specifications
11. pyDot11. https://github.com/ICSec/pyDot11/. Accessed 1 May 2019

Optimization on Metro Timetable Considering Train Capacity and Passenger Demand from Intercity Railways

Haiyang Guo, Yun Bai, Qianyun Hu, Huangrui Zhuang and Xujie Feng

Abstract Timetable optimization of metro lines connecting with intercity railway stations can reduce passenger waiting time and improve service quality. Based on arrival times of intercity trains and the entire process for passengers transferring from railway to metro, this paper develops a mathematical model to characterize the time-varying demand of passengers arriving at the platform of a metro station connecting with an intercity railway station. Provided the time-varying passenger demand and capacity of metro trains, a timetable model to optimize train departure time of a bi-direction metro line where an intermediate station connects with an intercity railway station is proposed. The objective is to minimize waiting time of passengers at the connecting station. A genetic algorithm is applied to solve the proposed timetable model. Real-world case studies show that the prediction accuracy of the proposed model on passenger demand at the connecting station is higher than 90%, and the timetable model can reduce waiting time of passengers at the connecting station by 23.57% with negligible influence on passengers at other stations.

Keywords Metro timetable · Intercity railway station · Train capacity · Passenger waiting time · Genetic algorithm

1 Introduction

Metro is the most effective way for passenger evacuation at intercity railway stations because of the superiorities of massive capacity, high speed and great reliability. In daily operation, most metro lines adopt peak/off-peak-based timetables. However,

---

H. Guo · Y. Bai (✉) · H. Zhuang
Key Laboratory of Transport Industry of Big Data Application Technologies for Comprehensive Transport, Ministry of Transport, Beijing Jiaotong University, Beijing 100044, China
e-mail: yunbai@bjtu.edu.cn

Q. Hu
Anhui Transport Consulting & Design Institute Co. Ltd., Anhui 230088, China

X. Feng
MOT, China Academy of Transportation Science, Beijing 100029, China

for metro lines connecting with intercity railway stations, whose inbound passenger flow varies significantly over a short period due to the discrete arrivals of intercity trains, regular timetables might increase waiting time of passengers [1]. Therefore, it is necessary to optimize timetable of such metro lines according to the time-varying passenger demand at the connecting station.

In the domain of demand-oriented metro timetable optimization, Berrena [2, 3] proposed timetable optimization model under dynamic passenger demand, Niu [4–6] analysed waiting behaviours of passengers at stations and constructed timetable optimization model with the aim of minimizing passenger waiting time. While above studies did not take transfer behaviours of passengers into account, Wu [7] put forward a model to minimize total waiting time of passengers including transfer passengers in a metro network. It only considered passengers transferring between different metro lines, however. Besides, passenger demands considered in above researches were all obtained through analysing historic data because passenger demands are similar in working days. For metro lines connecting with intercity railway stations, a slight change in arrival times of intercity trains can have a significant impact on passenger demand at the connecting station. Therefore, historic data of connecting stations is not universal and a passenger demand forecast model based on arrival times of intercity trains is called for.

Hu [8] built a train departure time optimization model for a metro line whose start station is connected to an intercity railway station, on the basis of characterizing the time-varying demand of passengers transferring from intercity trains to metros. Whereas the model proposed by Hu for the predication of the transfer passenger demand did not take the influence of the transfer facility layout into account. Also, the developed timetable model is only practical for single-direction metro lines where the start station is the connecting station and the capacity of metro trains can be neglected. It is not adaptable enough for a metro line where an intermediate station is connected to an intercity railway station.

To solve this problem, this paper puts forward a model to predict the number of passengers arriving at the platform of connection stations via analysing the entire process for passengers transferring from intercity trains to metros. Furthermore, a timetable optimization model aiming at minimizing passenger waiting time of a metro line where an intermediate station is connected to an intercity railway station is proposed. At last, a genetic algorithm is developed to find the optimal solution of the proposed model.

2 Model on Transfer Passenger Demand Predication

The entire process for passengers transferring from intercity trains to metros is shown in Fig. 1. According to arrivals of intercity trains and the transfer process of passengers, a model for calculating the number of passengers taking escalators and stairs, which are located at platforms of an intercity railway station, is proposed firstly, and the calculation result is regarded as the passenger flow input. Then, take the impact

Fig. 1 Process for passengers transferring from intercity trains to metros

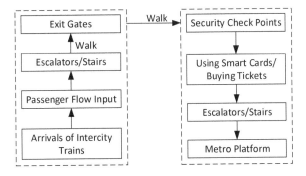

of each transfer facility (i.e. escalators/stairs, exit gates, etc.) into account and adjust the input passenger flow distribution orderly until the number of passengers arriving at the connecting station platform is obtained.

Transfer facilities considered of the transfer process are divided into two types: node facilities and facilities with branches. Facilities with branches are where two parallel facilities are provided for passengers to pass the same area, including escalators/stairs in addition to buying tickets at the station/using smart cards. Node facilities are those with capacity constraints, like exit gates of intercity railway stations and security checkpoints of metro stations. It is worth noting that escalators/stairs are also node facilities where passengers are influenced by capacity constraints after making choice between escalators and stairs.

2.1 Passenger Flow Input

It is very likely that more than one train get to the intercity railway station during the study period [0, T]. Therefore, the number of input passengers is calculated by the sum of passenger distribution of multiple trains:

$$C(t) = \sum_{k=1}^{K} A_k(t) \quad (1)$$

where $C(t)$ is the total number of input passengers at time t; $A_k(t)$ is the number of input passengers for train k at time t; K is the total number of intercity trains getting to the station during study period [0, T].

In general, transfer passengers spend different time walking from intercity trains to escalators/stairs which are located at platforms of an intercity railway station. According to Henderson's research that walking speed of passengers follows normal distribution whose mean is μ, standard deviation is σ [9]. With the distribution of passenger walking speed and walking distance of passengers, the distribution of passenger walking time can be calculated. Substitute the capacity of intercity trains, the number of input passengers for train k at time k is calculated by:

$$A_k(t) = Q_k H(\mu, \delta, t, l_k) \quad (2)$$

$$Q_k = P_k \times \varepsilon \quad (3)$$

where Q_k is number of transfer passengers from train k reaching the intercity railway platform; H is the distribution of passenger walking time; l_k is the average distance for passengers walking from train k to escalators/stairs on the intercity railway platform; P_k is the capacity of train k; ε is the load factor of intercity trains.

2.2 Facilities with Branches

Facilities with branches are where passengers need to make choice according to their conditions. For example, they need to decide whether to take escalators or stairs, whether to buy tickets at the station or use smart cards directly. Investigations on passengers using facilities with branches infer that it takes passengers nearly the same time to go through escalators and stairs, while the time they spent on buying tickets at the station is longer than using smart cards. As a result, the number of passengers choosing escalators and stairs is calculated, respectively, by:

$$L^1(t) = aL^0(t) \quad (4)$$

$$L^2(t) = (1 - \alpha)L^0(t) \quad (5)$$

where $L^1(t)$ is the number of passengers choosing stairs at time t; $L^2(t)$ is the number of passengers choosing escalators at time t; $L^0(t)$ is the number of passengers who intend to take escalators/stairs; α is the proportion of passengers who choose stairs.

The number of passengers passing AFC is calculated by:

$$S(t) = bS^0(t) + (1 - b)S^0(t - t_0) \quad (6)$$

where $S(t)$ is the number of passengers going through AFC at time t; $S^0(t)$ is the number of passengers intend to use AFC machines; b is the proportion of passengers passing AFC machines directly with smart cards; t_0 is the service lag time for passengers buying tickets at the station instead of using smart cards.

2.3 Node Facilities

Generally speaking, passengers who intend to take escalators will move to stairs when the entry of escalators is too crowded. Therefore, considering the capacity constraints of escalators/stairs, the number of passengers choosing stairs and escalators is re-calculated, respectively, by:

$$L^1(t) = L^1(t) + \max\{0, \eta(L^2(t) - c_2)\} \tag{7}$$

$$L^2(t) = L^2(t) - \max\{0, (1 - \eta)(L^2(t) - c_2)\} \tag{8}$$

where η is the number of passengers who change their choice and decide to take stairs rather than escalators; c_2 is service capacity of escalators.

Based on the re-calculated $L^1(t)$ and $L^2(t)$, the number of passengers going through stairs and escalators is expressed by:

$$L^3(t+1) = \min\{L^1(t+1) + \max\{0, L^1(t) - c_1\}, c_1\} \tag{9}$$

$$L^4(t+1) = \min\{L^2(t+1) + \max\{0, L^2(t) - c_2\}, c_2\} \tag{10}$$

where $L^3(t)$ is the number of passengers going through stairs at time t; $L^4(t)$ is the number of passengers going through escalators at time t; c_1 is the capacity of stairs; c_2 is the capacity of escalators.

Exit gates and security checkpoints have similar effect on the distribution of passenger flow, which is expressed, respectively, by:

$$J^1(t+1) = \min\{J^0(t+1) + \max\{0, J^0(t) - c_3\}, c_3\} \tag{11}$$

$$G^1(t+1) = \min\{G^0(t+1) + \max\{0, G^0(t) - c_4\}, c_4\} \tag{12}$$

where $J^1(t)$ is the number of passengers passing security checkpoints at time t; $G^1(t)$ is the number of passengers getting through exit gates at time t; $J^0(t)$ and $G^2(t)$ are the number of passengers who intend to be through security checkpoints and exit gates, respectively.

3 Timetable Optimization Model and Solution Methodologies

As shown in Fig. 2, this paper considers a bi-direction metro line with N stations which have been numbered sequentially from the up direction to the down direction. Although station 1 and station $2N$, station 2 and station $2N - 1$, ..., station $N + 1$

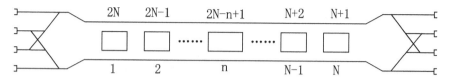

Fig. 2 Representation of a metro line

and station N refer to the same station in terms of geographic location, they are numbered separately to make the timetable model more understandable. Station n, that is station $2N - n + 1$, is the metro station which connects to an intercity railway station.

3.1 Objective Function

Divide the study period $[0, T]$ into a host of time intervals denoted by t ($t = 1, 2, 3, 4, \ldots$). Assume that all passengers arrive at metro stations at the end of each time interval, and all metro trains start their operation from the terminal station where the depot is located and turn around at the other terminal station. In order to evacuate passengers that get to the platform of connecting stations, this paper takes minimizing passenger waiting time at connecting stations as the objective of the timetable optimization model and it is calculated by:

$$\min W = W_1 + W_2 \tag{13}$$

$$W_1 = \sum_{j=1}^{K} \sum_{v=n+1}^{2N} \sum_{t \in \left(L_{j-1}^n, L_j^n\right]} P^{n,v}(t)\left(TD_j^n - t\right) \tag{14}$$

$$W_2 = \sum_{j=1}^{K} \sum_{v=2N-n+2}^{2N} \sum_{t \in \left(L_{j-1}^{2N-n+1}, L_j^{2N-n+1}\right]} P^{2N-n+1}(t)\left(TD_j^{2N-n+1} - t\right) \tag{15}$$

where W_1 is the waiting time of passengers at the connecting station when travelling towards up direction; W_2 is the waiting time of passengers at the connecting station when travelling towards down direction; $P^{n,v}(t)$ is the number of passengers travelling from connecting station n to station v; TD_j^n is the time when train j departs from station n; K is the total number of trains departing from the start terminal during the study period; L_j^n is the effective loading time of train j at station n.

Based on train departure times at the first station, running times at sections and dwell times at stations, train departure times at the connection station on up direction and down direction are expressed by:

$$TD_j^n = TD_j^1 + \sum_{u=1}^{n} d_j^u + \sum_{u=1}^{n-1} r_j^u \qquad (16)$$

$$TD_j^{2N-n+1} = TD_j^1 + \sum_{u=1}^{2N-n+1} d_j^u + \sum_{u=1}^{2N-n} r_j^u \qquad (17)$$

where d_j^u is the dwell time of train j at station u; r_j^u is the running time of train j from station u to station $u+1$.

3.2 Constraints

Whether a passenger can get on the oncoming train successfully depends on the remaining loading capacity of the train. In order to determine the number of passengers who can board the train, effective loading time is introduced, that is the critical time that the number of passengers onboard reaches the maximum loading capacity. The effective loading time for train j is at station u is calculated by:

$$L_j^u = \min\left\{ TD_j^u, \max\left\{ \tau \mid \sum_{t \in \left(L_{j-1,\tau}^u\right]} \sum_{v=u+1}^{2N} P^{u,v}(t) \leq C - R_j^{u-1} + \sum_{u'=1}^{u-1} B_j^{u',u} \right\} \right\} \qquad (18)$$

$$B_j^{u,v} = \sum_{t \in \left(L_{j-1}^u, L_j^u\right]} P^{u,v}(t) \qquad (19)$$

$$Q_j^u = R_j^{u-1} - \sum_{u'}^{u-1} B_j^{u',u} \qquad (20)$$

$$R_j^u = Q_j^u + \sum_{v=u+1}^{2N} B_j^{u,v} \qquad (21)$$

where $B_j^{u,v}$ is the number of passengers travelling from station u to v who board train j successfully; Q_j^u is the number of passengers left on train j after some passengers get off at station u; R_j^u is the number of onboard passengers after train j departs from station u; C is maximum loading capacity of metro trains.

To cover all passenger demand over the study period [0, T], departure times of the first train and the last train are pre-determined, which are denoted by:

$$TD_0^1 = 0 \qquad (22)$$

$$TD^1_{K+1} = T \tag{23}$$

Constraints of the headway between two adjacent trains are calculated by:

$$h_{\min} \leq TD^1_j - TD^1_{j-1} \leq h_{\max} \tag{24}$$

where $h_{\min}$ is the minimum headway; $h_{\max}$ is the maximum headway.

3.3 Solution Methodologies

The proposed timetable optimization model has a large solution space. In order to improve the efficiency of problem-solving, a genetic algorithm is adopted, the framework of which is shown in Fig. 3. In this study, chromosomes are binary coded, and the length of each chromosome depends on how many time intervals the study period is divided into. Each gene represents an alternative departure time of a train at the first station. If it equals "1", it means that a train departs at this time interval and "0" means no trains depart.

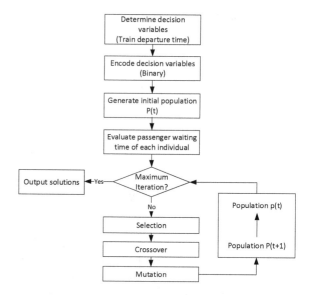

Fig. 3 Flowchart of the adopted generic algorithm

Table 1 Parameters of the passenger demand predication model

Parameters		Value
μ	Average walking speed	1.34 m/s
δ	Standard deviation	0.26
c_1	Capacity of stairs	54 Pax/10 s
c_2	Capacity of escalators	54 Pax/10 s
c_3	Capacity of security points	45 Pax/10 s
c_4	Capacity of exit gates	96 Pax/10 s
ε	Load factor of intercity trains	70%
α	Proportion of passengers taking stairs	0.25
b	Proportion of passengers using smart cards	0.60

4 Case Studies

The developed transfer passenger demand predication model and timetable optimization model are applied to Beijing Metro Line 9 where an intermediate station called Beijing West Metro Station connects to Beijing West Railway Station. The study period is 12:30–14:00 on a working day when intercity trains get to Beijing West Railway Station intensively.

4.1 Passenger Demand Predication of the Connecting Station

Based on investigations on the entire process for passengers transferring from intercity trains of Beijing West Railway Station to Beijing Metro Line 9, parameters of the passenger demand predication model are obtained, which are shown in Table 1.

As Beijing Metro Line 9 and Beijing Metro Line 7 are both connected with Beijing West Railway Station, this paper assumes that half of the passengers who intend to transfer from intercity trains to metros take Beijing Metro Line 9. Based on arrival times of intercity trains at Beijing West Railway Station over the study period and above parameters, the result of passenger demand predication is shown in Fig. 4.

4.2 Accuracy of Passenger Demand Predication

Compare the accuracy of passenger demand predication in this paper to that calculated by Hu in 2016 under different time intervals. As Table 2 shows, the predication model proposed in this paper has a smaller error and this advantage becomes more significant as the time interval becomes shorter.

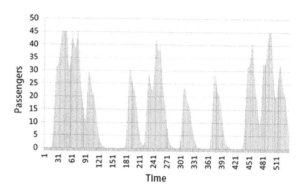

Fig. 4 Calculated passenger demand at Beijing West Metro Station

Table 2 Accuracy comparison of passenger demand predication models

Time interval (minutes)	Average error	
	This paper (%)	Hu (2016) (%)
0.5	8.36	10.12
1	7.96	10.07
2	7.52	8.25
5	5.07	5.66
10	4.20	4.55

4.3 Timetable Optimization of Beijing Metro Line 9

Input the results of passenger demand predication at the connecting station to the timetable optimization model and use GA to find the solutions. As the selected time interval is 10 s and the length of the study period is 90 min, the length of each chromosome is 540. Other parameters of GA are shown in Table 3.

Table 4 represents passenger waiting time of current timetable and the optimized timetable. It is found that the optimized timetable reduces passenger waiting time at the connecting station by 23.57%. Although the passenger waiting time at other stations increases by 1.61%, it is too low to affect riding experience of passengers. It is also noted that saving rate of passenger waiting time is higher when train capacity is neglected. However, the maximum load factor reaches 136.86% under this condition. If train capacity is considered, the maximum load factor is only 98.86%. As a result, the congestion on metro trains is relieved and service quality for passengers is improved.

Table 3 Parameters of GA adopted in the case study

Population size	Maximum iterations	Crossover probability	Mutation probability
50	3000	0.9	0.1

Table 4 Passenger waiting time of current timetable and the optimized timetable

Train capacity		Considered	Neglected
Passenger waiting time at the connecting station	Current timetable	24,490	23,943
	Optimized timetable	18,741	18,194
	Saving rate	23.57%	24.01%
Passenger waiting time at other stations	Current timetable	22,562	22,562
	Optimized timetable	22,925	22,957
	Saving rate	−1.61%	−1.75%
Average load factor		86.34%	95.93%
Maximum load factor		98.68%	136.86%

5 Conclusions

Based on the analysis on the entire process for passengers transferring from intercity trains to metro trains, a mathematical model is proposed to predict the number of passengers getting to the platform of a metro station which connects with an intercity railway station. Compared to the existing research, the passenger demand predication model developed in this paper is more accurate.

According to the calculated passenger demand, a timetable optimization model with the aim of minimizing passenger waiting time at a connecting station is established and solved by GA. Real-world case studies indicate that the optimized timetable can reduce passenger waiting time at the connecting station by 23.57% with negligible influence on passengers at other stations.

The timetable optimization model proposed in this paper takes train capacity into account, which improves service quality for passengers to some extent. However, this paper only considers the case that a metro line connects with an intercity railway station. However, some intercity railway stations connect with several metro lines. How to optimize their timetables coordinately will be introduced in the further work.

Acknowledgements This work was supported by the National Natural Science Foundation of China (71571016, 71621001).

References

1. Sun L, Jin JG, Lee DH et al (2014) Demand-driven timetable design for metro services. Transp Res Part C Emerg Technol 46:284–299
2. Barrena E, Canca D, Coelho LC et al (2014) Single-line rail rapid transit timetabling under dynamic passenger demand. Transp Res Part B Methodol 70:134–150
3. Barrena E, Canca D, Coelho LC et al (2014) Exact formulations and algorithm for the train timetabling problem with dynamic demand. Comput Oper Res 44(3):66–74
4. Niu H, Zhou X (2013) Optimizing urban rail timetable under time-dependent demand and oversaturated conditions. Transp Res Part C Emerg Technol 36:212–230

5. Niu H, Tian X, Zhou X (2015) Demand-driven train schedule synchronization for high-speed rail lines. IEEE Trans Intell Transp Syst 16(5):2642–2652
6. Niu H, Zhou X, Gao R (2015) Train scheduling for minimizing passenger waiting time with time-dependent demand and skip-stop patterns: nonlinear integer programming models with linear constraints. Transp Res Part B Methodol 76:117–135
7. Wu JJ, Liu MH, Sun HJ, Li TF, Gao ZY, Wang DZW (2015) Equity-based timetable synchronization optimization in urban subway network. Transp Res Part C Emerg Technol 51:1–18
8. Hu Q, Bai Y, Cao Y, Chen Y, Chen Z (2016) Optimization of subway timetable at the railway-subway transfer station. J Railway Sci Eng 13(12):2503–2507 (in Chinese)
9. Henderson LF (1971) The statistics of crowd fluids. Nature 229:381–383

Hybrid Intrusion Detection with Decision Tree and Critical State Analysis for CBTC

Yajie Song, Bing Bu and Xuetao Yang

Abstract Communication-based train control (CBTC) is considered as the main organ of urban rail transit systems, which is facing increasingly serious security threats. Intrusion detection systems (IDS) are crucial for security protection. This paper reports the design principles and evaluation results of a novel hybrid intrusion detection system which is suitable for CBTC systems. This hybrid method combines the advantages of the high true positive rate of network-based IDS (NIDS) and the ability of host-based IDS (HIDS) to monitor system behavior, where decision tree and critical state analysis are used, respectively. The proposed method is verified on a semi-physical simulation platform of CBTC and the experiments show that the designed scheme can detect intrusions accurately with a 97.8% detection rate.

Keywords CBTC · Intrusion detection · Decision tree · Critical state analysis

1 Introduction

Communication-based train control (CBTC) is a railway signaling system that makes use of the continuous, high-capacity, bidirectional train-to-wayside data communications between the train and track equipment for the traffic management and infrastructure control [1]. As a large number of information and communication technologies are applied, CBTC is exposed to more security threats. Intrusion detection systems (IDS) are effective at protecting systems from security issues by detecting network anomalies and monitoring system behavior [2].

However, the traditional IDS cannot be deployed to the CBTC systems directly since CBTC has obvious differences compared with traditional IT systems. Firstly, since the control object of CBTC is running trains, it is the primary goal to ensure the safe operation of trains. Secondly, there are many kinds of equipment in CBTC,

---

Y. Song (✉) · B. Bu
Beijing Jiaotong University, Beijing 100044, People's Republic of China
e-mail: 17120267@bjtu.edu.cn

X. Yang
Beijing Traffic Control Technology Company, Beijing, China

including some equipment with limited resources, which should be considered in detection methods. Thirdly, the device in CBTC is difficult to update or restart, resulting in numerous vulnerabilities in systems.

Since there are few studies on intrusion detection specifically for rail transit systems and CBTC is a typical industrial control system (ICS), we investigate intrusion detection technologies in ICS, which can be mainly classified into state analysis, packet classification, and model-based [3]. Zhu calculates whether the current state is close to or exceeds critical states to detect intrusions [4]. This method can effectively detect attacks that make the system significantly deviate from the normal state. Packet classification can be implemented by machine learning or data mining algorithms, where a normal model is constructed by training historical packets to detect abnormalities [5, 6]. The model-based detection mainly uses system parameters and physical features of ICS to establish a mathematical model, which is adopted to predict the current state [7, 8].

These existing IDS can be classified as network-based IDS (NIDS) and host-based IDS (HIDS). NIDS monitors traffic to and from all devices on the network and HIDS detects activities of one device only. In this paper, a hybrid method combining NIDS and HIDS is proposed. The main contributions are as follows:

- The network-based detection module adopts decision tree to implement protocol analysis and packet classification, which can improve the accuracy of IDS.
- The host-based detection module uses critical state analysis to detect anomalies which affect train operation. It can detect not only general intrusions, but also data integrity attacks (DIA), which tampers with the control signal.
- Experiments show that the proposed IDS can achieve higher detection accuracy than before as the principles of CBTC are fully considered.

The rest of this article is organized as follows. Section 2 introduces CBTC and the impacts of attacks. The construction of IDS is discussed in Sect. 3, including the network-based module and the host-based one. Section 4 analyzes the results of the proposed method and summaries are given in Sect. 5.

2 Attack Impacts on CBTC Systems

2.1 A Typical CBTC System

CBTC is a continuous automatic train control system as shown in Fig. 1, which utilize high-resolution train location determination; the continuous, high-capacity, bidirectional train-to-wayside data communication system (DCS); on-board equipment vehicle on-board controller (VOBC), including automatic train protection (ATP), automatic train operation (ATO), and mobile stations (MS) using for wireless communications; and wayside processors such as zone controller (ZC), automatic train supervision (ATS), computer interlock (CI), and data storage unit (DSU) [9].

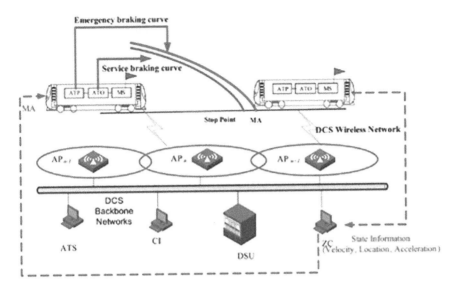

Fig. 1 Typical CBTC system

2.2 Impacts of Cyber Attacks on CBTC

Generally, Windows, universal protocols and information-based components are applied in critical signaling equipment, whose inevitable vulnerabilities introduce more threats into CBTC. As the high availability of CBTC systems is crucial for proper operation, an attack may result in serious impacts [10].

As shown in Fig. 2, an attacker takes a denial of service (DoS) attack on ZC by sending superfluous requests. Under normal circumstances, the state information of the former train is transmitted to ZC, which forms a movement authority (MA) and sends it to the back train. The ATP of the train calculates the emergency braking curve based on the received MA, as shown by the blue curve. However, when the ZC is attacked, the train cannot receive MA, which will take an emergency braking as

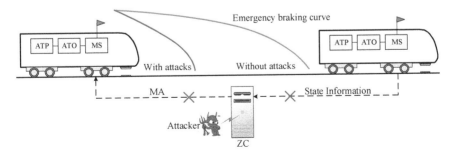

Fig. 2 Impact of attacks on CBTC

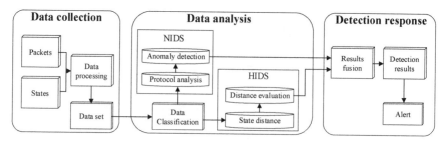

Fig. 3 Framework of the detection model combining NIDS and HIDS

shown by the red curve due to the fail-safety mechanism and if these interruptions are frequent enough it could seriously impact service. Other attacks may have the same impacts as the goal of attackers is to affect the train operation.

3 A Hybrid IDS Combining Network-Based and Host-Based Detection Module

3.1 Framework of the Proposed IDS

As shown in Fig. 3, the proposed IDS combining NIDS and HIDS which adopts decision tree and critical state analysis, respectively, is described. In the data collection phase, communication packets and device states are processed to form a data set, which is input to the data analysis phase. NIDS performs protocol analysis on packets and uses decision tree to implement anomaly detection. HIDS calculates the distance between the current state and the critical state and then evaluates the distance to determine whether the system is attacked. By fusing the results of NIDS and HIDS, the final detection results can be obtained. In this paper, an alert is generated if any detection module outputs an alarm.

3.2 The Detection Module Based on Network

Traditional NIDS constructs one model to classify all packets, where two problems exist. Firstly, as a system has a large number of packets, the training of a model may consume much computing resources. Secondly, due to the different protocols used in CBTC, a single model cannot extract valid features for each protocol, resulting in a low detection rate. Therefore, we identify the protocol type before classifying packets. A decision tree is constructed for each type of protocol to detect anomalies.

In this paper, the dedicated protocols used in railway are mainly analyzed, such as Railway Signal Safety Protocol-I (RSSP-I), whose format is shown in Fig. 4. By

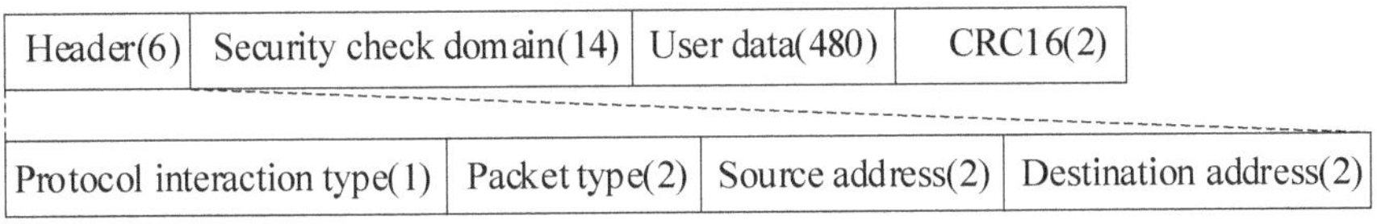

| Header(6) | Security check domain(14) | User data(480) | CRC16(2) |

| Protocol interaction type(1) | Packet type(2) | Source address(2) | Destination address(2) |

Fig. 4 Data payload format of RSSP-I

checking the length and header, a packet can be determined whether it conforms to RSSP-I. Thereby packets of CBTC are divided into three categories. They are RSSP-I, RSSP-II, and others.

Since the number of packets in CBTC is much smaller than that of an IT system, a tree-like structure machine learning method, decision tree is adopted to detect abnormal packets, which is suitable for the classification of small data sets. A packet starts from the root node of a tree, according to the attributes of the packet, passes through one of the different paths, finally reaches the leaf node to obtain the classification result. When building a decision tree, information gain is used to decide which attribute to split on at each node [11].

$$I(S, A) = E(S) - \sum_{v \in V(A)} \frac{|S_v|}{|S|} E(S_v) \qquad (1)$$

where I is the information gain; S is the input space; A is the attribute; v is the value of A; $V(A)$ is the number of A; S_v is a subset of S and $E(S)$ is the entropy of the system, which is calculated as

$$E(S) = \sum_{i=1}^{c} -p_i \log_2(p_i) \qquad (2)$$

where c is the number of classes of all samples after being classified according to attribute A; p_i is the probability of class i. The node order of a tree is determined according to I of an attribute A. The larger I is, the closer A is to the root node.

In this paper, three decision trees are constructed separately according to different protocols. Taking RSSP-I as an example, the attributes are denoted by L, including MAC, IP, port, length and the header information. Each address or port contains two attributes, such as the source MAC address represented by sMAC and the destination one represented by dMAC.

$$L = \{sMAC, dMAC, sIP, dIP, sPort, dPort, Len, T1, T2, sA, dA\} \qquad (3)$$

During the detection process, the decision tree obtained by training the historical data is used for packet classification, and results are input to the response phase.

3.3 The Detection Module Based on Host

As CBTC systems directly control trains, an attacker is likely to affect the normal operation of the train to maximize the impact of the attack. Therefore, the IDS also need to track and analyze the position and velocity of the trains. Due to the trains running strictly in accordance with timetables, the possible critical states of trains at different locations can be well documented by the host on trains. By monitoring the evolution of the states and tracking down when the train is entering a critical state, it would be possible to detect those attacks aiming at affecting the train operation, such as DIA. State analysis can detect anomalies by calculating the distance between the real-time state and the critical state [12].

The real-time state vector is defined as $R = (rs_1, rs_2, \ldots, rs_m)$ and the critical one is $C = (cs_1, cs_2, \ldots, cs_m)$, where rs_i is the current value of different states and cs_i is the value of critical states. Let $d(R, C)$ denotes the distance between R and C.

$$d(R, C) = \sqrt{\omega_1^2(rs_1 - cs_1)^2 + \omega_2^2(rs_2 - cs_2)^2 + \cdots + \omega_m^2(rs_m - cs_m)^2} \quad (4)$$

where ω_i is a weight coefficient determined by experts [13].

Taking the states of a running train as an example, the position and the speed of the train construct a state vector $R = (970, 54)$, while the critical state is $C = (1000, 60)$ according to the statistics of historical data. If the weight coefficients are 0.3 and 0.7, respectively, we can get the distance between the two states as $d(R, C) = 9.93$. When the distance is less than 10, the current state is close to the critical state and the HIDS module generates an alarm as follows.

$$R = (970, 54) \rightarrow \text{Abnormal} \quad (5)$$

4 Experimental Results and Discussion

The proposed IDS is verified on the semi-physical simulation platform of CBTC, where different attacks are launched to collect data, including DoS, Probing, User to Root (U2R), Remote to Local (R2L), and DIA. Since CBTC systems are connected to external networks through the ATS, an attacker may easily capture devices in the ATS, causing device failure, communication abnormality, and host-virus infection. If the goal of an attacker is to affect the operation of the train, the attacker needs to intrude the other equipment, where a critical node implements data transmission between the ATS and other equipment. Through the gateway, which is the critical node, invaders may implement attacks on ZC, CI, and DSU. The quantity of the collected data is shown in Table 1. 80% of the data are randomly selected as training sets to obtain the detection models and the rest as test sets to evaluate the accuracy.

Table 1 Quantity of the experimental data

Category	Times	Packets	States	Category	Times	Packets	States
Normal	50	1,023,700	1174	U2R	20	94,784	323
Dos	20	1,482,000	538	R2L	20	65,682	412
Probing	20	436,327	472	DIA	20	37,655	713

True positive rate (TPR) and false positive rate (FPR) are selected to measure the performance of the IDS [14]. Let normal → normal (M) denote the number of normal records that are detected true. Similarly, normal → attack (N), attack → attack (X), attack → normal (Y). Then TPR and FPR can be calculated as

$$\text{TPR} = \frac{M}{M+N} \quad (6)$$

$$\text{FPR} = \frac{Y}{X+Y} \quad (7)$$

Since research on intrusion detection for CBTC systems is limited, it is difficult to directly compare its performance with other IDS. In order to prove that the method of combining NIDS and HIDS is better than a single detection model, the TPR and FPR of NIDS, HIDS and hybrid IDS are calculated separately, as shown in Table 2, whose data is plotted as Fig. 5.

Due to the different principles of network-based and host-based detection modules, the detection performance for different attacks is also different. DIA does not change the header of a packet and it cannot be detected by NIDS. In addition, it can be seen from the results of all data that the TPR of hybrid IDS is 7.18 and 4.4% higher than that of the single detection module. Since the hybrid IDS alarms, once a single module detects an anomaly, the FPR of the hybrid IDS is 0.21% and 0.5% higher than NIDS and HIDS, respectively. In summary, the proposed method fully considers the characteristics of CBTC and effectively detects different attacks, which provides effective protection for CBTC systems.

Table 2 Results of different detection models

Attack	TPR (%)			FPR (%)		
	NIDS	HIDS	Hybrid IDS	NIDS	HIDS	Hybrid IDS
DoS	99.28	94.18	99.41	2.79	2.1	2.91
Probing	96.52	86.96	97.68	3.18	2.32	3.2
U2R	81.35	94.37	94.94	1.96	2.48	2.64
R2L	83.23	95.64	95.7	2.67	2.54	2.87
DIA	–	94.1	94.1	–	2.75	2.75
All	90.62	93.4	97.8	2.71	2.42	2.92

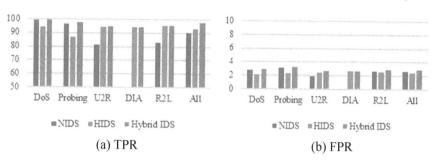

Fig. 5 Performance of different detection models

5 Conclusion

Intrusion detection is critical to the security of CBTC systems. Based on the characteristics of CBTC, we design the IDS combining NIDS and HIDS, which is verified on the platform of CBTC. Experimental results have shown that the designed IDS has better performance than other methods. Since the data of CBTC systems needs to be comprehensively processed, further analysis of the security risks and the impact of attacks are required to improve the existing detection scheme.

Acknowledgements This paper was supported by grants from the National Natural Science Foundation of China (No. 61603031), Beijing Natural Science Foundation (No. L181004), and projects (No. I19L00090), State Key Laboratory of Traffic Control and Safety of Beijing Jiaotong University, and projects (No. I18JB00110), and Beijing Laboratory for Urban Mass Transit.

References

1. Farooq J, Soler J (2017) Radio communication for communications-based train control (CBTC): a tutorial and survey. IEEE Commun Surv Tutorials 19(3):1377–1402
2. Barbará D, Jajodia S (eds) (2002) Applications of data mining in computer security (vol 6). Springer Science & Business Media
3. Mantere M, Sailio M, Noponen S (2014) A module for anomaly detection in ICS networks. In: Proceedings of the 3rd international conference on high confidence networked systems. ACM
4. Zhu B, Sastry S (2010) SCADA-specific intrusion detection/prevention systems: a survey and taxonomy. In: 40th COSPAR scientific assembly. Held 2–10 Aug 2014, in Moscow, Russia, Abstract F4.6-18-14
5. Buczak AL, Guven E (2016) A survey of data mining and machine learning methods for cyber security intrusion detection. IEEE Commun Surv Tutorials 18(2):1153–1176
6. Ponomarev S, Atkison T (2016) Industrial control system network intrusion detection by telemetry analysis. IEEE Trans Dependable Secure Comput 13(2):252–260
7. Pal S, Sikdar B, Chow J (2016) Detecting data integrity attacks on SCADA systems using limited PMUs. In: 2016 IEEE international conference on smart grid communications (SmartGridComm). IEEE

8. Valdes A, Cheung S (2009) Communication pattern anomaly detection in process control systems. In: IEEE Conference on IEEE technologies for homeland security, 2009. HST '09, pp 22–29
9. Zhu L, Yu FR, Wang F (2015) Introduction to communications-based train control. In: Advances in communications-based train control systems. CRC Press, pp 22–34
10. Ramdas, V et al (2010) ERTMS level 3 risks and benefits to UK railways. Transport Research Laboratory, Client project report CPR798 (2010)
11. Umanol M, Okamoto H, Hatono I et al (1994) Fuzzy decision trees by fuzzy ID3 algorithm and its application to diagnosis systems. In: IEEE international fuzzy systems conference
12. Carcano A et al (2011) A multidimensional critical state analysis for detecting intrusions in SCADA systems. IEEE Trans Ind Inf 7(2):179–186
13. Yang Y et al (2017) Multidimensional intrusion detection system for IEC 61850-based SCADA networks. IEEE Trans Power Deliv 32(2):1068–1078
14. Tsai C-F et al (2009) Intrusion detection by machine learning: a review. Expert Syst Appl 36(10):11994–12000

Research on Pedestrian Traversal Time in T-Shaped Passage

Mengdi Liang, Jie Xu, Yiwen Chen and Siyao Li

Abstract As the most common walking facilities in subway stations, passages usually have high density and complicated pedestrian flow. Different from previous researches on single passage, this paper mainly discusses T-shaped passage, a typical structure of combination of passages. In the event of an emergency, the T-shaped passage is likely to become a bottleneck point of traffic. Traversal time is an important indicator for evaluating the traffic efficiency. Firstly, we improved social force model based on the field survey. Secondly, we establish simulation scenes with different types of pedestrians and different widths of passage to describe pedestrians walking process in T-shaped passage. Finally, we analyze the influence of different factors on traversal time.

Keywords T-shaped passages · Improved social force model · Pedestrian walking process simulation · Traversal time

1 Introduction

Passages of urban railway station are important facilities for passenger's transfer, entering, and exiting station. The passage usually gathers a large number of passengers especially during peak hours [1–3]. T-shaped passage is a common transfer passage, which has the complex structure of flow streamline. Thus, pedestrian traffic congestion easily occurs, which causes safety hazard and deteriorated service of urban railway station [4–7]. Therefore, it is essential to improve pedestrian efficiency

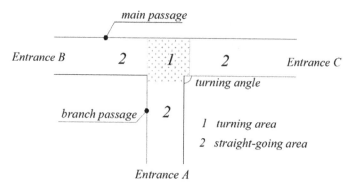

Fig. 1 Structure of a T-shaped passage

of passage to enhance the pedestrian travel experience and service degree. We divide a T-shaped passage into two distinct areas: the straight-going area and turning area (as shown in Fig. 1). The angle between the branch passage and main passage is defined as turning angle which is 90° in this paper, the angle at which a pedestrian turn is defined as steering angle. Similarly, the walking behavior of pedestrians can be classified into two processes: moving straight at the straight-going area and turning at the turning area. In this paper, based on the survey of pedestrian behavior, the pedestrian walking process model is developed. The proposed model is improved on basic social force model (BSFM) accommodating the pedestrian flow, types, and structure of passage. Furthermore, the traversal time of pedestrians under different scenes is obtained based on simulation to analyze the traffic efficiency of pedestrians in passages.

2 Field Survey

We selected two stations with typical characteristics pedestrians flow of the Beijing Subway, i.e., the Fuxingmen station and Xizhimen station, to collect walking behavior parameters at T-shaped passages. In the research of this paper, the manual follow-up counting method and video collection method were used, and the research work was carried out for 30 days during peak hours. The parameters included the pedestrian throughput, travel direction, speed, and density.

Walking speed of different pedestrians is evaluated by manual follow-up counting method. The recorder follows the pedestrians through the passage in per unit length; the speed is evaluated from the recorded traversal time and unit length. According to the field survey, the walking speed in T-shaped passage is 0.8–1.5 m/s for men and 0.6–1.3 m/s for women. When pedestrians take luggage in T-shaped passage, the walking speed is 0.6–1.25 m/s, and the pedestrians with company walk at a speed of 0.65–1.35 m/s. The walking speed in T-shaped passage of heads-down tribe is 0.6–1.3 m/s. We use SPSS software to do the regression analysis with speed

Table 1 Relationship between speed and the population density

Types of pedestrians	Relationship between speed and the population density
Men	$f(x)_{men} = 2.499 - 2.611x + 1.573x^2 - 0.345x^3$
Women	$f(x)_{women} = 2.221 - 2.099x + 1.076x^2 - 0.205x^3$
Pedestrians with luggage	$f(x)_{objects} = 2.051 - 1.963x + 1.097x^2 - 0.229x^3$
Pedestrians with company	$f(x)_{group} = 2.23 - 2.262x + 1.287x^2 - 0.276x^3$
Heads-down tribe	$f(x)_{phone} = 1.737 - 1.266x + 0.586x^2 - 1.222x^3$

Where x is the population density and the unit is p/m^2

under different densities of different types of pedestrians. The curves which adopted cubic spline function are in high fit degree. The relationship between speed and the population density of different pedestrians is shown in Table 1.

Based on the field survey, we found that pedestrians walk on the right side in the T-shaped passage. The rear pedestrians will follow the pedestrians in the same direction, and this phenomenon will be more obvious when density increases. When a pedestrian turns at the turning area, the form of movement of pedestrians can be seen as a 90° circular motion centered on the turning center. The pedestrian trajectory is similar to a quarter arc and tends to take the innermost path.

3 Model

3.1 Model Development

The BSFM which was proposed by Helbing et al. [8] is based on Newton's second law of motion. In the BSFM, pedestrians are driven by three forces: the desired force $\vec{F}_i^0$, which considers the destination of pedestrian i; the interaction force $\vec{F}_{ij}$ between pedestrians i and j, which prevents them from colliding; and the interaction force $\vec{F}_{i\omega}$ between pedestrian i and the wall, which is added to the model to keep pedestrians away from walls. The change in pedestrian speed over time t is given as follows:

$$m_i \frac{d\vec{v}_i(t)}{dt} = \vec{F}_i^0 + \vec{F}_{ij} + \vec{F}_{i\omega} + \xi_i \quad (1)$$

where m_i is the pedestrian's mass, $\vec{v}_i(t)$ is the actual walking speed, and ξ_i is a random variable.

On the basis of research data and document literatures, the pedestrian quality is uniformly distributed within the interval of (45, 80) kg [9]. The size of pedestrians without baggage is uniformly distributed between (0.4, 0.6) m, and the size of pedestrians carrying baggage is uniformly distributed within the interval of (0.6, 1.1) m [10–12].

Based on the field survey, pedestrians in the T-shaped passage can be seen as a 90° circular motion in which center is the midpoint of turning angle. We improved the BSFM with resistance and centripetal force:

$$m_i \frac{d\vec{v}_i(t)}{dt} = \vec{F}_i^0 + \vec{F}_{ij} + \vec{F}_{i\omega} + \xi_i + \varphi\left(\vec{F}_{res} + \vec{F}_{cen}\right) \qquad (2)$$

where $\vec{F}_{res}$ denotes the resistance, $\vec{F}_{cen}$ represents the centripetal force, and φ is a piecewise function.

$$\begin{cases} \varphi = 1 \text{ pedestrians in turning process} \\ \varphi = 0 \text{ pedestrians in non-turning process} \end{cases} \qquad (3)$$

The individual pedestrian force analysis diagram in improved social force model (ISFM) is shown in Fig. 2. During the turning process, the resistance force is related to the desired speed and the magnitude of the steering angle, and they are computed as follows:

$$\vec{F}_{res} = A \cdot \exp(\delta \cdot \beta)\vec{n} \qquad (4)$$

where A is a parameter related to the pedestrian speed and equals 2 m/s² [13], δ is related to the steering angle and equals 0.016 [11], and $\vec{n}$ is a vector with a direction opposite that of the desired speed.

In the ISFM, the pedestrian is subjected to the centripetal force pointing to the turning center. The specific calculation formula is as follows:

$$\vec{F}_{cen} = k \cdot |v_t|^2 \cdot \vec{n}' \qquad (5)$$

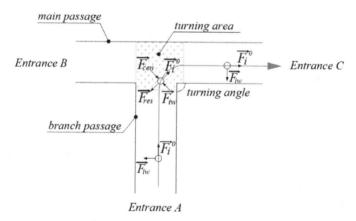

Fig. 2 Individual pedestrian force analysis diagram in ISFM

where k is the gravity coefficient and equals 24 N/m [11], v_t is the pedestrian speed at time t, and $\vec{n}'$ is a vector with a direction perpendicular to the walking direction and pointing toward the turning center.

The diversities between pedestrians will make the macro performance between pedestrians different. Pedestrians walking with company are attracted to each other and excluded from other pedestrians who are not in the group. We improved the ISFM with attraction for the pedestrians with company:

$$\int_{ai}^{x} = -O_{ai} \exp[(r_{ai} - d_{ai})/B_{ai}] \tag{6}$$

where O_{ai} is attraction intensity between pedestrians with company, B_{ai} is attracting scope, r_{ai} is the radius of the pedestrians who walked with company, d_{ai} is the distance of the pedestrians who walked with company.

4 Model Verification

To verify the model, we simulated a pedestrian walking in a Y-shaped passage with MATLAB 2016a. The width of the main passage is 4 m, and the width of branch passage is 5 m. Two pedestrians entered the main passage from Entrance A, after traversing the T-shaped passage, one left from Entrance B and another left from Entrance C. The simulation of the pedestrian trajectory with the ISFM deviated from that of the BSFM in the turning area (as shown in Figs. 3 and 4). Based on the ISFM, the simulated trajectory in the merging area is nearly a quarter arc, which is consistent with the actual pedestrian walking behavior.

The survey found that the pedestrians with company in the subway station are mostly two or three people walking side by side. The distance between the adults and the children is relatively small, and the distance between two men is large. We built a simulation scenario in which the length of main passage is 30 m and width is 10 m, the length of branch passage is 15 m and the width is 20 m. The group size of pedestrians walking together is evenly distributed between two and three.

Fig. 3 Pedestrians trajectory simulated with BSFM

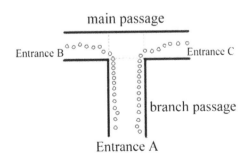

Fig. 4 Pedestrians trajectory simulated with ISFM

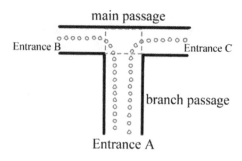

Fig. 5 Simulation diagram of different pedestrians with company

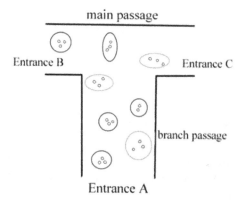

By increasing the attraction for the pedestrians with company, different types of pedestrians can be obtained, as shown in Fig. 5. The yellow circle indicates the group with the lager spacing between the pedestrians, the blue circle indicates the group with the small spacing.

5 Simulation

In this paper, we only discuss the unidirectional pedestrian flow in the T-shaped passage. These experiments contain diverging and merging scenarios. To evaluate the traversal time of pedestrians, we loaded different numbers of pedestrian in different scenarios. The simulation results obtained after each scene was run for 10 times were averaged.

The simulation Scene 1 and Scene 2 were grouped as Group 1 to explore the influence of different types of pedestrians on traversal time. Set the width of the main passage to 4 m, the length of main passage is 27 m, the width of the branch passage to 4.6 m. Scene 1 is a diverging-flow scenario, and the pedestrian arrival rate is 1000–10,000 p/h with taking as the interval of 1000 p/h. The pedestrian threshold is set to 1000 p/h. Scene 2 is a merging-flow scenario, the pedestrian arrival rate

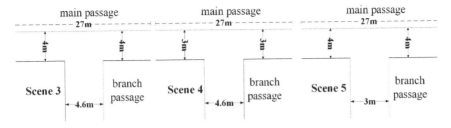

Fig. 6 Schematic of passage width influences on traversal time

of an entrance is 500–5000 p/h. The value is taken at intervals of 500 p/h. The pedestrian threshold is set to 1000 p/h. The pedestrians in Scene 1 and Scene 2 are homogeneous. According to the statistical analysis of the proportion and speed of different pedestrians, we consider the influence of different types of pedestrians on traversal time, such as pedestrian with different genders (p-genders), pedestrians with luggage (p-luggage), pedestrians with company (p-company), and heads-down tribe (p-phone). The simulation experiment of adding different types of pedestrians is compared with the simulation results of Scene 1 and Scene 2 for further analysis.

The scenarios in Group 2 focused on the influence of the width of passage on traversal time. The parameters and schematic diagram of the three scenarios are shown in Fig. 6. The way to load different pedestrian flows in diverging and merging scenarios is the same as the method adopted in Scene 1 and Scene 2.

6 Result Analysis

6.1 Discussion of Traversal Time Based on Different Pedestrian Types

Simulation results of diverging and merging flow of different pedestrian types are shown in Figs. 7 and 8. The traversal time of the scenes which considering with different genders is almost the same as that of Scene 1 and Scene 2.

By comparing the passing time of scenes which adding pedestrians with company, adding pedestrians with luggage, and adding heads-down tribe with the results of Scene 1 and Scene 2, it is found that the traversal time of each scene has little difference when the pedestrian flow is small. However, with the increase of pedestrian flow, the traversal time of the three scenes increases obviously. In the diverging scenes, it is easy for heads-down tribe to stop in the turning area, which causes great interference to the diverging flow. When the pedestrian flow reaches the passage capacity, the traversal time increases by 66% compared with Scene 1. In the merging scenes, there will be no pause for heads-down tribe, but the slower speed leads to a 13% increase in traversal time.

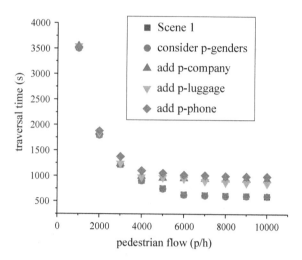

Fig. 7 Diverging flow simulation results scatter plot of different pedestrian types

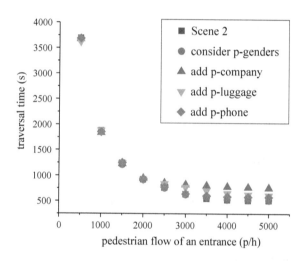

Fig. 8 Merging flow simulation results scatter plot of different pedestrian types

Pedestrians with luggage take up more space and move more slowly. When the pedestrian flow reaches the passage capacity, the pedestrian passage time in the diverging scenes increases by 42% compared with Scene 1, and the merging scenes reduces by 18% compared with Scene 2.

Simulation results show that the pedestrians with company are different from pedestrian with luggage. When pedestrian flow is large, a part of pedestrians with company will change from walking side by side to walking in file. When the pedestrian flow reaches the passage capacity, compared with Scene 1 and Scene 2, the traversal time in the diverging scenes increases by 56% and that of the merging scenes by 47%.

The results indicate that heads-down tribe will cause more interference to the traversal time in diverging scenes when pedestrian flow is large. In both diverging and merging scenarios, the higher the proportion of pedestrians with company and luggage, the greater the impact on traversal time.

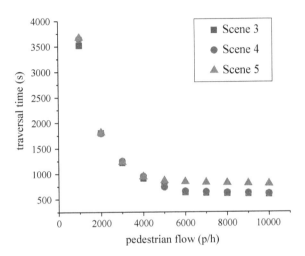

Fig. 9 Diverging flow simulation results scatter plot of different widths of passage

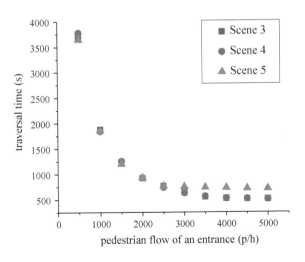

Fig. 10 Merging flow simulation results scatter plot of different widths of passage

6.2 Discussion of Traversal Time Based on Different Passage's Width

The simulation results of Group 2 are shown in Figs. 9 and 10. According to simulation results, there is no significant difference in the traversal time of the three scenarios in the diverging and merging scenes when pedestrian arrival rate is small.

In the diverging scenarios, when the pedestrian arrival rate exceeds 5000 p/h, the traversal time in Scene 5 is the longest, and the traversal time in Scene 3 is the shortest. The traversal time in Scene 5 is 34% more than that in Scene 3. The reason for this phenomenon is that with the increase of pedestrian flow, the force between pedestrians increases, and the pedestrians speed slows down, which causes great interference to pedestrians in the diversion of turning area. Since the width of passage is the largest in Scene 3, the traversal time is the shortest. Compared with Scene 5, the width of branch passage in Scene 4 is much wider, allowing pedestrians to complete the diversion more quickly, so the traversal time is less than Scene 5.

In the merging scenarios, the traversal time in Scene 5 is the longest when the pedestrian arrival rate exceeds 5000 people/h, which is 38% longer than that of Scene 3 with the shortest traversal time. With the increase of pedestrian flow, pedestrian which moving from the main passage merge into the branch passage, and the width of the branch passage of Scene 5 is the largest. The blockage probability is larger when pedestrians move into the branch passage. In Scene 3 and Scene 4, there is same width of branch passage, and the width of main passage in Scene 4 is less than that in Scene 3. This is because in the case of high pedestrian density, the small width of passage constrains the turning process, then pedestrians complete turning process in a relatively short time with small turning radius. However, as the flow increases, blockage probability will be caused when merging occurs, and the traversal time of two scenarios is close.

In conclusion, when pedestrian flow reaches the passage capacity, increasing the width of passage will help improve the pedestrian walking efficiency and reduce the traversal time. Reducing the width of the branch passage will increase the traversal time.

7 Conclusion

To depict the turning process and pedestrians walking with company, we introduced resistance, centripetal force, and attraction for the pedestrians with company to simulate pedestrian walking behavior in T-shaped passage based on the field survey. We explore the relationships among the pedestrian walking behavior parameters, such as the traversal time, types of pedestrians, and the width of passage. Simulation results show that diversion and merging have different characteristics. Heads-down tribe will cause more interference to the traversal time in diverging scenes. The increase of the proportion of pedestrians with luggage or company will increase the traversal time, whether it is in the diverging or merging scenarios. It is found that the width of the branch passage is the main factor affecting the traversal time. In the future, we will focus on the quantitative relationship among the width of passage, pedestrian flow, and traversal time in the T-shaped passage.

Acknowledgements The authors gratefully acknowledge the support provided by China National "13th Five-Year" key research project "Safety assurance technology of urban rail system" (Grant No. 2016YFB1200402).

References

1. Almeida JE, Rosseti RJF, Coelho AL (2013) Crowd simulation modeling applied to emergency and evacuation simulations using multi-agent systems. In: DSIE'11—6th Doctoral symposium in informatics engineering, pp 93–104
2. Haghani M, Sarvi M (2017) Stated and revealed exit choices of pedestrian crowd evacuees. Transp Res Part B Methodol 95:238–259. https://doi.org/10.1016/j.trb.2016.10.019
3. Currie G, Muir C (2017) Understanding passenger perceptions and behaviors during unplanned rail disruptions. Transp Res Procedia 25:4392–4402. https://doi.org/10.1016/j.trpro.2017.05.322
4. Tajima Y, Nagatani T (2002) Clogging transition of pedestrian flow in T-shaped channel. Phys A Stat Mech Appl 303:239–250. https://doi.org/10.1016/S0378-4371(01)00424-1
5. Zhang J, Klingsch W, Schadschneider A, Seyfried A (2011) Transitions in pedestrian fundamental diagrams of straight corridors and T-junctions. J Stat Mech Theory Exp 2011. https://doi.org/10.1088/1742-5468/2011/06/P06004
6. Zhang J (2013) Experimental study of pedestrian flow through a T-junction. https://doi.org/10.1007/978-3-642-39669-4
7. Shimizu T, Igakura T, Ishibashi R, Kojima A (2011) A modeling of pedestrian behavior based on hybrid systems approach—an analysis on the direction of confluence. In: Proceedings of SICE annual conference, 2011, pp 1442–1446
8. Helbing D (n.d.) Simulating dynamical features of escape panic, 1–16
9. Huifang S (2017) Observations and analysis of pedestrian walking microparameters in urban rail transit station passageway. Technol Econ Areas Commun 19:33–38 (in Chinese)
10. Parisi DR, Gilman M, Moldovan H (2009) A modification of the social force model can reproduce experimental data of pedestrian flows in normal conditions. Phys A Stat Mech Appl. https://doi.org/10.1016/j.physa.2009.05.027
11. Yu Q (2015) Improving pedestrian social force model based on pedestrians predicting theory in the rail transit station (in Chinese)
12. Ye Y (2015) Calculation and simulation on capacity of stairs in urban rail transit hub station (in Chinese)
13. Shan Q (2009) Modeling and simulation of pedestrian flow movement in urban transit (in Chinese)

Research on FN-Based MCVideo Service for Railway Communication System

Qingqing Wang, Mingchun Li, Bin Sun, Jianwen Ding and Zhangdui Zhong

Abstract With the emergence of the demand for railway multimedia services, the railway mobile communication system has begun to develop rapidly toward long-term evolution for railway (LTE-R). We analyze the railway video service requirements, compare the standardization processes of the 3rd Generation Partnership Project (3GPP) and International Union of Railways (UIC), and find that there is currently no solution for the rail-specific video services. Based on the above reasons, this paper designs and develops a video call scheme using Function Number (FN) for the railway system to enable railway multimedia services. First of all, the feasibility of the Mission Critical (MC) Service to the railway is discussed. Secondly, Mission Critical System (MCS) in LTE-R is introduced. Then, the Mission Critical Video (MCVideo) software architecture is designed, and the FN-based video service procedure is proposed in this paper, including FN registration and FN-based MCVideo

Q. Wang (✉) · Z. Zhong
State Key Laboratory of Rail Traffic Control and Safety, Beijing Jiaotong University, Beijing 100044, China
e-mail: 17120124@bjtu.edu.cn

Z. Zhong
e-mail: zhdzhong@bjtu.edu.cn

Q. Wang · M. Li
School of Electronic and Information Engineering, Beijing Jiaotong University, Beijing 100044, China
e-mail: 18111070@bjtu.edu.cn

Q. Wang · B. Sun · J. Ding · Z. Zhong
Beijing Engineering Research Center of High-Speed Railway Broadband Mobile Communications, Beijing 100044, China
e-mail: bsun@bjtu.edu.cn

J. Ding
e-mail: jwding@bjtu.edu.cn

M. Li
Jiaxun Feihong Intelligent Technology Institute, Beijing 100044, China

B. Sun · J. Ding
National Research Center of Railway Safety Assessment, Beijing 100044, China

© Springer Nature Singapore Pte Ltd. 2020
B. Liu et al. (eds.), *Proceedings of the 4th International Conference on Electrical and Information Technologies for Rail Transportation (EITRT) 2019*, Lecture Notes in Electrical Engineering 640, https://doi.org/10.1007/978-981-15-2914-6_18

call. Finally, the potential of the proposed scheme is verified on the laboratory static testbed. The results show that the proposed video call scheme has good performance.

Keywords LTE-R · Mission Critical · Function Number · Video call

1 Introduction

Recently, with the development of mobile devices with multimedia capabilities together with the deployment of high-capacity mobile network, the user data traffic has made an immense growth especially multimedia content [1], which also fully reflects the demand of the railway communication system services. The new demand of the railway multimedia services is gradually increasing, but it cannot be met by narrow-band mobile communication systems, and the industry support of the Global System for Mobile Communications-Railways (GSM-R) will be ended in 2030. Therefore, it is inevitable to develop the dedicated broadband mobile communication system LTE-R as the successor of GSM-R [2]. Video services are also applicative to complex railway application scenarios, to further ensure railway transportation safety and improve operational efficiency and service quality. Following this trend, it is necessary to study the implementation of video services based on LTE-R to meet new demands of railway services. Mission Critical Communication (MCC) is defined in 3GPP release 14, which is proposed for public safety [3]. It is highly valued by the telecommunications industry and has important significance [4]. MCVideo, proposed in the MCC series of standards, defines a service for MCVideo communication using 3GPP transport networks [5]. It is initially proposed for the LTE system, but with the potential to be applied in 5G. MCVideo can implement multiple functions such as video private call, video group call, and video surveillance. What is more, MCVideo can also provide a reference for railway services.

The future railway mobile communication system (FRMCS) functional working group of UIC has investigated and summarized the requirements for the next generation railway communication system in the future railway mobile communication user requirement specification (URS) [6]. They propose that MC Services cannot be directly applied to the railway communication system, because of the incompatibilities between the FRMCS requirements and MC Service specifications, such as functional alias for railway communication [7].

This paper proposes the optimization and improvement of the 3GPP MCVideo architecture to solve the problem of incompatibility with the railway services. All of the proposals are tested one by one on LTE-R laboratory testbed to verify the functionality and performance of the system.

The rest of this paper is organized as follows. Section 2 introduces the system architecture of LTE-R. Section 3 presents the design of the MCVideo software architecture and FN-based video call service procedure in the railway dispatching scenario. Section 4 describes the settings of the LTE-R testbed and the analysis of test results. Finally, Sect. 5 draws the conclusions.

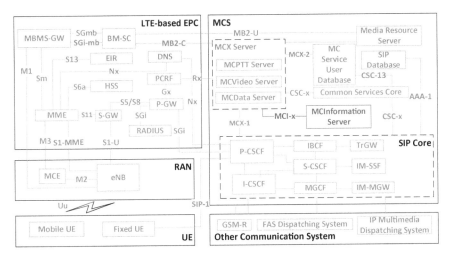

Fig. 1 Logical architecture of LTE-R

2 System Architecture

Considering 3GPP specifications and special service characteristics of the railway communication system, the video service system architecture for LTE-R should meet specific functional requirements for railway communication services [8].

As the best choice for the next generation railway communication system, LTE-R takes use of the advantages of large bandwidth, high transmission data rate, and low delay to support the railway-only services. Therefore, we introduce MCS based on LTE-R, to realize the multimedia dispatching services (MDS) for the railway system. The logical architecture of LTE-R is shown in Fig. 1.

In addition to the functional entities supported by 3GPP, a new entity is designed in our paper. Mission Critical Information Server (MCInformation Server, MCIS), created to support rail-specific services, mainly assists MCX Server (X refers to PTT, Video, or Data) in completing FN-based services and location-based services. And the MCI-x reference point exists between MCIS and MCX Server, providing MCX Server with data retrieval and call support related to FN services, as well as user location notifications.

3 Solutions to MCVideo Service

According to the above railway video service requirements and LTE-R logical architecture, we design MCVideo software architecture and propose the service procedure of FN-based MCVideo private call.

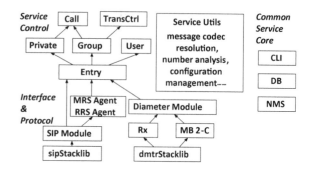

Fig. 2 MCVideo software architecture

3.1 MCVideo Software Architecture

The software architecture of MCVideo designed in this paper is shown in Fig. 2. Major parts are as follows:

- Interface and protocol part provides encapsulation of the stacks and interfaces.
- Service control part provides support for session initiation protocol (SIP), diameter protocol, and the service agent. Its function includes object selection for a private call or group call, and it is responsible for initiating a call to a call object. It also supports the media transmission and reception control, and other configurations for a user.
- Common service core part provides interfaces for configuration and management. command line module, database module (DB), and network management module (NMS) are included in this part.

We also design the state transition process for each functional module. Taking the call procedure as an example, a call object has four states as the called side, which are *Idle, IC_Offering, IC_Alerting, and Conversation*. The call object is *Idle* when there is no service transmission, and its state is changed to *IC_Offering* state when receiving a call request. Call module requests Media Resource Server (MRS) to create the session and distribute media resources for the call establishment. Then, call module sends a ring alert to the calling user, and the call object reaches *IC_Alerting*. When the called user connects the call and returns a response, the call object forwards 200 OK to other callers and switches to *Conversation*, and then, the call on the calling side is successfully established.

3.2 FN-Based Video Private Call Procedure

FRMCS mentions that in the next generation of railway communication systems, railway users should allow different functional aliases to distinguish identity roles and use this alias to establish voice calls or video calls [9]. To satisfy this service requirement, several solutions are also proposed in this work. Figure 3 shows an FN-

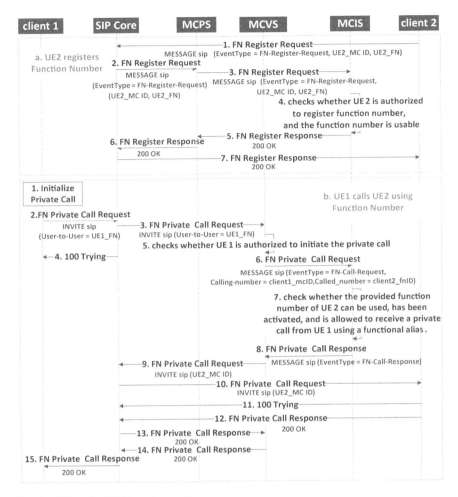

Fig. 3 FN-based MCVideo private call

based video private call procedure designed in this work. We divide it into two parts. The first part is that UE2 obtains the functional alias by registering FN to MCIS, and the second part is the process that UE1 initiates and establishes a call to UE2 using functional alias. There are some key fields carried in SIP MESSAGEs in each part, specified as follows:

- *FN-based registration*: SIP MESSAGE is FN-Register-Request sent by UE2 to MCIS. In its Message Body, the "EventType" value is set to "FN-Register-Request." MC ID and FN of UE2 for identifying the functional alias are also carried. It is worth noting that the registration request is forwarded and completed via MCPTT Server (MCPS) by default. Then, the mapping relationship between the user and its FN is sent to MCX Server. After completing FN registration and obtaining the function alias, the user can perform various FN-based services, such as FN-based voice, video, and data.

- *FN-based MCVideo call*: In Message Header of SIP INVITE, which is sent by UE1 to MCVideo Server (MCVS) to initiate a call, the value of field "User-to-User" is FN of UE1. "EventType," "calling-number," and "called-number" are carried in Message Body of SIP MESSAGE sent from MCVS to MCIS, whose values are "FN-Call-Request," "client1_mcID," and "client2_fnID," respectively. MCIS returns SIP MESSAGE to MCVS, sets "EventType" value to "FN-Call-Response," and carries the parsed MC ID of UE2. Then, MCVS completes the call request using MC ID of UE2.

4 Experimental Validation

To evaluate the performance of the proposed FN-based MCVideo service, a static LTE-R testbed is designed and built in the laboratory. Figure 4 shows the testbed framework and the explanations of system parameters. Test point configuration is also represented.

The frequency bandwidth configured for the testbed is specifically used for China railway. Multiple logical functional entities of the MC System are all integrated into the same server (MC Server) and have the same IP address. Different functional entities are distinguished by the port number. The terminals are divided into Mobile Equipment and Fixed Equipment, to simulate typical terminal roles in the railway system. For example, Train Police uses Handheld Terminal, Train Driver uses Cab Integrated Radio communication equipment (CIR), and Station Attendants or Dispatchers use FAS/MDS Dispatch Terminals. We repeat the FN-based registration and MCVideo private call tests using different terminals based on the testbed and use Wireshark for signaling tracking to verify the designed procedure and key fields. In

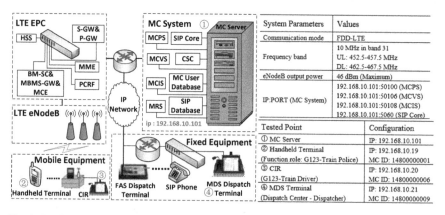

Fig. 4 Deployment of the LTE-based static testbed

addition, we also use MATLAB and other software to evaluate the performance of the call establishment time for FN-based MCVideo private call.

4.1 Session Establishment Procedure

Figure 5 is generated by the module of SIP stream analysis in Wireshark using testing data, which shows the FN registration procedure of a user (Fig. 5a) and the session establishment procedure of the FN-based video private call (Fig. 5b). The call procedure is used to simulate a train driver to initiate a video call to the train police of the train through Function Number.

Logical functional elements, including SIP Core, MCVS, and MCIS, are integrated into MC Server and have the same IP address (192.168.10.101). Hence, the details of the interaction between these network elements are omitted in the call stream on the left side in Fig. 5. The interactive messages are depicted in the comments on the right, including the call request SIP INVITE between SIP Core and MCVS, SIP MESSAGE between MCVS and MCIS, as well as the response message SIP 200 OK.

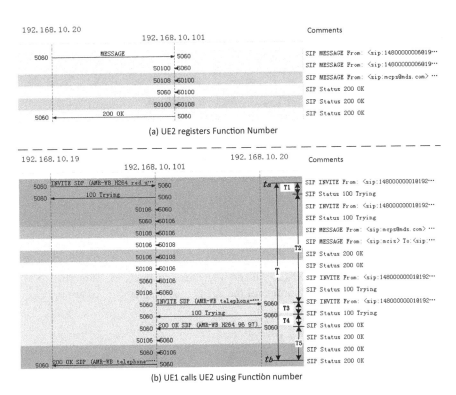

Fig. 5 SIP streams of FN registration and call establishment

```
Content-Type: application/mds+xml

<?xml version="1.0" encoding="UTF-8"?><MDS_XML EventType="FN-Register-Request"
RequestID="000001"><McpttId>14800000006</McpttId><FN>2007112331</FN></MDS_XML>
```

Fig. 6 "UDP stream" for FN-based registration (UE2 → MCPS)

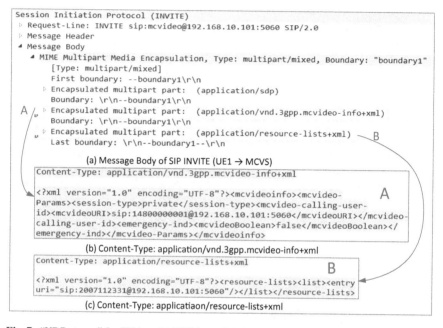

Fig. 7 "UDP stream" for FN-based MCVideo call (UE1 → MCVS)

For FN-based registration, the UDP stream of the SIP MESSAGE from the calling user (UE2) to MCPS is analyzed, as shown in Fig. 6. In XML content, what carried in the field are EventType (FN-Register-Request), MC ID (14800000006), and Function Number (2007112331) that is mapped to the user function alias.

Figure 7 represents the UDP stream of SIP INVITE for an FN-based MCVideo call initiated by the calling user (UE1). There are three encapsulated multiparts included in Message Body, where <session-type> and <mcvideo-calling-use-id> are included in mcvideo-info XML content (Fig. 7b) and in the resource-lists XML content (Fig. 7c), Function Number of the called user is carried in <entry uri>.

4.2 System Performance Evaluation

In addition, the end-to-end MCVideo access time is tested. As shown in Fig. 5, T indicates end-to-end MCVideo access time. It is defined as the time interval between when a calling user request a call (SIP INVITE, t_a) and when the user receives a response message (SIP 200 OK, t_b). 3GPP defines that end-to-end MCPTT access time is no more than 1 s, and there is no additional indicator of end-to-end access time for MCVideo service [10]. T is defined in Eq. (1).

$$T = t_b - t_a \qquad (1)$$

We choose two different terminal combinations for MCVideo communication, CIR-Handheld Terminal (Scenario 1) and CIR-MDS Terminal (Scenario 2). CIR is the calling user, and the called party is Handheld Terminal and MDS Terminal. We grab the timestamps t_a and t_b from CIR. Repeat tests are performed for each group of terminals, and in each test, T is calculated and collected. Then, we draw the cumulative distribution function (CDF) based on thousands of T values, as shown in Fig. 8.

In Scenario 1, 95.92% of calls are completed within 1 s, and the ratio is 97.96% in Scenario 2. Based on the key performance indicators (KPI) of MCPTT service and the test results in this paper, we propose a reasonable KPI for end-to-end MCVideo access time. End-to-end MCVideo access time should be no more than 1 s for 95% of MCVideo request in a static scenario, regardless of the type of terminal equipment used in a call.

On the other hand, the end-to-end access time in Scenario 1 is not very stable compared to Scenario 2. However, there should be a smaller latency in theory for Scenario 2 because MDS Terminal is accessed via IP data communication network, but the statistical results are inconsistent with the theory. Therefore, we divide its call establishment procedure into five stages for analysis and comparison referring to [11], as shown in Fig. 9. T1 represents the response time of the server to the call request initiated by UE1. T2 is the time for routing selection and session configuration by server as well as the reservation of media resources. T3 and T4 are processing delays of UE2 for the received call request. T5 is the time when UE2 forwards confirmation of the call establishment to UE1 via server. According to the statistical

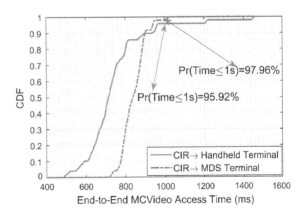

Fig. 8 Statistics of CDF for end-to-end MCVideo access time

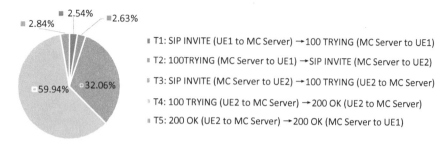

Fig. 9 Time distribution of the call establishment procedure for MDS Terminal

results, we find that MDS Terminal takes a longer time to process the call request received from the server, which is about 92% of the end-to-end latency. Therefore, the processing efficiency of the fixed terminal should be researched and improved as much as possible in the future system optimization.

5 Conclusions

In this paper, we propose a video call scheme based on function alias for the purpose of meeting the multimedia service needs of the railway communication system. The MCVideo software architecture and the state transition process for each functional module, as well as the detailed procedures of FN-based MCVideo service, are specifically designed. Then the implementation is also realized and validated based on the laboratory static testbed. The results show that FN-based MCVideo call is well implemented and meets the performance requirements. Proposals of a KPI for MCVideo call and optimized direction of call establishment time are also contributions of this paper. It is to be noted that although the proposed schemes are based on LTE-R in this paper, it can also be applied to 5G railway communication system.

Acknowledgements The authors express thanks to the support in part from the National Key R&D Program of China under Grant No. 2018YFB1201500, the Research on Development Planning of Next Generation Mobile Communication Technology for Railway under Grant N2018G025, the Research on Technology Application of LTE-R Mobile Communication under Grant P2018G052, and the Research on Application Technology of 5G Mobile Communications for Railway under Grant N2018G072.

References

1. Nguyen, ND, Knopp R, Nikaein N, Bonnet C (2013) Implementation and validation of multimedia broadcast multicast service for LTE/LTE-advanced in OpenAirInterface platform. In: IEEE 38th conference on local computer networks workshops, pp 70–76

2. Sun B, Ding J, Lin S et al (2019) Comparison analysis on feasible solutions for LTE based next-generation railway mobile communication system. ZTE Commun 17:56–62
3. 3rd Generation Partnership Project, Technical Specification (TS). Security of the mission critical service. 3GPP TR 33.180. http://www.3gpp.org/DynaReport/33180.htm. Last accessed 4 July 2019
4. Rajavelsamy R, Pattan B, Chitturi S (2015) Efficient registration procedure for multi-domain authentication for mission critical communication services. In: 2015 IEEE conference on standards for communications and networking (CSCN), pp 282–287
5. 3rd Generation Partnership Project, Technical Specification. Mission critical video services. http://www.3gpp.org/DynaReport/22281.htm. Last accessed 4 July 2019
6. UIC Future Radio Mobile Communication System. User requirement specification 2.0. http://www.uic.org/IMG/pdf/frmcs_user-requirements.pdf. Last accessed 4 July 2019
7. 3rd Generation Partnership Project, Technical Specification. Study on application architecture for the future railway mobile communication system (FRMCS). http://www.3gpp.org/DynaReport/23790.htm. Last accessed 4 July 2019
8. China Railway (2018) Interim specification of next generation mobile communication services and functional requirements for railway (in Chinese)
9. 3rd Generation Partnership Project, Technical Specification. Study on application architecture for the future railway mobile communication system (FRMCS) phase 2. http://www.3gpp.org/DynaReport/23796.htm. Last accessed 4 July 2019
10. 3rd Generation Partnership Project, Technical Specification Group Services and System Aspects. Mission critical push to talk (MCPTT). http://www.3gpp.org/DynaReport/22179.htm. Last accessed 4 July 2019
11. Huang J, Ding J, Zhong Z et al (2018) LTE-based MCPTT architecture for next generation railway dispatching communication system. In: 2018 16th international conference on intelligent transportation systems telecommunications (ITST), pp 1–7

End-to-End Communication Delay Analysis of Next-Generation Train Control System Based on Cloud Computing

Jiakun Wen, Lianchuan Ma and Yuan Cao

Abstract The communication system of the next-generation train control system (TCS) based on cloud computing is modeled in this paper. The end-to-end communication delay is analyzed as the main factor affecting the real-time performance of the system. To improve this performance, a time-sensitive network (TSN) is introduced into the system. We analyze and simulate the end-to-end communication delay of systems with the TSN and compare it with traditional Ethernet. It is concluded that applying a TSN, based on the nw-DRR scheduling algorithm and with optimized parameters, can significantly reduce the end-to-end communication delay decreases from tens of milliseconds to milliseconds.

Keyword End-to-end communication delay · Next-generation train control system · Time-sensitive network (TSN)

1 Introduction

With the development of a new generation train control system (TCS), the computing power requirements of the train control equipment are getting higher and higher. The device will inevitably become very complicated if TCS is still implemented based on the existing overlay method. However, the TCS based on cloud computing in [1] can perfectly solve these problems and achieve the performance requirements of safety, reliability, energy saving and consumption reduction while the safety and reliability of the TCS in [1] are affected by the real-time nature of the system. A method of using FPGA to reduce the processing delay is proposed in [2], but this approach only accelerates the encryption and decryption process. Another TSN solution based on RSC is introduced in [3]. Besides, TSN is introduced into the fronthaul network in

J. Wen (✉) · L. Ma · Y. Cao
School of Electronic and Information Engineering, Beijing Jiaotong University, Beijing, China
e-mail: 18120268@bjtu.edu.cn

L. Ma · Y. Cao
National Engineering Research Center of Rail Transportation Operation and Control System, Beijing Jiaotong University, Beijing, China

[4, 5]. Nevertheless, the above documents all only focus on TSN theory research, do not combine theory with practical application scenarios. Therefore, this paper is aimed to analyze the factors affecting the real-time performance of the system and proposes a TSN based on nw-DRR (a classification-based scheduling method based on deficit cycle) to improve the real-time performance. At last, we carry on a simulation to verify this method.

2 Next-Generation Train Control System Network Structure

It is shown in Fig. 1 that a communication network topology structure of a cloud-based next-generation TCS is suitable for [1].

At present, the railway signal safety network in CTCS using the redundant ring wide-area network architecture which is based on traditional Ethernet network switches and Fibre Channel is a typical network architecture of the rail transit backbone network. Therefore, the backbone of the next-generation TCS based on cloud computing in this paper still adopts this structure. In terms of vehicle-to-ground wireless communication, 4G has been identified as the core technology of the high-speed railway next-generation communication system. As a result, the wireless communication of the next-generation TCS based on cloud computing in this paper adopts the BBU and RRU fronthaul network architecture. The onboard equipment accesses the backbone network through the air interface and the fronthaul network, and the ground device and the cloud computing device access the backbone network directly.

In order to facilitate the following analysis, the data transmission end-to-end delay t_D is divided into backbone network delay t_{BN} and access network delay t_{AN} (There is no node between the ground device and the backbone switch, so there is no t_{AN}), and access network delay can be divided into vehicle access network delay

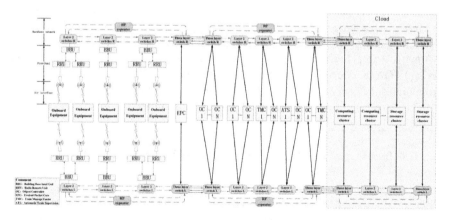

Fig. 1 Network topology of next-generation TCS-based cloud computing

for fronthaul network t_{ANF}, vehicle access network delay for cyberspace delay t_{ANI}, so the end-to-end delay can be expressed by the following equation.

In order to facilitate the following analysis, the data transmission end-to-end delay t_D is divided into backbone network delay t_{BN} and access network delay t_{AN} (There is no node between the ground device and the backbone switch, so there is no t_{AN}), and access network delay can be divided into vehicle access network delay for fronthaul network t_{ANF}, vehicle access network delay for cyberspace delay t_{ANI}, so the end-to-end delay can be expressed by the following equation.

$$t_D = t_{BN} + t_{AN} \qquad (1)$$

$$t_{AN} = t_{ANF} + t_{ANI} \qquad (2)$$

The traditional Ethernet-based contention-based transmission method will cause large delay and jitter under heavy load conditions, which will affect the real-time performance of the system. For this purpose, the system delay will be analyzed in the next section.

3 Maximum End-to-End Delay Analysis

3.1 Maximum End-to-End Delay in Ethernet

End-to-end delay in Ethernet can be divided into processing delay, transmission delay and propagation delay [6].

The transmission delay at each node in traditional switched Ethernet can be expressed by the following Eq. (3)

$$t_{Di} = \frac{\sum \sigma_i - \sigma_i}{C} + \frac{L_{max}}{R_i} \qquad (3)$$

where σ_i is the backlog of the ith data stream, C is the upstream port rate, and L_{max} is the maximum packet length. In the scenario described in [7], the transmission delay of each node is approximately 1.3 ms.

Propagation delay in the physical link, the propagation delay of the signal in the fiber can be expressed by.

$$t_p = l \times \frac{n}{c} \qquad (4)$$

where l is the physical link length, n is the refractive index of the signal in the fiber, and c is the propagation speed of the signal in the fiber. That is, the propagation rate is 5 us/km.

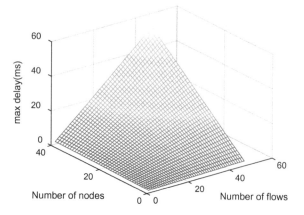

Fig. 2 End-to-end delay in traditional Ethernet

The processing delay in the traditional processing mode is about 40 us, and after using the hardware acceleration technology is approximately 1–2 us.

Obviously, the delay generated by the store-and-forward mechanism accounts for the largest proportion, and the transmission delay is larger under heavy load conditions. Therefore, the following focuses on the analysis of the transmission delay. The traditional Ethernet end-to-end delay simulation is shown in Fig. 2.

It can be concluded from the figure that as the number of streams accessed by each node and the number of nodes increases, the end-to-end delay increases linearly. In vehicle-to-ground communication, the transmission delay of the air interface transmission and the fronthual network will be larger. If the cloud-based next-generation TCS still adopts traditional Ethernet as the backbone network, the large delay is not expected. Therefore, the time-sensitive network is introduced into the backbone network and the vehicle access network to optimize it.

4 A TSN Based on nw-DRR Scheduling

TSN realizes accurate time synchronization between each node through *IEEE802.1AS* timing and synchronization protocol [8], achieves data stream sender and receiver service request transmission and transmission path management through *IEEE802.1Qat* stream reservation protocol [9], enforces priority queuing, fast-forwarding of data and scheduling algorithms in protocols such as *802.1Qbv* and *802.1Qbu* by *IEEE802.1Qav* queue and forwarding protocol [7]. The above protocols jointly implement low-latency, the low-jitter transmission of real-time data in the local area. *nw-DRR* is a classification-based scheduling method based on deficit cycle [3]. The basic principle is shown in Fig. 3.

At the output port of the relay node (switch or router), the convection is divided according to the priority, then the N inputs are divided into N high-priority queues and one low-priority queue, that is, divided into $N + 1$ queues. These queues are

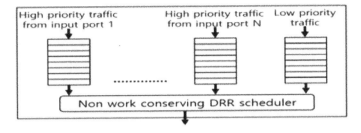

Fig. 3 nw-DRR output queue structure

scheduled using the following mobilizer, but are subject to these following scheduler principles:

Step 1: Assign a deficit value and a quantum value to each queue. The deficit value is variable, the quantum value is fixed, the quantum value is proportional to the flow rate of the queue, and the smaller the value of the sum of the quantum values, the smaller the delay, but the more the scheduler processes.
Step 2: Each queue is executed in turn and all queues are executed once as one frame.
Step 3: The quantum value and the deficit value of each queue at the beginning of each frame are the same.
Step 4: If the deficit value is greater than or equal to the packet length of the queue header, the packet enters the encapsulation of the frame, and the deficit value decreases by a corresponding amount after the data packet enters the frame encapsulation. As long as the conditions are met, the loop is executed in Step 4.
Step 5: If the queue is empty, the deficit value is set to zero and the next queue is processed. And jump back to Step 3 until the frame is full.

In the case of using the *nw-DRR* classification scheduling algorithm, if the data stream is greater than or equal to zero at the node, no queue contention will occur, and no queue burst delay will occur. If there is a negative deficit value, the maximum burst delay of the relay node, i.e., the transmission delay, can be expressed by the following formula:

$$\Theta_i^{DRR} = \frac{1}{r}\left[(F - \phi_i)\left(1 + \frac{L_i}{\phi_i}\right) + \sum_{n=1}^{N} L_n\right] \quad (5)$$

where Θ_i^{DRR} is the transmission delay of node i, r is the link transmission rate L_i is the maximum data packet of stream I length, N is the number of data streams received by the node, ϕ_i is the quantum value, and F is the sum of the quantum values.

4.1 Maximum End-to-End Delay for Next-Generation TCS Using TSN

To ensure the stability of information transmission in a multipoint-to-multipoint communication network, the network must meet the following two constraints [10].

- The arrival process of each flow conforms to the arrival curve of the flow, i.e.,

$$A_i(t_0, t) \leq \rho_i(t - t_0) \tag{6}$$

where ρ_i is the arrival rate of flow i.
- The sum of the traffic arrival rates on each link in the network is less than or equal to the bandwidth of the link, i.e.,

$$\sum \rho_i \leq r \tag{7}$$

Under the premise of satisfying the above two conditions, the delay of the flow i in the node through k nodes satisfies the following constraints.

$$D_i \leq \frac{\sigma_i - L_i}{\rho_i} + \sum_{j=0}^{k} \Theta_j^{S_j} \tag{8}$$

where $A_i(t_0, t)$ is the arrival amount of stream i at the moment (t, t_0), ρ_i is the stream bursty traffic of arrival rate.

Because the topology of the backbone network and the access network are inconsistent, separate analysis is needed.

- **Access network transmission delay**

The ground control device and the cloud computing device directly access the backbone network and only generate delays at the access points of the backbone network. Therefore, it is only necessary to analyze the access network delay of the onboard device [11, 12].

The access time of the onboard equipment to the backbone network needs to be transmitted through the air interface and the fronthual. The maximum transmission delay in the fronthual from Eqs. (5) and (8) can be expressed by the following formula:

$$t_{\text{ANF}} \leq \frac{\sigma_j - L_j}{\rho_j} + \frac{1}{r}\left[(F - \phi_j)\left(1 + \frac{L_j}{\phi_j}\right) + \sum_{n=1}^{N} L_n\right] \tag{9}$$

where σ_j is maximum stream bursty traffic of j in the fronthual. The transmission delay of the air interface, that is, the delay of data transmission between the onboard device and the ground receiving device is represented by t_{Air}.

- **Backbone transmission delay**

The backbone network has a ring structure and a long link length. Compared with the access network, the propagation delay of the signal in the physical link needs to be considered. The propagation delay of the signal in the optical fiber can be expressed by (4). The maximum queuing delay in the backbone network from (5) can be expressed by:

$$\Theta_i^{BN} \leq \sum_{i=0}^{k} \left\{ \frac{1}{r} \left[(F - \phi_i)\left(1 + \frac{L_i}{\phi_i}\right) + \sum_{n=1}^{N} L_n \right] \right\} \quad (10)$$

where k is the number of nodes that the stream i in the backbone network has experienced in the backbone network.

Consequently, the transmission delay in the backbone network from Eqs. (8), (4), and (10) can be expressed by:

$$t_{BN} \leq l \times \frac{n}{c} + \frac{\sigma_i - L_i}{\rho_i} + \sum_{i=0}^{k} \left\{ \frac{1}{r} \left[(F - \phi_i)\left(1 + \frac{L_i}{\phi_i}\right) + \sum_{n=1}^{N} L_n \right] \right\} \quad (11)$$

The existing LTE-R does not disconnect during handover, so the vehicle-to-ground communication delay only considers the delay variation due to the change of the wireless communication rate. Subsequently, the end-to-end transmission delay of the onboard equipment from (9) and (11) can be expressed by the following formula:

$$t_{DV} \leq t_{Air} + \frac{\sigma_j - L_j}{\rho_j} + \frac{1}{r}\left[(F - \phi_j)\left(1 + \frac{L_j}{\phi_j}\right) + \sum_{n=1}^{K} L_n\right] + \frac{\sigma_i - L_i}{\rho_i} + l_i$$

$$\times \frac{n}{c} + \sum_{i=0}^{m} \left\{ \frac{1}{r} \left[(F - \phi_i)\left(1 + \frac{L_i}{\phi_i}\right) + \sum_{n=1}^{N} L_n \right] \right\} \quad (12)$$

where m is the number of nodes that the stream i transmitted by the onboard device has experienced in the backbone network, l_i is the physical link length of the backbone network, K is the number of RRU accessing the BBU, and ρ_i is the rate of wireless communication of the vehicle, which is related to the distance between them and base station.

Similarly, the end-to-end transmission delay of the ground equipment obtained by (11) can be expressed by:

$$t_{DG} \leq \frac{\sigma_i - L_i}{\rho_i} + l \times \frac{n}{c} + \sum_{i=0}^{m} \left\{ \frac{1}{r} \left[(F - \phi_i)\left(1 + \frac{L_i}{\phi_i}\right) + \sum_{n=1}^{N} L_n \right] \right\} \quad (13)$$

where m is the number of nodes that the flow sent by the ground device passes through the backbone network, and l is the physical link length of the flow sent by the ground device on the backbone network.

5 Simulation and Comparative Analysis

For the purpose of analysis, the end-to-end delay of TSN is compared with traditional Ethernet, and it is obviously that the former is significantly smaller than the latter. The simulation results are shown in Fig. 4, where (a) is the maximum end-to-end delay of traditional Ethernet, and (b) is the maximum end-to-end delay of applying TSN.

The analysis of the mathematical model shows that the end-to-end transmission delay of the data is mainly affected by factors such as quantum value, link length, maximum packet length, link capacity, number of data streams received by each node, and number of switches which the data amount passes through. In the orbital traffic scenario, each calculation is periodic, so the data transmission required for the calculation is also periodic, and the length of the data transmitted each time is also regular, which make it possible that through analyzing the relationship of those related parameters and finding the appropriate combination to minimize the end-to-end delay

It can be seen from Fig. 5 that the number of nodes and flows accessing each node are approximately proportional to the maximum end-to-end delay, but the maximum packet length, link capacity, flow arrival rate, maximum end-to-end delay and the slope of the curve increases as the maximum packet length increases, the link capacity decreases and the flow arrival rate decreases. Therefore, the number of nodes and flows accessed by each node should be controlled as much as possible; meanwhile, a secondary switch can be used to share the burden of the switches in the backbone network. In addition, the link capacity should be increased as much as possible without changing the maximum packet length.

It can be concluded from Fig. 6 that the slope of the curve of the maximum end-to-end delay of the onboard device and the wireless communication rate increases as

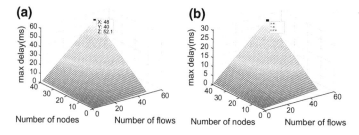

Fig. 4 Maximum latency comparison between traditional Ethernet and TSN

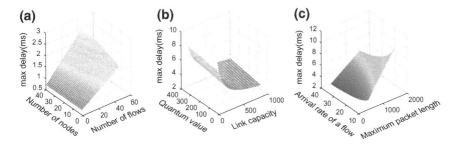

Fig. 5 Relationship between various parameters of end-to-end delay of ground equipment with TSN

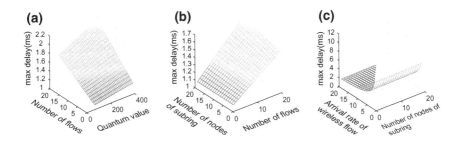

Fig. 6 Relationship between various parameters of end-to-end delay of onboard equipment with TSN

the communication rate decreases. Therefore, we should try our best to ensure that the wireless communication rate of the car is kept at around 10 Mbps.

Figure 7 is a comparison of the maximum end-to-end delay in the case where the parameters are optimized, that is, the quantum value takes 80 bit and link capacity takes 1000 Mbps. It can tell us that the maximum end-to-end delay is significantly reduced after parameter optimization.

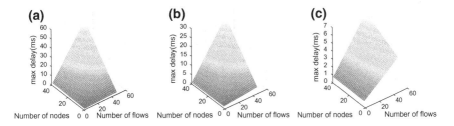

Fig. 7 End-to-end maximum delay comparison of TSN with optimized parameters

6 Conclusion

In summary, the TSN based on the *nw-DRR* scheduling algorithm can reduce the maximum end-to-end delay of real-time data transmission from 60 to 30 ms. Furthermore, after optimizing the quantum value (80 bit) of the *nw-DRR* scheduling algorithm and increasing the link capacity the maximum end-to-end delay is reduced to about 7 ms, when the link capacity is 1000 Mbps. Consequently, the real-time performance of the cloud-based TCS is effectively improved by the TSN introduced into the backbone network and the fronthual. More information on how to optimize the network layout, improve the wireless communication rate of the vehicle, and reduce the rate loss of the handover would help us to further improve the real-time performance of the system to achieve millisecond delay.

Acknowledgements This work was supported by National Key R&D Program of China under Grant No. 2016YFB1200602; and Beijing Laboratory of Urban Rail Transit.

References

1. Chinese patent in progress A cloud-based rail transit train operation control system
2. Peng W, Kaicheng L (2007) Research on implementation of Rijndael algorithm based on FPGA. Railway Comput Appl 3:46–48 (in chinese)
3. Joung J (2019) Regulating scheduler (RSC): a novel solution for IEEE 802.1 time sensitive network (TSN). Electronics 8:189
4. Guangdong S (2018) Research and simulation of clock synchronization and scheduling algorithm in time sensitive networking. Beijing University of Posts and Telecommunications
5. Puye W (2018) Research and simulation of schedule algorithm in time sensitive networking. Beijing University of Posts and Telecommunications
6. Zhen X, Lianchuan M, Jiancheng M (2013) Design and real-time performance analysis on switched ethernet based train communication network. Railway Comput Appl 22(6):51–56. https://doi.org/10.3969/j.issn.1005-8451.2013.06.015
7. IEEE P802. IQav forwarding and queuing enhancements for time-sensitive streams, 2009. http://www.Ieee802.Orglllpages/802.1av.html
8. IEEE 802.1 AVB TG, IEEE P802. IAS-timing and synchronization for time sensitive applications in bridged local area networks, 2011. http://www.Ieee802.Orthollpages/802.1as.html
9. IEEE P802. IQat virtual bridged local area networks stream reservation protocol, 2010. http://www.Ieee802.Orglllpages/802.1at.html
10. Braden R, Clark D, Shenker S (2018) Integrated services in the internet architecture: an overview. https://tools.ietf.org/html/rfc1633. Accessed on 13 Dec 2018
11. Assimakopoulos P, Al-Hares MK, Gomes NJ (2016) Switched ethernet fronthaul architecture for cloud-radio access networks. Opt Commun Netw 8:B135–B146
12. Weimin T (2018) The research and design of 5G new radio numerology. Beijing University of Posts and Telecommunications

A New Method of High-Speed Railway Station Planning Based on Network Evolution

Chengchen Liu, Yanhui Wang, Limin Jia and Shuai Lin

Abstract With the development of high-speed rail industry in various countries, it is very important to plan the network properly. The planning of the high-speed rail station affects the development of the high-speed rail network. This paper combines complex networks with high-speed rail stations as network nodes. Based on the gravity model, we established a network evolution rule to study the impact of new stations on overall network performance. At the end of the article, we use some Chinese high-speed rail stations as examples to verify the method. We hope to use this method to provide guidance for high-speed rail station planning.

Keywords High-speed railway · Network planning · Complex network · Unauthorized evolution

1 Introduction

In this article, we analyze the high-speed rail network topology based on the attraction of nodes to analyze the relevant parameter indexes when it reaches the equilibrium state of attraction, so as to optimize the high-speed railway topology network structure in China.

In recent years, due to the increasingly prominent traffic problems, many experts and scholars at home and abroad have done a lot of research on network structure optimization, and many new methods and heuristic algorithms have appeared. Aleksander K uses the network design to improve the model to study the optimal structure

C. Liu · Y. Wang · L. Jia (✉) · S. Lin
State Key Laboratory of Rail Traffic Control and Safety, Beijing Jiaotong University, Beijing 100044, China
e-mail: 18125717@bjtu.edu.cn

C. Liu
School of Traffic and Transportation, Beijing Jiaotong University, Beijing 100044, China

Y. Wang · L. Jia · S. Lin
Research and Development Center of Transport Industry of Technologies and Equipments of Urban Rail Operation Safety Management, MOT, PRC, Beijing, China

of the transport network [1]. Guo Chonhui et al. discovered the evolutionary community structure in dynamic weighted networks, providing a new method for network evolution theory [2]. Madleňák Radovan and others studied the optimization of postal transport network in Slovakia and provided new ideas for us to study the problem from the perspective of the network [3]. Wu Tao et al. predicted the evolution of complex networks via similarity dynamics, which provide us with ideas for studying evolutionary issues from a mechanical perspective [4]. In addition, the most classic one is that Newman et al. used the renormalization group method to analyze the average distance of the network [5]. At present, the domestic and international methods for network structure optimization have made certain breakthroughs, and a relatively complete theoretical system has been formed.

However, none of the above studies have analyzed the high-speed rail station planning from the perspective of complex network evolution. We will analyze whether it is necessary to build a high-speed railway station in a certain place and guide the planning of the high-speed railway station.

We consider that the connection between nodes is determined by the attraction between nodes. The attraction between nodes is determined by the degree of attraction. We use network evolution to study the planning problems of high-speed rail stations.

2 Node Attraction Calculation Model

2.1 Comprehensive Value of Attraction—Comprehensive Evaluation of Entropy Weight

2.1.1 Entropy Weight Method

We hold that factors such as population, government policies, and accessibility will affect the planning of high-speed rail stations in a region. Therefore, we use the entropy weight method to calculate the attraction of nodes through these data.

The principle of entropy weight method is to quantify and synthesize the information of each unit to be evaluated in the evaluation; using entropy weight method to weight each factor can simplify the evaluation process. Therefore, this paper uses the entropy method to determine the weight of the indicator.

In the multi-index evaluation system, in order to ensure the reliability of the results, to overcome the unreasonable influence of the original data on the clustering analysis results due to the difference of the measurement unit and the order of magnitude, the original indicator data needs to be transformed.

This article uses the normalization changes.

The normalization processing formula is as follows:

$$r_{ij} = \frac{x_{ij} - \min_{1 \le i \le m}(x_{ij})}{\max_{1 \le i \le m}(x_{ij}) - \min_{1 \le i \le m}(x_{ij})}. \tag{2.1}$$

2.1.2 Index Entropy Weight Determination

For r_{ij}, follow the steps below to do the transformation.

$$P_{ij} = r_{ij} / \sum_{i=1}^{m} r_{ij} \tag{2.2}$$

When $r_{ij} = 0$, the above formula should be modified to

$$P_{ij} = (1 + r_{ij}) / \sum_{i=1}^{m} (1 + r_{ij}) \tag{2.3}$$

We improve the calculation formula of the entropy value E_j of the index F_j as follows:

$$E_j = -em^{-1} \sum_{i=1}^{m} p_{ij} \ln(p_{ij}) \tag{2.4}$$

At this time, the range of entropy values can be limited to the range of (0, 1), and then the solution of the entropy weight has no negative value, and thus, the negative value of the attraction degree calculation does not occur.

Entropy weight of indicator F_j

$$W_j = (1 - E_j) / \sum_{j=1}^{N} (1 - E_j) \tag{2.5}$$

We can calculate the node degrees of each node. Its calculation formula is

$$z_j = 100 \sum_{j=1}^{N} r_{ij} - W_j \tag{2.6}$$

2.2 Attractive Model

Using the basic form of the gravity model [6], we give a new parametric form for node attraction. Combined with the actual situation of high-speed rail passenger transport, the following improved models are now available:

$$F_{ij} = \frac{GKZ_iZ_j}{D_{ij}^2} \tag{2.7}$$

where

- F_{ij} The attraction between nodes i and j
- G The coefficient of attraction, the dimension of the equilibrium equation, which is usually 1
- K The correlation coefficient of passenger flow is related to the correlation of passenger flow between the two stations
- Z_i, Z_j Node attraction obtained by component analysis by establishing high-speed railway station node attraction evaluation index system
- D_{ij} Spatial distance between nodes i and j of high-speed rail station.

Next, we can use the node attraction to build an evolution model.

3 Network Topology Evolution Model

3.1 Basic Idea of the Model

This paper is based on the BA model. The construction algorithm of the BA model is as follows [7]:

a. Growth: starting with a network with nodes n_0, each time introducing a new node and connecting to m existing nodes, where $m \leq n_0$;
b. Priority connection: the connection probability between a new node i and an original network node j and the degree of node j satisfy the following relationship:

$$\prod(k_{ij}) = k_i / \sum_{j=1}^{N-1} k_j \tag{3.1}$$

Based on the integrated BA network model and the obtained node attraction, we give the algorithm formula of the modified priority connection probability ϕ_{ij}:

$$\phi_{ij} = \frac{k_j^q F_{ij}^p}{\sum_j k_j^q F_{ij}^p} \tag{3.2}$$

The p and q in the formula are related to the level of the node.

3.2 Evolution Model Establishment and Algorithm Steps

In order to better complete the evolution process of the real network, we must not only consider the practicability and accuracy in the evolution process, but also consider the variables of the real network added to the established network model [8]. Only in this way can we avoid the evolutionary deviation or the wrong result caused by the difference between the virtual network and the real network.

The evolution model of high-speed railway network based on node attraction is constructed as follows:

(1) Build an initial network model:

Before the network evolution begins, it is necessary to determine an initial network with n_0 nodes.

(2) Introducing a new node:

As a starting point, the initial network introduces a new node into the network each time, connecting the new node to the existing network node through the connection probability between the node and the node. (The node cannot be connected to itself.)

(3) Priority connection rules for nodes:

According to the degree distribution of the obtained nodes, according to the order of the probability distribution, m points are required according to the rules, then the first m original network nodes are connected with the new node.

The specific algorithm steps are as follows:

Step 1: Initialization. Determine the number of initial nodes of the network (n_0) and the degree of each node (k_j), the degree of attraction of any node (Z_i), determine the number of nodes (m) to be connected preferentially, and determine the saturation of the node (K_{max});

Step 2: Determine the node priority connection rule. Randomly select a new node i to calculate the connection probability of nodes in the network to newly joined nodes ϕ_i, and prioritize the join sequence in descending order $\{\phi_1, \phi_2, \phi_3, \ldots, \phi_{n_0}\}$

Step 3: Connect the node. Selecting the first m nodes with high connection probability from the priority connection sequence to connect with the new node;

Step 4: Update the network. Update the network based on node connection status;

Step 5: When the node satisfies $K_i = K_{max}$, the node exits the subsequent evolution process.

Step 6: Evolution termination judgment. When the number of network nodes $t < n_0$, jump to Step 2 to continue; otherwise, the evolution process terminates and jumps to Step 6.

Step 7: Result analysis. Calculate the basic parameters that reflect the characteristics of the network and analyze the status of the network operation.

Table 1 Network evaluation index

Network parameters	Symbol	Calculation formula	Acquisition method
Average degree	$<k>$	$\frac{1}{N}\sum \lambda$	Software Gephi
Average clustering coefficient	C	$\frac{2e_i}{k_i(k_i-1)}$	Software Gephi
Average path length	L	$\frac{2}{N(N-1)}\sum_{i>j} d_{ij}$	Software Gephi
Global validity	E	$1/L$	Simple calculation
Local validity	E_0	$1/C$	Simple calculation
Total system cost	TC	K	Simple calculation
Network connectivity	Q	$2M/N$	Formula calculation

3.3 Network Performance Analysis

In order to evaluate the overall performance of the network, this paper selects the following evaluation indicators for the overall performance of the network: (1) the average path length (using the symbol L); (2) the average clustering coefficient (using the symbol C); (3) the network connectivity (using the symbol Q); (4) the total system cost (using the symbol TC); (5) the validity (using the symbol E) [9]. The specific calculation method is shown in Table 1.

The above network indicators reflect the performance of the network itself. By comparing the network performance before and after planning, it is possible to analyze whether the planning is effective.

4 Examples of Model Validation

This paper selects some high-speed railway stations in China as network nodes to verify the above methods.

4.1 Node Degree Calculation

Combined with the actual situation of China's high-speed railway network, the seven indicators that affect the attraction of high-speed railway stations are

Table 2 Node degree of each station

Station	Node degree	Station	Node degree
Dalian	59.4	Xiamen	35.5
Beijing	97.40	Yueyang	22.43
Guangzhou	88.23	Foshan	23.10
Qingdao	72.10	Chongqing	42.92
Tianjin	77.01	Qinhuangdao	23.35
Shanghai	87.42	Yantai	25.71
Wuhan	91.46	Weihai	18.22
Ningbo	39.45	Fuzhou	35.65
Shenzhen	78.41	Zhongshan	23.67
Zhuhai	35.56	Haerbin	30.07
Nanjing	58.77	Xuzhou	57.92
Hangzhou	58.43	Hefei	60.01
Nanchang	42.93	Yichang	23.81
Changsha	40.23	Nanning	42.43
Jinan	42.82		

(1) The population size of the administrative area where the station is located (unit: 10,000 people)
(2) The level of economic development of the administrative area where the station is located
(3) Traffic accessibility (taking the mileage of the administrative area of the station as an indicator, unit: km)
(4) Radiation radius of the station (unit: km)
(5) Location of the station (10-point system)
(6) Development level of transportation facilities (10-point system)
(7) Government policy control and control (10-point system).

Using the improved entropy weight model given above, the node degrees of each station can be obtained, as shown in Table 2.

4.2 Node Attraction Calculation

The calculation of the attraction degree of the selected high-speed railway station node has been carried out above, so the programming according to formula 1-1 is carried out, and the above-mentioned numerical values are brought into the program to obtain the calculation result of the attraction of the high-speed railway station in China.

Fig. 1 Final evolution image

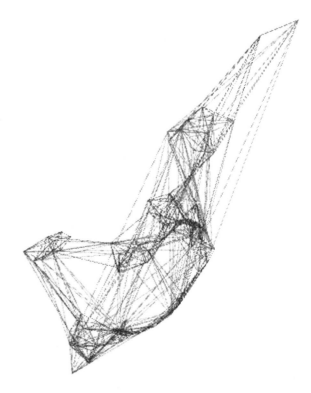

4.3 Unauthorized Evolution of the Network

Based on the original nodes, we use the attractiveness to calculate the connection probability and apply the network evolution model proposed in this paper to evolve. In order to take into account the connection between stations as much as possible, we make $K_{max} = 7$.

Import the final topology map into Gephi to get Fig. 1.

We assume that we plan to build a high-speed rail station in Binzhou, Shandong. Use the model to evolve to get Fig. 2.

4.4 Calculation and Comparison of Network Performance Indicators

It can be seen that the network structure has changed after joining the new node. The network average decreased from 5.915 to 2.966, which indicates that the overall attraction of the network is weakened. The clustering coefficient has changed from 0.571 to 0.572. Generally, the change has not changed much, and the network diameter has changed from 7 to the present 13, the scope of this network has become

Fig. 2 Network diagram after the new station

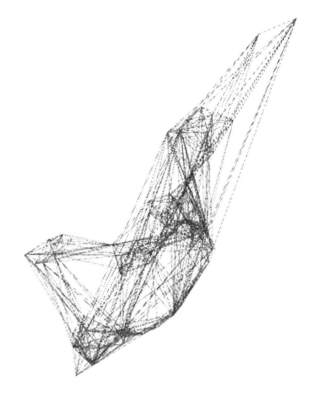

larger, the average path length has changed from 3.631 to 3.813, the network path has become longer, the global effectiveness and partial effectiveness have declined, and the overall efficiency of the network has deteriorated. The network is worse than the previous network.

In summary, it is not recommended to plan high-speed rail stations in Binzhou, Shandong. Since the current high-speed rail station has not been built in Binzhou, the verification by this model is basically consistent with the current plan.

5 Conclusion

This paper analyzes the rationality of planning from a new perspective. In this paper, a lot of models have been established for the evolution process, which is obtained by reviewing the literature, discovering problems, thinking about problems, solving problems, and finally improving the model. Through the research on the evolution of high-speed railway transportation network, we hope to find out the development direction of future transportation network. For the selection of parameters in the model, there may be deviations in the results, and the model itself has room for improvement. We also hope that the reader can provide valuable suggestions on this basis.

Finally, the evolutionary model is used to carry out the powerless evolution of the whole network. The advantages and disadvantages of the transportation network are analyzed, and the guiding suggestions for the future high-speed railway station planning are given.

Acknowledgements The authors gratefully acknowledge the support from "the Fundamental Research Funds for the Central Universities" (Grant No. 2017JBZ103).

References

1. Król A (2016) The application of the artificial intelligence methods for planning of the development of the transportation network. Transp Res Procedia 14:4532–4541
2. Guo C, Wang J, Zhang Z (2014) Evolutionary community structure discovery in dynamic weighted networks. Physica A 413:565–576
3. Madleňák R, Madleňáková L, Štefunko J (2015) The variant approach to the optimization of the postal transportation network in the conditions of the Slovak Republic. Transp Telecommun J 16(3)
4. Wu T, Chen L, Zhong L et al (2017) Predicting the evolution of complex networks via similarity dynamics. Physica A 465:662–672
5. Newman MEJ, Watts DJ (1999) Renormalization group analysis of the small-world network model. Phys Lett A 263:341–346
6. Yin G, Chi K, Dong Y et al (2017) An approach of community evolution based on gravitational relationship refactoring in dynamic networks. Phys Lett A 381(16):1349–1355
7. A-L Barabási, Albert R, Emergence of scaling in random networks. Science 286:509–512
8. Liu Q, Liu C, Wang J et al (2017) Evolutionary link community structure discovery in dynamic weighted networks. Physica A 466:370–388
9. Huo L, Cheng Y, Liu C et al (2018) Dynamic analysis of rumor spreading model for considering active network nodes and nonlinear spreading rate. Physica A: Stat Mech Appl

A Rail Train Number Identification Algorithm Based on Image Processing

Shuangyan Yang, Ruifeng Xu, Zhihui Zhou, Zongyi Xing and Yong Zhang

Abstract The identification of the rail train numbers is of great significance for the daily inspection and maintenance of trains. In this paper, the image processing technology is used to locate, segment, and identify the vehicle number, thus completing the online detection system for the rail train identification. First, we use a high-speed industrial camera to obtain an image containing the rail train number; secondly, a series of operations such as rough interception, smoothing, binarization, and morphological processing are performed on the rail train number image to locate the rail train number; then, according to the connected domain characteristics of the single character of the car number, a single character is segmented and normalized. Finally, the artificial neural network (ANN) is used to realize the identification of the rail train number character.

Keywords Image processing · Background color · Morphological processing · ANN · Character recognition

1 Introduction

Rail trains are an important part of urban rail transit. The record of train number plays a very important role in rail transit. Therefore, accurate rail train number identification makes it a crucial process to enter the information system [1]. The rail train number record can be completed by manual transcription, but it consumes a lot of manpower and material resources and is prone to unnecessary errors, and the rail train number data cannot be provided in real time.

At present, the existing train number identification system is an automatic identification system for railway number based on RFID technology. In general, the system can realize the function of automatically copying the rail train number. However, since the accuracy of ATIS relies on the RFID tag installed at the bottom of the train,

S. Yang · R. Xu · Z. Zhou · Z. Xing (✉) · Y. Zhang
School of Automation, Nanjing University of Science and Technology, Nanjing, China
e-mail: xingzongyi@163.com

the RFID tag is easily damaged and lost; therefore, the accuracy of the system for identification of the car number will be greatly reduced.

In recent years, with the rapid development of image processing technology, the train number recognition system based on image processing has gradually shown its advantages [2]. This paper proposes and implements an online detection system for rail train number identification based on image processing. The first is to obtain the image containing the rail train number through the image acquisition hardware system and then use the image processing-related algorithm to preprocess the rail train number image, locate the rail train number area, segment and identify the single character, etc., and finally achieve the identification of urban rail train number.

2 Algorithm

There are four steps: First, in order to be able to locate the rail train number area more accurately, we need to process the original image in advance; and then, according to the background color of the region where the character is located, a series of operations such as binarization and morphological processing are performed to locate the rail train number; then, according to the connected domain characteristics of the single character of the car number, a single character is segmented and normalized; finally, through the artificial neural network (ANN), the identification of the rail train number characters is achieved. The image processing algorithm flow of the detection system is shown in Fig. 1.

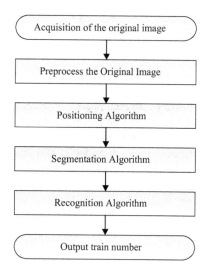

Fig. 1 Image processing algorithm flow of the detection system

2.1 Preprocess the Original Image

At the scene where the train travels, due to the influence of illumination and environment, it is necessary to preprocess the original image collected [3], thereby improving the quality of the image of the car number and achieving accurate positioning. The original image is shown in Fig. 2.

First, the original image is filtered using a Gaussian low-pass filter to obtain the filtered image I_f. Then, according to the background color of the area where the car number is located [4], the filtered image is processed by setting the range of the three components in the RGB space, the area outside the threshold range is removed, and the area where the car number is located is divided by the background color area. The outer area turns black, and the image $I_b(i, j)$ is obtained, namely:

$$\begin{cases} I_b(i, j) = I_f(i, j), & \begin{cases} R_{\min} < I_f(i, j)_R < R_{\max} \\ G_{\min} < I_f(i, j)_G < G_{\max} \\ B_{\min} < I_f(i, j)_B < B_{\max} \end{cases} \\ I_b(i, j) = 0, & \text{others} \end{cases} \quad (1)$$

where (i, j) represents the pixel point coordinates of the image, $I_f(i, j)$ represents the value on the image I_f after the filtering where the coordinates (i, j) are located, $I_f(i, j)_R$ represents the R component in (i, j) of $I_f(i, j)$, $I_f(i, j)_G$ represents the value of the G component in $I_f(i, j)$, and $I_f(i, j)_B$ represents the B component of $I_f(i, j)$, $R_{\max}$ and $R_{\min}$ represent the maximum and minimum values of the background color R component of the area where the car number is located, $G_{\max}$ and $G_{\min}$ represent the maximum and minimum values of the background component G color component of the area where the car number is located, $B_{\max}$ and $B_{\min}$ represent the maximum and minimum values of the background component B color component of the area where the car number is located. The processed image is shown in Fig. 3.

Then, perform binarization and morphological processing on the processed image [5, 6]. Delete the connected domain of the binary image with the area less than the entire image area by 5–10%. The processed image is shown in Fig. 4.

Fig. 2 Original image

Fig. 3 After preprocessing according to background color

Fig. 4 After binarization and morphological processing

2.2 Positioning Algorithm

After analyzing the preprocessed image data, according to the aspect ratio of the circumscribed rectangle of the connected area [7], the ratio of the number of pixels in the connected area to the number of contiguous rectangular pixels in the connected area, and the number of horizontal hops in each line of the connected area [8], we can find the area with the rail train number characters in the remaining connected areas, as shown in Figs. 5 and 6.

2.3 Segmentation Algorithm

Perform a negation operation on the binarized image containing the track train number characters and analyze the inverted image. According to the aspect ratio of the circumscribed rectangle of the connected area and the ratio of the number of pixels in the connected area to the number of pixels in the outer rectangle, we can locate the single characters in the inverted image. As shown in Figs. 7 and 8.

Fig. 5 Located rail train number area

Fig. 6 Segmented car number area

Fig. 7 Located single character in the rail train number area

Fig. 8 Split single character

2.4 Recognition Algorithm

Through the character segmentation algorithm, we will get a single car number character, and then, we need to normalize them, unify them into uniform size images [9], and perform binarization operations.

This stage also needs to train the ANN model. The ANN is set to three layers, which are input layer, hidden layer, and output layer, respectively. The neuron excitation function is sigmoid, and the sigmoid function expression is:

$$f(\sigma) = \frac{1}{1 + e^{-\sigma}} \qquad (2)$$

where σ represents the input of the neuron and $f(\sigma)$ represents the output of the neuron.

We use n as the number of input layer nodes, use l as the number of hidden layer nodes, use m as the number of output layer nodes, initialize the connection weights ω_{ij} and ω_{jk} between the input layer, the hidden layer, and the output layer neurons, and initialize the hidden layer function threshold φ_j, the output layer function threshold is θ_k, i represents the input layer node number, j represents the hidden layer node number, and k represents the output layer node number. The training steps are as follows:

Step 1: According to the input x_i, the connection weight ω_{ij} between the input layer and the hidden layer, and the hidden layer function threshold φ_j, the output of the hidden layer h_j is obtained as follows:

$$h_j = f\left(\sum_{i=0}^{n} \omega_{ij} x_i - \varphi_j\right) \quad j = 1, 2, \ldots, l \tag{3}$$

where f is the excitation function of the neuron.

Step 2: According to the output h_j of the hidden layer, the connection weight ω_{jk}, the output layer function threshold θ_k, and the output y_k of the output layer are obtained as:

$$y_k = f\left(\sum_{j=1}^{n} \omega_{jk} h_j - \theta_k\right) \quad k = 1, 2, \ldots, m \tag{4}$$

where f is the excitation function of the neuron.

Step 3: Set o_k to the desired output, and calculate the error δ_k between the actual output and the expected output:

$$\delta_k = o_k - y_k \tag{5}$$

Step 4: Update the network weights ω_{ij} and ω_{jk} according to the error δ_k:

$$\omega_{jk} = \omega_{jk} + \eta \delta_k h_j \tag{6}$$

$$\omega_{ij} = \omega_{ij} + \eta h_j (1 - h_j) x_i \sum_{k=1}^{m} \omega_{jk} \delta_k \tag{7}$$

where η is the learning rate.

In the ANN, the information in the input layer is first forwarded to the hidden layer, and the output is sent to the output layer through the excitation function. The weight of each layer is unchanged during forward propagation. If the result is different from the expected value, the back-propagation is performed and the weight of each layer network is modified. After the back-propagation is completed, the forward

Fig. 9 Final identified result

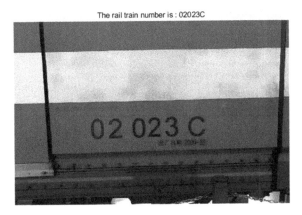

propagation is performed again until the difference value is extremely small. Finally, the processed image information is loaded into the trained labor. The identification of the car number character is performed in the ANN model, as shown in Fig. 9, and the identified car number is output.

3 Conclusion

The identification of the rail train number is one of the key technologies in the train online detection system. It is of great significance for the daily inspection and maintenance of trains. Therefore, accurate rail train number identification is a crucial process.

First, we use a high-speed industrial camera to get an image containing the car number; then, a series of operations such as rough interception, smoothing, binarization, and morphological processing are performed on the car number image to locate the car number; According to the connected domain characteristics of the single character of the car number, a single character is segmented and normalized. Finally, the ANN is used to realize the identification of the car number character.

Compared with the current train number recognition technology, this algorithm has the following advantages: (1) It uses image processing to identify the train number to determine the relevant information of the train, improves the accuracy of the vehicle number information collection, and reduces the vehicle number collection system. (2) The algorithm has good real-time performance, can quickly get the car number, get the car number area based on the body and character color, and the car number area positioning accuracy is higher.

Acknowledgements This work is supported by National Key R&D Program of China (2016YFB1200401).

References

1. Wang J, Zhao Y, Ge L, Niu Z (2015) Video identification system for train numbers. Comput Technol Dev 25(03):28–31 (in Chinese)
2. Zheng Q (2016) Image-based EMU identification and pantograph detection. Southwest Jiaotong University (in Chinese)
3. Yang J (2017) Research on high-speed train number recognition algorithm based on image processing. Southwest Jiaotong University (in Chinese)
4. Cheng L (2013) Automatic recognition system for locomotive car number based on image processing. Nanjing University of Aeronautics and Astronautics (in Chinese)
5. Wang C, Yang Z, Yu Y et al (2017) Friction surface crack identification of brake disc based on edge detection and mathematical morphology. Railway Roll Stock 37(01):10–13+24 (in Chinese)
6. Huang L (2018) Application research of image recognition system for locomotive vehicle number. Chongqing University of Technology (in Chinese)
7. Zhao R (2011) Research on algorithm for identification of railway freight car number. Hebei University of Technology (in Chinese)
8. Gu X, Li Z (2013) License plate location and character segmentation based on mathematical binary morphology. J Hebei United Univ (Nat Sci Ed) 35(02):85–89 (in Chinese)
9. Wang J (2015) Research on truck number recognition technology based on LabVIEW. Shijiazhuang Railway University (in Chinese)

Multi-stage Attack and Proactive Defense for CBTC

Xiang Li, Bing Bu and Li Zhu

Abstract CBTC is one of the most advanced signal systems at present. The security of the CBTC systems are crucial for its close relation with people's livelihood, thus the defense methods study is necessary. The information attacks in the CBTC systems have two main attributes: multi-stage and cyber-physical fusion which makes the consequence of failed defense enormous, and more importantly, the traditional passive defense that only statically defense against network domain attacks that can hardly achieve effective defense anymore. Therefore, we adopt proactive defense to solve this problem. In this paper, an in-depth analysis of the multi-stage feature of attacks in the CBTC systems has been made, based on which, we propose a multi-stage game model framework to describe the attack–defense interaction in each stage and obtain the optimal defense strategies by calculating the game equilibrium. Finally, the result reveals the effectiveness and superiority of the proposed method.

Keywords CBTC · Proactive defense · Multi-stage game

1 Introduction

Urban rail transit has become one of the most important means to solve urban traffic congestion problems because of its high efficiency and environmental protection. The communication-based train control system (CBTC) is an advanced type of signal system. CBTC introduces information technology into the train control field. By using generic software and hardware for signal equipments and universal computer network for data transmission, the efficiency of the signal system is enhanced remarkably, and it also makes the system construction cost and complexity reduced

---

X. Li (✉) · B. Bu · L. Zhu
State Key Lab of Rail Traffic Control & Safety, Beijing Jiaotong University, Beijing, China
e-mail: 17120253@bjtu.edu.cn

B. Bu
e-mail: bbu@bjtu.edu.cn

L. Zhu
e-mail: lizhu@bjtu.edu.cn

greatly. At present, CBTC has gradually become the approved solution for urban rail transit signal system, while the CBTC systems also face more security risk under the pressure of greater system openness and the complicated and abundant APT attacks such as the Stuxnet virus [1] invading the Iranian nuclear facility in 2010.

Due to it's characteristic of cyber-physical fusion, the CBTC systems are facing effect from both network and physical aspects while under information attacks, and there will be significant reduction of safety and efficiency for the signal system, together with the multi-stage feature of attacks, the traditional passive defense, such as packet filter firewall, antivirus software, is no longer suitable satisfying protection. Proactive defense is becoming more popular as it aims at in-validating attacks by deploying defense before attack happens or causes serious impacts, which matches the security demand of CBTC [2, 3]. The mainstream proactive defense method is the attack–defense model-based proactive defense, and this method generates optimal strategies base on the attack–defense interaction analysis and feasible quantitative scheme.

There has been some researches on the attack–defense model of CBTC; Bao [4] proposed an attack countermeasure tree-based model to calculate optimal strategies for both attacker and defender; Peng [5] designed a defense mechanism especially for man-in-the-middle attack in CBTC system by using Bayesian game; Li [6] develops a risk assessment method on the top of attack tree model and attack path prediction; and Yu [7] models CBTC system under DoS attack and data deceive attack separately with the consideration of random packet loss and build an intrusion detection system base on the model. Game theory has been used to quantify the attack and defense interaction in many researches. Luo et al. [8] developed a game theoretical approach to analyze the attack impacts on the IDS guarded systems; Orojloo et al. [9] use game theory as a quantification scheme to evaluate the security of cyber-physical systems; Zhu et al. [10] divide the APT attack into stages and phases and calculate optimal defense strategy by game theory. Most present studies focus on specific attacks, others treat the defense as a attack route prediction problem and the attack analyzation is overly simplified, by choosing neither of the above two ways can we take both attacks' multi-stage and network physical fusion features into consideration.

In this paper, we divide a whole attack process in CBTC into three stages: intrusion, penetration, and disruption, for their distinct attack–defense scenarios. In this paper, we proposed a multi-stage game model to describe the three stages of interactions between attacker and defender, where a reasonable quantification structure is used for the aid of equilibrium calculation and optimal strategy generation. And especially, we design a game-based defense method for the unguarded disruption stage.

2 Security Analyzation of CBTC System

In this section, we will give a brief introduction of the CBTC systems and analyze the security risks the systems face.

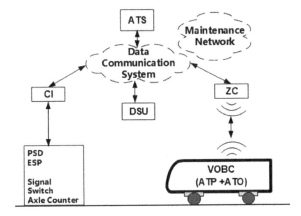

Fig. 1 CBTC system architecture

2.1 CBTC System Introduction

CBTC system is the most advanced train control system at present thanks to the more accurate train positioning technology compared with traditional track circuit and high-speed, two-way, large-capacity vehicle-to-ground wireless communication, and CBTC can operate in moving block style which improves system efficiency greatly. Figure 1 shows the structure of CBTC. CBTC consists of ground equipments and vehicle equipments. The vehicle on-board controller (VOBC) sends positioning and speed information to the ground, the zone controller (ZC) integrates trains' positioning information and the safety route information that the computer interlock (CI) delivered to calculate the moving authority (MA) and send it back to VOBC; VOBC works out a driving curve base on MA. According to the massages transmitted, the CBTC networks are divided into ATC network, ATS network, and maintenance network. The ATC network is safety-critical and carries data which are directly related to the train control process, like the aforementioned MA; ATS network is responsible for monitoring and scheduling data transmission; maintenance network connects crucial devices and obtains operation and error logs. Some of the signal equipments are based on redundancy architecture such as double 2-vote-2 system, so CBTC system also contains a large amount of device intranets.

2.2 CBTC Security Concerns

CBTC systems mainly have three types of security concern: access, equipment, and communication protocol.

Access vulnerabilities. Although CBTC networks have no direct connection with public network, there are possible ways for attackers to invade. At present, most CBTC systems still use WLAN for train-ground communication. The access and authentication process is dominated by pre-shared key (PSK) and the 4-way handshake in WLAN, while the password brute force attack is very mature and can be carried out offline, so if the CBTC systems do not regularly perform full-line key update, with certain automatic tools and large computing capacity, the attacker can crack PSK while it is still valid.

Another possible way is using other urban rail subsystems as springboard to invade. CBTC has direct or indirect interfaces with other urban rail subsystems such as the passenger information system (PIS) and the closed-circuit television (CCTV). Some of the interfaces lack of valid data transmission control measurement and can cause the outsider to use other urban rail subsystems as springboard.

Finally, the most popular method used by external attackers is social engineering attacks. The outsider can apply spear phishing, waterhole, etc., to infect the CBTC staffs' personal computer and deliver the malicious code into the CBTC systems by the means of portable storage devices.

Equipment vulnerabilities. CBTC employs general-purpose operating systems such as Windows and Linux which are known as insecure. A large number of vulnerabilities exist in these operating systems and the patch installation in the CBTC systems are far behind normal level.

Most of the key equipments in CBTC systems are based on redundancy architecture, such as double 2-vote-2 system, and use industry control operating systems which are highly customized such as VxWorks. This approach greatly improves the reliability and safety of the devices, but often lacks consideration of information attacks, so that the device tolerance is even reduced under attack, for example, a simple port probe against ZC's open port could cause ZC to shutdown because the defined timing of the double 2-vote-2 platform is disturbed and the device's counter measure for unexpected behavior is shutdown.

Communication protocols vulnerabilities. There is no mandatory provision about the communication protocol used in CBTC systems although the interoperation process is ongoing. At present, commonly used protocols in CBTC are RSSP-1 and venders' private protocols, some of the venders' private protocols lack of security protection methods, and the integrity and authenticity of data transmitted this way can hardly be guaranteed. And in our experiments, the RSSP-1 protocol has been verified to have security vulnerabilities, and data tampering attacks can be implemented under certain conditions.

3 The Multi-stage Attack–Defense Game Model

In this section, we will propose a full-sized multi-stage game model (shown in Fig. 2) to describe the attack–defense interactions in the CBTC systems. The model proposed is a scalable game framework, in which exist multi types of the game according to the different attack scenarios.

Two players exist in the whole game: the attacker and system manager or defender expressed as R_A and R_D, respectively. The whole game model consists of three

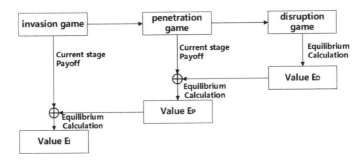

Fig. 2 Multi-stage game model's sequential structure

subgames: the intrusion game G_I, the penetration game G_P, and the disruption game G_D. In the rest of this section, we will define the action space and payoffs of every subgame as well as the equilibrium calculation method.

3.1 Intrusion Game

In this subgame, the attackers choose which intrusion method to invade into the CBTC systems, and attacks from insider and outsider should both be considered, and it is reasonable to use Bayesian game to capture the lack of attacker role information for defenders.

As described in the last section, the outsiders mainly have three possible ways to invade, while the insiders have two means to get unexpected access. The intrusion point has to be considered too, because the different convenience they may offer in the subsequent stages. The defenders' actions are general network defense methods like reconfiguration, adding firewall, password update, etc. The attackers' payoff can be calculate by Eqs. (1), (2), and (3), where P_I^A is the payoff for attackers in this subgame, E_I is the gain and C_I is the cost, V_P is the game value of the penetration stage game, and L_I represents the intrusion points' secure level; the ATC network has the highest score because if the attack succeed, the subsequent attack can directly work on the critical equipments such as ZC and make disruption immediately. I_M indicates the intrusion method level and has two options: network access and code implementation which can offer different convince for subsequent attacks. $(W_L \ W_M)$ is the weight vector of $(I_L \ I_M)$. The attacking costs consist of attack difficulty and detected risks, and for certain vulnerabilities, the difficulty score can be replaced by generic evaluation method like CVSS.

$$E_I = (I_L \ I_M) \cdot (W_L \ W_M) \tag{1}$$

$$C_I = (I_D \ I_E) \cdot (W_D \ W_E) \tag{2}$$

Table 1 Intrusion gain construction elements

Intrusion network level	Score	Intrusion method level	Score
ATC network	1	Network access	1
Equipment intranet	0.8	Code implementation	0.5
Maintenance network	0.6		
ATS network	0.4		

Table 2 Intrusion cost construction elements

Intrusion difficulty	Score	Detected risks	Score
Ready-made tools	0.3	Code implementation	0.9
Detailed attack tutorial	0.6	Physical access	0.6
Completely self design	0.9	Virtual link access	0.3

$$P_I^A = E_I + C_I + V_P \tag{3}$$

The defenders' gain is the loss avoided by successful defense and it equals the attackers' gain. The defenders' cost can be calculated in the same manner as attackers. The actual calculation refers to Tables 1 and 2. We use λ to indicate the uncertainty of attacker role, $\lambda = 1$ means the attacker is a outsider and $\lambda = 0$ means the attacker is a insider.

3.2 Penetration Game

The two main goals of attackers in penetration stage is to get the access authority of the attacking targets and collect information necessary for the disruption stage, and in this stage of game, the attackers will penetrate among different CBTC networks extensively and deeply, and the defender will try to deploy defensive countermeasures to prevent the proliferation of attack authority.

After the intrusion stage we defined, the attacker will have access authority of ATS subnet or maintenance subnet, and in our case, we assume that the disruption stage has to be in the ATC networks. For attackers start from ATS networks, the only way to penetrate into ATC subnet is hecking through the gateway computer which is responsible for the data transmission between ATS and ATC. If this staged attack starts from maintenance network, the penetration process will be hecking through the key equipment intranet to the ATC subnet. We use sequential game for this stage, and the payoff quantification scheme is the same with intrusion stage. The payoff of this stage of game consists of the current stage payoff and the gaming value of disruption game as shown in Eq. (5) where I_d is the value of disruption game and $E(d_s)$ is the output of the current game payoff matrix given certain action, $(p_0^* \ldots p_i^*)$ is the mixed equilibrium if satisfy Eq. (4), and vise versa for $(q_0^* \ldots q_j^*)$.

$$\forall\, p_m^*, p_n^* in (p_0^* \ldots p_i^*) n \neq m \quad E_m(d_0 \ldots d_j) = E_n(d_0 \ldots d_j) \tag{4}$$

$$E(d_0 \ldots d_j) = \sum_{s=0}^{j} P(d_s) + I_d. \tag{5}$$

3.3 Disruption Game

In this stage, the attackers will implement attacks that can affect the system in a visual style. According to the attack effect, attacks in the CBTC systems can be divided into three categories: data leakage, denial of service (DoS), and data tampering. In this paper, we consider one type of data tampering attacks: the MA biased attack. The attack scenario is shown in Fig. 3.

The attackers intercept the communication between ZC and VOBC, and add a random offset X to the origin MA, both the efficiency and safety can be affected by this attack. There is no built defense, the "fail-safe" principle only guarantee safety while failure exists, but cannot be triggered by biased attack. Here, we design a simplified sequence anomaly detector F whose input is the MA sequence VOBC received $\{MA_{k-n}, MA_{k-n+1}, \ldots, MA_k\}$, and output the predicted attack offset X' as well as the attacking probability p. Based on the detection result, we design a operation level switch mechanism, where the defenders chose to switch between the CBTC level and the intermittent ATP level (IATP). The timing of switch is guided by

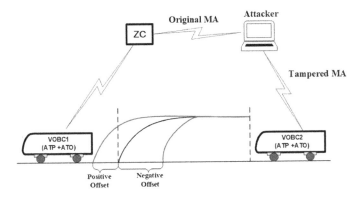

Fig. 3 MA biased attack scenario

Table 3 Defenders' payoff functions in disruption game

	Defense actions	Observation X'
Negative X'	Remain CBTC level	$\left(\frac{X'}{S_{z1}}, 0\right) \times (W_e, W_S)$
	Switch to IATP level	$\left(\frac{S_{z2}-S_{z1}}{S_{z1}}, 0\right) \times (W_e, W_S)$
Positive X'	Remain CBTC level	$\left(0, \frac{X'}{S_{p1}}\right) \times (W_e, W_S)$
	Switch to IATP level	$\left(\frac{S_{z2}-S_{z1}}{S_{z1}}, 1\right) \times (W_e, W_S)$

the game result. It should be noted that the attack offset can be positive or negative, so the game payoff matrix are as appropriate. The defenders' payoffs are indicated in Table 3 where S_{z1} and S_{z2} are the optimal tracking distance in CBTC and IATP level, respectively, and S_{p1} is the safety margin of trains in CBTC level. (W_e, W_S) is the weigh vector of efficiency and safety.

4 Model Solution and Result Analyze

As shown in Fig. 2, the equilibrium calculation of penetration stage takes both the current stage game parameters and the equilibrium of disruption game into consideration, and it is the same for the intrusion stage, so the result calculation will be in a backward manner.

Disruption game. First, we determine some parameters needed. We consider the situation where the train's emergency brake acceleration is 1 m/s² and the top speed is 70 km/h and one block is 300 m, so the theoretical optimal tracking distance for CBTC level and IATP level are 190 m and 300 m, respectively, and the weight vector is (0.1, 0.9).

When the detector gives negative offset prediction, the switching point can be determined by $\min\left\{\frac{X'}{S_{z1}} \times W_e, \frac{S_{z2}-S_{z1}}{S_{z1}} \times W_e\right\}$, so until the detector outputs estimation smaller than -110 m, the defender will choose to remain CBTC level. If the detector estimation is positive, the switch point can be determined by $\min\left\{\frac{X'}{S_{p1}} \times W_s, \frac{S_{z2}-S_{z1}}{S_{z1}} \times W_e+W_s\right\}$, and the result indicates that once the detector outputs estimation larger than 8.5 m, system will switch to IATP level to maintain relatively high level of safety.

Penetration game. Because the disruption stage only has one attack method in our case, the equilibrium calculation of penetration stages can be simplified by ignoring the effect of the last stage game. We use the penetration start from ATS subnet as an example to illustrate the equilibrium calculation. We set mix strategy equilibrium meaning that the attackers choose $A10$ and $A11$ with certain probability distribution, and defenders choose $D8$ and $D9$ with certain probability distribution, we can calculate the equilibrium by applying Eqs. (6), (7), finally, we get equilibrium ((0.58,0.42),

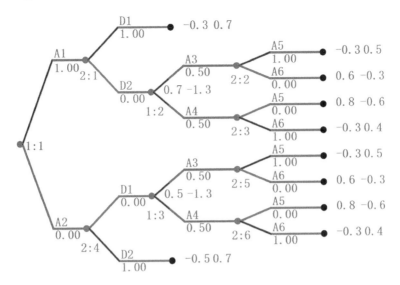

Fig. 4 intrusion game

(0.4,0.6)), and the game value (0, −0.31) which will be passed to the intrusion game.

$$C_{A10} \times q_8 + (E_{A10} - C_{A10}) \times q_9 = (E_{A11} - C_{A11}) \times q_8 + C_{A11} \times q_9 \quad (6)$$

$$(E_{A10} - C_{D8}) \times p_{10} + (-E_{A11} - C_{D8}) \times p_{11}$$
$$= (-E_{A10} - C_{D9}) \times p_{10} + (E_{A11} - C_{D9}) \times p_{11} \quad (7)$$

Intrusion game. Since this is a Bayesian game, defenders do not know against whom to play with, so the role information λ is a random choice; and in our case, we set $\lambda = 0.3$, and we plug in the game value obtained in penetration stage and calculate the Bayesian Nash equilibrium with the software Gambit, the game tree structure is shown in Fig. 4, and the equilibrium obtained is ({(0.4068, 0.5932), (0.5, 0.5)}, (0, 0, 0.2609, 0.7391)).

5 Conclusion

In this paper, we proposed a multi-stage game model based on the detailed analyzation of the attack–defense scenario in the CBTC systems to guide the defense deployment. We divided the information attack in CBTC into three unequal stages: intrusion, penetration, and disruption. We selected different kinds of game model according to the distinct attack scenes, and especially we designed a game-based operation level switch mechanism to defend the data tampering attack in disruption stage. We adopted a backward style equilibrium calculation method which passes the impact of subsequent stages to the pre-order ones. We evaluated the model proposed by analyzation and simulation. The result proved that the multi-stage game model is effective in directing proactive defense.

For future work, we plan to construct a more adequate and reasonable attack classification system for CBTC and polish the multi-stage game model furthermore.

Acknowledgements This paper was supported by grants from the National Natural Science Foundation of China (No. 61603031), Beijing Natural Science Foundation (No. L181004), and projects (No. I19L00090, I18JB00110), and Beijing Laboratory for Urban Mass Transit.

References

1. Farwell JP, Rohozinski R (2011) Stuxnet and the future of cyber war. Survival 53(1):23–40
2. Huang L, Zhu Q (2018) Adaptive strategic cyber defense for advanced persistent threats in critical infrastructure networks
3. Hu P, Li H, Fu H, Cansever D, Mohapatra P (2015) Dynamic defense strategy against advanced persistent threat with insiders. In: Computer communications. IEEE
4. Bao Z (2017) Research on active defense of information security risk in train control system. Doctoral dissertation (in Chinese)
5. Peng Y (2018) Research on detection and defense of urban rail CBTC system man-in-the-middle attack. Doctoral dissertation (in Chinese)
6. Li L (2018) Research of the penetration test on the security of the train control system. Doctoral dissertation (in Chinese)
7. Yu S (2017) Research on information security detection technology of urban rail transit control system. Doctoral dissertation (in Chinese)
8. Luo Y, Szidarovszky F, Alnashif Y, Hariri S (2009) A game theory based risk and impact analysis method for intrusion defense systems. In: IEEE/ACS international conference on computer systems & applications. IEEE
9. Orojloo H, Azgomi MA (2017) A game-theoretic approach to model and quantify the security of cyber-physical systems. Comput Ind 88:44–57
10. Zhu Q, Rass S (2018) On multi-phase and multi-stage game-theoretic modeling of advanced persistent threats. IEEE Access (99):1–1

Design and Development of a General Equilibrium Calculation System for Rail Vehicles

Yuwen Liu, Fengping Yang, Lin Jin, Feng Liu and Jian Wu

Abstract Train safety is greatly affected by the vehicle equilibrium calculation. Firstly, this paper discusses the physical structure of the vehicle and analyzes the derivative process of calculation formulas, which are deduced by force balance and moment balance. Secondly, based on the software structure design of three-tier architecture, the detailed requirement analysis and software design of vehicle equilibrium calculation system are carried out. Finally, the engineering verification is carried out with the actual data of vehicle, and the difference between the calculated results and the actual measurement results is within a reasonable range. The system can realize mass equilibrium calculation of rail vehicles, manage vehicle data effectively, and improve the efficiency of actual calculation. The software system has already been applied in a rail vehicle designing department of a company and will be a good value of engineering application.

Keywords Rail vehicles · Equilibrium calculation · Three-tier structure · System design

1 Introduction

In order to alleviate the traffic congestion caused by urbanization, the demand of multiple vehicles, such as electric multiple units (EMU) and urban rail vehicles, is increasing, which inevitably requires to ensure the safety and efficiency of vehicle

Y. Liu · J. Wu
CRRC Qingdao Sifang Co., Ltd., Qingdao 266111, Shangdong, China
e-mail: sf-liuyuwen@cqsf.com

F. Yang (✉) · F. Liu
School of Electrical and Automation Engineering, East China Jiaotong University, Nanchang 330013, China
e-mail: yangfp@ecjtu.edu.cn

L. Jin
Zhuzhou CRRC Electric Co., Ltd., Zhuzhou 412005, Hunan, China
e-mail: jin7223309@163.com

design. From the perspective of vehicle body structure, under certain conditions, the greater the axle load of a vehicle is, the greater the carrying capacity of a vehicle is [1]. However, the larger the axle load of a vehicle is, the greater imbalance of the axle load is. When the imbalance of the axle load becomes higher, it will affect the traction and braking performance of the vehicle, thus endangering the safety of the vehicle [2]. Taking the EMU as an example, the EMU is composed of bogies, vehicle body, and a lot of equipments hanged on the vehicle. The weight of the whole vehicle is all sustained by two bogies. For the manufacture of the EMU, all the weight is evenly distributed on the four shafts of two bogies. However, it is impossible to achieve the full equilibrium distribution of vehicle weight as a result of manufacturing errors and other factors. The balance of the vehicle is required so as to achieve a reasonable weight distribution. Because of so many different types of rail vehicles and different methods of vehicle calculation, it is inefficient to meet the design demand of vehicles by conducting a large number of manual calculations. Document [3] introduces a quality management tool and a computing tool for vehicle equilibrium calculation under the Pro/E environment. The quality management is used to input the information of parts and components, and the calculation results are displayed in the equilibrium calculation interface. This method increases the calculation efficiency, but it needs to establish a large number of models. Since the process of modeling is tedious, it is not conducive to massive calculations. Based on LabVIEW and MATLAB, document [4] adopts mix-programming and uses a graphical software as the computing interface to achieve the overall import of data and parameters modified, which provides multiple inputting windows to meet the import of different types of data. The results are displayed on the interface after calling MATLAB computing scripts. This method solves the problem of batch calculation. Document [5] solidifies the calculation formula in Excel and establishes workbooks for inputted data and calculation result, with calculation results displaying in real time. This method can modify the inputted data in real time without additional software installed, but it only supports output format of Excel. All of the above documents take one type of vehicle as an analytical object to deduce the calculation formula and realize the equilibrium calculation. These methods can only calculate a single type of vehicle and thus are unable to achieve the overall calculation of vehicles for multiple types and multiple structures in the same system.

In view of the above problems, this paper firstly introduces the formula deduction process of rail vehicle. Secondly, it describes the analyzing and designing procedure of a set of application software for vehicle equilibrium calculation, which can satisfy the equilibrium calculation of the EMU, the metro, and other different types of a tram. Finally, the practicability of the software is verified by comparing the software calculation results with the actual measured data.

2 Procedure of Equilibrium Calculation

The components of the vehicle, as well as mounted equipment, include the equipment under the vehicle, the outside equipment, the vehicle body, the inside equipment, the electric equipment, and the wiring in the vehicle. A large deal of equipment is distributed in the three-dimensional space of the vehicle body, which requires a uniform distribution of weight on the axle.

2.1 Deduction Process of the Calculation Formula

First, the vehicle is divided into the upper part and the lower part, and the weight of the two parts is calculated, respectively. The upper part includes the vehicle body, the inside and outside installation equipments, the undercarriage suspended equipment, and passengers. The other part includes the bogie and its auxiliary equipment. The equilibrium calculation of the vehicle is divided into three steps:

(1) A coordinate system for each vehicle is established, which determines the center of gravity coordinates for each component. Then, the center of gravity coordinates in each vehicle body are calculated.
(2) The force analysis of the secondary springs under each vehicle is carried out. Then, a single bogie is used as a model to carry out the force analysis. Finally, the axle load under each bogie is calculated.
(3) The corresponding wheel load is calculated according to axle load and distribution of vehicles' center of gravity [6, 7].

2.2 Calculation Formula of Tram

2.2.1 Center of Gravity Calculation

On the empty vehicle condition, the total weight of the upper part of the vehicle is equal to the sum of the parts' mass,

$$M_{sum} = \sum m_i \tag{1}$$

Among them, m_i is the weight of a single component, M_{sum} is the total weight (excluding bogies) on an empty vehicle; the unit is kg, the center of gravity coordinate is (X_e, Y_e, Z_e), and take the X-axis gravity center as an example,

$$X_e = \frac{\sum W_{xi}}{M_{sum}}. \tag{2}$$

2.2.2 Calculation of Secondary Springs

Taking an articulated tram as an example, in a no-articulated vehicle, based on the formula of force balance and torque balance, the force analysis of a tram is shown in Fig. 1.

Among them, F_1 and F_2 represent pressure of the two line springs, X_i represents the center of gravity on the X-axis, F_g represents the gravity applied by the vehicle, L represents the fixed distance of the vehicle, and a represents the longitudinal span of the two line springs of the articulated bogie,

$$F_1 = \left(\frac{L + 2X_i - a}{2L - a}\right) F_g \tag{3}$$

$$F_2 = \left(\frac{L - 2X_i}{2L - a}\right) F_g \tag{4}$$

Similarly, in articulated vehicles, force analysis is shown in Fig. 2. The following formulas can also be obtained.

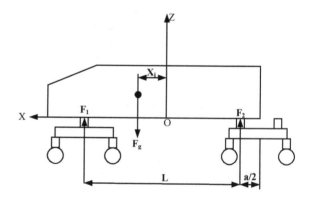

Fig. 1 Stress analysis over secondary spring (no-articulated vehicle)

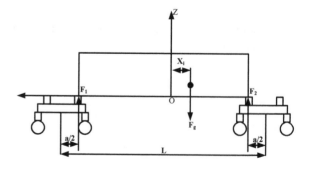

Fig. 2 Stress analysis over secondary spring (articulated vehicle)

$$F_1 = \left(0.5 + \frac{X_i}{L - a}\right) F_g \tag{5}$$

$$F_2 = \left(0.5 - \frac{X_i}{L - a}\right) F_g. \tag{6}$$

2.2.3 Axle Load

In a no-articulated vehicle, the force analysis of a no-articulated bogie is carried out separately, as shown in Fig. 3.

F_{a1} and F_{a2} represent the first and second axes axle load, F_{b1} is the self-gravity of the bogie, F_1 is the pressure on the two line springs applied by articulated vehicle upper part, X_z is the center of gravity of the no-articulated bogie, and L_1 represents the axle distance of the articulated bogie. The calculation formulas of the axle load are as follows,

$$F_{a1} = \frac{F_1}{2} + \frac{L_1 + 2X_z}{2L_1} F_{b1} \tag{7}$$

$$F_{a2} = \frac{F_1}{2} + \frac{L_1 - 2X_z}{2L_1} F_{b1} \tag{8}$$

The axle load analysis of articulated bogies is shown in Fig. 4, and the following formulas can also be obtained,

$$F_{a3} = \left(0.5 + \frac{X'_z}{L_2}\right) F_{b2} + \left(0.5 + \frac{a}{2L_2}\right) F_2 + \left(0.5 - \frac{a}{2L_2}\right) F'_2 \tag{9}$$

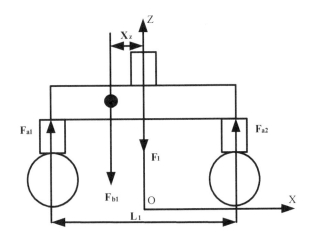

Fig. 3 Stress analysis of bogie (no-articulated bogie)

Fig. 4 Axle load analysis of articulated bogie

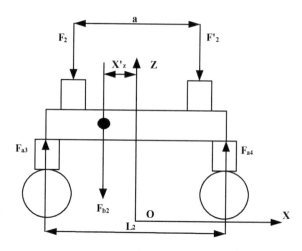

$$F_{a4} = \left(0.5 - \frac{X'_z}{L_2}\right)F_{b2} + \left(0.5 - \frac{a}{2L_2}\right)F_2 + \left(0.5 + \frac{a}{2L_2}\right)F'_2 \quad (10)$$

F_{a3} and F_{a4} represent the axle load of the third and fourth axes, F_{b2} is the self-gravity of the bogie, F_2 and F'_2 are the pressure on the two springs applied by articulated vehicle upper part, and X'_z is the center of gravity of the articulated bogie.

2.2.4 Wheel Load

The force balance and moment balance analysis of each pair of wheelsets is carried out to derive the following formulas,

$$F_{wdl} = F_a * \left(0.5 + \frac{Y_i}{dg}\right) \quad (11)$$

$$F_{wdr} = F_a * \left(0.5 - \frac{Y_i}{dg}\right) \quad (12)$$

F_{wdl} is the left wheel load, F_{wdr} is the right wheel load, Y_i is the distance between the center of gravity and the Y-axis of the reference system, and F_a is the axle load. For the standard gauge of the 1435 mm, dg represents the distance between the wheel and the rail contact point on the two sides of the wheel, equal to the 1500 mm.

3 Analysis and Design of the Software

3.1 Software Requirement Analysis

The goal of the vehicle equilibrium calculation system is to design a set of standardization process to achieve the weight balance of the rail vehicle. While ensuring the accuracy of the calculation results, it can store all related data of the project and finally form a large database that can be referred to and called for future projects. The specific use case diagram of the system is shown in Fig. 5.

(1) Input data: In order to meet various requirements, three kinds of data-input methods are designed: manual input, Excel input, and historical database called. Then, many data values, such as basic data of the vehicle, the weight of each component, the coordinate, and moment of inertia, are introduced into the database. Manual input is mainly used for inputting basic parameters, such as length, width, height, and manual operation of parts. Excel input and historical database called are used for simultaneous inputting of vehicle parts data and increasing computing efficiency.

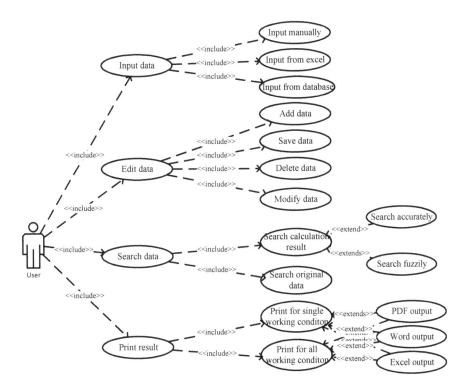

Fig. 5 Software use case diagram

(2) Edit data: A new equilibrium calculation can save the input data and modify or delete the calculation results when they do not meet the requirements.
(3) Search data: The query function is designed on the main interface, and the historical calculation results can be viewed and searched by keyword searching.
(4) Print result: According to the customer's demand, crystal report is used to customize the content and typeset of the output results. Three formats of output are provided: Word, Excel, and PDF.

3.2 Software Detailed Design

3.2.1 Software Architecture Design

The overall architecture of the system is based on the three-tier structure under the *C/S* mode. The system is designed with flexible software architecture. It ensures the high cohesion and low coupling characteristics of the system while it meets the functional requirements of the system and reduces the difficulty of design and development. The architecture design of the equilibrium calculation system is shown in Fig. 6.

(1) Database access layer (DAL)

The database access layer includes the balanced-number information table, the vehicle parameter information table, the working condition information table, the component classification table, the calculation result table, and the information table of each model part.

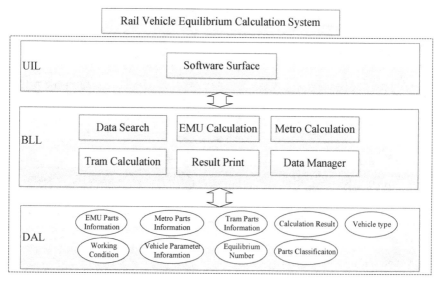

Fig. 6 Software architecture

(2) Business logic layer (BLL)

The business logic layer includes the function modules, such as equilibrium calculation, report output, information query, and the modification of calculation results. The layer is between the user interface layer and the database access layer. The database access layer is used to access the business data through the data operation object of the data access layer, and the data support is provided for user interface layer [8].

(3) User interface layer (UIL)

The user interface layer is located at the outermost level, which is the nearest to users. It is used to display data, receive data from users, and provide users with an interactive operation interface. It mainly completes client view display and data validation, including displaying data in a user-friendly interface, validating the integrity and effectiveness of the inputted data, and saving the data in a specified format [9].

3.2.2 Software Procedure Design

The activity diagram is a kind of Unified Modeling Language (UML) dynamic model. It clarifies the workflow of the business use case, as shown in Fig. 7.

(1) Firstly, the new vehicle equilibrium calculation project is set up when a user opens the interface. Then, the user inputs the vehicles' basic information and clicks save button.
(2) User can choose among three input ways to input the basic parameters of the vehicle and then click the save button. A large number of repetitive input tasks can be avoided by historical database called. It can reduce data redundancy and save storage space.
(3) When the system recognizes that all data is inputted, the user can click the save button. If the data were saved successfully, clicking the calculation button, the system will calculate the result and display it in the result interface. Otherwise, the data must be inputted again. If the result meets the requirement, the crystal report will be called to print. Otherwise, the system will go back to the previous layer to modify the input data until the calculation result meets the requirement.
(4) When users want to search data, they can enter the keywords for fuzzy query or accurate query. If the results do not meet the requirement, click the modification button to recalculate.

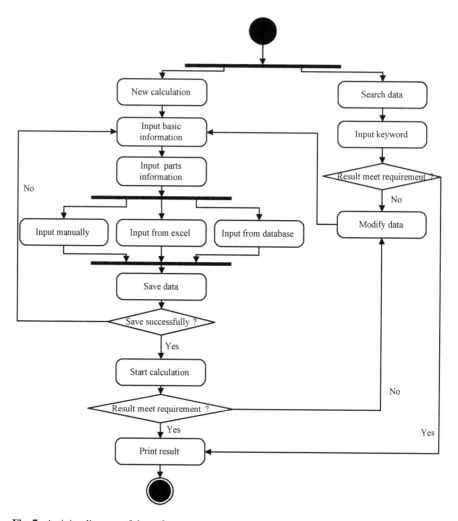

Fig. 7 Activity diagram of the software

4 Software Implementation and Result Verification

4.1 Software Implementation

Take EMU as an example, in the EMU management interface, new computation or query for historical computation can be created, as shown in Fig. 8.

After inputting the basic parameters of the vehicle, the user can choose among three input modes to input the vehicle components: manual input, Excel input, and historical database called. This is selected from the Excel input, and the interface is shown in Fig. 9.

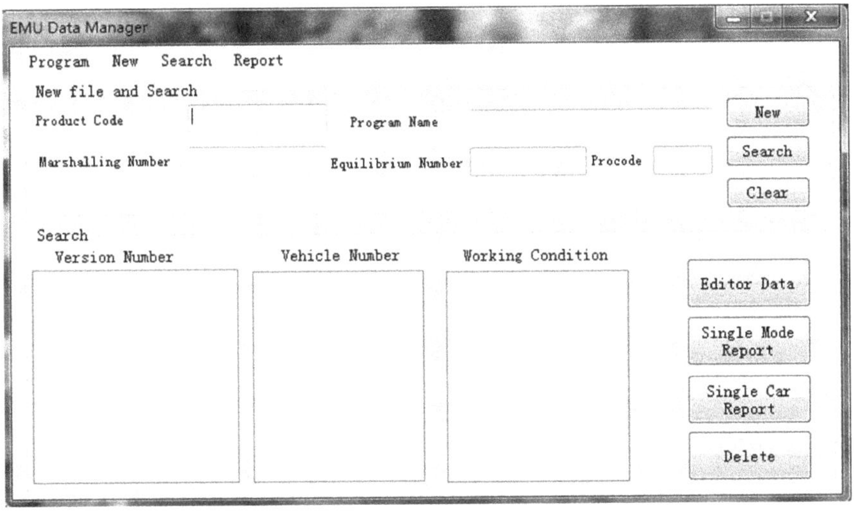

Design and Development of a General Equilibrium Calculation ... 235

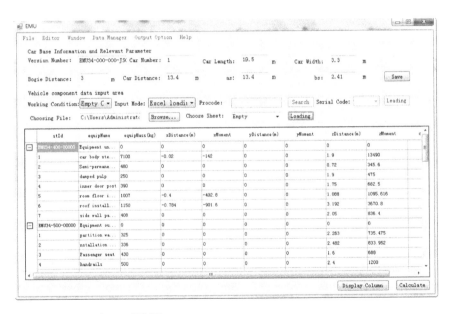

Fig. 8 Main interface of EMU

Fig. 9 Component input of EMU

After clicking "calculate" button, the user will get the calculation results, which is shown in Fig. 10. The maximum axle load difference is 0.09%, and the maximum wheel load difference is 1.33%, which meets the technical standard.

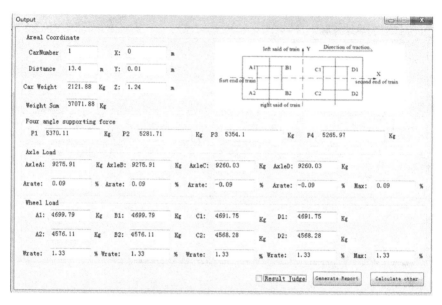

Fig. 10 Calculation results of EMU

4.2 Result Verification

In terms of data management, the system stores the accumulated original data and the calculation results for each calculation. Therefore, it forms a large database, which can be referred to and called by subsequent projects. It will not only achieve an orderly and reliable historical data but also avoid the waste of data resources and improve the utilization rate of the original data. In terms of computing efficiency, the system supports calculations of different vehicles at the same time. The rich Structured Query Language (SQL) statements greatly increase the speed of each calculation and shorten the design period.

In order to verify the correctness of the results calculated by software, the calculation results of four axle-loads and the calculation results of wheel load on the two sides in the same bogie are analyzed and compared with the actual measured data of axle load and wheel load, which are shown in Figs. 11 and 12. Under the same original data, the following figure is a column contrast diagram of the axle load and wheel load, which are, respectively, calculated by software and actually measured.

After analyzing and comparing with the measured results, the maximum deviation rate between the data calculated by the software and the measured data is only − 0.06%. Considering the deviation of component manufacturing, the accumulation of measurement error, and the difference of equipment installation accuracy, the result obtained from the software calculation is coinciding with the actual measurement result. It shows that the calculated result is accurate and the software system has a good application prospect.

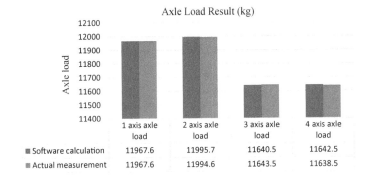

Fig. 11 Comparison diagram of axle load results

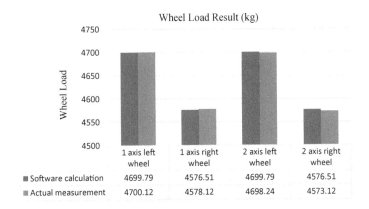

Fig. 12 Comparison diagram of wheel load results

5 Conclusion

Aiming at the problems of low-efficiency and non-standard equilibrium calculation of railway vehicle, a general software system for vehicle equilibrium calculation of EMU, metro, and tram is designed in this paper. Through force balance and moment balance, the force analysis of rail vehicles is carried out, and the relevant calculation formulas are deduced. Using three-tier architecture as the software architecture of the system, the system design and detailed design of the software are completed. Through practical engineering verification, the difference between the software calculated results and the actual weighing results is within a reasonable range, which meets the requirements of engineering design. The system supports data transfer between different vehicles and data output in different formats. It can greatly improve the efficiency of designers and meet the design requirements of different vehicles under different working conditions.

References

1. Yin Ren FA (2013) Research on type selection of rail transit vehicle in metropolitan area. Railway Stand Des 04:135–139 (in Chinese)
2. Fan JF (2015) Design of Guangzhou rail transit line 3 north project vehicle. Electr Locomot Mass Transit Veh 38(01):31–36 (in Chinese)
3. Liao Y, Yu J (2012) The development of rail vehicle design weight equilibrium calculation tool. Mod Manuf Technol Equip 06:14–15 (in Chinese)
4. Wang X, Yue Y, Lei ZW, Liu Y (2013) Weight calculation and adjustment of CG of metro vehicle based on LabVIEW. Electr Locomot Mass Transit Veh 36(02):36–39 (in Chinese)
5. Wang Y, Xie H, Lin W, Xia D, Lv Z (2017) Calculation method of axle load and wheel load for articulated EMUs. Electr Locomot Mass Transit Veh 40(01):39–42+64 (in Chinese)
6. Wang W, Yan Y (2006) Discussion of the calculation of locomotive's weight, axle load and wheel load. Electr Locomot Mass Transit Veh (02):67–69 (in Chinese)
7. Yasin Y, Gunay A (2008) Investigation of the effect of counterweight configuration on main bearing load and crankshaft bending stress. Adv Eng Softw 40(2)
8. Tan Q (2014) Server database performance optimization research. Silicon Valley 7(08):54+48 (in Chinese)
9. Xiao YL, Jian Z (2015) Software quality evaluation for vehicle based on set pair theory. Appl Mech Mater 3785(738) (in Chinese)

Research on Vehicle Number Localization of Urban Rail Vehicle Based on Edge-Enhanced MSER

Ruifeng Xu, Jiandong Bao, Shuangyan Yang, Chen Zhang and Zongyi Xing

Abstract In order to reduce the impact of shooting angle, distance and motion blur on vehicle number localization, and improve the accuracy of the localization in complex environments, a new vehicle number localization method based on edge-enhanced maximally stable extremal region (MSER) is proposed. This method first performs edge enhancement on the vehicle number image and extracts the character candidate region in the edge-enhanced image by using MSER. Then, the candidate region is filtered according to the vehicle number character features using the heuristic rule and the stroke width rule, and the remaining candidate regions are merged. The final step is completed by combining the geometric features of the vehicle number. The experimental results show that the comprehensive localization accuracy of this method is as high as 98.50%. High accuracy and robustness can be achieved by using this method in vehicle number localization with images acquired in complicated scenes with various shooting distances and angles.

Keywords Edge-enhanced maximally stable extremal regions · Stroke width · Vehicle number localization

1 Introduction

With the development of train maintenance automation, the installation process of vehicle identification equipment based on radio frequency identification technology is complicated, and the problems of electromagnetic interference are becoming more and more prominent. The identification technology based on image processing has become a research hotspot. This technology consists of two steps: vehicle number localization and vehicle number identification. Commonly used urban rail vehicle number localization methods are mainly based on vehicle number color features [1, 2], texture features [1], edge information [2], transform domain analysis [3], and morphological processing [4]. Commonly, localization of vehicle number

R. Xu · J. Bao · S. Yang · C. Zhang · Z. Xing (✉)
Nanjing University of Science and Technology, Nanjing 210094, China
e-mail: xingzongyi@163.com

images in specific scenes with certain shooting angle and shooting distance can be achieved. However, there are fewer effective localization methods for the vehicle number images in complex situations with multiple scenes, different shooting angles, and distances.

In this paper, an algorithm based on edge-enhanced MSER is proposed. Images of the vehicle number in complex situations with different scenes, shooting angles, and distances are taken as the object. Firstly, the affine invariance, strong stability, and the adaptability of illumination extract the appropriate vehicle number candidate region, which solves the MSER sensitivity problem to blurred image and improves the adaptability to the threshold. Then, the extracted MSER region is filtered according to the vehicle number character feature, and the stroke width of the filtered region is calculated, which is used to rescreen the candidate regions before they are aggregated, and the final vehicle number localization is completed by combining the vehicle number geometric features. The entire algorithm flow is shown in Fig. 1.

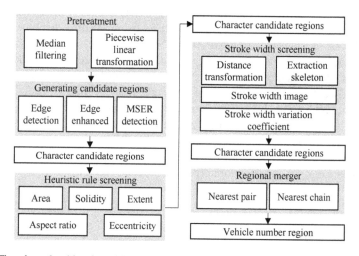

Fig. 1 Flowchart algorithm for vehicle number localization based on edge-enhanced MSER

Fig. 2 Original image

2 Vehicle Number Text Localization Algorithm

2.1 Vehicle Number Picture Preprocessing

Because of the high speed of the urban rail vehicle, the camera that captures its images is usually set with a short shutter time and a high gain, which leads to inevitable noise interference and uneven brightness in the images of the vehicle number. In order to reduce the noise interference, the original image taken is filtered by a 3 × 3 template. In order to reduce the influence of light and the unevenness caused by the camera, the contrast between the character region and the background region is enhanced, the vehicle number character region [5] is highlighted, and the image is processed by the piecewise linear gradation transformation as given in Eq. (1).

$$f(x) = \begin{cases} \frac{c}{a}x, 0 \leq x \leq a \\ \frac{d-c}{b-a}(x-a) + c, a \leq x \leq b \\ \frac{N-d}{M-b}(x-b) + d, b \leq x \leq M \end{cases} \quad (1)$$

where x is the gray value of the original image, and $f(x)$ is the gray value after the piecewise linear gray transformation. The gray level of the original input image is between 0 and M, while the gray level of the transformed image is between 0 and N. a, b, c, and d are the control points of the gradation transformation. After tests with a large number of pictures, the suitable values of highlighting vehicle number region and restraining non-vehicle number region are obtained. Figure 3 is a preprocessed image of the original image in Fig. 2, where it can be seen that the contrast of the vehicle number region and its background is increased in the processed image.

Fig. 3 Piecewise linear gray-scale transformation

2.2 Vehicle Number Character Candidate Region Detection

After preprocessing the original image, a gray image with better contrast has been obtained, and then, the candidate region of the vehicle number character is extracted by the maximally stable extremal regions (MSERs) algorithm [6]. The MSER algorithm was first proposed by Matas et al. in 2004, and its characteristics such as scale rotation invariance and affine invariance are considered to be the best region detection operators. The method takes the value of Δ in the gray interval of [0–255] and uses these values as the threshold to binarize the image. All the connected domains obtained in this process are extremal regions (ER). For a certain extreme region Q under the threshold i, it is assumed that the connected domain region of Q under the thresholds $i - \Delta$, i, and $i + \Delta$ are, respectively, $Q_{i-\Delta}$, Q_i, and $Q_{i+\Delta}$. If the region change rate $q(i) = |Q_{i+\Delta} - Q_{i-\Delta}|/Q_i$ when it is less than a certain threshold, the extreme value region Q is called the maximally stable extremal region.

The MSER extraction process shows that the algorithm is sensitive to image blur. When the image quality is low, a large number of non-text candidate regions are easily generated, resulting in a decrease in the accuracy of the results. Therefore, an edge-enhancement-based MSER algorithm is designed. The steps of the method are as follows:

1. Using the Canny edge detection operator to perform edge detection on the preprocessed image P to obtain an edge image E, as shown in Fig. 4;
2. Combining image P with image E to generate edge-enhanced image B, where $B = P - \lambda B$, $0 \leq \lambda \leq 1$, and normalize B to [0, 255], as shown in Fig. 5;
3. Using the MSER algorithm for the edge-enhanced image B, all the detected maximally stable extremal regions are used as candidate regions for the vehicle number characters, as shown in Fig. 6.

Figure 7 shows the MSER region extracted from the image of the vehicle number without edge enhancement. As can be seen from the comparison of Figs. 6 and 7, the MSER region with edge enhancement is increased, forming many new MSER regions with a large number of enhanced edges. However, the cases where the vehicle number character region is not extracted due to edge blurring are also reduced.

Fig. 4 Edge detection map

Fig. 5 Edge-enhancement map

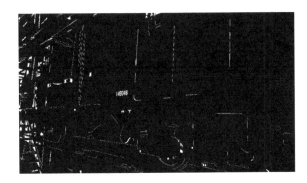

Fig. 6 MSER diagram of edge-enhanced image

Fig. 7 Original image of MSER

2.3 Vehicle Number Character Region Filtering Based on Heuristic Rules

After detection of the vehicle number images using the edge-enhanced MSER, there are still a large number of non-character regions in the candidate region. In order to get the exact position of the vehicle number character, a preliminary screening is needed to remove candidate regions that are clearly not characters. In this paper, the region of the candidate connected region, the aspect ratio (the aspect ratio of the

Fig. 8 MSER diagram

circumscribed rectangle of the connected region), the extent (the ratio of the region of the connected region to the region of the circumscribed rectangular region), the solidity (the minimum region of the connected region and its external minimum convex polygon), the eccentricity (the eccentricity of the ellipse having the same standard second-order central moment to the connected region), and other features are used to filter out regions that are clearly not the vehicle number characters. The detected MSER region is drawn in the vehicle number picture. As shown in Fig. 7, after filtering the vehicle number using heuristic rules, the remaining MSER regions are as shown in Fig. 8. It can be seen that in this step, a large number of MSER regions formed by the edges from the edge-enhancement step are also filtered out.

2.4 Vehicle Number Character Region Filtering Based on Stroke Width

It can be seen from Fig. 9 that there are still many non-character regions in the candidate region after the heuristic rule filtering, so it is necessary to filter out some non-character regions according to the characteristics of the vehicle number characters themselves. It can be seen from the analysis that the vehicle number characters

Fig. 9 Heuristic rule filter

generally have similar stroke widths, so the stroke width of the candidate region can be obtained, and the non-character regions can be filtered out according to the mean value and variance of the stroke width of the candidate region.

Epshtein et al. [7] propose the concept of stroke width transformation (SWT) based on that texts in adjacent regions usually have roughly equal stroke widths, which has been applied by many scholars in the field of text localization. Since the vehicle number character itself is a set of 6 characters, the stroke width is roughly similar, and in this paper, a simpler algorithm for obtaining the stroke width is designed. The steps of this algorithm are as follows:

1. In order to prevent the boundary effect, a circle of background edges is added to the candidate region image R (Fig. 10a) to obtain a region image Re (Fig. 10b);
2. Performing Euclidean distance transformation on the image Re to obtain a distance image Rd (Fig. 10c);
3. Extract the skeleton image Rs (Fig. 10d) of the image Re using the Zhang–Suen refinement algorithm;
4. Using the skeleton image Rs as a mask, perform an AND operation on the distance image Rd, obtaining a result as the stroke width image Rw (Fig. 10e);
5. Calculate the average value μ and the standard deviation σ of all foreground pixel points in the stroke width image Rw as the average and standard deviation of the stroke width of the candidate region.

After obtaining the stroke width values of all candidate regions, the connected domains are filtered according to the stroke width characteristics of the vehicle number characters, as shown in Figs. 11 and 12. For the same scene, shooting angle, and distance of the same vehicle number picture, the mean and standard deviation of the stroke width is a good filtering standard, while for different situations, the coefficient of variation of the stroke width, as the standard deviation of the stroke width divided by the mean, is a better filtering criterion. The candidate region whose stroke width variation coefficient is larger than the threshold value is filtered out, and the threshold size is related to the font number printed by the vehicle number and is obtained according to data of a large number of regions during the application process.

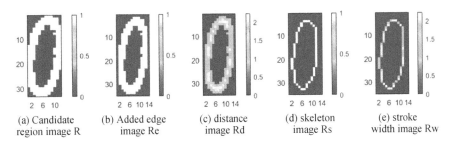

(a) Candidate region image R
(b) Added edge image Re
(c) distance image Rd
(d) skeleton image Rs
(e) stroke width image Rw

Fig. 10 Calculation of stroke width

Fig. 11 Stroke width filter

Fig. 12 Vehicle number region map after stroke width filtering

2.5 Vehicle Number Character Region Merge

After the heuristic rule and the stroke width are filtered, a single vehicle number character candidate region is obtained and aggregated, and the final vehicle number is located on the aggregated region according to the geometric feature of the vehicle number. The steps for vehicle number localization designed in this paper are as follows:

1. For all remaining vehicle number character candidate regions, obtain the height, width, and center point coordinates of the circumscribed rectangle. For any two connected regions i and j, assume that the width of the circumscribed rectangle is w_i and w_j, the height is h_i and h_j, and the coordinates of the center point are (x_i, y_i) and (x_j, y_j), if Eq. (2) is satisfied:

$$\begin{cases} 0.6 \leq \frac{h_i}{h_j} \leq 1.6 \\ |x_i - x_j| \leq 1.2 \max(w_i, w_j) \\ |y_i - y_j| \leq 0.3 \max(h_i, h_j) \end{cases} \quad (2)$$

Then, the connected regions i and j form a nearest pair [8];

Fig. 13 Final localization number

2. The nearest neighbor pairs are aggregated and constructed into the nearest neighbor chain [8]. Assume that the number of character candidate regions in the nearest neighbor chain is N, and the aspect ratio of the circumscribed rectangle is r. If Eq. (3) is satisfied:

$$\begin{cases} N \geq 6 \\ 2 \leq r \leq 8 \end{cases} \quad (3)$$

It is considered that the region where the nearest neighbor chain is located is where the train number is located (Fig. 13).

3 Results

In order to verify the effectiveness of the algorithm, 600 vehicle number images were collected using the equipment installed in the depot. Gallery 1 and Gallery 4 were shot right in front of the vehicle number region. Gallery 2 and Gallery 3 were offset by 30° and 60° shooting, respectively (Fig. 14). The Galleries 1, 2, and 3 were shot in the Guangzhou Metro, and the Gallery 4 was shot in the Nanjing Metro, which has a more serious uneven illumination problem. Each gallery has 150 images. In order to verify the validity and robustness of the location detection method, Algorithm 1 (fixed color search and morphology-based vehicle number localization algorithm [4]) and Algorithm 2 (vehicle number localization algorithm based on maximally stable extremal region) are used [9]. The test results of the vehicle number localization test on the four galleries on the MATLAB platform are given in Table 1.

It can be seen from the experimental results that Algorithm 1 has lower localization accuracy for the pictures in the case of uneven illumination in the library 4; Algorithm 2 has a gradual decrease in the localization accuracy for the pictures with large shooting angles; the impact is small; and the accuracy of localization in complex environments is still as high as 98.50%, which is highly robust.

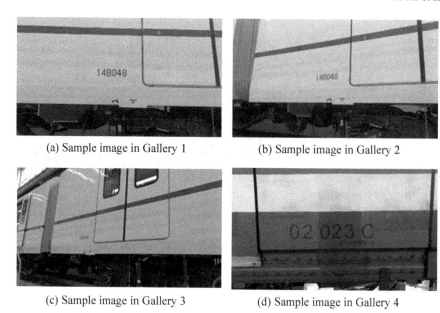

Fig. 14 Sample image in gallery

Table 1 Comparison of vehicle number localization algorithms

Algorithm	Gallery 1 localization accuracy (%)	Gallery 2 localization accuracy (%)	Gallery 3 localization accuracy (%)	Gallery 4 localization accuracy (%)	Total localization rate (%)	Localization time (ms)
Algorithm 1	97.33	95.33	96.67	85.33	93.67	367
Algorithm 2	96.67	94.67	91.33	92.67	93.83	244
Algorithm in this paper	99.33	98.67	98.67	97.33	98.50	497

4 Summary

In this paper, we analyze the difficulties in the localization of urban rail vehicle numbers and propose a series of solutions:

- Preprocessing the vehicle number image using median filtering and piecewise linear gray-scale transformation, reducing noise points in the vehicle number picture, and increasing the contrast between the vehicle number character region and the background;

- Using edge-enhanced MSER algorithm to extract the candidate region of the vehicle number characters, which reduces the problems of motion blur and uneven illumination;
- Using heuristic rules and stroke width rules to filter out most of the non-vehicle number characters improving the accuracy of the detecting;
- Using the vehicle number character region merging method based on the nearest pair and the nearest neighbor chain, significantly increasing the robustness of the entire localization algorithm.

Experiments show that the proposed algorithm remains high accuracy and robustness in vehicle number localization in complex situations with various scenes, shooting distances, and angles. The introduced algorithm can meet the needs of practical applications with a comprehensive accuracy of localization as high as 98.5%.

Acknowledgements This work is supported by National Key R&D Program of China (2016YFB1200402).

References

1. Wan Y, Xu QY, Huang MM (2013) On license plate location in complex background based on texture and colour. Comput Appl Softw 30(10):259–262 (in Chinese)
2. Li L Q, Peng JY, Feng XY (2012) New approach for precise license plate locating based on edge analysis and color statistics. Appl Res Comput 29(1), 336–339, 343 (in Chinese)
3. Rajput H, Som T (2015) An automated vehicle license plate recognition system. Comput Pract 48(8):56–61
4. Zhou L, Wang L, Zhou YM et al (2013) License plate location based on fixed color assortment search and morphology. J Highw Transp Res Dev 30(5):118–125 (in Chinese)
5. Cheng SH, Gao X, Cheng SC (2017) Moving vehicle detection based on computer vision. Acta Metrol Sin 38(3):288–291 (in Chinese)
6. Matas J, Chum O, Urban M et al (2004) Robust wide-baseline stereo from maximally stable extremal regions. Image Vis Comput 22(10):761–767
7. Epshtein B, Ofek E, Wexler Y (2010) Detecting text in natural scenes with stroke width transform. In: Computer vision and pattern recognition. IEEE, pp 2963–2970
8. Miao LG (2011) License plate detection algorithm based on nearest neighbor chains. Acta Autom Sin 37(10):1272–1280 (in Chinese)
9. Li B, Tian B, Yao Q et al (2012) A vehicle license plate recognition system based on analysis of maximally stable extremal regions. In: IEEE international conference on networking, sensing and control. IEEE, pp 399–404

Urban Rail Transit Power Monitoring System Techniques Based on Synchronous Phasor Measurement Unit

Shize Huang, Lingyu Yang, Ji Xue and Kai Yu

Abstract Failures in power systems of rail transit systems lead to the disruption of transportation, which causes significant economic losses and profound social effects. We consider the application of smart energy system to power fault monitoring of rail transit. The causes of these failures are often not immediately determined in the existing rail transit power monitoring system, and this situation will prolong the repair time. To analyze accidents and immediately determine faults in the rail transit power systems, the entire power data acquisition system of a rail transit is built on the basis of the existing power system. Data acquisition, processing, and computing of power system AC phasor are achieved through precision sampling of phasor measurement units and advanced algorithms. Furthermore, problems of decentralized power equipment and data transmission delay are solved through clock synchronization technology. At the same time, the shortage of power data acquisition system is effectively compensated by demonstration project, which is beneficial to the safe operation of power supply system of Shanghai urban rail transit. In this case, equipment failure of the rail transit power system, fault location, and repair time are reduced.

S. Huang
Shanghai Key Laboratory of Rail Infrastructure Durability and System Safety, Sichuan Railway Industry Investment Group Co., Ltd., Tongji University, 4800 Caoan Road, Shanghai, People's Republic of China
e-mail: hsz@tongji.edu.cn

L. Yang (✉)
Key Laboratory of Road and Traffic Engineering of Ministry of Education, Tongji University, 4800 Caoan Road, Shanghai, People's Republic of China
e-mail: 1572062220@qq.com

J. Xue
Shanghai Electrical Apparatus Research Institute, 505 Wuning Road, Shanghai, People's Republic of China
e-mail: xueji@vip.sina.com

K. Yu
China Railway Second Hospital Engineering Group Co., Ltd., No. 3, Tongjin Road, Chengdu, Sichuan, People's Republic of China
e-mail: ekyukai@qq.com

Keywords Phasor measurement unit (PMU) · Mass traffic · Power monitoring system

1 Introduction

Electrical power is the energy of urban rail transit systems. Once the rail power system fails, transportation is disrupted. In rail transit network operation, the fault line soon affects other relevant lines, and the situation affects ground traffic and causes traffic barriers in a large area. Thus, the urban rail transit power system should be safe, uninterrupted, and characterized by high quality. It is important to realize intelligent management of rail transit electric energy system.

Urban rail transit lines have adopted power monitoring systems. As the operating parameters of 10 kV power system for railway vary quickly and the transmission and processing of massive data are much tough, the SCADA system with synchronous real-time data transmission is developed [1]. The supervisory control and data acquisition (SCADA) focuses on operating, monitoring, and controlling the power supply load [2]. The implementation of SCADA and energy management systems took place from 2002 to 2006 in Indian electricity grid [3]. However, these smart grid and SCADA systems have efficiency constraints when it comes to the vast amount of computing resources with enormous amount of data [4]. And the current SCADA system is faced with more and more severe challenges in terms of data throughput capacity, scalability, reliability, real-time performance, and whatnot [5]. The new technique based on synchronized phasor measurement technology (PMU) with fast communication network and GPS system for time transfer has a dynamic adaptation for back-up protection of power system. It can provide effective monitoring tools for safe and stable operation of power system and can improve the monitoring and stable operational level of power system. Prediction of the power on a grid using PMU data shows great promise, and it will improve grid reliability and efficiency [6]. Gopakumar et al. [7] proposed a remote monitoring system for real-time detection and classification of transmission line faults in a power grid using PMU measurements. Karpe S. U. [8] speculated that PMU can be used to identify faults, protect power system, and eliminate faults. Oleinikova I et al. [6] provides the possibility of phasor measurement unit implementation for the estimation of power transmission line parameters in order to solve problems related to transmission line temperature, sag, and clearance. Luigi Vanfretti et al. [9] took advantage of real-time data obtained from PMU devices for monitoring wind-farm-to-grid sub-synchronous dynamics. PMU is mainly applied in power stations, substations, and power grid of 500 and 220 kV. However, the railway is a 10 kV power system and the operating parameters change rapidly, and PMU cannot be directly applied to rail transit.

Although voltage and current monitoring is conducted by the power monitoring systems, this method cannot be used for microanalysis because it belongs to the average category used for macro-control. Thus, the existing power monitoring system is short of operation safety management system. Most alarms of existing power monitoring systems do not provide information on faulty and risky equipment. In addition, the SCADA is designed and managed in line and is unable to adapt the fault diagnosis and quick-fault point location requirements of the regional professional maintenance unit under network operation. Given the lack of information-sharing mechanisms, the SCADA not only has an effect on emergency measures but also prevents the implementation of contingency plans. By installing the synchronized PMU in the rail transit power system, the system not only collects AC data but also collects DC data in order to provide criterion when the power dispatchers diagnose power fault. Meanwhile, the real-time information can help determine the origin and development of various faults, help dispatchers, solve problems, identify the root cause of alarms, and achieve intelligent alarm.

The present study is based on the existing rail transit power monitoring system. The front-end data acquisition device is applied to obtain data for achieving intelligent fault location and fault warning alarm, as well as to transfer these data resources to the upper node to provide data support for upper applications. PMU is used to sample the current as well as voltage data for AC and DC circuits; in this setup, each measurement data time stamp requires microsecond precision. The system can achieve synchronous sampling of the voltage and current data of the distribution system, as well as solve the problems of data scattering, asynchronous sampling, and transmission delay. This study presents original data of intelligent fault location and fault alarm for the rail traffic power system maintenance unit, which facilitates real-time monitoring, diagnosis and evaluation, emergency response, and integrated dispatch throughout the system.

2 General Design Scheme

To conduct fault analysis and quickly determine the location of a failure in the rail power system, the new system based on the rail transport power system increases the measuring device to gather electricity data. A time delay occurs during data transfer because of the characteristics of distribution in rail transit power system measurement devices. To analyze the rail power system state and possible accidents, the system collects data synchronously and attaches a precise time stamp for accurately recovering the system state during centralized computing and processing.

To achieve the aforementioned functions, we must investigate high-precision measurement technology for accurate and reliable measurement data, study clock synchronization technology for the same time reference between dispersion measurements, and consider high-reliability and real-time network transmission technology for timely and accurate data transfer. Therefore, the PMU is used in this study to sample current, voltage, phase data of AC circuit and voltage, and current data of the DC circuit [9, 10]. To meet the requirements of high-precision time mark, the GPS clock system in each front-end data acquisition station (substation) is set up, a high-accuracy electric energy data measurement unit controlled by synchronous pulse is used, and then, the time stamp is marked at the same time that the data are

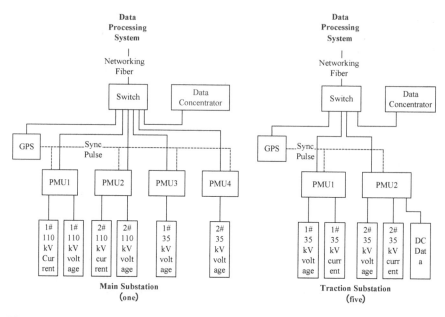

Fig. 1 System diagram

collected. The data concentrator (DC) is used to package the PMU data to constitute the entire rail power data acquisition system [11].

The system diagram is shown in Fig. 1.

The design program of the main substation consists of the following:

Data collection content

(1) Three-phase voltage of 110 kV side in the two main transformers, with a total of six AC voltage datasets.
(2) Three-phase voltage and current of 35 kV side in the two main transformers, with a total of six AC voltage and current datasets.

Signal source

(1) The three-phase voltage of 110 kV side in the main transformer is retrieved from the transformer voltage signal of the 126 kV GIS cabinet through micro-switches and fuses that lead to the PMU.
(2) The three-phase voltage of 35 kV side in the main transformer is retrieved from the transformer voltage signal of the 35 kV incoming cabinet through micro-switches and fuses that lead to the PMU.

The current transformer is installed separately inside the cable sandwich of the 35 kV main transformer, and the current signal leads to the remote PMU.

The design program of the traction substation consists of the following:

Data acquisition content

(1) Three-phase voltage and current of 35 kV in the two main transformers with a total of three AC voltage and six current datasets.
(2) DC voltage of a positive cabinet, DC current of two positive cabinets, DC current of four feeder cabinets, one DC voltage dataset, and six DC current datasets.

Signal source

(1) The three-phase voltage of 35 kV in the rectifier transformer is retrieved from transformer voltage signal of the incoming cabinet through micro-switches and fuses that lead to the PMU.
(2) The current transformer is installed separately inside the cable sandwich of the 35 kV rectifier transformer and the current signal leads to the remote PMU.
(3) DC data acquisition is achieved by installing an isolation amplifier on the shunt, and a signal is retrieved from the original signal.

To pack the data on the PMU and transfer them to the upper system, the data concentrator is placed on each master substation and traction substation.

3 Design of Phasor Measuring Device

3.1 Requirements and General Planning

In the power systems based on 50 Hz sinusoidal power, the vector of amplitude and phase can occur on behalf of all information of the basic signal. Thus, the phasor measurement indicates accurate measurement of amplitude and phase in the signal. To reflect all information included in the signal, the signal information is demarcated according to time and space, and a phasor measurement method is used to collect phasor information.

To measure grid parameters, a PMU is developed. The unit conforms to the requirements of IEEE Standard for Synchrophasor Measurements for Power Systems. The unit can measure amplitude, phase, frequency, and so on; AC amplitude error is within 0.2%, phase error is within 0.5°, and frequency error is within 0.02% (no more than ±0.01 Hz). The unit can also measure the DC analog signal, digital input interface, and Ethernet communication interface; the communication protocol is standard industrial Ethernet/IP protocol, and it further encapsulates the IEEE C37.118 protocol based on the Ethernet/IP protocol, thereby meeting the requirements of high-speed, real-time transmission of phasor data.

Design of the solution of the phasor measurement unit is shown in Fig. 2.

The measurement program includes three parts according to the function requirements of the phasor measurement unit, namely sampling module, communication

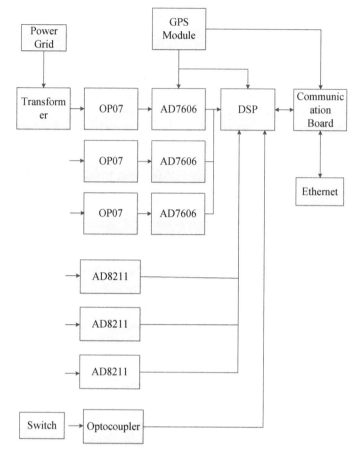

Fig. 2 PMU measurement program

module, and synchronous clock module. The sampling module is made up of transformers, amplifiers, processor unit, and related circuit. The communication module is made up of the communication interface, processor unit, and related circuit. The synchronous clock module aims to provide a high-accuracy synchronous second pulse.

As shown in Fig. 2, DSP is the processor unit of the sampling module, and it has data processing functions such as acquisition and computing of the AC power phasor and analog. The communication board is the processor unit of the communication module, which is used for adding standard space of measurement data, package packets, and sending data to the data concentrator through industrial Ethernet. The GPS module synchronous clock module is used to provide a high-accuracy synchronous second pulse, collect real-time data on the measuring unit for the phasor, and add a time tab.

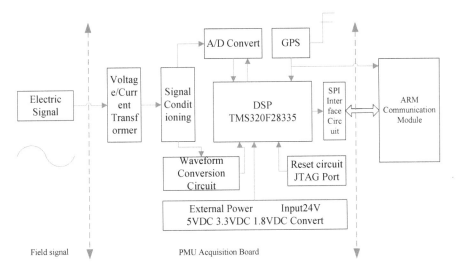

Fig. 3 PMU hardware block of acquisition module

3.2 Hardware Design for Phasor Measuring Device

For hardware design, the measuring device changes external measurement information into a board-level signal and follows the zero crossing of the signal through the squaring circuit. The measuring device also introduces the synchronous clock second pulse and Ms pulse signals, as well as completes the exchange of measurement data with the communication board through the SPI interface.

The phasor measuring device includes function modules such as acquisition, synchronization, waveform conversion, frequency tracking, and communication. It also completes the precision acquisition and computing of electrical signals in the field, as well as communicates with the communication module through the SPI interface. The hardware block diagram of the PMU sampling module is shown in Fig. 3.

The PMU sampling module is powered by a 24 V power source, and the power supply voltage can be converted to 5, 3.3, and 1.8 V depending on the requirement of each chip. The main CPU uses DSP chip TMS320F28335 and has strong data processing capabilities. It also communicates with the communication module through the SPI interface.

3.3 Software Design for Phasor Measuring Device

For software design, phasor measurement device software is divided into two parts, namely automatic calibration procedure and main features procedure. The automatic

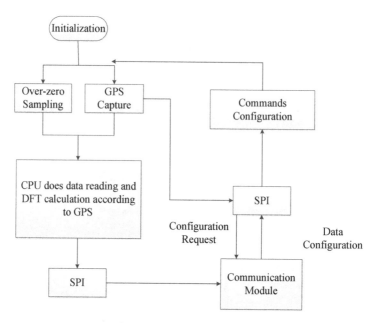

Fig. 4 PMU software flowchart

calibration procedure calculates the error coefficient of each channel from the measuring device. It corrects the deviation of corresponding channels based on the command status of the DIP switch and indicates the completion status by LED. After completing the correction of deviation for each channel, the corrected deviation coefficients are stored in the PMU flash to compensate for sampling data.

Main features procedure is the program operated in normal mode. The program contains data acquisition, FFT computation, SPI communication, and the communication module as main function modules. DSP controls the process to collect and calculate data, as well as to send the processed key variable information to the communication module by SPI. The main design process is shown in Fig. 4.

The communication module mainly reads the power data and configuration for both the acquisition part and time of the GPS. This module achieves data exchange with the data concentrator in the network as a slave of Ethernet/IP before being packaged.

3.4 Implementation of Time Synchronize

As phasor measurement units are distributed all over the power grid, high-precision time synchronization technology is needed to enable all phasor measuring devices to collect data at the same time [12].

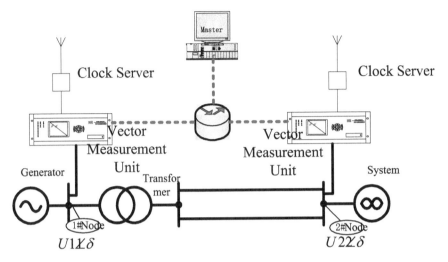

Fig. 5 PMU time synchronization diagram

The PMU supports IEEE 1588 Precision Clock Synchronization Protocol. GPS clock synchronization system composed of PMU, clock server, and satellite [13].

The solution design is shown in Fig. 5.

The time synchronization unit is an important part of the PMU; it consists of NTP synchronous message transceiver unit, system time synchronization unit, time difference calculation unit, clock adjustment unit, and DSP signal generation unit. The accuracy of the DSP acquisition signal pulse depends on NTP time synchronization accuracy, and it does not produce the DSP acquisition signal pulse when time synchronization accuracy is low. The pulse width of the DSP acquisition signal also changes when the clock shakes; thus, time synchronization accuracy can affect the output of the DSP acquisition signal pulse [14].

4 Application and Testing

4.1 Test System Design

A test system is built and is illustrated in Fig. 6.

The test system compares the measured values at the same time as two or more PMUs under the same input signal. The accuracy is considered to be acceptable if the values can maintain the same precision and do not change under other conditions (time, amplitude, phase, different MPU, and so on).

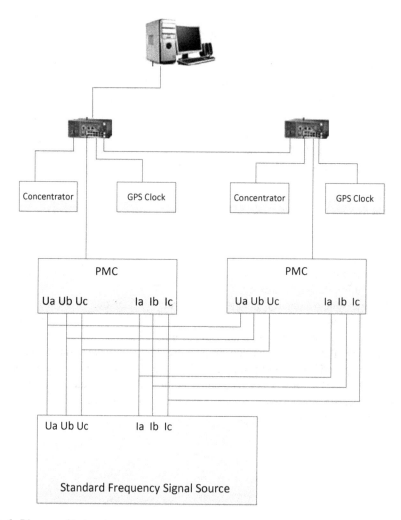

Fig. 6 Diagram of test system

Two PMUs are connected to the host computer via the concentrator, and both of them have three AC voltage inputs and three AC current inputs. A standard signal source is connected parallel to them, and each has one GPS clock source. PMU test is the PMU capture and data analysis software in the host computer. The software repeatedly counts the times of mission and receiving messages and calculates the amplitude and phase errors of the AC voltage and current. It also calculates amplitude errors when frequency changes. Phase error value is replaced by the phase errors of two PMUs; the phase error value is considered to be correct as long as the phase between the two PMUs is the same.

The tests of amplitude errors and phase errors of the AC voltage and current, as well as the test of amplitude errors when frequency changes are conducted under the condition that the temperature is 25° and the humidity is 50% RH, requiring amplitude errors to be within 0.2% and phase errors to be within 0.5°.

4.2 Amplitude Error Testing

A parallel signal source is connected to a PMU under the aforementioned conditions, and a data pack is obtained through the test software. Input channels are tested under the frequencies of 45, 50, and 55 Hz, and practical measured values of PMU channels are recorded to compare it to the rational values of the signal source. The amplitude errors are then analyzed. The maximum percentage errors of the voltage and current amplitudes are shown in Tables 1 and 2, respectively.

According to Tables 1 and 2, the maximum percentage of errors of the voltage and current amplitudes are within 0.2%; thus, the test is validated.

4.3 Amplitude Error Testing When Frequency Changes

A parallel signal source is connected to a PMU under the preceding condition, and the data pack is obtained through the test software. Input channels are tested under different frequencies, and the practical measured values of PMU channels are recorded to compare with the rational values of the signal source. The amplitude errors are then analyzed. Table 3 shows the maximum percentage errors of amplitudes when the frequency changes.

According to Table 3, the maximum percentage errors of amplitudes when the frequency changes are within 0.2%; thus, the test is validated.

Table 1 Maximum percentage errors of voltage amplitudes

Frequency (Hz)	Input 10U 0.1I	30U 0.3I	50U 0.5I	70U 0.7I	90U 0.9I	100U 1.0I	110U 1.1I	120U 1.2I	Voltage channel
	Error (%)								
45	0.079	−0.088	−0.1014	−0.1212	−0.1435	−0.1514	0.1577	−0.1796	Channel 1
	0.079	−0.088	−0.1414	−0.1498	−0.1546	−0.1514	0.1486	−0.163	Channel 2
	0.179	−0.088	−0.1014	−0.1212	−0.1324	−0.1414	0.1486	−0.163	Channel 3
50	0.179	0.0453	−0.0414	−0.0642	−0.088	−0.1013	−0.0577	−0.1296	Channel 1
	0.079	−0.0546	−0.0814	−0.0927	−0.088	−0.1013	−0.0759	−0.1046	Channel 2
	0.179	0.0453	−0.0414	−0.0642	−0.0768	−0.0914	−0.0395	−0.1046	Channel 3
55	−0.121	−0.1213	−0.1014	−0.0784	−0.0991	−0.1114	−0.1213	−0.1213	Channel 1
	−0.121	−0.088	−0.0814	−0.0784	−0.1213	−0.1414	−0.1577	−0.163	Channel 2
	−0.121	−0.088	−0.0614	−0.0642	−0.0991	−0.1214	−0.1304	−0.138	Channel 3

Table 2 Maximum percentage errors of current amplitudes

Frequency (Hz)	Input									Current channel
	10U 0.1I	30U 0.3I	50U 0.5I	70U 0.7I	90U 0.9I	100U 1.0I	110U 1.1I	120U 1.2I		
	Error (%)									
45	−0.2	0.167	−0.1	−0.1	−0.0889	−0.08	−0.0727	−0.075		Channel 1
	−0.1	−0.1	−0.1	−0.114	−0.1	−0.09	−0.0818	−0.0833		Channel 2
	−0.1	0.133	−0.06	−0.1	−0.1	−0.09	−0.0818	−0.1		Channel 3
50	−0.2	−0.133	−0.1	−0.1	−0.0889	−0.08	−0.0727	−0.0667		Channel 1
	−0.1	−0.1	−0.08	−0.1	−0.0889	−0.09	−0.0727	−0.075		Channel 2
	−0.1	0.1	−0.06	−0.0714	−0.0889	−0.09	−0.0818	−0.0833		Channel 3
55	−0.2	0.1	−0.04	0.0429	−0.0667	−0.08	−0.0818	−0.0917		Channel 1
	−0.1	0.1	−0.04	0.0429	−0.0667	−0.07	−0.0727	−0.0833		Channel 2
	−0.1	0.1	−0.04	0.0429	−0.0667	−0.07	−0.0727	−0.075		Channel 3

Table 3 Maximum percentage error of amplitude for frequency variation

Input	Frequency							AC channel
	45 Hz	47 Hz	49 Hz	50 Hz	51 Hz	53 Hz	55 Hz	
	Error (%)							
100U 1.0I	−0.141	−0.141	−0.161	−0.111	−0.141	−0.161	−0.131	AC1
	−0.151	−0.161	−0.171	−0.101	−0.161	−0.141	−0.131	AC2
	−0.111	−0.121	−0.131	−0.081	−0.121	−0.091	−0.111	AC3
	0.07	0.06	0.05	0.07	0.04	0.02	0.09	AC4
	0.05	0.05	0.04	0.06	0.04	0.03	0.08	AC5
	0.06	0.05	0.05	0.06	0.04	0.03	0.07	AC6

4.4 Phase Error Testing

A parallel signal source is connected to two PMUs under the aforementioned condition, and the data pack is obtained through the test software. Input channels are tested under the frequency of 45, 50, and 55 Hz. The results are recorded, and the phase errors of these two PMUs are compared and analyzed. The phase error maximum values are shown in Table 4.

According to Table 4, the phase error maximum values are within 0.5°; thus, the test is validated.

According to the test results, the three test systems can achieve accurate amplitude error and phase error, thereby realizing the synchronized functions. The DC error and AC frequency error can also be tested through this system and they can meet the accuracy required.

4.5 Application

The researchers measured the power system of the urban rail transit and demonstrated the transmission project application at six stations, namely Jiangpu main transformer station, Hongkou Football Stadium Station, Anshan Village Station, Yanji Road Station, City Light Road Station, and Yin-bound traffic stop in Metro Line 8. Accurate power data on the main substation and traction substation main circuit are within the scope of the data collection, which was uploaded to a data maintenance processing center through fast Ethernet.

Jiangpu main substation is configured with two metering cabinets, which were equipped with a PMU, a scanner, a satellite synchronized clock, an Ethernet switch, and so on. The PMU was used to collect AC voltage at the 110 kV side of the main transformer and AC voltage as well as current at the 35 kV side. The satellite synchronized clock was used to provide precise satellite clock signals to synchronous data. The scanner was used to collect massive data and upload them, and then, the

Table 4 Phase error maximum values

Frequency (Hz)	Input										AC channel
	10 U 0.1 I	30 U 0.3 I	50 U 0.5 I	70 U 0.7 I	90 U 0.9 I	100 U 1.0 I	110 U 1.1 I	120 U 1.2 I			
	Error (%)										
45	0.464	0.378	0.349	0.366	0.344	0.332	0.343	0.338			AC 1
	0.476	0.378	0.349	0.366	0.344	0.326	0.343	0.338			AC 2
	0.436	−0.373	0.332	0.349	0.327	0.321	0.338	0.338			AC 3
	0.476	0.39	0.361	0.378	0.349	0.332	0.343	0.338			AC 4
	0.453	0.361	0.338	0.349	0.327	0.315	0.326	0.326			AC 5
	0.453	−0.373	0.332	0.343	−0.327	0.303	0.315	0.315			AC 6
50	0.464	0.396	0.378	0.395	0.373	0.366	0.372	0.366			AC 1
	0.47	0.396	0.389	0.401	0.378	0.372	0.372	0.372			AC 2
	0.436	0.361	0.361	0.372	0.343	0.349	0.361	0.355			AC 3
	0.459	0.401	0.389	0.401	0.366	0.366	0.372	0.361			AC 4
	0.459	0.389	0.373	0.366	0.343	0.349	0.355	0.349			AC 5
	0.482	0.389	0.367	0.361	0.338	0.338	0.349	0.349			AC 6
55	0.487	0.491	0.476	0.452	0.447	0.435	0.435	0.424			AC 1
	0.476	0.499	0.458	0.435	0.435	0.429	0.447	0.435			AC 2
	0.476	0.453	0.412	0.401	0.412	0.412	0.429	0.424			AC 3
	0.489	0.459	0.493	0.481	0.492	0.475	0.481	0.464			AC 4
	−0.49	0.447	0.395	0.407	0.424	0.424	0.435	0.429			AC 5
	−0.482	0.453	0.401	0.407	0.418	0.418	0.441	0.429			AC 6

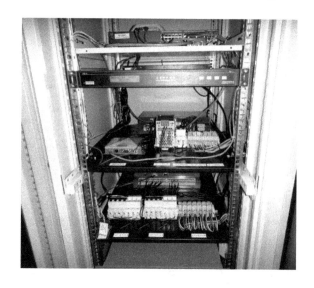

Fig. 7 Demonstration of project site map

real-time data were uploaded to the maintenance central control room through fiber fast Ethernet for analysis and application. The demonstration project site map is shown in Fig. 7.

As Hongkou Football Stadium Station, Anshan Village Station, Yanji Road Station, City Light Road Station, and Yin-bound traffic stop are traction substations, they collect not only AC voltage and current at the 35 kV side but also DC voltage and current signals that are used for pulling trains. Every station has one metering cabinet to complete the acquisition and transmission of synchronous vector data. The construction and commissioning phase of the project are complete, and the system is operating normally in accordance with the desired effect.

5 Conclusions

Existing urban rail transit power monitoring system cannot diagnose DC fault due to lack of DC data. So, the power dispatcher only performs remedial measures after an accident and does not diagnose fault real time. In the current study, key technologies of precision measurement and time synchronization of rail transit distribution system were investigated. A network data system was built, which included a synchronous acquisition device, a data concentrator, a core algorithm, and a GPS clock device. The PMU solves problems that the traditional fault recorder and the existing power monitoring system called SCADA can only record the transient waveform for a few seconds before and after the fault, and no common time stamp is recorded between different sites. A powerful guarantee for the analysis of accidents is provided by using PMU real-time recorded waveform data with accurate time scales. At the same time, through its real-time information, it can realize online judgment of various faults occurring in the power grid and the origin and development process of complex faults. The time synchronization module receives time synchronization signals from the GPS and precisely corrects time. It ensures the stability of the time of the acquisition module through an FM debugging algorithm to ensure that its clock deviation from the GPS clock device is less than 10 uS. Meanwhile, certain sizes of application of this system have been conducted on the Shanghai rail transit AC and DC power distribution system, thereby providing a technical means to avoid or reduce protective trips caused by high load.

This study applies the synchronous data acquisition to a rail transportation and distribution system for the first time. The system data will be used in metro operation scheduling management and can warn about faults, simultaneously avoiding overload trips caused by multiple locomotive starts, and ensuring the safe operation of rail traffic. The system can also be used to locate failure in the power system, assist in quick troubleshooting, and restore running traffic if the power failure occurs.

The rail power system is the entry point of the project. According to the actual operation of rail transit, the research will obtain more and more PMU monitoring data, so as to further study and realize the accurate location of the fault, and carry on the early warning to the fault to ensure the safe operation of the rail transit.

Acknowledgements This research is supported by the Fundamental Research Funds for the Central Universities and National Natural Science Foundation of China, National Key R&D Program of China (2016YFB1200402), and National Natural Science Foundation of China (61703308). The authors are grateful for the reviewer of initial drafts for their helpful comments and suggestions.

References

1. Zhijian QU, Jiang S, Wang J et al (2011) Distributed SCADA system with asynchronous data transmission for 10 kV power supply of railway. Electr Power Autom Equip 31(2):129–133
2. Roop DW (2015) Power system SCADA and smart grids [book reviews]. IEEE Power Energ Mag 14(1):115–116
3. Soonee SK, Agrawal VK, Agarwal PK et al (2015) The view from the wide side: wide-area monitoring systems in India. Power Energy Mag IEEE 13(5):49–59
4. Bitzer B, Kleesuwan T (2015) Cloud-based smart grid monitoring and controlling system. In: Power engineering conference
5. Zheng Z, Zhai M, Peng H et al (2017) Architecture and key technologies of distributed SCADA system for power dispatching and control. Autom Electr Power Syst 41(5):71–77
6. Oleinikova I, Mutule A, Grebesh E et al (2015) Line parameter estimation based on PMU application in the power grid. In: IEEE international conference on power engineering
7. Gopakumar P, Mallikajuna B, Reddy MJB et al (2018) Remote monitoring system for real time detection and classification of transmission line faults in a power grid using PMU measurements. Prot Control Mod Power Syst 3(1):16
8. Karpe SU, Kalgunde MN (2017) Power system backup protection in smart grid using synchronized PMU. In: International conference on signal processing
9. Vanfretti L, Baudette M, Domínguez-García J-L et al (2016) A phasor measurement unit based fast real-time oscillation detection application for monitoring wind-farm-to-grid sub-synchronous dynamics. Electr Mach Power Syst 44(2):12

10. Shahsavari A, Sadeghi-Mobarakeh A, Stewart E et al (2017) Distribution grid reliability versus regulation market efficiency: an analysis based on micro-PMU data. IEEE Trans Smart Grid PP(99):1–1
11. Chow JH (2016) Online monitoring of dynamics with PMU data. In: Smart grid handbook
12. Renjie D, Yong M, Weihua X, Power system dynamic monitoring device and dynamic simulation experiment based on GPS. J Tsinghua Univ (Sci Technol)
13. Li P, Wang Y, Synchronous clock and the usage of TD series of satellite. Autom Electr Power Syst
14. DP83640 Precision PYTER—IEEE 1588 precision time protocol transceiver. National Semiconductor Corporation, 2008
15. Bhuiyan SMA, Khan JF, Murphy GV (2017) Big data analysis of the electric power PMU data from smart grid. In: Southeastcon
16. Rihan M, Kumar V (2016) GIS aided PMU placement for dynamic state measurement in power grid: a case study. In: International conference on computing

Railway Capacity Calculation in Emergency Using Modified Fuzzy Random Optimization Methodology

Li Wang, Min An, Limin Jia and Yong Qin

Abstract Accurate estimated capacity of the railway section can provide reliable information to railway operators and engineers in decision-making, particularly, in an emergency situation. However, in an emergency, the optimization of capacity of a railway section is usually involved to study, for example, the characteristics of dynamic, fuzziness, randomness, and non-aftereffect properties. This paper presents a proposed capacity calculation method based on the modified fuzzy Markov chain (MFMC). In this method, the capacity of a railway section in an emergency can be expressed by a fuzzy random variable, which remains the randomness of capacity changing according to the impact of emergencies and the fuzziness of the driving behavior and other factors. A case study of a high-speed line from Beijing to Shanghai is used to show the process of the proposed methods for optimization of section capacity calculation in an emergency.

Keywords Railway section capacity · High-speed railway · Fuzzy Markov chain · Emergency

---

L. Wang
School of Traffic and Transportation, Beijing Jiaotong University, Beijing, China

L. Jia · Y. Qin
State Key Laboratory of Rail Traffic Control and Safety, Beijing Jiaotong University, Beijing 100044, China

M. An (✉)
School of Science, Engineering and Environment, University of Salford, Salford M5 4WT, UK
e-mail: M.An@salford.ac.uk

L. Wang · L. Jia · Y. Qin
Beijing Engineering Research Center of Urban Traffic Information Intelligent Sensing and Service Technologies, Beijing 100044, China

1 Introduction

Railway emergency can be defined as the railway system will be affected by different types and levels of failures or incidents [1, 2], such as track system failure, signal system failure, and even an accident. Capacity calculation of a railway section and rerouting path generation in an emergency are usually the basis for the planned subsequent adjustment scheme which includes service planning [3], rolling-stock planning, and timetabling [4–6]. Railway section capacity refers to the maximum number of the trains running on the railway section in a contain time (day, or night or even if several hours) under the conditions of certain train types and the traffic organization. Its influence factors mainly include the number of sections, the condition of railway signal systems, speed of trains, length of block sections, and train sequences [7]. The current methods used to assess capacity of the railway section include graphic methods [8], analytical methods [9], and optimization methods [10]. Since the section capacity calculation involves many factors, the mathematical model or computer simulation only considered the key factors that cannot be completed in conformity with the real situation. Especially, in emergency conditions, the railway system will be affected by different types and levels of failures or incidents, so that the transport will be interrupted or a train will be limited to a low speed in a certain section [11]. Therefore, it is essential to develop new methods that can take these factors into consideration to provide more accurate capacity estimation in emergency conditions, so that optimization and rerouting path can be undertaken to find the best solution.

2 Factor Analysis of Railway Section Headway

The headway of railway section is very important in safe railway operation, which controls the capacity of railway sections. Usually, there are two headway block modes, i.e., the quasi-moving block mode and the fixed block mode, which are described in the below sections.

2.1 Train Headway Under Fixed Block Mode

Fixed block mode defines that a section can be divided into a number of sub-sections, and each sub-section has a fixed distance space. As can be seen L'_s, L''_s, L'''_s as shown in Fig. 1 are three-fixed blocks, which are recognized by signals. The train operation depends on the signal control modes, such as three- and four-patterns that are mostly used in control operation in railway systems.

In the three-pattern fixed block section, the headway interval of the train usually separates in three-pattern block areas (see Fig. 1), in which Trains *A* and *B* have a

Fig. 1 Three-pattern fixed block mode

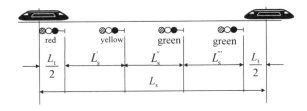

same length; L_t is length of *train*; L'_s, L''_s, L'''_s are three-block section intervals length, respectively, (m); the calculation of the train headway expression is shown in Eq. 1.

$$I = (L_t + L'_s + L''_s + L'''_s)/v \times 3.6 \tag{1}$$

In the four-pattern fixed block section, the headway interval of the train usually separates in four-pattern block areas, as shown in Fig. 2. Assume $L'_s = L''_s = L'''_s = L_s$, the train headway can be calculated by Eq. 2.

$$I = (4L_{\text{sunsection}} + L_{\text{train}})/v \times 3.6 \tag{2}$$

In fixed block mode, the train headway mainly depends on the length of the block section and the train operation speed. Normally, the three-pattern block section requires whose length of one block section to satisfy one train barking distance. And the four-pattern block section requires two- or three-block sections. The length of the block sections is determined by the train braking distance, signaling indication system, and the safety redundancy. However, the main factors that affect the braking distance are the traction machines, traction weight, route speed restriction, route slope, etc. These factors increase the uncertainty of the train headway under the fixed block mode. The length of the block section can choose from the range 1600–2600 m under the three-pattern mode and the 700–900 m under the four-pattern mode. In this case, the parameter L_s is a fuzzy value and denoted as $\widetilde{L}_s$ (in this study, superscript "~" expresses that a parameter is a fuzzy parameter). The average train speed v is affected by, for example, rail conditions, the driver behavior, the states of equipment, and the external environmental influence. During an emergency period, the average train speed v is varied fuzzily and randomly, which can be denoted as $\hat{v}$ (in this study,

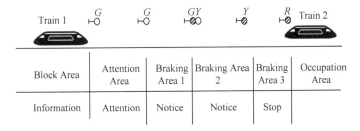

Fig. 2 Four-pattern fixed block mode

the superscript "∼" expresses a parameter is a fuzzy random parameter). Therefore, under the three-pattern mode and four-pattern mode, the calculation function of the train headway with uncertainty can be transformed into Eqs. 3 and 4, respectively.

$$\tilde{I} = (L_t + 3\tilde{L}_s)/\hat{v} \times 3.6 \tag{3}$$

$$\tilde{I} = (L_t + 4\tilde{L}_s)/\hat{v} \times 3.6. \tag{4}$$

2.2 Train Headway Under Fixed Block Mode

Quasi-moving block mode defines the distance between two trains which is controlled by signals using digital technologies such as GPS and mobile communications. In other words, the distance between two trains is varied depending on train speeds, and rail conditions, the time use of signal system reaction, etc. Currently, high-speed railways adopt the quasi-moving block mode to control the distance between *Trains A* and *B* (see Fig. 3). In this case, the time of the train headway can be calculated as Eq. 5.

$$I = \left((L_a + L_d + L_b + L_p + L_c + L_t)/v\right) \times 3.6 \tag{5}$$

where I is the total time of headway that is needed for *Train A* stopping in order to prevent clash with *Train B*, v is the average speed (km h^{-1}) of *Train A*, L_a is the distance related to response time, for example, when *Train A* receives a braking signal, the time is needed for a train driver to take an action, L_d is the distance related to the time needed to launch the brake system, L_b is the distance from break launch to *Train A* stopping, L_p is the safety distance depending on rail conditions and train speeds, L_c is the distance of normal operation, and L_s is the total of the distance as shown in Fig. 3.

Equation 5 can be transferred to the speed-time expression as Eq. 6.

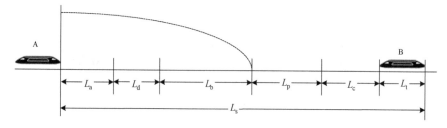

Fig. 3 Quasi-moving block mode

$$I = t_a + t_d + (v/3.6)/2a + \left(L_p + L_c + L_t\right)/v/3.6 \tag{6}$$

As can be seen from Eq. 6, to determine the parameters depends on, for example, rail conditions, train speeds, types of train control systems, types of the braking systems, and also the driver behaviors, etc. The t_a is affected by many factors such as the time of the train control system to response, the time of information transmission, and the reaction time of the driver. Landex [12] studied the t_a and believes that t_a is around 9.5 s. The t_d is the response time of the train brake system. Again, t_d is also a fuzzy value, which is generally considered as around 2.5 s [12]. The a is the braking deceleration speed and is determined based on the braking performance of the train and route conditions. Xianming [13] studied the corresponding deceleration speeds when a train is running on a downhill slope with 20‰, plain section and approaching to the station, and a can choose 0.565 m s^{-2}, 0.75 m s^{-2}, and 0.5 m s^{-2}, respectively. Because of the differences of the train braking system and the route conditions, the value of a is fuzziness with the range from 0.5 to 0.75 m s^{-2}. If the safety margin was considered, the value of L_p can choose with the range of 80–150 m; The value of L_c changes with the differences of the control mode, the longer the value of L_c, the drivers' behavior shows more calm down. However, it will increase the train headway at the meantime. Therefore, it will keep the comfortable driving and high usage of the ability by chosen 1–2 block length. Equation 6 can be expressed as Eq. 7

$$\tilde{I} = \tilde{t}_a + \tilde{t}_d + \hat{v}^2/2\tilde{a} + \left((\tilde{L}_p + \tilde{L}_c + L_t)/\hat{v}\right) \times 3.6. \tag{7}$$

3 Factor Analysis of Railway Section Headway

Based on the analysis of uncertainty of fuzzy and stochastic, the railway section capacity status changing process in emergency is described as a fuzzy Markov chain. Then, the calculation of section capacity is changed to the fuzzy status transfer processing. Finally, the parameters sensitivity analysis is given in the section.

3.1 Uncertainty Analysis of the Railway Section Capacity

The severity of the incident influences and the railway department repair ability are changing randomly, which cause a random variation in train operation speed correspondingly. Therefore, the section capacity changes randomly in a period of time. Moreover, in a certain time of future, the railway network capacity is associated with the current equipment instead of the former condition. As the conclusion, the transformation of the section capacity has the non-aftereffect property for different rail lines and emergency types. The changing process of the section capacity status can be described as Markov chain (MC) [14]. Due to the fuzziness property of the

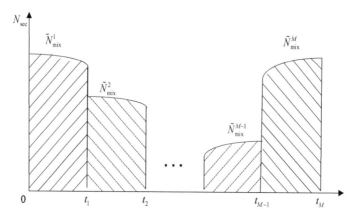

Fig. 4 Section capacity changing process in an emergency

section capacity itself, the capacity status changing process is considered as the fuzzy Markov chain (FMC) in this paper. The definition of FMC is given in Sect. 3.2.

As shown in Fig. 4, the time axis is divided into M periods. Because of the different effect level of the special events in each period, the different emergency grade, the different degree of the maintenance, and the train speed limitation are different. So, the capacity of each time period is also different. Then, the whole section capacity can be computed by Eq. 8. Where, $\widetilde{N}_{mix}^i$ is fuzzy value of the section capacity in the ith period, by setting $\Delta t^{(i)} = t_i - t_{i-1}, i = 1, 2, \ldots, M$.

$$N_{\sec} = \sum_{i=1}^{M} \widetilde{N}_{mix}^i \tag{8}$$

The dividing of the time periods depends on the influence degree of the incidents and the railway line recovery levels. Therefore, the dividing has property of fuzziness. In reality, it can be determined by the development of the incident and the speed limited strategy.

3.2 Fuzzy Markov Chain

The fuzzy random variable (FRV) is a measurable function, which is the set from the probability space mapping to the fuzzy space. For example, for a given probability space $\{\Theta, \Lambda, \Pr\}$, $\tilde{u}_1, \tilde{u}_2, \ldots, \tilde{u}_n$ are the fuzzy variable, if $\xi(\omega_i) = \tilde{u}_i, i = 1, 2, \ldots, n$, then, $\xi(\omega_i)$ is the fuzzy random variable [15]. Figure 5 shows the relationship among an FRV, a random variable, and a fuzzy variable.

Definition 1 Fuzzy Markov Chain (FMC) For a given probability space (Ω, Λ, P), Θ is nonempty set, Λ is the σ algebra of Γ, and P is the probability. The number

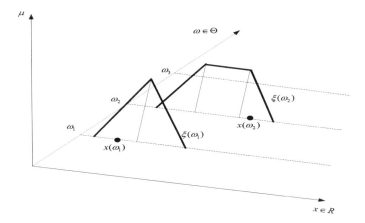

Fig. 5 FRV graph

of the fuzzy random variable $\{\tilde{X}(t), t = 0, 1, 2, \ldots\}$ is limited or countable. For the whole possible fuzzy conditions of all $\tilde{X}(t)$, there is one group of fuzzy set $\tilde{A} = \{\tilde{A}_0, \tilde{A}_1, \tilde{A}_2, \ldots\}$ corresponding to $\tilde{X}(t)$, make S to the possibility of one fuzzy event, if the possibility of $\tilde{X}(n+1) = \tilde{A}_{n+1}$ (variable $\tilde{X}(n+1)$ at the moment $n+1$ stay in the $\tilde{A}_{n+1}$ status) can only relate to $\tilde{X}(n)$, but not the status before the n, expressed as $S(\tilde{X}(n + 1) = \tilde{A}_{n+1}|\tilde{X}(n), \tilde{X}(n - 1), \ldots, \tilde{X}(0)) = S(\tilde{X}(n + 1) = \tilde{A}_{n+1}|\tilde{X}(n))$, thus, the sequence $\tilde{X}(t)$ is named as fuzzy Markov chain (FMC).

Signed S as one-step transfer possibility matrix, which is denoted by Eq. 9.

$$S = \begin{bmatrix} s_{0,0} & s_{0,1} & \cdots & s_{0,j} & \cdots \\ \cdots & \cdots & \cdots & \cdots & \cdots \\ s_{i,0} & s_{i,1} & \cdots & s_{i,j} & \cdots \\ \cdots & \cdots & \cdots & \cdots & \cdots \end{bmatrix}. \tag{9}$$

3.3 Calculation of Section Capacity Fuzzy Value

Assuming in the continuous emergency time T, the speed type of the train is Q_{spe}, denoted as $v^{(1)}, v^{(1)}, \ldots, v^{(Q_{spe})}$. Divide the continuous time T into Q_c time periods, the correspond section capacity has Q_c kind of situations, denoted as $\tilde{c}_1, \tilde{c}_2, \ldots, \tilde{c}_i, i = 1, 2, \ldots, Q_c$, consisting all the possible fuzzy sets $\tilde{C}$ of the capacity changing Markov chain procession.

Firstly, the section capacity can be calculated in one time period by Eq. 10.

$$c = (60 \times 60 - t_{\text{mai}})/I \tag{10}$$

Here, c is the section capacity in each time period. t_{mai} is the comprehensive maintenance time. If different trains' speeds exist, the capacity can be calculated by the deduction coefficient method.

The status transfer matrix of train speed can be denoted by Eq. 11.

$$P = \begin{bmatrix} P_{1,1} & P_{1,2} & \cdots & P_{1,Q_{speed}} \\ P_{2,1} & P_{2,2} & \cdots & P_{2,Q_{speed}} \\ \vdots & \vdots & \ddots & \vdots \\ P_{Q_{speed},1} & P_{Q_{speed},2} & \cdots & P_{Q_{speed},Q_{speed}} \end{bmatrix} \quad (11)$$

Here, $P_{i,j}$ is the probability of the train speed from $v^{(i)}$ to $v^{(j)}$. Then, set the status transfer matrix of the section capacity under the possible measurement can be calculated by Eq. 12.

$$S = \begin{bmatrix} s_{1,1} & s_{1,2} & \cdots & s_{1,Q_c} \\ s_{2,1} & s_{2,2} & \cdots & s_{2,Q_c} \\ \vdots & \vdots & \ddots & \vdots \\ s_{Q_c,1} & s_{Q_c,2} & \cdots & s_{Q_c,Q_c} \end{bmatrix} \quad (12)$$

Here, $s_{i,j}$ is the condition probability of the section capacity from $\tilde{c}_i$ to $\tilde{c}_j$ under the possible measurement [16], denoted as Eq. 13.

$$s(\tilde{c}_j/\tilde{c}_i) = s(\tilde{c}_j, \tilde{c}_i)/s(\tilde{c}_i) \quad (13)$$

For the convenient of calculation, define matrix Q_1 and Q_2 by Eqs. 14 and 15.

$$Q_1 = \begin{bmatrix} \mu_{\tilde{c}_1}(0) & \mu_{\tilde{c}_2}(0) & \cdots & \mu_{\tilde{c}_N}(0) \\ \mu_{\tilde{c}_1}(1) & \mu_{\tilde{c}_2}(1) & \cdots & \mu_{\tilde{c}_N}(1) \\ \vdots & \vdots & \ddots & \vdots \\ \mu_{\tilde{c}_1}(N) & \mu_{\tilde{c}_2}(N) & \cdots & \mu_{\tilde{c}_N}(N) \end{bmatrix} \quad (14)$$

$$Q_2 = \begin{bmatrix} p_0\mu_{\tilde{c}_1}(0)/P(\tilde{c}_1) & p_1\mu_{\tilde{c}_1}(1)/P(\tilde{c}_1) & \cdots & p_N\mu_{\tilde{c}_1}(N)/P(\tilde{c}_1) \\ p_0\mu_{\tilde{c}_2}(0)/P(\tilde{c}_2) & p_1\mu_{\tilde{c}_2}(1)/P(\tilde{c}_2) & \cdots & p_N\mu_{\tilde{c}_2}(N)/P(\tilde{c}_2) \\ \vdots & \vdots & \ddots & \vdots \\ p_0\mu_{\tilde{c}_N}(0)/P(\tilde{c}_N) & p_1\mu_{\tilde{c}_N}(1)/P(\tilde{c}_N) & \cdots & p_N\mu_{\tilde{c}_N}(N)/P(\tilde{c}_N) \end{bmatrix} \quad (15)$$

Therefore, we can get Eq. 16.

$$s(\tilde{c}_j/\tilde{c}_i) = Q_2 P_{ml} Q_1 \quad (16)$$

For all periods, section capacity can be generalized as Eq. 17.

$$\left(\widehat{N}_{\text{mix}}^{(l)}\right)^{\text{T}} = \left(\tilde{c}_1, s_{h,1}, \tilde{c}_2, s_{h,2} \cdots \tilde{c}_{Q_c}, s_{h,Q_c}\right), \quad l = h+1 \tag{17}$$

Therefore, the section capacity of the divide-period in limited speed condition $\widehat{N}_{\text{mix}}^{(i)}$ is the fuzzy random variable.

The section capacity in the whole time T is the sum of all periods of capacity, denoted by Eq. 18.

$$N_{\text{sec}} = \sum_{i=1}^{Q_c} \widetilde{N}_{\text{mix}}^i = \begin{cases} \sum_{i=1}^{Q_c} \tilde{c}_1, \sum_{i=1}^{Q_c} s_{i,1} \\ \vdots \\ \sum_{i=1}^{Q_c-1} \tilde{c}_{Q_c} + \tilde{c}_{Q_c-1}, \sum_{i=1}^{Q_c-1} s_{i,Q_c} \cdot s_{Q_c,Q_c-1} \\ \sum_{i=1}^{Q_c} \tilde{c}_{Q_c}, \sum_{i=1}^{Q_c} s_{i,Q_c} \end{cases} \tag{18}$$

It is noticeable to see that in possible measurement the section carrying capacity is still a fuzzy random value in the continuous duration T under the emergency condition.

4 Case Study

4.1 Emergency Scenario

In this paper, the real-world case study is based on Shanghai to Nanjing section. Figure 6 describes the network of this section and its nearby railways. The Shanghai to Nanjing intercity line and the Beijing to Shanghai existing line are both parallel to Beijing to Shanghai high-speed rail line. In Fig. 6, these three routes are distinguished by the imaginary line, the dash-dot line, and the solid line, respectively. The distance between any two stations is given in Fig. 6. The details of design and train schedules of Shanghai to Nanjing section can be found in [17].

Suppose that the broken-down situation occurs in Suzhou to Kunshan section in Beijing to Shanghai high-speed rail line at 8 a.m., i.e., no train service can provide. Recovery is expected for 3 h. In Zhenjiang West to Wuxi East section, due to the heavy wind situation, speed restriction is implemented. The wind speed is expected to keep at about 20 m s^{-1} for 2 h. Then, the wind speed would increase at about 24 m s^{-1} for 5 h. At last, the wind speed would decrease and back to normal operation environment in 2 h. Besides, other sections and stations are not affected by the emergency.

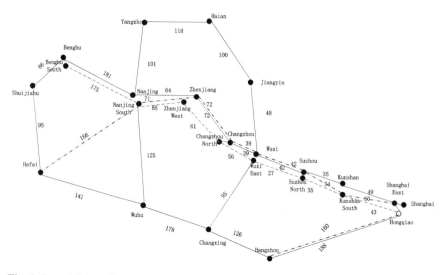

Fig. 6 Part of China railway network

4.2 Section Capacity Calculation

In the broken-down section, the capacity can be set to zero directly. Therefore, the capacity calculation is focused on the speed restriction conditions. According to emergency development, the section capacity changing process can be divided into three stages based on the wind speed variation. Based on the railway traffic safety specification, in the above scenario, there are three kinds of train operation speed, 160 km h^{-1}, 70 km h^{-1}, and normal speed. Firstly, calculate the section capacity in different limited speeds. In this period, it only involved of the medium and high-speed trains, therefore, when the train recovered to normal operation, the average speed of the train can be expressed as the fuzziness value $\tilde{v}_0 = $ (330 km h^{-1}, 340 km h^{-1}, 350 km h^{-1}). In addition, the average speed of the train changes due to the environment, the operation equipment conditions, and the drivers' driving skills [18]. Normally, the average speed changes in one range and smaller than the limited speed. Therefore, when the limited speed is 160 km h^{-1}, the average speed of a train can be set as $\tilde{v}_1 = $ (130 km h^{-1}, 140 km h^{-1}, 155 km h^{-1}). When the limited speed is 70 km h^{-1}, the average speed can be set as $\tilde{v}_2 = $ (50 km h^{-1}, 60 km h^{-1}, 65 km h^{-1}). **Calculate the section capacity value in different speed restriction conditions**. We can set t_a as 9.5 s, t_d as 2.5 s, a as 0.7 m s^{-2}, L_p as 150 m according to Sect. 2. The value of L_c relates to the train speed, when the train is with high speed, it can take the length of two-block section, which is about 4000 m; when the train is running in lower speed, L_c can take the length of one block section, which is about 2000 m. L_t can be set to the conservative value 400 m. Then, put all of the parameters into Eq. 2, we getting the following results.

When the train operated normally, the train headway $\widetilde{I}_0$ is (124, 127, 131). Section capacity $\tilde{c}_0$ can be calculated by using Eq. (10) is (27, 28, 29). When the limited speed is 160 km h^{-1}, the train headway $\widetilde{I}_1$ is (143,156, 168), $\tilde{c}_1$ is (21, 23, 25). When the limited speed is 70 km h^{-1}, $\widetilde{I}_2$ is (163, 187, 208), $\tilde{c}_2$ is (17, 19, 22).

Calculate the capacity transfer conditional probability in emergency condition. Assume the first stage of the train's limited speed is 160 km h^{-1}. We can get the first stage section capacity $\widehat{N}_{mix}^{(1)}$ by Eq. 19.

$$\widehat{N}_{mix}^{(1)} = 2\tilde{c}_1 = (42, 46, 50) \tag{19}$$

In order to make the calculation convenience, assume $v_0 = 330$, $v_1 = 340$, $v_2 = 350$, $v_3 = 130$, $v_4 = 140$, $v_5 = 155$, $v_6 = 50$, $v_7 = 60$, $v_8 = 65$. The probabilities of different speed p_{v_i} are set to the same, $p_{v_i} = 1/9$, $i = 1, 2, \ldots, 8$. The capacity transfer condition probability of the second stage and the third stage can be obtained based on Eq. 12. The transfer matrixes are denoted as S^1 and S^2.

$$S^1 = \begin{bmatrix} 0.2000 & 0.7833 & 0.0167 \\ 0.1248 & 0.7891 & 0.0861 \\ 0.0079 & 0.1566 & 0.8355 \end{bmatrix} \tag{20}$$

$$S^2 = \begin{bmatrix} 0.8667 & 0.1317 & 0.0016 \\ 0.6097 & 0.3758 & 0.0145 \\ 0.0207 & 0.9017 & 0.0776 \end{bmatrix} \tag{21}$$

Sum divided-period capacities. According to section capacity calculation method based on FMC as described in Sect. 3, the carrying capacity of the second stage and third stage can be expression as the fuzziness random variable, particular form shown as:

$$\widehat{N}_{mix}^{(2)} = \begin{cases} 5\tilde{c}_0, s_{1,0}^1 \\ 5\tilde{c}_1, s_{1,1}^1 \\ 5\tilde{c}_2, s_{1,2}^1 \end{cases} = \begin{cases} (135, 140, 145), & 0.1248 \\ (105, 115, 125), & 0.7890 \\ (85, 95, 110), & 0.0861 \end{cases} \tag{22}$$

$$\widehat{N}_{mix}^{(3)} = \begin{cases} 2\tilde{c}_0, s_{1,0}^2 \\ 2\tilde{c}_1, s_{1,1}^2 \\ 2\tilde{c}_2, s_{1,2}^2 \end{cases} = \begin{cases} (54, 56, 58), & 0.6097 \\ (42, 46, 50), & 0.3758 \\ (34, 38, 44), & 0.0145 \end{cases} \tag{23}$$

Therefore, the total capacity for nine hours is calculated by Eq. 18.

$$\widehat{N}_{\text{sec}} = \sum_{i=1}^{3} \widehat{N}_{\text{mix}}^{(i)} = \begin{cases} (231, 242, 253), 0.0761 \\ (219, 232, 245), 0.0469 \\ (211, 224, 239), 0.0018 \\ (201, 217, 233), 0.4811 \\ (189, 207, 225), 0.2965 \\ (181, 199, 219), 0.0114 \\ (181, 197, 218), 0.0525 \\ (169, 187, 210), 0.0324 \\ (161, 179, 204), 0.0013 \end{cases} \quad (24)$$

It can be found from Eq. 24, the expectancy value of the comprehensive capacity is 215.197. The calculation process of the comprehensive capacity expectation explained that in the current scenario the section capacity is about 215. The possible maximum and minimum values are 242 and 180, respectively. This provides alternative offers for the policymaker to make the decision depended on the different preferences. If the radical arrangement was chosen, the value can set to 242. However, if the operator preferred to the conservative strategy, the main target is to meet the basic demand of the transportation, the value can set to 180. Generally, the compromise strategy is adopted, which means the value sets to 215.

This paper focuses on the railway section capacity calculation, without the station capacity. In this case, the heavy wind affects train operation speed running on section mainly, but less effect on the station. Therefore, the section capacity decides the line capacity at current scenario.

5 Conclusion and Future Work

Section capacity calculation involves multiple factors and complicated relationships. In emergency conditions, the factors of the calculation capacity present the characteristics of dynamic, fuzziness, randomness, no aftereffect, etc. In the reality, the capacity calculation cannot include all of the factors. FMC-based capacity calculation method can be more fault-tolerant of all kinds of sensitive factors and the uncertainties. In addition, this method provides variable choices for the policymakers, so that it can suit to radical or conservative strategies.

The research in the future is mainly concentrated in two aspects: ① the train speed transfer matrix is the key factor to realize the section capacity calculation, the value of the transfer matrix need to be confirmed in the practical operation environment; ② there is a big difference of train operation objectives, strategies, and principles between in the emergency and in the normal conditions. Line planning, timetable rescheduling, and rolling-stock rebalancing are yet to consider together in emergency condition. Our ultimate goal is to design and develop a real-time decision support framework in the future, which will decrease the influence of the emergency effectively and recover the train operation quickly.

Acknowledgements This study is funded by the National Key Research and Development Program of China (2016YFB1200401) and National Natural Science Foundation of China (71701010).

References

1. Xu XN, Zhuang L, Xing B et al (2018) An ultrasonic guided wave mode excitation method in rails. IEEE Access 6:60414–60428
2. Kou L, Qin Y, Zhao X et al (2019) Integrating synthetic minority oversampling and gradient boosting decision tree for bogie fault diagnosis in rail vehicles. Proc Inst Mech Eng Part F: J Rail Rapid Transit 233(3):312–325
3. Meng XL, Qin Y, Jia LM (2014) Comprehensive evaluation of passenger train service plan based on complex network theory. Measurement 58:221–229
4. Meng XL, Jia LM, Xiang WL (2018) Complex network model for railway timetable stability optimisation. IET Intell Transp Syst 12(10):1369–1377
5. Zhang HR, Jia LM, Wang L et al (2019) Energy consumption optimization of train operation for railway systems: algorithm development and real-world case study. J Clean Prod 214:1024–1037
6. Wang MM, Wang L, Xu XY et al (2019) Genetic algorithm-based particle swarm optimization approach to reschedule high-speed railway timetables: a case study in China. J Adv Transp
7. Abril M, Barber F, Ingolotti L, Salido M, Tormos P, Lova A (2008) An assessment of railway capacity. Transp Res Part E: Logist Transp Rev 44:774–806
8. Riejos FAO, Barrena E, Ortiz JDC, Laporte G (2016) Analyzing the theoretical capacity of railway networks with a radial-backbone topology. Transp Res Part A: Policy Pract 84:83–92
9. Zhang X, Nie L (2016) Integrating capacity analysis with high-speed railway timetabling: a minimum cycle time calculation model with flexible overtaking constraints and intelligent enumeration. Transp Res Part C: Emerg Technol 68:509–531
10. Petering MEH, Heydar M, Bergmann DR (2016) Mixed-integer programming for railway capacity analysis and cyclic, combined train timetabling and platforming. Transp Sci 50:892–909
11. Zhang YG, Zeng QF, Lei DY, Wang XY (2016) Simulating the effects of noncrossing block sections setting rules on capacity loss of double-track railway line due to the operation of out-of-gauge trains. Discret Dyn Nat Soc
12. Landex A, Kaas AH, Schittenhelm B, Schneider-Tilli J (2006) Evaluation of railway capacity. In: Annual transport conference at Aalborg University, Aalborg, pp 1–22
13. Xianming S (2005) Research on the headway time of passenger train in China. China Railway Sci 7:32–35 (in Chinese)
14. Wang L, Qin Y, Xu J et al (2013) Section carrying capacity calculation in emergency for high-speed railway. J Beijing Inst Univ S1:18–21
15. Liu B (2007) Uncertainty theory. Springer, Berlin
16. Pardo MJ, de la Fuente D (2010) Fuzzy Markovian decision processes: application to queueing systems. Comput Math Appl 60:2526–2535
17. Wang L, Jia LM, Qin Y et al (2011) A two-layer optimization model for high-speed railway line planning. J Zhejiang Univ: Sci A 12:902–912
18. Wang L, Qin Y, Xu J, Jia L (2012) A fuzzy optimization model for high-speed railway timetable rescheduling. Discret Dyn Nat Soc

Research on FlexRay Bus Communication Protocol Stack of Rail Vehicle Electronic Control System Based on AUTOSAR Standard

Zhi yuan Wang, Yong Liu, Yong liang Ni and Yan ming Li

Abstract Based on the automotive open system architecture (AUTOSAR) specification of FlexRay bus, this paper designs the FlexRay bus communication protocol stack. With the development of rail vehicles, the driving safety, driving comfort and passenger comfort of rail vehicles have been significantly improved. Accordingly, the number and the complexity of electronic control units (ECUs) are increasing, and the amount of software code is rising rapidly. However, the bandwidth and data volume of controller area network (CAN) bus are relatively limited, and its fault tolerance and reliability are relatively low. It will be difficult to meet the requirements of future vehicle network control. FlexRay bus replaces CAN bus, which has higher bandwidth and better fault-tolerant performance. At the same time, the communication protocol stack conforming to AUTOSAR specification can shorten the software development cycle of electronic control system for railway vehicles, reduce the labor cost of products and improve the software portability rate. Experimental results verify the feasibility and validity of the protocol stack.

Keywords Rail vehicle · AUTOSAR · FlexRay · Protocol stack

1 Introduction

With the wide application of advanced technologies, including electronic information technology and Internet technology, the electronic system of rail vehicles has been supplemented and improved to a large extent [1, 2]. Efficient communication between vehicular ECUs has a higher requirement to the bandwidth of vehicular bus, the response speed to emergency tasks and the fault tolerance ability. The bandwidth and data volume of CAN bus are relatively limited, and its fault tolerance and reliability are relatively low. It will be difficult to meet the requirements of future vehicle network control such as large data volume, real-time and flexibility [3, 4]. In this

Z. Wang (✉) · Y. Liu · Y. Ni · Y. Li
China North Vehicle Research Institute, Beijing, China
e-mail: Wangzy_1995@163.com

case, the mature FlexRay bus standard applied in the automotive field becomes a good choice.

AUTOSAR standard is a new set of criteria for automotive electronic system development. The development of automotive electronic system based on AUTOSAR standard separates the application software design from the underlying hardware and improves the reusability and portability of the upper application software.

2 FlexRay Bus

FlexRay bus has high bandwidth and good fault tolerance. And it has prominent advantages in real-time, reliability and flexibility [5].

The maximum rate of traditional CAN bus is 1 Mbps. FlexRay has two channels by comparison. The highest rate of traditional CAN bus can reach 10 Mbps and the total data rate can reach 20 Mbps. Because FlexRay has two routes, it can achieve redundancy better and make messages fault-tolerant [6].

FlexRay is a time-triggered bus system, which conforms to the principle of time division multiple access. Therefore, in the time control area, time slots are allocated to certain messages. Time slots are repeated periodically, that is to say, the time of information on the bus can be predicted, thus ensuring its certainty. For other messages with low time requirement, they can be transmitted in the event control area. Therefore, the flexibility of time and event triggering are formed [7, 8].

3 AUTOSAR

AUTOSAR is mainly divided into three layers, as shown in Fig. 1. From top to bottom, there are application layer (APP), runtime environment (RTE) layer and basic

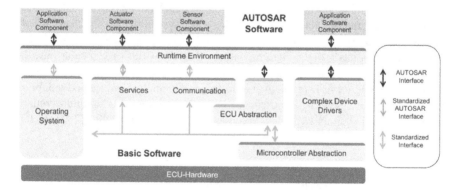

Fig. 1 AUTOSAR

software layer (BSW). The basic software layer is subdivided into service layer, ECU abstraction layer, microcontroller abstraction layer (MCAL) and complex driver [9].

The functions in the application layer are implemented by various software components and are not connected to the automobile hardware system. RTE is a virtual function bus, which summarizes all the interfaces that need to interact with external software. The basic software layer provides basic software services, including standardized system functions and functional interfaces, and consists of a series of basic service software components.

4 FlexRay Communication Module Design

This paper develops FlexRay transceiver driver module, FlexRay interface module and FlexRay transport protocol module of FlexRay communication module and realizes basic communication function based on Freescale microcontroller MPC5646C.

4.1 FlexRay Transceiver Driver Module (FrTrcv)

The FlexRay transceiver is a hardware device which transforms the logical 1/0 signals of the microcontroller ports, or the information given by the serial peripheral interface (SPI) connection to the bus compliant electrical voltage, current and timing. The FlexRay transceiver driver abstracts use FlexRay transceiver hardware. It offers a hardware independent interface to the higher layers. It also abstracts from the ECU layout by using application programming interfaces (APIs) of the MCAL layer to access the FlexRay transceiver hardware. The FrTrcv module in the AUTOSAR layered architecture is shown in Fig. 2.

4.1.1 File Structure of FrTrcv Module

Table 1 lists main core files of the module.
Table 2 lists header files of the module.
Table 3 lists configuration files of the module.

4.1.2 Interface Provided by FrTrcv Module

Table 4 lists the interfaces of FrTrcv, which are provided to other modules.

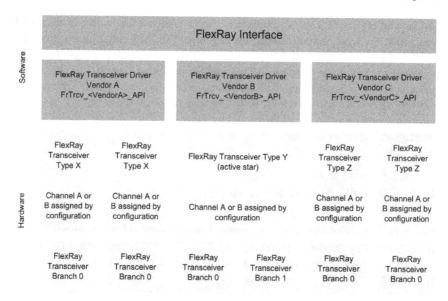

Fig. 2 FrTrcv module in AUTOSAR layered architecture

Table 1 Names and functions of core files

File name	Function
FrTrcv_CheckWakeupByTransceiver.c	FrTrcv_CheckWakeupByTransceiver ()
FrTrcv_ClearTransceiverWakeup.c	FrTrcv_ClearTransceiverWakeup ()
FrTrcv_DisableTransceiverBranch.c	FrTrcv_DisableTransceiverBranch ()
FrTrcv_EnableTransceiverBranch.c	FrTrcv_EnableTransceiverBranch ()
FrTrcv_GetTransceiverError.c	FrTrcv_GetTransceiverError ()
FrTrcv_GetTransceiverMode.c	FrTrcv_GetTransceiverMode ()
FrTrcv_GetTransceiverWUReason.c	FrTrcv_GetTransceiverWUReason ()
FrTrcv_GetVersionInfo.c	FrTrcv_GetVersionInfo ()
FrTrcv_Init.c	FrTrcv_Init ()
FrTrcv_MainFunction.c	FrTrcv_MainFunction ()
FrTrcv_SetTransceiverMode.c	FrTrcv_SetTransceiverMode ()

4.2 FlexRay Interface Module (FrIf)

The FlexRay interface executes jobs as described in the ECU configuration and connects the upper layer's modules to the underlying FlexRay driver and FlexRay

Research on FlexRay Bus Communication Protocol ...

Table 2 Names and description of header files

File name	Description
FrTrcv.h	Contains the provided function declarations and type definitions to be used by other modules
FrTrcv Types.h	Contains configuration specific type definitions
FrTrcv Internal.h	Contains internal function declarations and variable declarations
FrTrcv Cbk.h	Call-back function declaration

Table 3 Names and description of configuration files

File name	Description
FrTrcv Cfg.h	FrTrcv configuration header file
FrTrcv LCfg.h	FrTrcv Link-Time configuration header file symbolic names
FrTrcv LCfg.c	FrTrcv Link-Time configuration file

Table 4 Interfaces of FrTrcv

Service name	Description
FrTrcv_Init	This service initializes the FrTrcv
FrTrcv_GetVersionInfo	This service returns the version information of this module
FrTrcv_SetTransceiverMode	This service sets the transceiver mode
FrTrcv_GetTransceiverMode	This service returns the actual state of the transceiver
FrTrcv_GetTransceiverWUReason	This service returns the wakeup reason
FrTrcv_ClearTransceiverWakeup	This service clears a pending wakeup event
FrTrcv_GetTransceiverError	All mandatory errors can be accessed via this API. In addition to errors on the physical layer and local to the ECU hardware, a global error flag is provided
FrTrcv_MainFunction	This cyclically executed API service of the FlexRay interface serves the following purposes: detects for wake-up-by bus events and informs higher layers

transceiver. This implementation of the FlexRay interface handles many types of FlexRay driver and FlexRay transceiver, with one or more controlled physical hardware.

4.2.1 File Structure of FrIf Module

Table 5 lists main core files of the module.
Table 6 lists header files of the module.
Table 7 lists configuration files of the module.

Table 5 Names and functions of core files

File name	Function
FrIf_CancelTransmit.c	FrIf_CancelTransmit ()
FrIf_CheckWakeupByTransceiver.c	FrIf_CheckWakeupByTransceiver ()
FrIf_ClearTransceiverWakeup.c	FrIf_ClearTransceiverWakeup ()
FrIf_ControllerInit.c	FrIf_ControllerInit ()
FrIf_DisableLPdu.c	FrIf_DisableLPdu ()
FrIf_DisableTransceiverBranch.c	FrIf_DisableTransceiverBranch ()
FrIf_GetChannelStatus.c	FrIf_GetChannelStatus ()
FrIf_GetTransceiverError.c	FrIf_GetTransceiverError ()
FrIf_GetTransceiverMode.c	FrIf_GetTransceiverMode ()
FrIf_GetTransceiverWUReason.c	FrIf_GetTransceiverWUReason ()
FrIf_GetVersionInfo.c	FrIf_GetVersionInfo ()
FrIf_GetWakeupRxStatus.c	FrIf_GetWakeupRxStatus ()
FrIf_Init.c	FrIf_Init ()
FrIf_MainFunction.c	FrIf_MainFunction ()
FrIf_SetState.c	FrIf_SetState ()
FrIf_SetTransceiverMode.c	FrIf_SetTransceiverMode ()
FrIf_SetWakeupChannel.c	FrIf_SetWakeupChannel ()
FrIf_StartCommunication.c	FrIf_SetWakeupChannel ()
FrIf_Transmit.c	FrIf_StartCommunication ()

Table 6 Names and description of header files

File name	Description
FrIf.h	Contains the provided function declarations and type definitions to be used by other modules
FrIf_Types.h	Contains configuration specific type definitions
FrIf_Private.h	Contains internal function declarations and variable declarations

Table 7 Names and description of configuration files

File name	Description
FrIf_Cfg.h	FrIf configuration header file
FrIf_Lcfg.h	FrIf Link-Time configuration header file symbolic names
FrIf_Lcfg.c	FrIf Link-Time configuration file
FlexRay_Fibex.xml	Field bus exchange format (FIBEX) database XML file, contains only cluster parameters

4.2.2 Interface Provided by FrIf Module

Table 8 lists the main interfaces of FrIf, which are provided to other modules.

4.3 FlexRay Transport Protocol Module (FrTp)

The FlexRay transport protocol executes jobs as described in the ECU configuration and it provides the following functions:

- Segmentation of data in transmit direction;
- Reassembling of data in receive direction;
- Control of data flow;
- Error detection in segmentation sessions;
- Transmit cancellation.

Table 8 Main interfaces of FrIf

Service name	Description
FrIf_ControllerInit	Initialized a FlexRay communication controller (CC)
FrIf_GetState	Get current FrIf state
FrIf_GetTransceiverError	Wraps the FlexRay transceiver driver API function FrTrcv_GetTransceiverError. The enum value "FR_CHANNEL_AB" shall not be used
FrIf_GetTransceiverMode	Wraps the FlexRay transceiver driver API function FrTrcv_GetTransceiverMode (). The enum value "FR_CHANNEL_AB" shall not be used
FrIf_GetVersionInfo	Returns the version information of this module
FrIf_Init	Initializes the FlexRay interface
FrIf_SetState	Requests FrIf state machine transition
FrIf_SetTransceiverMode	Wraps the FlexRay transceiver driver API function FrTrcv_SetTransceiverMode (). The enum value "FR_CHANNEL_AB" shall not be used
FrIf_Transmit	Requests transmission of a protocol data unit (PDU)
FrIf_CancelTransmit	Requests cancellation of an ongoing transmission of a PDU in a lower layer communication module
FrIf_MainFunction	This cyclically executed API service of the FlexRay interface serves the following purposes: • Program the absolute timer interrupt in order to start the execution of FrIf_JobListExec_<ClstIdx> () if the CC does not support asynchronous buffer access • Monitoring the proper (in time) execution of the FrIf_JobListExec_<ClstIdx> () and resynchronize the joblist if necessary

4.3.1 File Structure of FrTp Module

Table 9 lists main core files of the module.
Table 10 lists header files of the module.
Table 11 lists configuration files of the module.

Table 9 Names and functions of core files

File Name	Function
FrTp_CancelReceive.c	FrTp_CancelReceive ()
FrTp_CancelTransmit.c	FrTp_CancelTransmit ()
FrTp_ChangeParameter.c	FrTp_ChangeParameter ()
FrTp_CancelTransmit.c	FrTp_CancelTransmit ()
FrTp_GetVersionInfo.c	FrTp_GetVersionInfo ()
FrTp_Init.c	FrTp_Init ()
FrTp_MainFunction.c	FrTp_MainFunction ()
FrTp_Receiver.c	FrTp_Receiver ()
FrTp_RxIndication.c	FrTp_RxIndication ()
FrTp_Receiver.c	FrTp_Receiver()
FrTp_Sender.c	FrTp_Sender()
FrTp_Shutdown.c	FrTp_Shutdown()
FrTp_Transmit.c	FrTp_Transmit()
FrTp_TriggerTransmit.c	FrTp_TriggerTransmit()
FrTp_TxConfrmation.c	FrTp_TxConfrmation()

Table 10 Names and description of header files

File name	Description
FrTp.h	Includes the declarations and type definitions that need to be exported for use in other modules; that is, function prototypes
FrTp_Private.h	Contains the module specific declarations and macro definitions
FrTp_Types.h	Contains the module specific type definitions
FrTp_Cbk.h	Contains Call-back functions declarations

Table 11 Names and description of configuration files

File name	Description
FrTp_Cfg.h	Contains the pre-compile time configurable parameters for the FrTp module
FrTp_LCfg.c	Contains the link-time configurable parameters for the FrTp module
FrTp_PBCfg.h	Post build configuration header file for the FrTp module
FrTp_PBCfg.c	Contains the post build configurable parameters for the FrTp module

Table 12 Main interfaces of FrTp

Service name	Description
FrTp_Init	This service initializes all global variables of a FlexRay Transport Layer instance and set it in the idle state. It has no return value because software errors in initialization data shall be detected during configuration time (e.g., by configuration tool)
FrTp_Shutdown	This service closes all pending transport protocol connections by simply stopping operation, frees all resources and stops the FrTp Module
FrTp_Transmit	Requests transmission of a PDU
FrTp_RxIndication	Indication of a received PDU from a lower layer communication interface module
FrTp_TxConfrmation	The lower layer communication interface module confirms the transmission of a PDU, or the failure to transmit a PDU
FrTp_GetVersionInfo	Returns the version information
FrTp_CancelTransmit	Requests cancellation of an ongoing transmission of a PDU in a lower layer communication module
FrTp_CancelReceive	Requests cancellation of an ongoing reception of a PDU in a lower layer transport protocol module
FrTp_MainFunction	Maintains each channel state and checks for timer expirations

4.3.2 Interface Provided by FrTp Module

Table 12 lists the main interfaces of FrTp, which are provided to other modules.

5 Results and Discussion

Using Vector's CAN open environment (CANoe) VN7600, the hardware MPC5646C is tested on the vehicle electronic simulation platform. The framework of simulation platform is shown in Fig. 3. The communication between Panel ECU and Head ECU is through local interconnect network (LIN) bus. The communication between Head ECU (MPC5646C), Trail ECU, Engine ECU and Drive ECU is through FlexRay bus. FlexRay module code of Head ECU conforms to AUTOSAR standard and is the

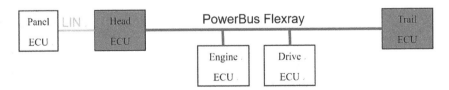

Fig. 3 Framework of simulation platform

tested ECU. Engine ECU and Driver ECU are simulated by CANoe. Gear status is sent by Drive ECU through FlexRay. Engine speed is sent by Engine ECU through FlexRay. The Head ECU sends the steering light status and the heating control command to the Trail ECU through FlexRay bus.

The Simulation nodes are shown in Fig. 4. The database file is shown in Fig. 5. The detail Flexray messages will be described as Table 13. The detail Flexray testing messages will be described as Table 14.

First, set the gearbox gear is D4, the speed is 150 km/h, and the engine speed is 4000 rpm on the panel. The frame of ID61 shows the gear is 4 and the speed is 150, as shown in Fig. 6. The frame of ID63 shows that the engine speed is 4000, as shown in Fig. 7. Considering the two frames, the FlexRay bus receiving function of the Head ECU is normal.

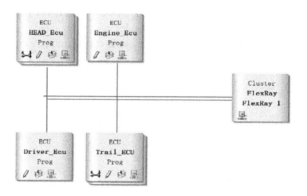

Fig. 4 Simulation nodes

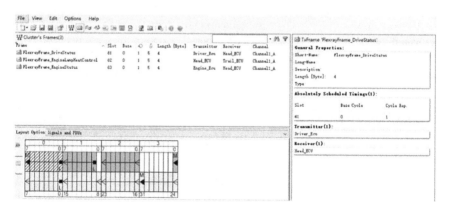

Fig. 5 Database file

Research on FlexRay Bus Communication Protocol ...   293

Table 13 Flexray messages detail descriptions

Signal name	Period (MS)	ID (Hex)	Start position (Update bit)	DLC	Frame
Gears	10	0×03D (61)	0 (8)	8	Drive status
Vehicle speed			9 (25)	16	
Engine speed	10	0×03F (63)	0 (16)	16	Engine status
Exhaust and brake			17 (25)	8	
Engine shut down command	5	0×03E (62)	0 (1)	1	Engine lamp heat control
Clearance lamp switch status			2 (3)	1	
Left turn signal switch status			4 (5)	1	
Right turn signal switch			6 (7)	1	
Power supply voltage			8 (24)	16	
Heating switch status			25 (26)	1	

Table 14 Flexray testing messages detail descriptions

Name	Type	The source ECU	Value
Gears	FlexRay bus signal	Drive ECU	0—N; 1—D1; 2—D2; 3—D3; 4—D4; 5—D5; 6—D6; 7—R
Vehicle speed	FlexRay bus signal	Drive ECU	0–150 km/h
Engine speed	FlexRay bus signal	Engine ECU	0–4000 rpm
Left rear turn signal switch	FlexRay bus signal	Head ECU	1—open; 0—close
Right rear turn signal switch	FlexRay bus signal	Head ECU	1—open; 0—close
Heating switch status	FlexRay bus signal	Head ECU	1—open; 0—close

Fig. 6 Frame of ID61

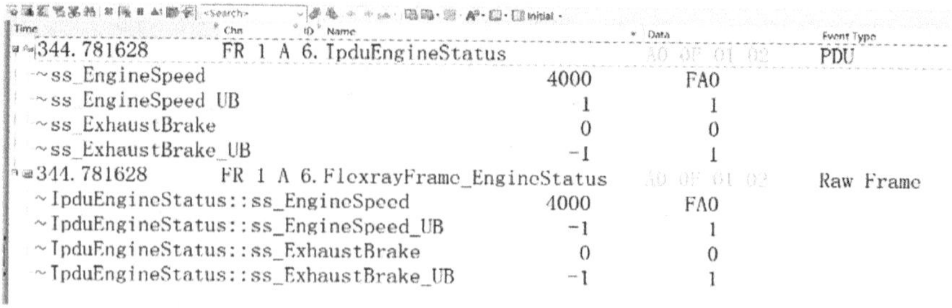

```
Time                    Chn      ID  Name                              Data        Event Type
344.781628              FR 1 A 6. IpduEngineStatus          A0 0F 01 02 PDU
 ~ss_EngineSpeed                                    4000        FA0
 ~ss_EngineSpeed_UB                                   1           1
 ~ss_ExhaustBrake                                     0           0
 ~ss_ExhaustBrake_UB                                 -1           1
344.781628              FR 1 A 6. FlexrayFrame_EngineStatus  A0 0F 01 02 Raw Frame
 ~IpduEngineStatus::ss_EngineSpeed                 4000        FA0
 ~IpduEngineStatus::ss_EngineSpeed_UB               -1           1
 ~IpduEngineStatus::ss_ExhaustBrake                  0           0
 ~IpduEngineStatus::ss_ExhaustBrake_UB              -1           1
```

Fig. 7 Frame of ID63

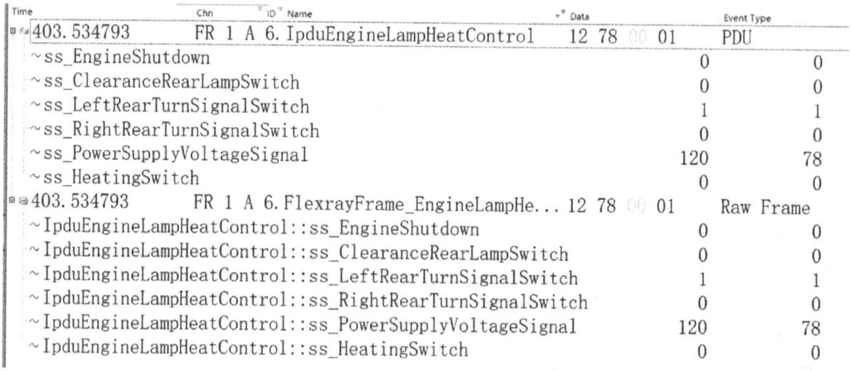

```
Time                    Chn      ID  Name                              Data        Event Type
403.534793              FR 1 A 6. IpduEngineLampHeatControl  12 78 00 01 PDU
 ~ss_EngineShutdown                                  0           0
 ~ss_ClearanceRearLampSwitch                         0           0
 ~ss_LeftRearTurnSignalSwitch                        1           1
 ~ss_RightRearTurnSignalSwitch                       0           0
 ~ss_PowerSupplyVoltageSignal                      120          78
 ~ss_HeatingSwitch                                   0           0
403.534793              FR 1 A 6. FlexrayFrame_EngineLampHe... 12 78 00 01 Raw Frame
 ~IpduEngineLampHeatControl::ss_EngineShutdown       0           0
 ~IpduEngineLampHeatControl::ss_ClearanceRearLampSwitch 0        0
 ~IpduEngineLampHeatControl::ss_LeftRearTurnSignalSwitch 1       1
 ~IpduEngineLampHeatControl::ss_RightRearTurnSignalSwitch 0      0
 ~IpduEngineLampHeatControl::ss_PowerSupplyVoltageSignal 120    78
 ~IpduEngineLampHeatControl::ss_HeatingSwitch        0           0
```

Fig. 8 Frame of ID62

The left turn switch is turned on the panel. As shown in Fig. 8, the frame of ID62 shows that the left turn light is turned on. The right turn and heating switch are turned on the panel. As shown in Fig. 9, the frame of ID62 shows that the right turn light is turned on and the heating switch is turned on. Considering the two frames, the FlexRay bus transmission function of the Head ECU is normal.

6 Conclusion

The FlexRay bus communication protocol stack based on MPC5646C hardware platform, which conforms to AUTOSAR specification, can realize normal communication function. It contributes to solving the problems of longer software development cycle, higher labor cost and lower software portability caused by the increasing number and complexity of electronic control units of railway vehicles and the rapid increase of software codes.

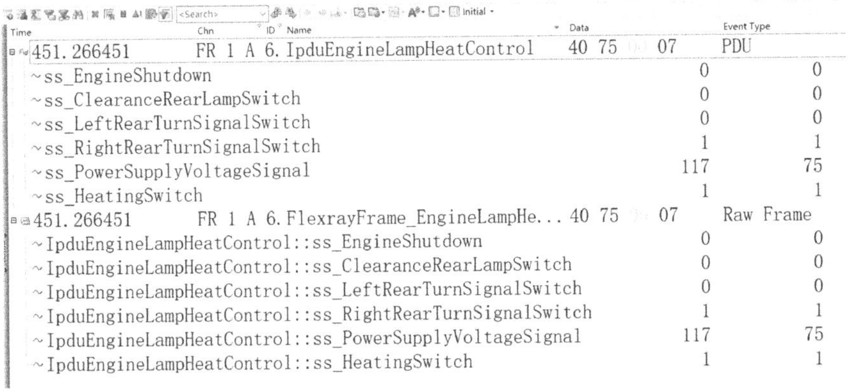

Fig. 9 Frame of ID62

References

1. Wang H (2018) Brief analysis of the characteristics and development trend of the braking system of urban rail transit vehicles. People's Transp 12:50–51 (in Chinese)
2. Guo QY, Kang JS, Feng JH, Zhou GF (2017) Development status and trend of traction control for rail transit vehicles. J Power Supply 15(02):40–45 (in Chinese)
3. Bai WW (2018) Research on the optimization of communication mechanism for in-vehicle FlexRay. Beijing Jiaotong University (in Chinese)
4. Zhang YS (2016) Research of automotive FlexRay network communication management based on AUTOSAR. Hefei University of Technology (in Chinese)
5. Wang YF, Zhang YS, Liu HJ, Huang WK (2016) Fault detection and management for automotive FlexRay network communication based on AUTOSAR. Trans Chin Soc Agric Eng (Trans CSAE) 32(4):105–112 (in Chinese)
6. Wu X, Wang W, Xu DH (2015) Research of vehicle FlexRay network management based on AUTOSAR specification. Automot Sci Technol 03:32–37 (in Chinese)
7. Wu X (2015) Research on AUTOSAR network management of automobile FlexRay network. Hefei University of Technology (in Chinese)
8. Li B (2012) Research and implementation of FlexRay bus communication protocol based on AUTOSAR. University of Electronic Science and Technology (in Chinese)
9. Li YM, Ni YL, Li S, Qiao FP (2017) Research on CAN communication protocol stack of vehicle E/E system based on AUTOSAR criterion. Comput Meas Contr 25(11):239–243 (in Chinese)

Research on Indoor Location Technology in Metro Station

Zongyi Xing, Hang Yang and Yuan Liu

Abstract In order to assist blind people to travel independently and solve the problems of jumping of location points and poor real-time location caused by traditional indoor location method, a method based on region classification and electronic map fusion is proposed. In this method, the idea of position fingerprint matching is adopted. In the off-line stage, the best Gauss filtering template is searched for RSSI sequence generated by Bluetooth sensor by iteration optimization. The filtered sequence mean is used to construct position fingerprint database, and the support vector machine model is used to classify the position fingerprint database at the first level. In the online stage, the idea of sliding window is used to classify the location area in two levels. In the window range, KNN algorithm based on Euclidean distance is used to calculate the position coordinates, and the path layer information of electronic map is used to correct the position coordinates, so as to further control the error range and improve the location efficiency. Experiments in the subway station hall show that the filtering method improves the positioning accuracy by nearly 4%, and the positioning accuracy can reach 1.59 m by using this positioning algorithm.

Keywords Indoor location · Gauss filtering · Support vector machine · Sliding window · Electronic map

1 Introduction

In recent years, the rise of the concept of smart city and the large-scale application of cloud computing technology have led to the rapid development of location-based service (LBS) industry [1]. The indoor positioning market represented by Bluetooth, WiFi and UWB positioning technology is becoming more and more mature, and it brings great convenience to human activities in airports, hospitals, warehouses and other scenarios. The number of blind people in China is huge and growing rapidly. The travel problems of these traffic vulnerable groups are receiving more and more

Z. Xing (✉) · H. Yang · Y. Liu
School of Automation, Nanjing University of Science and Technology, Nanjing, China
e-mail: xingzongyi@163.com

attention [2]. With the rapid advancement of China's modernization process and the continuous improvement of urban infrastructure, metro, as a representative of modern social travel tools, has been widely favored by visually impaired people with its advantages of safety, speed and convenience [3].

Because most of the metro stations do not provide power supply interface, the operators restrict the changes of facilities in the station (such as ground embedded receivers, roof, and gallery installation receivers hanger) [4]. The low power Bluetooth positioning technology based on iBeacon has the characteristics of simple networking mode, independent of external power supply at the device access end, and high precision positioning effect in medium and long distance. Therefore, the technology is very suitable for positioning in metro stations [5]. At present, Bluetooth-based indoor location algorithms are mainly based on received signal strength indication (RSSI) indoor location and matching location based on signal fingerprint library. Fingerprint matching is used to get rid of the dependence on path loss model, effectively reduce the multipath effect of Bluetooth signal and the influence of non-line-of-sight propagation on location accuracy [6]. At the same time, some efficient machine learning algorithms can be used in off-line training stage, which can often achieve the desired location accuracy. Therefore, this method has been widely studied and used [7].

In this paper, firstly, support vector machine (SVM) algorithm is used to classify the positioning area off-line, then sliding window is used to classify the positioning area twice in the process of online matching and positioning, and the error range is further controlled. Finally, the passenger positioning coordinates are corrected by combining the electronic map in the station. Experiments in laboratory and subway stations show that the method has better positioning effect [8].

2 Algorithm

2.1 Position Matching Algorithm Based on SVM and Sliding Window

The iBeacon information received in Android application includes the Universal Unique Identifier (UUID) corresponding to the module, the area number Major identifying a certain location area, the Minor, RSSI value identifying each iBeacon in the area, and the distance between the receiver and the Bluetooth module, where RSSI meets the logarithmic path distance loss model, as shown in formula (1).

$$P(\mathrm{d}B) = P(d_{EF}) + 10n \lg\left(\frac{d}{d_{EF}}\right) + \xi \tag{1}$$

In the formula, $P(d_{EF})$ refers to the receiving power at the reference distance d_{EF} (usually 1 m), $P(dB)$ refers to the receiving power at the reference distance d, and the path loss index n, generally between 2 and 4, ξ is the path loss factor.

The implementation of the location fingerprint location algorithm can be divided into off-line training phase and online matching phase. In the off-line training phase, firstly, the Bluetooth signal intensity collected from each observation point is analyzed, and then the adaptive Gauss algorithm is used to de-noise to establish the position fingerprint features of each observation point. Finally, the data objects are clustered by SVM algorithm. In the online positioning stage, a dynamic window is constructed by selecting the matched location points around the cluster, which further reduces the time complexity of position matching in the moving process. At the same time, the navigation interface provided by ArcGIS is used to correct the position coordinates at the path level and display them. The detailed algorithm flow of this method includes data preprocessing, off-line training and online position matching and display.

2.2 Data Processing

Bluetooth signal is affected by multipath effect and non-line-of-sight transmission in the transmission process. RSSI value fluctuates greatly with time and has a span of up to 20 dBm. Using Gauss function fitting method to process the measurement data can effectively reduce the impact of some small probability, large interference events on the overall measurement, and significantly improve the ranging error. Adaptive Gaussian Filtering (AGF) has better denoising and smoothing effect than the traditional Gauss filtering with fixed filter template value by iterative optimization and selecting the optimal filter template suitable for RSSI sequence characteristics [9].

The adaptive Gauss filter uses the Gauss function, which is the probability density function of the normal distribution.

$$f(x) = \frac{1}{\sigma\sqrt{2\pi}} \exp\left(-\frac{(x-\mu)^2}{2\sigma^2}\right) \quad (2)$$

where σ represents RSSI Sequence Standard Deviation of the Point to be Measured, μ represents the size of one-dimensional signal filtering template.

Mean errors corresponding to different μ are detected in the preprocessing stage. The definition of mean error is as follows:

$$\text{error} = \sum_{k=1}^{N} (\text{rssi}'_{ik} - \overline{\text{rssi}_i}) \quad (3)$$

where N is the total number of measurements at the point with serial number i, rssi'_{ik} is the value of rssi_{ik} after Gaussian filtering, and $\overline{\text{rssi}_i}$ is the mean of iBeacon for the

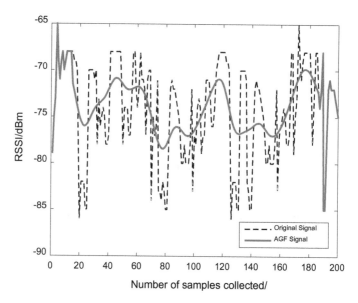

Fig. 1 Effects of AGF algorithm before and after filtering

corresponding ID in N measurements. Through iterative optimization, the μ value corresponding to error$_{min}$ is selected as the optimal template for Gauss filtering.

In the laboratory environment, AGF algorithm is used to filter the collected sample data. Figure 1 shows the effect before and after filtering. The fingerprint database is established by using the data filtered by GF and AGF respectively, and the same algorithm is used for online position matching. The accuracy of position matching is calculated within 1.5 m at each location point. The location effect diagram is shown in Fig. 2. The average matching accuracy of the GF filtering algorithm at each location is 83.7%, which is 87.5% when the AGF algorithm is used [10].

2.3 Support Vector Machine Model

Support Vector Machine (SVM) is a two-class classification model. Its basic model is defined as a linear classifier with the largest spacing in feature space. Its learning strategy is to maximize the spacing. For the m-dimensional eigenvector composed of sample signal strength set, the input space data can be mapped to the high-dimensional eigenvalue space by using the non-linear function $\varphi(x)$, and the optimal classification hyperplane can be constructed in the space:

$$f(x) = \omega\varphi(x) + b = 0 \qquad (4)$$

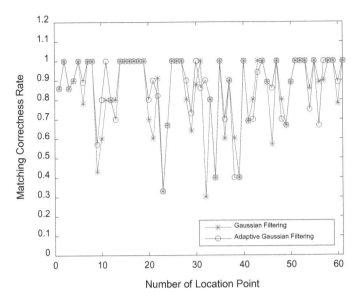

Fig. 2 Contrast map of location accuracy of two filtering methods

where ω is the weight vector, b is the deviation, ω and b determine the location of the classification surface. In order to find the optimal classification hyperplane corresponding to formula (4), the maximum problem with constraints has to be solved.

$$\begin{cases} \min\left(\dfrac{1}{2}\|\omega\|^2 + C\sum_{i=1}^{n}\xi_i\right) \\ \text{s.t.} y_i(wr_i^T + b) \geq 1 - \xi_i, \quad i = 1, 2, \ldots, n \\ C > 0 \\ \xi_i \geq 0 \end{cases} \quad (5)$$

where C is the penalty factor, which is proportional to the importance of SVM classification to the wrong samples. To solve this kind of constrained minimum value problem, Lagrange equation needs to be applied. After a series of transformations, the quadratic programming problem can be transformed into a corresponding dual problem, namely:

$$\begin{cases} \max\left(\dfrac{1}{2}\sum_{i=1}^{N}\sum_{j=1}^{N}\alpha_i\alpha_j y_i y_j(x_i \cdot x_j) + \sum_{i=1}^{N}\alpha_i\right) \\ \text{s.t.} \sum_{i=1}^{N}\alpha_i y_i = 0, \alpha_j \geq 0 \quad (j = 1, 2, \ldots N) \end{cases} \quad (6)$$

By calculating the optimal solution $\tilde{\alpha} = (\tilde{\alpha}_1, \tilde{\alpha}_2, \ldots, \tilde{\alpha}_N)$, the optimal weight vector ω and the optimal offset $\tilde{b}$ separated by hyperplanes, the following decision functions can be used to determine the type of samples:

$$f(x) = \text{sgn}\left(\sum_{i=1}^{N} \tilde{\alpha}_i y_i (x_i \cdot x) + \tilde{b}\right) \tag{7}$$

2.4 Online Fingerprint Matching Based on Sliding Window

The reciprocal of the RSSI change norm is used as an estimate of the current state:

$$w = \frac{1}{\|\mathbf{x} - \mathbf{rssi}\|_p} \tag{8}$$

Generally we take $p = 2$, the reciprocal of Euclid distance is taken to represent the similarity between the observed eigenvector **x** and the position fingerprint vector. Since SVM multi-classifiers train classifier models in M sub-regions, **x** is substituted into each sub-region. The sub-region corresponding to the target value $c_i = +1$ is the sub-region of the current location.

The idea of sliding window can be understood as that in the time interval of two online matches, the distance of blind passengers is limited, so they can choose the location point adjacent to the previous location point as the next matching area. As passengers move, the window range is also changing. The shape of the window is defined as follows:

$$W = \begin{bmatrix} 1 & 1 & 1 & 1 & 1 \\ 1 & 1 & 1 & 1 & 1 \\ 1 & 1 & 0 & 1 & 1 \\ 1 & 1 & 1 & 1 & 1 \\ 1 & 1 & 1 & 1 & 1 \end{bmatrix} \tag{9}$$

The size of the window can be adjusted and some adjacent points can be discarded at the boundary of the sub-region. Finally, $K(K \geq 2)$ locations with smaller w values are selected as the reference points for final location in the window, and the average values of corresponding coordinates are calculated as the result of position estimation, that is, KNN algorithm. The calculation formula is as follows:

$$(x, y) = \frac{1}{K} \sum_{i=1}^{K} (x_i, y_i) \tag{10}$$

3 Indoor Electronic Map and Map Correction

In this system, mixed vector and raster layers are used to represent the geographic information of indoor maps. There are three main types of layers: points, lines and planes. Each layer is a collection of certain elements and represents a kind of map information, such as blind path, elevator and inquiry desk. In addition, there are some extended application layers such as grid element layer, navigation connection point layer and annotation layer.

The positioning system is implemented on Android application, and APP will plan its path (along the blind road as far as possible) according to the destination chosen by blind passengers. When the location coordinates of blind passengers deviate from the path, the grid layer "pulls" them back to the established path. The model is shown in Fig. 3 For the position coordinates of the blind passengers at the previous moment and the actual coordinates of the current moment, the online position matching results are used to determine whether the passengers deviate from the established path or not.

The electronic navigation map (partial) of the first floor of the underground station hall of Suyuan Station of Guangzhou Metro is made by ArcGIS software, as shown in Fig. 4.

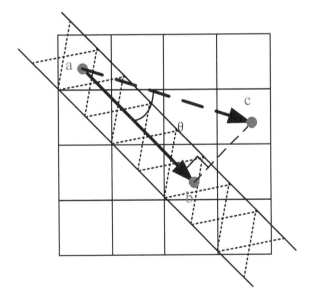

Fig. 3 Model diagram for error correction of location points

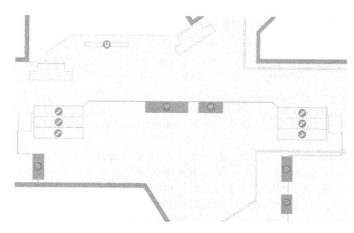

Fig. 4 Electronic map of the hall layer of SuYuan station of Guangzhou metro (Part)

4 Experiments and Conclusions

In the experimental scenario, the underground floor of SuYuan Station of Guangzhou Metro is selected, and the location area is 2600 m^2. In this area, iBeacon(supporting BLE4.0 protocol) is deployed evenly in the horizontal and vertical direction at a distance of 3 m. The Android mobile phone with Honor 6X is used for data acquisition, fingerprint matching and position display at the application end.

The location area is divided into 2436 reference points for signal acquisition. The Bluetooth fingerprint data format after acquisition is: position point serial number, position point relative coordinates, sensor type identification and RSSI array. The iBeacon broadcasting frequency is set to 10 Hz, and 200 sets of data are collected at each reference point to construct training set. Table 1 is the statistical information of positioning error distance obtained from the experimental data of navigation and positioning by three positioning methods.

It can be seen from the table above that the location errors of traditional fingerprint matching algorithm based on KNN are concentrated in the range of more than or equal to 2 m, and there are few observation points with an error of about 1 m. After classifying the location area by SVM, the proportion of observation points with error distance of about 2 m is increased, but the overall positioning effect is not improved.

Table 1 Statistical information of location error distance for several location algorithms

Location algorithm	Location error probability (%)				Average error (m)
	1 m	2 m	3 m	4 m	
KNN	25.3%	73.1%	94.2%	98.7%	2.087
KNN + SVM	27.1%	79.5%	95.0%	98.7%	1.997
Our algorithm	55.0%	87.1%	98.6%	99.5%	1.598

In contrast, the algorithm described in this paper improves the positioning accuracy to 55% within 1 m, which obviously improves the overall positioning accuracy. At the same time, using the idea of two-level classification, the time complexity of online location is reduced from $O(n)$ to $O\left(\frac{n}{M \cdot |W|}\right)$, where n is the summary of data objects, M is the number of sub-regions and $|W|$ is the size of sliding window. In summary, the fast and accurate characteristics of this algorithm make it very suitable for real-time location of blind passengers in metro stations.

Acknowledgements This work is supported by National Key R&D Program of China (2017YFB1201203).

References

1. Zhan J, Wu L-X, Tang Z-J (2010) Ranging method and accuracy analysis based on RSSI of wireless sensor network. Telecommun Eng (in Chinese)
2. Lu H, Liu X, Zhang C et al (2010) Compare the triangle location algorithm and the fingerprint recognition algorithm based on WiFi positioning. Mob Commun 34(10):72–76 (in Chinese)
3. Min Z (2008) Analysis of indoor location technologies. Mod Comput (in Chinese)
4. Chao-Lan X, Jun-Li G, Xiao-Hua Z et al (2019) Algorithm of bluetooth indoor localization based on K-means and SVM. Transducer and Microsystem Technologies (in Chinese)
5. Wang Hongyu, Yan Yu, Jin Minglu et al (2012) Toward robust indoor localization based on Bayesian Filter using chirp-spread-spectrum ranging. IEEE Trans Industr Electron 59(3):1622–1629 (in Chinese)
6. Zhi-Qiang LI (2016) Research on indoor positioning technology based on the fusion of particle filter and electronic map. University of Electronic Science and Technology of China, Sichuan (in Chinese)
7. Yu Y-P (2014) A new indoor map matching algorithm fusing map and sensor information. Telecommun Eng (in Chinese)
8. Sang N, Xing-Zhong Y, Zhou R (2014) Method of WiFi indoor location based on SVM. Appl Res Comput (in Chinese)
9. New patterns for mobile indoor maps designing based on spatial cognitive. J Syst Simul (9):2097–2102 (2014) (in Chinese)
10. Qian MO, Shuo X (2014) Close degree classification of indoor positioning algorithm based on bluetooth 4.0. J Astronaut Metrol Meas (in Chinese)

Arrival Train Delays Prediction Based on Gradient Boosting Regression Tress

Rui Shi, Jing Wang, Xinyue Xu, Mingming Wang and Jianmin Li

Abstract Delay prediction based on real-world train operation records is an essential issue to the delay management. In this paper, we present the first application of gradient boosting regression tress (GBRT) prediction model that can capture the relation between train delays and various characteristics of a railway system. Delayed train number (DN), station code (SC), scheduled time of arrival at a station (ST), time travelled (TT), distance travelled (DT), and percent of journey completed distance-wise (PC) are selected as the explanatory variables, and the delay time (WD) is the target variable. The model can evaluate various impact factors on train delays, which can assist dispatchers to make decisions. The results demonstrate that the GBRT model has a higher prediction precision and outperforms the support-vector machine (SVR) model and the random forest (RF) model.

Keywords High-speed railway · Real-world data of train operation · Arrival train delay prediction · Gradient boosting regression tress

1 Introduction

By 2018, China has had more than 25,000 km of high-speed railways in operation, and more than 65% of the high-speed trains are operated every day. The construction and operation of high-speed railway have achieved remarkable achievements in improving the scale and quality of railway network, easing the shortage of transport capacity, and improving the quality of railway transport service [1]. However, statistics from China railway corporation (CRC) shows that the punctuality of the high-speed trains in China is only 90% in 2015, despite the fact that their starting and

---

R. Shi · X. Xu (✉) · M. Wang · J. Li
State Key Laboratory of Rail Traffic Control and Safety,
Beijing Jiaotong University, Beijing, China
e-mail: xxy@bjtu.edu.cn

School of Traffic and Transportation, Beijing Jiaotong University, Beijing, China

J. Wang
China Waterborne Research Institute, Beijing, China

finishing times have reached 98.8% and 95.4%, respectively. Train delays cause huge losses for railway operators and passengers. Thus, it is extremely critical to study and predict train delays to improve the punctuality and the quality of train services.

Various techniques have been used to predict train arrival delays. Different random distributions are considered, such as Weibull, chi-squared, Erlang, and log-normal, while exponential distributions are mostly used [2, 3]. However, these distributions are hypothetical and difficult to adapt to different scenarios of railway systems. With the development of big data technology, machine-learning model becomes popular. These models use the presented data to learn a general rule that maps inputs to outputs. Artificial neural networks (ANN) are mostly used to predict delays of passenger trains [4–7]. They typically provide better fit but less straightforward interpretation than traditional models. Marković presents the support-vector regression in analysis of train delays and explores an expert opinion as an approach for estimating influence of infrastructure on train arrival delays [8]. Xu et al. give a short-term passenger flow prediction model under passenger flow control using a dynamic radial basis function network [9]. A random forest regression model is also established to study the arrival time prediction [10]. However, there is plenty of room of these models for accuracy improving.

Clearly, there is still a need for better predictive models that account for massive real-world train operation data. Thus, based on the real-world train operation records of Guangzhou Railway Bureau database, the influencing factors of high-speed train delay are analyzed in this paper, and a machine-learning model is proposed. To our best acknowledge, this is the first application of the gradient boosting regression tress (GBRT) in the analysis of train delays. The case study shows that it has good data fitting and predictive properties.

2 Materials and Methods

2.1 Description and Analysis of the Data Set

The data used in this study come from train operation records on Wuhan–Guangzhou (WH–GZ) high-speed rail (HSR) in China. The WH–GZ HSR connects Wuhan in Hubei province to Guangzhou in Guangdong province with a 1096 km double-track HSR line. A total of 15 stations and 14 sections, from GuangzhouSouth (GZS) to ChibiNorth (CBN), are managed by the Guangzhou Railway Bureau. All train operation records were obtained from the CTCS system, including all delay train records from GZS railway station to CBN railway station. The data recorded the train number, station, the planned and actual arrival and departure time of each train at each station, the maximum train running speed and the daily traffic volume, etc. Some data are shown in Table 1.

Due to the different redundant time distribution, timetables of up and down train need to be considered separately. This paper only considers the data of up train for

Table 1 Original train operation data

Train	Date	Station code	Scheduled arrival time	Scheduled departure time	Actual arrival time	Actual departure time
G1152	2018-8-11	361	13:35	13:37	13:40	13:42
G1505	2018-8-11	361	14:50	14:53	14:55	14:57
G537	2018-8-11	361	15:12	15:14	15:16	15:17
G840	2018-8-11	361	15:44	15:46	15:49	15:51
G9691	2018-8-11	361	22:32	22:34	22:37	22:38

analysis. Between July 1, 2018 and December 31, 2018, a total of 330,787 arrival and departure events were recorded in the CTCS system, among which 63,119 were data of train arrival delaying exceeding 2 min. Data analysis and dispatcher interviews revealed several factors that could affect HSR train delays. The first obvious factor is the category of trains, since high-level trains have priority in the conflict event. Secondly, station is a resource for dispatchers to adjust train operation and recover train delay, which can absorb the delay time caused by random factors in train operation to some extent [11]. The arrival delay rate of each station is shown in Table 3. Moreover, the scheduled arrival time could also be taken into account as we expect more delays during peak hours. In addition, the travel distance is also closely related to train delay since the farther the distance, the higher the expected delay. Finally, the travel time and the percentage of travel completed are also considered to improve the prediction accuracy.

2.2 Statistics of the Variables

Based on the analysis of available data, we have selected seven factors that might influence train arrival delays on railways. The six independent input variables are as follows:

- High-speed train number (DN)
- The station code (SC)
- Scheduled time of arrival at station (ST)
- Percent of journey completed distance-wise (PC)
- Distance travelled (DT)
- Time travelled (TT).

The correlation between train delays and every input variable is shown in Fig. 1. While there may be linear correlation between some variables, for example, a negative linear correlation between DN and DT may exists, the dependent variable WD has a complex correlation with each independent variable which is difficult to determine. The complex relationships between the above variables indicate that traditional

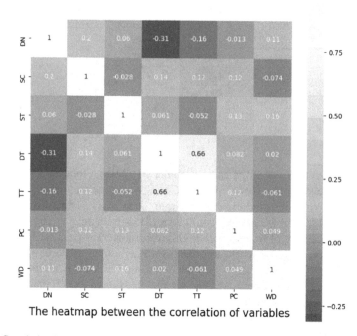

Fig. 1 Correlation between train delays and input variables. The lighter the color of the square, the stronger the positive correlation between the variables corresponding to the *x*-axis and *y*-axis, respectively. Conversely, the darker the color, the higher the negative correlation

statistical models (such as multiple linear regression model) are not suitable for solving this problem, and the prediction outcomes of these models will be inaccurate. Therefore, the machine-learning models which can solve the complex relationship are considered in this paper.

2.3 Gradient Boosting Regression Tress

GBRT is a boosting algorithm improvement. The original boosting algorithm gives an equal weight value to each sample in the first place, and all the initial basic learners have the same importance. The model obtained in each training will make the estimation of data points slightly different, so at the end of each step, the weight value needs to be processed by increasing the weight of the wrong classification points and reducing the weight of the correct classification points. This makes certain points "seriously concerned" if they are misclassified for a long time, giving them a high weight. N simple basic learners will be obtained after N iterations. Finally, they will be combined and weighted or voted to get a final model.

GBRT is greatly different from traditional boosting. The kernel of GBRT lies in that every calculation is to reduce the last residual, and in order to reduce these residuals, a new model can be established in the direction of the gradient of residual reduction. In GBRT, each new model is built to reduce the residuals of previous models in the gradient direction, which is greatly different from the traditional boosting algorithm in weighting correct and wrong samples.

The specific generation process of GBRT model is as follows: the dataset is $D = \{(x_1, y_1), (x_2, y_2), \ldots, (x_N, y_N)\}$, and the loss function is $L(y, f(x))$. The number of leaf nodes of each regression tree is J, and its input space is divided into J disjoint regions $R_{1m}, R_{2m}, \ldots, R_{jm}$, and a constant value b_{jm} is estimated for each region. The regression tree $g_m(x)$ is expressed by the formulation as follows:

$$g_m(x) = \sum_{J=1}^{J} (b_{jm} I), \, x \in R_{jm} \tag{1}$$

$$I(x \in R_{jm}) = \begin{cases} 1, x \in R_{jm}; \\ 0, \text{others}. \end{cases} \tag{2}$$

Step 1: Initialization of the model:

$$f_0(x) = \arg\min_{\rho} \sum_{i=1}^{n} L(y_i, \rho) \tag{3}$$

Step 2: Iteratively generate M regression trees:

Calculate the negative gradient value of the loss function, and use it as the estimated value r_{im} of the residual:

$$r_{im} = -\left[\frac{\partial L(y_i, f_{m-1}(x_i))}{\partial f_{m-1}(x_i)}\right]_{f(x)=f_{m-1}(x)} \tag{4}$$

A regression tree $g_m(x)$ is generated according to the residual generated in the previous step, and the m tree is divided into J disjoint regions $R_{1m}, R_{2m}, \ldots, R_{jm}$, the step size of gradient descent is calculated as follows:

$$\rho_m = \arg\min_{\rho} \sum_{i=1}^{n} L(y_i, f_{m-1}(x_i)) + \rho g_m(x_i)) \tag{5}$$

Step 3: Update the model, where lr is expressed as learning rate:

$$f_m(x) = f_{m-1}(x) + \text{lr} * \rho_m g_m(x) \tag{6}$$

Step 4: Output model $f_M(x)$

GBRT is suitable for well handling unclean and noisy data, supporting different loss function. And it also has strong predictive ability for nonlinear data [12], which avoids the overfitting problem in decision tree learning by stopping tree growth as early as possible. Through grid search, we finally determine the optimal parameters, and the main parameters are as follows: learn_rate = 0.01, n_estimators = 200, max_depth = 30, and alpha = 0.9. The other two models SVR and RF will not be introduced due to limited space.

2.4 Statistical Measurements

To evaluate the performance of the models, some criteria are used in this paper. These criteria include mean average error (MAE), squared error (MSE), root-mean-squared error (RMSE), and R^2. These statistical parameters are defined as follows:

$$\begin{cases} \text{MAE} = \frac{\sum_{i=1}^{m} |y_i - \bar{y}|}{m} \\ \text{MSE} = \frac{\sum_{i=1}^{m} (y_i - \bar{y})^2}{m} \\ \text{RMSE} = \sqrt{\frac{\sum_{i=1}^{m} (y_i - \bar{y})^2}{m}} \\ R^2 = 1 - \frac{\sum_{i=1}^{m} (y_i - f(x_i))^2}{\sum_{i=1}^{m} (y_i - \bar{y})^2} \end{cases} \quad (7)$$

where m denotes the number of target values $y = (y_1, y_2, \ldots, y_m)^T$, $\bar{y}$ is the prediction value, and $f(x_i)$ represent the regression function for feature vector x_i. All statistical analyses are done using Python 3.6 programming.

3 Case Study

Because, too small delays may be absorbed by the buffer time in station or near section directly, it is reasonable to consider them to be unobvious. In this paper, 654 trains with delay time longer than 2 min were extracted, and trains with delay time increasing due to secondary or multiple disturbances were deleted. After noise reduction and other pre-processing process of data, the remaining data sample size for modeling was 1306.

To compare the GBRT model with SVR and RF model, 80% of the data in the dataset are randomly selected to train the models, and the remaining 20% are as the test set. The training set is applied to calculate various criteria of different models. The results are shown in Table 2. The R^2 value (i.e., coefficient of determination) of GBR is 0.9009, higher than other two models, and MAE, MSE, *RMSE* are all the lowest, which indicates that GBR model has the best fitting effect among all models. The comparison of prediction results of each model is also visualized in Fig. 2, which is complementary to the above results. To verify the validity of each

Table 2 Comparing the criteria of GBRT model with the other two methods on training set

Methods	R^2	MAE	MSE	RMSE
SVR	0.8730	0.6356	2.5791	1.6059
RF	0.8866	0.9957	2.1628	1.4706
GBR	0.9063	0.4628	1.8845	1.3727

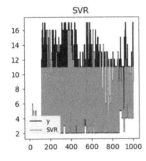
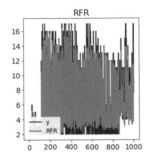
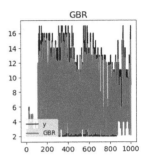

Fig. 2 Comparison between the real delay and the predicted delay by three models in the training dataset

model, the test set is input into the trained model. Although these criteria of GBRT for test date are a little lower than those for training data, these are still the highest of all models shown in Table 4. The R^2 value of SVR is only 0.6348, which indicates the effect of overfitting to noise. The results show that the generalization ability and the prediction accuracy of the GBR are better than the other two models. Visualization of the renderings (Fig. 3) is helpful to better understand the regression capabilities of each model. As expected, the GBRT outperforms other two models (Table 3).

4 Conclusion

Delay prediction is an essential issue to the delay management. In this paper, train operation records on WH–GZ HSR line is obtained, and the running data of delayed train (delay time is more than 2 min) are extracted from train operation recorded data. According to the correlation between each explanation variable, the first application of GBRT prediction model is proposed in this paper. Delayed train number, station code, scheduled time of arrival at a station, time travelled, distance travelled, percent of the journey completed distance-wise are selected as the explanation variables, and the delay time is the target variable. The performance of the model is compared with the SVR and the RF model which were used for similar problems in the literature. The main conclusions can be summarized as follows:

- GBRT model can well fit the delay data of high-speed trains under different spatial and temporal distributions;

Table 3 Rate of delays at each station on the line studied

Station code	Number of arrival trains	Number of arrival delayed trains	Rate of delays (%)
369	64,762	6605	12.13
368	71,586	27,266	26.83
367	35,792	14,550	21.12
366	35,794	11,573	8.00
365	35,788	15,338	39.02
364	35,612	16,199	58.96
363	35,422	16,384	47.51
362	35,350	14,887	42.11
361	60,920	28,946	46.25
360	26,112	15,396	45.49
359	25,454	9933	42.86
357	48,712	3898	32.33
356	31,394	6632	40.65
355	22,711	6085	38.09
354	30,446	3669	10.20

Table 4 Comparing the criteria of GBRT model with the other two methods on test set

Methods	R^2	MAE	MSE	RMSE
SVR	0.6349	1.7440	9.4515	3.0743
RF	0.7257	1.5157	4.4284	2.1043
GBR	0.8353	1.1299	3.5952	1.8961

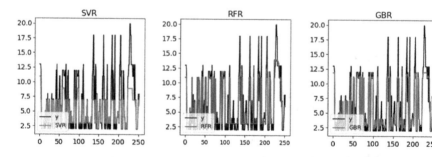

Fig. 3 Comparison between the real delay and the predicted delay by three models in the test dataset

- GBRT model can capture the relation between train delays and different characteristics of a railway system;
- The comparison between GBRT model, SVR model, and RF model shows that GBRT model has better prediction accuracy.

In future work, characteristic of the infrastructure as input variables can be integrated into this model to increase the prediction accuracy of the GBRT. These variables would include the number of junctions and stations, type of signalization, and other characteristics of the infrastructure, which can be quantified by capturing these influences. It would require extensive data collection to calibrate the delay models.

Acknowledgements The authors gratefully acknowledge the support provided by the national key research project "Railroad Comprehensive Efficiency and Service Level Improvement Technology with High Speed Railway Network (Grant No. 2018YFB201403)" in China.

References

1. Peng Q, Li J, Yang Y, Wen C (2016) Influences of high-speed railway construction on railway transportation of China. J Southwest Jiaotong Univ 51(2/3):525–533 (in Chinese)
2. Yuan J (2006) Stochastic modelling of train delays and delay propagation in stations. Ph.D. thesis, TU Delft
3. Goverde RMP, Corman F, D'Ariano A (2013) Railway line capacity consumption of different railway signalling systems under scheduled and disturbed conditions. J Rail Transport Plan. Manage 3(3):78–94
4. Malavasi G, Ricci S (2001) Simulation of stochastic elements in railway systems using self-learning processes. Eur J Oper Res 131(2):262–272
5. Peters J, Emig B, Jung M, Schmidt S (2005) Prediction of delays in public transportation using neural networks. In: International international conference on computational intelligence for modelling. IEEE Computer Society
6. Yaghini M, Khoshraftar MM, Seyedabadi M (2013) Railway passenger train delay prediction via neural network model. J Adv Transp 47(3):355–368
7. Pongnumkul S, Pechprasarn T, Kunaseth N, Chaipah K (2014) Improving arrival time prediction of Thailand's passenger trains using historical travel times. In: 11th International joint conference on computer science and software engineering (JCSSE). IEEE, pp 307–312
8. Marković N, Milinković S, Tikhonov KS, Schonfeld P (2015) Analyzing passenger train arrival delays with support vector regression. Transp Res Part C Emerg Technol 56:251–262
9. Li H, Wang Y, Xu X et al (2019) Short-term passenger flow prediction under passenger flow control using a dynamic radial basis function network. Appl Soft Comput J. https://doi.org/10.1016/j.asoc.2019.105620
10. Yuan Z, Zhang Q, Huang K et al (2016) Forecast method of train arrival time based on random forest algorithm. Railway Transport Econ 38(05):60–63+79 (in Chinese)
11. Wen C, Peng Q (2011) Mechanism of train operation conflict on high-speed rail. J Transp Eng 12(2):119–126 (in Chinese)
12. Hong W (2015) Wavelet gradient boosting regression method study in short-term load forecasting. Smart Grid 05:189–196

The Subway Passenger Flow Macroscopic State Analysis

Yanhui Wang, Hao Shi and Suizheng Zhang

Abstract Recently, the ridership explosion brings great challenges to the service of subway companies such as crowdedness in trains and the insufficient capacity of subway facilities. Thus, understanding, describing, and controlling the subway passenger flow state is necessary to mitigate estimated or forecasted traffic congestion. In this context, we incorporated the passenger volume of the station which includes inbound, outbound and transfer process and a load of section into the exploration of system element states. Thus, the network can be examined in the context of operation, which is the ultimate purpose of the existence of a subway network. Utilization rate and load factor were proposed to describe the state of station and section. Based on the basic states of elements, system passenger flow macroscopic states which can describe the subway system operation condition from a global perspective are presented with the entropy. Finally, Beijing subway system is implemented to validate the accuracy and superiority of this method.

Keywords Subway passenger flow · Element state · Macroscopic states

1 Introduction

Subway transportation systems are world widely adopted as effective countermeasures to mitigate the adverse effects of rapid urbanization and traffic congestion [1]. Compared with other public transport systems like bus system, taxi system, the subway system provides frequent more safe trips for large numbers of passengers in a short period of time. However, the expanding subway networks have stimulated a sharp rise in ridership; for example, there are 19 operating lines in Beijing subway systems by the end of 2017 and the passengers can on average reach 10 million each

---

Y. Wang
State Key Laboratory of Rail Traffic Control and Safety, Beijing Jiaotong University, Beijing 100044, China

H. Shi (✉) · S. Zhang
Beijing Hongdexin Zhiyuan Information Technology Co., LTD, Beijing 101102, China
e-mail: beyondshihao@163.com

weekday, which could increase to 11.5 million for peak periods [2]. The ridership explosion brings great challenges to the service of subway companies such as crowdedness in trains and the insufficient capacity of subway facilities [1]. To reduce the burden of high passenger demands and congestion, passengers are routinely restricted at some subway stations and have to line up for boarding or are even prohibited from entering the stations during peak hours [1]. The reasons for congestion include the mismatch between passenger flow and capacity, the mismatch between station infrastructure and passenger flow, and the untimely regulation of passenger flow [3]. Thus, fully understanding, describing and controlling the subway passenger flow state is necessary to mitigate estimated or forecasted traffic congestion.

The indirect method has been used in the traditional research to carry out an analysis of the changing trend of the passenger flow. Gao et al. [4] obtained the travel time distribution of passengers within a period of time between each pair of origin–destination and estimated the proportion of passengers on each path in this period. Wang et al. [5] established a disaggregate method based on the disaggregate model approach to forecast the passenger flow distribution. We et al. [6] proposed a dynamic simulation model of passenger flow distribution on schedule-based transit networks with train delays. Li et al. [7] introduced four radius fractal dimensions and two branch fractal dimensions by combining a fractal approach with passenger flow assignment model. These dimensions quantified the spatiotemporal distribution of passenger flow and described the change of characteristics.

Subway transport is a complex sociotechnical system, and its function is to transfer passenger from origin station to destination station through trains. Lots of valuable insights into the structure and dynamics of subway networks had been in-deed shown in previous studies, such as scale-free patterns with relatively high survivability and robustness and small world effects which including node degree, average shortest path, clustering coefficient, and efficiency [8, 9]. Nevertheless, these studies focus more on complex network topological properties of subway system, and with data collection challenges few of them considered the characters of passenger flow through network which is fundamental for a full description of the real-life network [10]. Few of studies have focused on the passenger flow macroscopic state which provides a new view to describe of control subway system safety.

2 Subway Elements and State Description

In this section, the states of station and section are presented after analyzing the evolution process of the stations and sections.

2.1 Subway System Elements Illustration

The network can be represented as a bi-directed passenger flow network. Figure 1

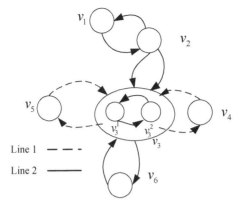

Fig. 1 Passenger flow network description

shows a simplified example of passenger flow network based on a topology network. The nodes' set $V = \{v_i | i = 1, 2, \ldots, N_v\}$ (Nv denotes the number of nodes) represents metro stations, and the edges' sets $L = \{l_{ij} | i, j \in V\}$ represent sections between these adjacent stations and l_{ij} means section from v_i to v_j.

In practice, a station could have several platforms, and if passengers want to transfer from one line to the other, they need to walk from one platform to the other. Hence, the transfer station is divided into multiple nodes $v_i = \{v_i^k | k = 1, 2, \ldots\}$ and k is determined by the number of operation lines. For example, in Fig. 1, station v_3 is split into two nodes $\{v_3^1, v_3^2\}$, because both line 1 and line 2 go through this station.

2.2 States Description of Passenger Flow

2.2.1 States of Station Passenger Flow

The activities process of passengers at stations can be divided into three procedures: inbound process, outbound process, and transfer process. Similar classifications are obtained by other authors. As shown in Fig. 2, inbound passengers arrived at a station from the outside network according to a time-dependent pattern during a given period and then moved from the entrance to the platform. The process contains these critical facilities: gates, staircases, escalators, walkways and hall. The outbound process is similar to the inbound process, but different in direction. The transfer process demonstrates the flow of transfer passengers within the station, i.e., the passengers' transfer from the current platform to other platforms of lines. In fact, passengers may need to make several interchanges between different lines to arrive at their destinations.

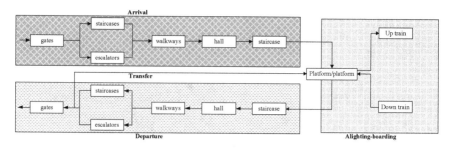

Fig. 2 Active process of passenger in subway system

At halls, platforms, and in transfer corridors, it is not realistic to observe the passenger flow in any place of station at present and this inspired us to study passenger flow state from the global perspective.

Station capacity can be defined as the maximum passenger volume that a station can accommodate in a given period of time under with fixed facilities of the station [11]. The capacity takes into account station services and passengers' security. With increasing passenger volume, control strategies such as boarding limiting were used to dynamically adjust passenger flow volume and relieve some demand pressure in crowded stations. For example, inbound passenger flow would be restricted by station managers once the total passenger volume reached 70% of the station capacity in Beijing subway [11]. Thus, the station capacity is the threshold to measure station state. According to the Code for design of the metro [12], the station capacity $P(i)$ is equal to the peak hour passenger volume $P^{\text{hour}}(i)$, multiply the peak factor λ ($\lambda = [1.1, 1.4]$). This can be formulated as follows:

$$P(i) = P^{\text{hour}}(i) \times \lambda \tag{1}$$

Considering the passenger flows moving into, moving out, and internal exchange of station v_i, we define the throughflow $P_T^{\text{th}}(i)$ as:

$$P_T^{\text{th}}(i) = P_T^{\text{in}}(i) + P_T^{\text{out}}(i) + P_T^{\text{tr}}(i) \tag{2}$$

where $P_T^{\text{in}}(i)$, $P_T^{\text{out}}(i)$, $P_T^{\text{tr}}(i)$ represent the inbound, outbound, and transfer passenger volume, respectively. For non-transfer station, the transfer passenger volume is zero.

The utilization rate of station, which is determined by $P_T^{\text{th}}(i)$ and $P(i)$, denoted by $u_T(i)$, is given as:

$$u_T(i) = P_T^{\text{th}}(i)/P(i) \tag{3}$$

$u_T(i)$ is the hybrid scalar depicting the passenger flow situation of station from the perspective of time and space. The higher of the value $u_T(i)$, the higher utilization rate of the station and the more congestion.

Table 1 Establishment of passenger flow state set of stations

X	Condition	Physical character description
1	$u_T(i) < 0.6$	There are few passengers through the station, and station is free
2	$0.6 \leq u_T(i) < 0.9$	The station is busy but not congested
3	$0.9 \leq u_T(i) \leq 1.1$	The station is relatively congested, and it is necessary to take measures to limit the flow
4	$u_T(i) > 1.1$	Stations are heavily congested and unsafe, requiring diversion

Use vector $X(t) = (x_1(t), x_2(t), \ldots, x_{Nv}(t))$ for representing all station states at t, where $x_i(t) \in X(t)$ denotes the state of station v_i. Notes that, for a station, the states are independent that means the station can only belong to one state at a time. The state set of each station is established and elucidated in Table 1.

2.2.2 States of Section Passenger Flow

In subway system, the transit lines are separate from each other and each direction of a line has a separate rail track. Moreover, the trains of different lines are operated separately, which means that trains are not shared between different lines.

The connection between stations is realized through the transport capacity of the train. Under normal circumstances, when the train leaves the previous station and enters the next station, the passenger flow of the station will be exchanged between the two, so as to ensure the safe and effective transportation of the passenger flow of the station. Thus, the capacity of train is a critical factor to affect the state. However, the exact number of people on the train is difficult to obtain and it is usually estimated by load factor. The load factor of section l_{ij}, denoted by $w_T(l_{ij})$, is given as:

$$w_T(l_{ij}) = P_T(l_{ij})/P_T^{design}(l_{ij}) \qquad (4)$$

$P_T(l_{ij})$ and $P_T^{design}(l_{ij})$ mean the single-track passenger volume and the design transport capacity for a section l_{ij} per unit time T, respectively. $P_T(l_{ij})$ and $P_T^{design}(l_{ij})$ can be calculated by the following formulas:

$$P_T(l_{ij}) = \sum_{k=1}^{m} P_T(tr_k) \qquad (5)$$

$$P_T^{design}(l_{ij}) = \sum_{k=1}^{m} P^{design}(tr_k) \qquad (6)$$

where m is the number of trains obtained through the schedule in unit time T. $P_T(tr_k)$ and $P_T^{design}(tr_k)$ represent the actual train capacity and the design train capacity of train k, respectively.

Table 2 Establishment of passenger flow state set of section

Y	Condition	Physical character description
1	$w_T(l_{ij}) < 0.6$	The section is free, and it means the utilization rate of trains is low
2	$0.6 \leq w_T(l_{ij}) < 0.9$	The section transport capacity is higher, and passengers are congestion in train
3	$0.9 \leq w_T(l_{ij}) \leq 1.1$	The section transport capacity is higher, and passengers are congestion in train
4	$w_T(l_{ij}) > 1.1$	The section transport capacity is close to maximum, and passengers are severely congested

Vector $Y(t) = (y_{12}(t), x_{23}(t), \ldots, y_{ij}(t))$ is used to represent the states of all sections, where $y_{ij}(t) \in Y(t)$ denotes the state of section l_{ij}. The state collection of each section is proposed and elaborated in Table 2.

3 The System Passenger Flow Macroscopic States

The variation of passenger flow on the network is the deduction of the passenger flow along with the train running in time and space dimensions. In the spatial dimension, the evolution of passenger flow has a sequential feature because the passenger flow takes the train to pass through different sections successively. In the time dimension, if the inbound passenger flow in the current period is not outbound, the inbound passenger flow will have an impact on the interval full-load rate in the following period with the passage of time. The system state indicator needs to demonstrate the macroscopic feature which is emerged from the elements (i.e., station and section).

Entropy was usually used to calculate disorder in a system. The amount of entropy is also a measure of the molecular disorder, or randomness, of a system. The concept of entropy provides deep insight into the direction of spontaneous change for many everyday phenomena which includes passenger flow state. Thus, entropy theory can be used to solve this problem.

For state of stations, the higher the entropy, the worse of the station state. Based on this, the entropy of stations $H(X)$ is given as

$$H(X) = -\sum_{i=1}^{CX} (N(X_i)/N_V) \log(N(X_i)/N_V) \tag{7}$$

where CX is the total number of elements in the set of states of stations. N_V means the number of stations. $N(X_i)$ represents the number of stations in state X_i. Note that, $N(X_{CX})$ usually are zeros in the early and late stages of operation, which will make log unsolvable. To overcome this problem, Formation (7) is improved as follows:

$$H(X) = -\sum_{i=1}^{CX}((N(X_i)+1)/(N_V+1))\log((N(X_i)+1)/(N_V+1)) \quad (8)$$

Similarly, the entropy of section state $H(Y)$ is defined as

$$H(Y) = -\sum_{j=1}^{CY}((N(Y_j)+1)/(N_L+1))\log((N(Y_j)+1)/(N_L+1)) \quad (9)$$

where CY is the total number of elements in the section state collection. N_L gives the number of sections. $N(X_j)$ represents the number of stations in state Y_j.

The system passenger state entropy $H(S)$ defines the sum of station entropy and section entropy, i.e.,

$$H(S) = H(X) + H(Y) \quad (10)$$

4 Case Studies

In this paper, we studied partial operation lines of Beijing subway system. It included 102 stations and 100 sections. June 6, 2018, operation data is used to analysis. Simulation real time is from 4:30 am to 00:00 pm. All of the data was divided into half-hour segments.

(1) Station states

After determining the through passenger flow, we investigated the utilization rate and states of Fuxingmen station in real time, as Fig. 3 shows. In the early peak and late peak, the utilization rate and state value were high. This is consistent with the actual situation and represents that the state can accurately describe the station passenger flow. From station, we knew that, at 8:00–9:00, the station needs to take measures to limit the flow.

(2) Section states

Figure 4 describes the Guomao–Dawanglu up and down section states. At down section, the high state appeared at early peak, while the high state of up section appeared at late peak and 21:30–22:00. This reflected the distribution of residence and business.

Subsequently, we obtain all section load factor and states of line 1 at 8:00–8:30, as illustrated in Fig. 5. For line 1, the up direction is from Pingguoyuan to Sihuidong. As apparent from the figure, there is significant difference between the up direction and down direction. The up sections from Wukesong to Fuxingmen have higher state while the down sections from Sihui to Guomao have higher state. Fuxingmen and Guomao both are transfer station from which passengers transfer to work area lines.

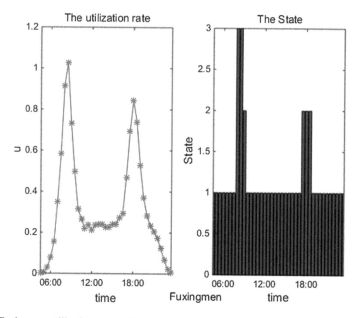

Fig. 3 Fuxingmen utilization rate and state

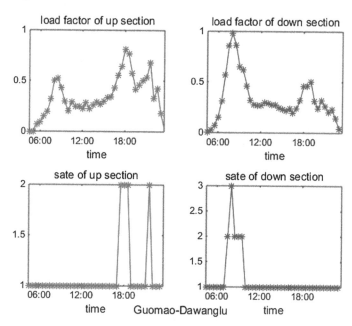

Fig. 4 Guomao–Dawanglu section load and states

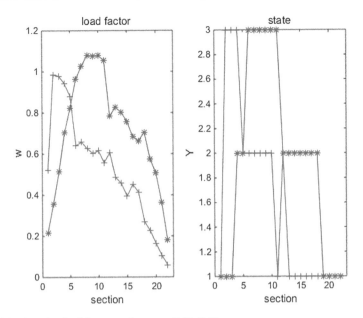

Fig. 5 Line1 section load factors and states at 8:00–8:30

(3) System macroscopic state

Lastly, we calculated the system entropy to analyze the system passenger states based on the station and section states obtained according to time-of-day and considering the train and station capacity constraint, as shown in Fig. 6. As apparent from the figure, there are significant morning and evening peaks at 7:30–9:00 and 17:00–19:00, respectively. With the morning peak more prominent than the evening peak in entropy, but in the temporal duration the evening peak has a larger span.

5 Conclusion

In this study, we presented a method to analyze the system states based on the research of station flow states and section flow states. According to the analysis and discussion, we know that the subway system can meet the demand of passenger flow, but the morning peak and evening peak will put enormous service pressure. The macroscopic state of the system can intuitively show the change of passenger flow in the whole network and provide a basis for staff to make decisions. Our results provide a general framework for studying macroscopic flow in transportation systems.

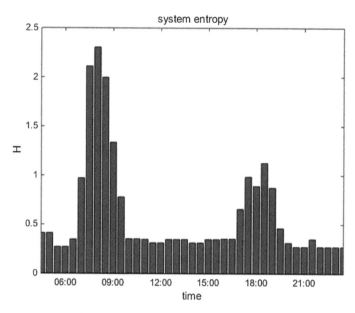

Fig. 6 System macroscopic state entropy of operation time

Acknowledgements This work was supported jointly by the National key research and development program "urban rail transit system security technology" of China (2016YFB1200402).

References

1. Li Y, Wang X, Sun S et al (2017) Forecasting short-term subway passenger flow under special events scenarios using multiscale radial basis function networks. Transp Res Part C Emerg Technol 77:306–328
2. Su S, Tang T, Wang Y (2016) Evaluation of strategies to reducing traction energy consumption of metro systems using an optimal train control simulation model. Energies 9(2):105
3. Li M, Wang Y, Jia L (2017) The modeling of attraction characteristics regarding passenger flow in urban rail transit network based on field theory. PLoS ONE 12(9):1–17
4. Gao S, Wu Z (2011) Modeling passenger flow distribution based on travel time of urban rail transit. J Transp Syst Eng Inf Technol 11(6):124–130
5. Wang D, Yao E, Yang Y et al (2014) Modeling passenger flow distribution based on disaggregate model for urban rail transit. In: Foundations and practical applications of cognitive systems and information processing. Springer, Berlin, Heidelberg, pp 715–723
6. Li W, Zhu W (2016) A dynamic simulation model of passenger flow distribution on schedule-based rail transit networks with train delays. J Traffic Transp Eng (English Edition) 3(4):364–373
7. Li X, Chen P, Chen F et al (2018) Passenger flow analysis of Beijing urban rail transit network using fractal approach. Mod Phys Lett B 32(10):1850001–18500012
8. Yang Y, Liu Y, Zhou M et al (2015) Robustness assessment of urban rail transit based on complex network theory: a case study of the Beijing subway. Saf Sci 79:149–162

9. Zhu L, Luo J (2016) The evolution analysis of Guangzhou subway network by complex network theory. Procedia Eng 137:186–195
10. Xu Q, Mao BH, Bai Y (2016) Network structure of subway passenger flows. J Stat Mech Theor Exp 3(033404):1–17
11. Xu X, Liu J, Li H et al (2016) Capacity-oriented passenger flow control under uncertain demand: algorithm development and real-world case study. Transp Res Part E Logistics Transp Rev 87:130–148
12. GB 50157-2013. Code for design of metro. China National Standard (in Chinese)

Real-Time Monitoring System for Passenger Flow Information of Metro Stations Based on Intelligent Video Surveillance

Xiang ling Yan, Zheng yu Xie and Ai li Wang

Abstract The vigorous development of urban rail transit provides an important guarantee for people's travel, and it also brings potential safety hazards to people. Metro operation often faces peak passenger flow pressure. Although the application of video surveillance can reduce the occurrence of such events, there are still some shortcomings in the existing video surveillance system. This paper first analyzes the shortcomings of the video surveillance system of the existing metro station and then introduces the function and advantages of the real-time monitoring system of the passenger flow information of the metro station based on intelligent video surveillance. The system can realize the real-time detection of passenger flow basic information and the statistical analysis of passenger flow basic information in each monitoring area of the subway station, fully consider the monitoring of key areas of metro stations, effectively reducing or avoiding the occurrence of emergencies and ensuring the safety of subway operations.

Keywords Intelligent video surveillance · Metro station · Passenger flow information · Real-time monitoring

1 Introduction

With the rapid development of the subway, it has exerted more and more influence on people's travel modes, travel environment and social development, and has become an indispensable transportation mode for citizens to travel. By the end of 2018, 35 cities across the country had opened subways and operated 185 lines, with a total operating mileage of 5761.4 km. There are 16 cities that realize the network operation of the subway with four or more lines and at least three interchange stations [1]. As of January 2019, Shanghai's rail transit construction is the largest in China's important cities, with a mileage of 705 km (676 km of subway and 29 km of magnetic levitation). There are 415 stations (413 subways and 2 maglevs) and 17 lines, of which the number of stations and operating mileage is the highest in the country; Beijing opened 21 lines, built 389 stations, operating mileage reached 626 km, the number of lines ranked first in the country; Guangzhou has a mileage of 454 km, built 240 stations and opened 14 lines (Table 1).

With the continuous deepening of metro network operation and the rapid increase of network passenger traffic in major cities in China, daily operations often face the pressure of large passenger flow under normal or unexpected events. Real-time

Table 1 Ranking of subway passenger traffic in Chinese cities

Ranking	City	Passenger capacity (ten thousand)
1	Beijing	912.002
2	Shanghai	885.003
3	Guangzhou	731.604
4	Shenzhen	392.46
5	Chengdu	343.346
6	Xian	267.097
7	Nanjing	262.1
8	Wuhan	261.649
9	Chongqing	259.701
10	Tianjin	127.1611
11	Hangzhou	124.6512
12	Shenyang	82.5013
13	Zhengzhou	74.8714
14	Suzhou	73.1
15	Changsha	63.28
16	Naning	57.01
17	Kunming	54.41
18	Dalian	49.29
19	Qingdao	44.64
20	Hefei	42.79

monitoring of passenger flow information in various areas of metro stations is the key to improving network operation efficiency and safety. In recent years, some vision-based surveillance system prototypes [2] have been developed to monitor real-world security environments: airports, highways, metro stations, railway infrastructure, etc. Video surveillance has been widely used in both public and private environments, such as homeland security, crime prevention, traffic control, accident prediction and detection. At present, there are certain operational risks in the subway passenger flow organization, the video surveillance demand is gradually increasing and the number of cameras in the subway station video surveillance system is also increasing. According to incomplete statistics [1], there were 1303 incidents with a delay of more than 5 min in 2018, and the trains exited the mainline for a total of 8372 times. The goal of intelligent video surveillance [3] is to effectively extract useful information from a large number of videos collected by surveillance cameras by automatically detecting, tracking and identifying objects of interest. In the subway station, intelligent video surveillance is used to detect, count and analyze the passenger flow information in real time, which guarantees the passengers' personal safety and the safe operation of the train and effectively avoids the occurrence of security incidents to a certain extent. At the same time, after an emergency, the staff can play back and analyze the video recording and can find out the cause of the problem, so as to better solve the problem and avoid the same event in the future [4]. Therefore, the real-time passenger flow analysis system based on intelligent video surveillance is particularly important for the safe operation of rail transit.

2 The Insufficient of Existing Video Surveillance System

2.1 Poorly in Real-Time Passenger Flow Information Detection

At present, the video monitoring system of the subway station [5] only processes information by checking video in real time, which has high requirements for the attention and vigilance of monitors. In most cases, false negatives and false positives will occur. If the station staff can not monitor the video screen in real time, it will not be effective in dealing with the event in the event of an emergency, such as a passenger flow congestion. Usually, after an emergency occurs, video playback is used to understand the passage of the event and then the corresponding processing work. This is a great hindrance to the timely and effective handling of sudden security incidents. With the increasing requirements for subway safety, the current video surveillance system has more prominent defects in real-time detection of passenger flow information.

2.2 Poorly Comprehensiveness in Passenger Flow Information Detection

The comprehensiveness of passenger flow information detection is mainly reflected in two aspects: First, whether it can obtain passenger flow information in various areas inside the station, such as entrances and exits, stairs, escalators, transfer passages, station halls, platforms, etc. The second is whether it can fully obtain different parameters of passenger flow status, such as passenger flow, passenger flow density, passenger flow speed and so on.

The AFC passenger flow information acquisition system can obtain the passenger flow data of the passengers in and out of the station, but the distribution of passenger flow in each area of the station can not be comprehensively calculated [6]. The current video surveillance system for station applications often only provides passenger flow and lacks comprehensiveness in the acquisition of different passenger flow information parameters.

2.3 Weakly Ability in Passenger Flow Information Analysis

At this stage, although the video surveillance system has been applied in the field of rail transit, its analysis of passenger flow information has obvious deficiencies, such as the lack of analysis of congestion in various metro stations. With the development of rail transit video surveillance and the increase of passenger traffic, massive data has appeared, which requires video surveillance systems to improve the analysis and processing capabilities of data [7].

3 Passenger Flow Information Real-Time Monitoring System

Aiming at the shortcomings of a video monitoring system in existing metro stations, this paper designs a real-time monitoring system for passenger flow information of metro stations based on intelligent video surveillance. The system mainly includes two modules: real-time detection module of passenger flow basic information and statistical analysis module of passenger flow basic information.

The system can accurately display the three-dimensional map of the station and all the passenger flow information collection points, and the station staff can view the real-time passenger flow status of each monitoring area by clicking the icon. The use of green, yellow, orange and red colors corresponds to the smooth flow, mild congestion, moderate congestion and severe congestion. The staff can visually see the current passenger flow when querying the monitoring areas. Take the National Library Station as an example (Fig. 1).

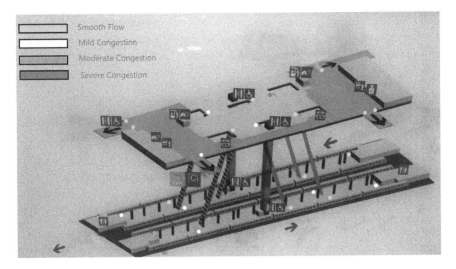

Fig. 1 Real-time passenger flow status at each point of the station

3.1 Real-Time Detection Module of Basic Information of Passenger Flow

The module can use the data of the passenger flow information collection device to perform intelligent analysis based on artificial intelligence algorithm, extract accurate passenger flow parameters and perform statistical analysis on the data. For example, according to the monitoring video image information of the passenger flow, real-time passenger traffic, passenger flow density, passenger flow speed and other information of each area of the station can be obtained in real time, and the corresponding passenger flow information can be visualized and displayed in real time (Fig. 2).

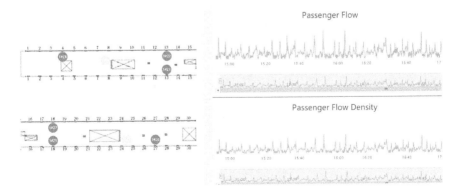

Fig. 2 Real-time detection of passenger flow basic information

Based on intelligent video surveillance, the real-time monitoring system of passenger flow information of metro stations provides real-time passenger flow information for subway station staff by displaying information such as passenger flow, passenger flow density and passenger walking speed in various areas of the subway station in real time. It is convenient for the station staff to understand the passenger flow in each area, and it is not necessary to rely on manual inspection to check the passenger flow inside the station and reduce the burden on the staff.

3.2 Passenger Flow Basic Information Statistical Analysis Module

Based on the real-time data of the passenger flow information detection module, the module calculates the arrival characteristics, congestion degree, accumulated passenger flow and passenger real-time walking speed of each station, and realizes the analysis functions of passenger flow evolution law, congestion status, safety status and other information. Taking the key facilities of the station as the evaluation object, the criteria for the classification of congestion levels of these key facilities are formed and quantified by grading, and the threshold of passenger flow density alarms with different congestion levels is clarified. According to the standard of crowding degree [8], the output is quantified according to the congestion degree of each key facility of the station according to the smooth, light congestion, moderate congestion and severe congestion levels (Fig. 3).

In order to ensure the safety of the operation of the subway station, when the passenger flow density of each station in the station reaches a crowded state, it is necessary to control the number of passengers. By setting the passenger flow

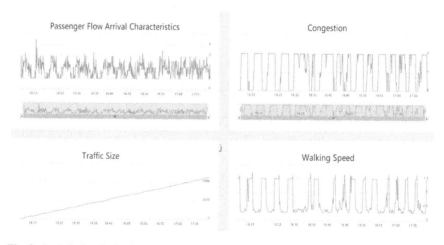

Fig. 3 Statistical analysis of passenger flow basic information

threshold at each level, the alarm can be triggered when the passenger flow density of the key facility exceeds the threshold so that the station personnel can take passenger flow control in time to humanize the passengers and disperse the passenger flow. It can not only improve the efficiency of subway operation but also effectively take corresponding measures in time before the peak of passenger flow [9].

4 Conclusions

Aiming at the problems existing in the existing video surveillance system, a real-time monitoring system for passenger flow information of metro stations based on intelligent video surveillance designed in this paper can realize real-time detection, statistics and analysis of passenger flow information. The application of the intelligent video surveillance system can effectively ensure the safety of the subway operation, reduce the occurrence of emergencies and provide guarantee for passengers to travel comfortably and safely. Instead of the on-site inspection of the staff, it saves a lot of manpower and financial resources so that the subway station staff can guide the passenger flow more scientifically and effectively improve the operation level of the subway.

Acknowledgements The study is sponsored by National Engineering Laboratory project for the Safety Technology of Urban Rail Transit System (Development and Reform Office High Technology [2016] No. 583); 2018 China Railway Information Technology Co., Ltd. Science and Technology Research and Development Program Major Project JGZG-CKY-2018013 (2018A01).

References

1. China Urban Rail Transit Association. 2018 annual statistical and analysis report of urban rail transit. (2019-3-30) [2019-06-03]. http://www.camet.org.cn/index.php?m=content&c=index&a=show&catid=18&id=16219. (in Chinese)
2. Foresti GL (1998) Real-time system for video surveillance of unattended outdoor environments. IEEE Trans Circ Syst Video Technol 8(6):97
3. Wang X (2013) Intelligent multi-camera video surveillance: a review. Pattern Recogn Lett 34:3
4. Zheng Y (2017) Application analysis of video surveillance system in rail transit. Technol Innov 18:152 (in Chinese)
5. Nong Xu (2012) Analysis and design of intelligent video surveillance system. Beijing University of Posts and Telecommunications, Beijing (in Chinese)
6. Chen J (2018) Characteristics and application analysis of urban rail transit passenger flow detection technology. Urban Rail Transit Res 1:37–138 (in Chinese)
7. Guo Z, Liu C, Xu Z (2018) Application of intelligent video analysis technology in rail traffic monitoring. Sci Technol Econ Guide 26(32):6 (in Chinese)
8. Cao Z, Qin Y, Xie Z (2018) Research on classification methods of LOS for urban rail transit stations passages. In: 2018 Prognostics and system health management conference
9. Chen W, Cai S, Qi J (2017) Dynamic monitoring and control decision system for passenger flow of subway station based on multi-source data. Urban Public Transport 4:4 (in Chinese)

Application of Target Detection Algorithms in Railway Intrusion

Xingwei Jia, Yumeng Sun and Zhengyu Xie

Abstract China's railway construction level is relatively high, and the level of operation and maintenance needs to be developed. The development of railway video technology is of great significance for improving the safety of railway operations and reducing the occurrence of accidents. This paper mainly studies the application of target detection algorithm in railway scenarios. For the railway video detection, this paper adopts four kinds of target detection algorithms: unsupervised frame difference method, background difference method, supervised class based on deep learning YOLOv3, Faster-RCNN algorithm. By collecting the video image data of the railway scene, the data set training deep learning algorithm is created, and then the four kinds of target detection algorithms are used to process the collected railway video image data, respectively, and the test algorithm is used for the perimeter intrusion detection effect in the railway scene. By comparison, it points out its advantages and disadvantages in the railway perimeter invasion.

Keywords Target detection · Perimeter invasion · Deep learning · Railway · Application · Compared

1 Introduction

Rail transit takes into account the advantages of air transportation and road transportation. It has the characteristics of strong economy, high timeliness and long transportation distance. It is one of the most popular modes of transportation. As of 2018, the railway network of China's high-speed rail eight vertical and eight horizontal has been built four vertical and four horizontal, the total operating mileage of the railway reached 127,000 km. The degree of rail transit construction is so high, and how to ensure its safe operation has become an important issue we have to solve. The

X. Jia · Y. Sun · Z. Xie (✉)
School of Transportation, Beijing Jiaotong University, Beijing, China
e-mail: xiezhengyu@bjtu.edu.cn

development of railway video detection technology will help prevent and reduce the occurrence of railway intrusion and destruction events and ensure the safe operation of China's rail transit.

At present, most of the intrusion detection of railway video in China adopts unsupervised target detection algorithms, such as background difference method and interframe difference method. However, such detection algorithms have the problem of high false negative rate. With the development of artificial intelligence and machine learning in recent years, the target detection algorithm based on deep learning has also emerged. The use of supervised and deep learning target detection algorithm can effectively solve the problem of high false negative rate. The target detection algorithms based on deep learning are representative of the YOLOv3 algorithm of the YOLO series and the Faster-RCNN target detection algorithm of the RCNN series.

2 Introduction to Target Detection Algorithm

We briefly introduce the principle of interframe difference method, background difference method, YOLOv3 algorithm and Faster-RCNN algorithm detection.

2.1 Interframe Difference Method

The interframe difference method uses the absolute value of the difference between the current frame of the video and the grayscale image of the previous frame to distinguish the foreground from the background [1]. When the difference exceeds the threshold, the connected is determined as the foreground. Otherwise, it is considered to be background. The nth frame and the $n-1$th frame image in the tagged video are F_n and F_{n-1}, and the grayscale images are denoted as f_n and f_{n-1}, respectively, and the sitting mark of each pixel in the image is (x, y). Then the difference image D_n:

$$D_n(x, y) = |f_n(x, y) - f_{n-1}(x, y)| \tag{1}$$

D_n is binarized according to the set threshold T to obtain a binary graph r_n:

$$r_n(x, y) = \begin{cases} 255, & D_n(x, y) > T \\ 0, & \text{else} \end{cases} \tag{2}$$

Finally, the foreground R_n is obtained through the connectivity calculation [2] (Fig. 1).

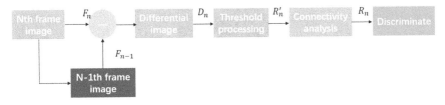

Fig. 1 Interframe difference method flowchart

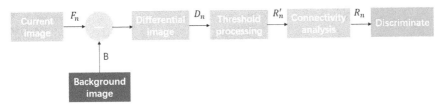

Fig. 2 Background difference method flowchart

2.2 Background Difference Method

The basic idea of the background difference method is similar to the interframe difference method. The difference is that the background difference method does not use the adjacent frame to make a difference, but uses the current frame to subtract from a real-time updated background model [3]. The background model is directly or inferior and affects the effect of target detection. Compared with the interframe difference method, the background difference method has two steps of background modeling and background update, which can reduce the influence caused by illumination and climate change in the scene. Common background modeling methods include mixed Gaussian, statistical average and median filtering [4] (Fig. 2).

2.3 YOLOv3

The YOLO algorithm has been updated to YOLOv3. The YOLOv3 algorithm is improved on the basis of YOLOv2 [5]. Based on the idea of ResNet-101, the Darknet53 network structure is proposed, which significantly improves the detection accuracy. The most prominent feature of the YOLO algorithm is end to end, which directly predicts the border and category information of the image through a single stage, which greatly improves the detection speed, and can reach a rate of 45 frames on the Titan X GPU. The main detection process: YOLO algorithm divides the picture into S × S grid cells. Each grid cell is responsible for predicting the confidence of B bounding boxes and C categories, and determining a bounding box needs to be predicted (x, y, w, h) four information, respectively, the bounding box's center point

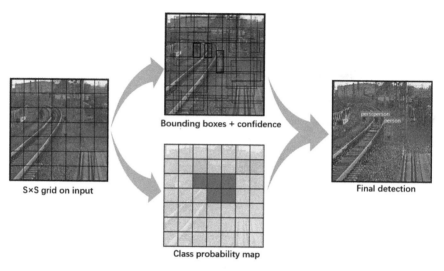

Fig. 3 YOLOv3 testing process

horizontal and vertical coordinates and length and width, the network finally uses the fully connected layer for category output, so the dimension of the fully connected layer is S × S × (5 × B + C) [6]. For the Pascal VOC2007 dataset, S takes 7 and B takes 2 because it contains 20 types of detectors C (Fig. 3).

2.4 Faster-RCNN

Faster-RCNN is proposed based on Fast-RCNN. It proposes to use region proposal networks (RPN) instead of selective search in Fast-RCNN to speed up the extraction process of region proposal, thus improving the overall detection speed of the algorithm. Unlike YOLOv3, the target detection of the Faster-RCNN algorithm is implemented in two steps [7]. First, the region proposal networks are used to extract about 2000 region proposals, and then the feature map of the region proposal is converted into a fixed-length fully connected layer. Finally, Softmax is used to perform specific categories. The classification, meanwhile, uses loss function to complete the bounding box regression to obtain the exact location of the object [8] (Fig. 4).

3 Application of Target Detection Algorithm in Railway Scenes

We use the captured video data to test the applicability of some mainstream target detection algorithms in railway scenarios. The video data collection was completed in the experimental loop of the eastern suburbs of Beijing. By simulating the intrusion

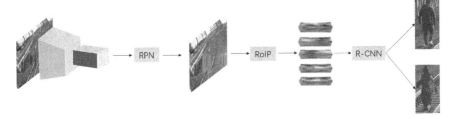

Fig. 4 Faster-RCNN testing process

behavior of the railway, such as crossing the track and walking along the track, the live video data of the railway test line was collected to study the applicability of these algorithms in the railway scene.

3.1 Application of Interframe Difference Method in Railway Scenes

The interframe difference method is used to monitor the video data of the intrusion behavior of the railway scene. The algorithm effect is shown in the following Fig. 5.

The railway camera has a fixed viewing angle and conforms to the premise of the interframe difference method. By processing the collected field data, it is found that the interframe difference method can detect the intrusion behavior in the scene, but it can hardly find objects with small motion amplitude or static motion and is affected by factors such as video noise and illumination. High can not identify intrusion object category information, so the false positive rate is high. However, the data processing speed is fast and real time, and the performance requirements of the processor are low.

Fig. 5 Interframe difference method detection

Fig. 6 Background difference method detection

3.2 Application of Background Difference Method in Railway Scene

The background data method of hybrid Gaussian background modeling is used to monitor the video data of the intrusion behavior of the railway scene. The algorithm effect is shown in the following Fig. 6.

The background difference method also has the problems of false positives and high false negative rate. However, because the algorithm performs background modeling, the detection effect of the intrusion target with less static and motion amplitude is better than the interframe difference method. The same background difference method has lower processor performance requirements.

3.3 Application of YOLOv3 Algorithm in Railway Scene

We use the trained YOLOv3 model for railway scene intrusion detection. The results are shown in the following Fig. 7.

It can be found that the use of the trained YOLOv3 algorithm for intrusion detection is very effective, and the type and position information of the object can be identified. The speed of monitoring can be monitored in real time, and the recognition effect on distant small targets is ideal and even accurately monitors the person at the 150 m pole. However, collecting video data and making data sets consume a lot of time and require high performance on the processor.

Fig. 7 YOLOv3 detection

3.4 Application of Faster-RCNN Algorithm in Railway Scenes

We use the trained Faster-RCNN algorithm for railway scene intrusion detection. The results are shown in the following Fig. 8.

The Faster-RCNN target detection algorithm has a good detection effect, and the false negative report rate is low. The category and position information of the object can be identified, but the detection rate is slow, and the real-time monitoring effect cannot be achieved, and the performance requirement of the processor is high.

Fig. 8 Faster-RCNN detection

Table 1 Comparison of four algorithms on Titan X graphics

Target detection algorithm	Interframe difference method	Background difference method	YOLOv3	Faster-RCNN
The time it takes to process a single image (s)	0.01	0.01	0.02	0.12
Used internal memory (mib)	68	146	1150	1387
MAP			Person 0.79	Person 0.78

4 Comparison and Summary

4.1 Algorithm Comparison

For the railway perimeter intrusion detection, the background difference method is more suitable in the target detection algorithm of unsupervised learning. Compared with the frame difference method, the background difference method can detect targets with a small amplitude or a small motion range, which greatly reduces the false negative rate. For the target detection algorithm of YOLOv3 and Faster-RCNN combined with deep learning, YOLOv3 is more suitable because the MAP of the two algorithms is similar, and the detection speed of YOLOv3 is much faster. We compare the processing speed of the above four algorithms on the Titan X graphics card and occupy the memory. The results are as follows (Table 1).

For unsupervised algorithms and supervised target detection algorithms combined with deep learning, unsupervised algorithms have faster processing speeds and lower processor performance requirements, but in actual applications, false positives have a high false negative rate. For the supervised and deep learning target detection algorithm, the detection effect is better, the false negative report rate is low, and the category position information of the object can be accurately identified, but the performance requirement of the processor is high, and the data is marked. Making data sets takes time and effort (Table 2).

4.2 Summary

In general, in the railway perimeter intrusion detection, YOLOv3 has better detection effect and fast processing speed, which meets the requirements of railway perimeter intrusion detection. It can be used to replace the commonly used background difference method to further improve detection accuracy and reduce errors, reporting the rate of underreporting and maintaining railway safety operations.

Table 2 Comparison of advantages and disadvantages of four methods in railway scene application

Category	Target detection algorithm	Advantage		Disadvantage	
Unsupervised learning	Interframe difference method	• Sensitive to moving targets	• Fast detection and high real-time performance • Low-performance requirements on the processor	• Slow motion or unable to extract feature pixels when similar to background color	• False positive false negative rate • Influenced by conditions such as light climate
	Background difference method	• Sensitive to shadow		• Need to carry out background modeling, high requirements for background model update	
Supervised learning	YOLOv3	• Fast processing speed and high real-time performance	• Can identify the category location information of the object • False positive false negative rate		• Need to make a data set • High-performance requirements on the processor
	Faster-RCNN			• Slow processing speed and poor real-time performance	

Acknowledgements The study is sponsored by Multi-source information fusion and big data analysis technology for air-conditioned vehicles-02(T17B500040).

References

1. Ramya P, Rajeswari R (2016) A modified frame difference method using correlation coefficient for background subtraction. Procedia Comput Sci 93:478–485
2. Kazanskiy NL, Popov SB (2015) Integrated design technology for computer vision systems in railway transportation. Pattern Recogn Image Anal 25(2):215–219
3. Zhu S, Guo Z (2012) Ma L (2012) Shadow removal with background difference method based on shadow position and edges attributes. EURASIP J Image Video Process 1:4
4. Lu X, Xu C, Wang L, Teng L (2018) Improved background subtraction method for detecting moving objects based on GMM. IEEJ Trans Electr Electron Eng 13(11):1540–1550
5. Redmon J, Farhadi A Yolov3: an incremental improvement [Z/OL]. Computer Vision and Patern Recognition (CVPR) [2018-04-01]. https://arxiv.org/abs/1804.02767v1
6. Redmon J, Divvala S, Girshick R, Farhadi A. You only look once: unified, real-time object detection. In: 2016 IEEE conference on computer vision and pattern recognition. Las Vegas, IEEE, pp 779–788
7. Yang L, Sang N, Gao C (2018) Vehicle parts detection based on Faster—RCNN with location constraints of vehicle parts feature point. In: International symposium on multispectral image processing and pattern recognition
8. Ren SQ, He KM, Girshick R, Sun J (2017) Faster R-CNN: towards real-time object detection with region proposal networks. IEEE Trans Pattern Anal Mach Intell 39(6):1137–1149

A Method for Pedestrian Intrusion Detection of Railway Perimeter Based on HOG and SVM

Yumeng Sun and Zhengyu Xie

Abstract In order to ensure the safety of railway operation, it has become urgent to strengthen the detection and protection of railway perimeter safety. In this paper, a method for pedestrian intrusion detection of railway perimeter based on histograms of oriented gradients (HOG) plus support vector machine (SVM) is proposed. By establishing the perimeter intrusion image sample set of the railway scene, Gaussian filtering is performed on the sample image to obtain the training set of positive and negative samples, and then, the HOG features are extracted for training. Finally, the field image is used to test this algorithm. The results show that the proposed algorithm is better than other algorithms, such as the inter-frame difference method and the mixed Gaussian background modeling method in the detection of pedestrian intrusion detection of railway perimeter. This paper describes in detail the above process and puts forward the deficiencies for further research in the future.

Keywords HOG · SVM · Gaussian filtering · Railway perimeter

1 Introduction

In recent years, the railway industry has developed exponentially in China. However, with the rapid development of railway in China, the traffic accidents in railway operation are getting more and more attention. In addition to the faults of railway equipment that can be avoided by strengthening the examination, traffic accidents caused by the external environment of railway are also important reasons for traffic accidents along railway lines. Therefore, it can be seen that strengthening the detection and protection of railway perimeter safety is the direction worthy of our attention in the railway construction process.

Y. Sun · Z. Xie (✉)
School of Traffic and Transportation, Beijing Jiaotong University, No. 3 Shangyuancun, Haidian District, Beijing, People's Republic of China
e-mail: xiezhengyu@bjtu.edu.cn

Y. Sun
e-mail: sunyumeng@bjtu.edu.cn

For such a large-scale railway system, research on railway pedestrian intrusion detection algorithms has become an indispensable part of railway security operations; video image processing technology is widely used due to its easy installation and visualization of results. This makes railway video monitoring system become one of the important means of railway foreign body intrusion detection [1]. Furthermore, pedestrian is the main factor with strong subjectivity and great threat to railway perimeter in the process of railway safety operation. Therefore, this paper takes pedestrian as the intrusion target of the railway perimeter. In the object detection algorithms [2], frequently used algorithms include background difference method, inter-frame difference method, optical flow method, and other improved algorithms. Guo et al. [3] studied detection and extraction algorithm of foreign matter in railway clearance based on inter-frame difference accumulation, but the installation equipment in the middle of the track bed and objects such as shadows and spots under illumination would cause false detection. Apart from the above-improved traditional algorithms, the emerging deep learning in recent years has shown excellent results in the field of image processing. Zhang [4] proposed a detection system based on deep learning to achieve the automatic detection function of invasion objects, which only remained in the theoretical target. Considering the cost of railway scene and the problem that warning image would be processed on the computer of monitoring center along railway line, it could not be used on a large scale, so it is not suitable for field application.

Since the HOG feature has an excellent effect in describing the local area of an image, many relevant researchers have found that the HOG feature is more effective in pedestrian detection [5]. Dalal et al. [6] first proposed the HOG feature extraction algorithm at the International Conference on Computer Vision and Pattern Recognition and combined with the SVM classifier to achieve the purpose of detecting pedestrians. Saika et al. [7] proposed a method for improving the accuracy of human detection using HOG features on train-mounted camera. Pang [8] proposed an effective HOG pedestrian detection, which reduced the computational cost and improved the detection accuracy.

2 Railway Perimeter Pedestrian Intrusion Detection Based on HOG and SVM

Because the railway scenario is mostly in the outdoor environment, it is greatly affected by light and haze. Therefore, according to the characteristics of the scene image of the railway and the characteristics of the current application, this paper proposes a pedestrian invasion based on the HOG plus SVM. The detection method, combined with Gaussian filtering, achieves the purpose of detecting pedestrians. The overall architecture is shown in Fig. 1.

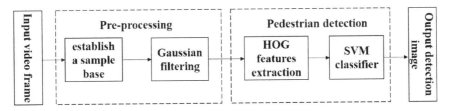

Fig. 1 Overall architecture

Firstly, Gaussian filtering is applied to the sample database. And then, the HOG features of the samples are extracted. Finally, the sample is trained in conjunction with the linear SVM classifier to achieve the purpose of detecting pedestrians and boxing pedestrians.

2.1 HOG Features Extracting Process

The HOG forms a feature by counting the histogram of oriented gradient of the local region of the image. The flow chart is shown in Fig. 2.

Image preprocessing. Firstly, select an image $P(x, y)$ and make the original picture into the grayscale image to reduce the calculation time in subsequent feature extraction. Then, it is processed by gamma normalization which not only improves its robustness under illumination conditions, but also reduces the influence of shadow on detection and effectively suppresses some noise interference simultaneously. The preprocessing image is shown in Fig. 3a. At last, the target in the dashed box is

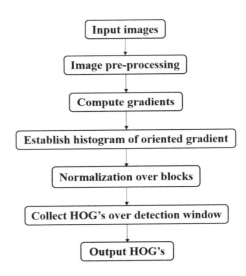

Fig. 2 HOG features extraction flow chart

Fig. 3 Preprocessing image is shown in (**a**) and **b** is extracted from (**a**)

extracted from Fig. 3a, and the resolution is adjusted to 64 * 128, as shown in Fig. 3b.

Compute gradients. The convolution operation of Fig. 3b is carried out by Formula (1) to obtain the gradient $G_x(x, y)$ in the horizontal direction of the image, and the convolution operation of Fig. 3b is performed by Formula (2) to obtain the gradient $G_y(x, y)$ of the vertical direction of the image, where $H(x, y)$ represents the pixel value of the input pixel. The horizontal and vertical gradients of the image are shown in Fig. 4a, b, respectively.

$$G_x(x, y) = H(x + 1, y) - H(x - 1, y) \tag{1}$$

$$G_y(x, y) = H(x, y + 1) - H(x, y - 1) \tag{2}$$

The gradient magnitude calculated at the pixel of the image is shown in Eq. (3), and the gradient direction calculated at the pixel of the image is shown in Eq. (4). The gradient is shown in Fig. 4c.

$$G(x, y) = \sqrt{[G_x(x, y)]^2 + [G_y(x, y)]^2} \tag{3}$$

$$\alpha(x, y) = \arctan\left(\frac{G_y(x, y)}{G_x(x, y)}\right) \tag{4}$$

Establish histogram of oriented gradients. Figure 3b (64 × 128) is divided into small areas which are called cell and each cell contains 8 × 8 pixels which can produce 128 cells. Then, regardless of the positive and negative directions in

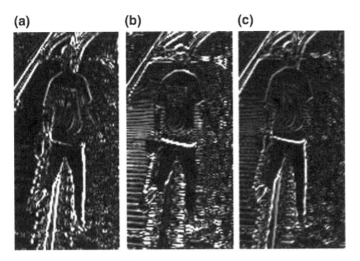

Fig. 4 Horizontal gradients image is shown in (**a**). The vertical gradients image is shown in (**b**) and the gradients image is shown in (**c**)

the plane, 360° can be divided into nine parts, each of which is called bin, that is, nine histograms are used to count the feature information in a cell and finally a 9-dimensional HOG feature vector can be obtained.

Normalization over blocks. By combining four cells into one block, it can slide 7 times in the horizontal direction and 15 times in the vertical direction, so there are 105 blocks in total. This makes a lot of information overlap, so it needs to be normalized to make the detection window more robust to the illumination and shadow of the image. In this paper, Formula (5) is adopted for normalization calculation, where V represents the vector before normalization, $\|V\|_2$ is the second-order norm of V, ε is a tiny quantity.

$$\|V\|_2 = V/\sqrt{\|V\|_2^2 + \varepsilon^2} \qquad (5)$$

Collect HOG's over detection window. The HOG feature descriptors of all the cells in a block are concatenated to obtain the HOG feature descriptors of the block, and the HOG feature descriptors of all the blocks are concatenated to obtain the HOG feature descriptor of Fig. 3b, as shown in Fig. 5. There are a total of $4 \times 9 \times 105 = 3780$ features in the 64×128 sample image, which is the feature vector that can be finally used for classification.

Fig. 5 HOG feature descriptor of all blocks

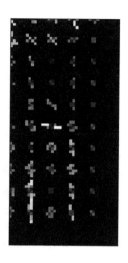

2.2 SVM Classifier

Linear classifier SVM is the simplest and most effective classifier. It is a classification problem with only two types of samples in a two-dimensional space. For example, R_1 and R_2 are two categories to be distinguished. For this article, R_1 is a pedestrian category and R_2 is a non-pedestrian category. You can find a line separating the two types of samples, which is the classification function. Label 1 indicates that the pedestrian sample belongs to category R_1, and label 0 indicates that the non-pedestrian sample belongs to category R_2.

3 Implementation Details

3.1 Establish the Sample Database

In order to improve the accuracy of the algorithm, we carry out the railway perimeter pedestrian intrusion experiment at Beijing eastern suburban loop testing base and construct a sample database of perimeter intrusion images of the railway scenario, which contains 1000 positive samples and 1500 negative samples. The pedestrian sample in the railway scene is taken as a positive sample, and the non-pedestrian sample is added to the negative sample, such as the sky, rail, stone, utility pole, and the like. The positive and negative samples partial sample sets are shown in Figs. 6 and 7, respectively.

Fig. 6 Partial sample sets of positive samples

Fig. 7 Partial sample sets of negative samples

3.2 Experimental Process

Firstly, Gaussian filtering is needed for the perimeter intrusion image sample database of railway scenario, and then, HOG feature extraction is performed on it. Label 1 is added to positive samples, and label 0 is added to negative samples to distinguish them.

Secondly, the HOG feature and labels obtained are sent to the linear SVM for training to generate the basic classifier. Then, the basic classifier is used to identify the negative sample, extract the error identification sample area, and add it to the original negative sample set as a new negative sample to form the final sample database.

Finally, extract HOG feature from the updated sample set, add labels, send it to the linear SVM for secondary training, and generate the final classifier to detect pedestrians. The training flow chart is shown in Fig. 8.

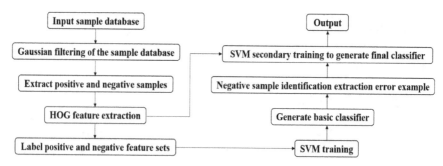

Fig. 8 Training flow chart

3.3 Experimental Result

In this paper, two original images of different time instants are randomly selected in the railway perimeter intrusion image sample database, and the pixels are adjusted to 640 × 480 size. We select the inter-frame difference method, the mixed Gaussian background modeling method, and the algorithm in this paper to detect them, of which Fig. 9a is the original picture. The detection effect is as follows: In the inter-frame difference method as Fig. 9b shows that the detection of the target is incomplete, the illumination is greatly affected, and the leaf swing is detected; the mixed Gaussian background modeling method as the Fig. 9c shows that it will detect the distant vehicle which is a false alarm; and the algorithm in this paper as Fig. 9d shows that it can detect the target more accurately but when the distance is far away the false negative occurs.

The experimental results show that the proposed algorithm is better than the inter-frame difference method and the mixed Gaussian background modeling method. However, when the targets overlap, only one detection frame appears. Such situations require further research. In addition, due to the small number of training samples in

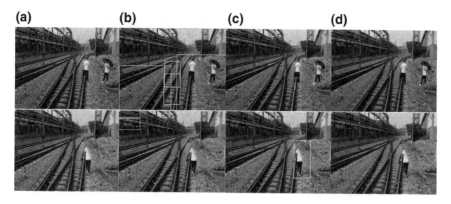

Fig. 9 Detection results of different methods in two moments

this paper, there is a high rate of missed detection in the detection of small targets. Therefore, in the future research and study, we should focus on the detection of small targets.

4 Conclusions

According to the requirements of on-site processing of railway video images, this paper proposes a pedestrian perimeter intrusion detection method based on HOG and SVM. Through experimental comparison, this method has achieved good detection results in the target detection of railway perimeter intrusion, and it has higher detection rate and lower false negative rate than inter-frame difference method and hybrid Gaussian background modeling method. It is more suitable for railway scenes to implement object detection.

Acknowledgements The study is sponsored by multi-source information fusion and big data analysis technology for air-conditioned vehicles-02(T17B500040).

References

1. Dong H, Ge D, Qin Y, Jia L (2010) Research on railway invasion detection technology based on intelligent video analysis. China Railway Sci 31(2):121–125 (in Chinese)
2. Wang C, Liu M, Qi F (2018) Summary of dynamic target detection and recognition algorithm in intelligent video surveillance system. Electr Eng 19(9):20–25 (in Chinese)
3. Guo B, Ding C (2019) Detection and extraction algorithm of foreign matter in railway clearance based on inter frame difference accumulation. Railway Stand Des 63(9):1–6 (in Chinese)
4. Zhang C (2017) A detection system based on deep learning to achieve the automatic detection function of invasion objects. Doctoral dissertation, North China University of Technology (in Chinese)
5. Wang M, Li X, Chen Q, Li L, Zhao Y (2016) Surveillance event detection based on CNN. ACTA Autom Sin 42(6):892–903 (in Chinese)
6. Dalal N, Triggs B (2005) Histograms of oriented gradients for human detection. In: IEEE computer society conference on computer vision and pattern recognition (CVPR), pp 886–893
7. Saika S, Takahashi S, Takeuchi M, Katto J (2016) Accuracy improvement in human detection using HOG features on train-mounted camera. In: IEEE global conference on consumer electronics. IEEE, pp 1–2
8. Pang Y, Yuan Y, Li X, Pan J (2011) Efficient hog human detection. Sig Process 91(4):773–781

Timetabling, Platforming, and Routing Cooperative Adjustment Method Based on Train Delay

Yinggui Zhang, Zengru Chen, Min An and Aliyu Mani Umar

Abstract This paper presents a methodology in which three sub-models of train timetabling, platforming, and routing are combined by studying the real-time adjustment and optimization of high-speed railway in the case of train delay in order to produce a cooperative adjustment algorithm, so that the general plan of train operation adjustment can be obtained. The results of a case study show that the proposed method can quickly adjust the train operation plan under the condition of train delay, restore the normal train operation order, and reduce the impact of train delay effectively.

Keywords Train delay · Railway timetabling · Train platforming · Optimization model

1 Introduction

Trains in a railway network are usually operated by following systematic predetermined schedules. If a train is delayed, it would affect other trains' arrivals and departures at the station. However, the change of timetabling because of the delays will also influence the platforming and routing plans in the railway network [1].

Some studies have been conducted on the optimization of train timetable in order to managing train delay. For example, [2 and 3] investigated the problems of railway transportation interference and train delay management. Cacchiani et al. [4] studied the recovery of real-time interference and interference management of railways. Yang et al. [5] developed a method for minimizing the total delay time at the departure station and the residence time of intermediate station as objective functions and established a collaborative optimization model for the integration of stop planning

Y. Zhang · Z. Chen · A. M. Umar
School of Traffic and Transportation Engineering, Central South University, Changsha 410004, China

M. An (✉)
School of Science, Engineering and Environment, University of Salford, Manchester M5 4WT, UK
e-mail: M.An@Salford.ac.uk

in order to solve train dispatching problems. Zhang et al. [6] combined train route, interlocking, train platform compilation rules, and processing theory, and designed a heuristic algorithm based on classical and combined scheduling rules. Samà et al. [7] proposed a method for real-time adjustment and re-arrangement of train platforms and train routes. However, the current study only addresses the optimization of train timetable and the adjustment of train delay, but the adjustment of platforming and routing plans is not considered, which is important in train operation management.

This paper presents a cooperative adjustment method in the case of train delay based on the synergistic relationship among train timetabling, platforming, and routing plans in order to restore normal train operation order and reduce the impact of train delay effectively.

2 Model Development

The proposed timetabling, platforming, and routing models are described in this section, respectively, which will be used to establish a cooperative adjustment model.

2.1 Notations

i—Trains, $i \in \mathbf{I}, \mathbf{I} = \{1, 2, \ldots, k\}$;
s—Stations, $s \in \mathbf{S}, \mathbf{S} = \{1, 2, \ldots, m\}$;
u_s—Arrival/departure track in a station s, $u_s \in \mathbf{U}_s$, $\mathbf{U}_s = \{1_s, 2_s, \ldots, n_s\}$;
t_{is}^a—Original scheduled time for train i to arrive at station s;
t_{is}^d—Original scheduled time for train i depart from station s;
$t_{is}^{\prime a}$—Adjusted time for train i to arrive at station s;
$t_{is}^{\prime d}$—Adjusted time for train i depart from station s;
h_u^s—The route to the track u in station s, $h_u^s \in \mathbf{H}_u^s$, $\mathbf{H}_u^s = \{1, 2, \ldots, g_u\}$;
$x_{iu}^s = \begin{cases} 1, & \text{if train } i \text{ occupies track } u \text{ at station } s \text{ in the original plan} \\ 0, & \text{otherwise} \end{cases}$;
$x_{iu}^{\prime s} = \begin{cases} 1, & \text{if train } i \text{ occupies track } u \text{ at station } s \text{ in the adjusted plan} \\ 0, & \text{otherwise} \end{cases}$;
X_{iu}^s—when $x_{iu}^s = 1$, $X_{iu}^s = u$; $X_{iu}^{\prime s}$—when $x_{iu}^{\prime s} = 1$, $X_{iu}^{\prime s} = u$;
$y_{iuh}^{\prime s} = \begin{cases} 1, & \text{if train } i \text{ passes through route } h \text{ to track } u \text{ in the adjusted plan} \\ 0, & \text{otherwise} \end{cases}$;
$y_i = \begin{cases} 1, & \text{if train } i \text{ is delayed} \\ 0, & \text{otherwise} \end{cases}$; $\theta_i = \begin{cases} 1, & \text{if } (t_{is}^{\prime d} - t_{is}^d) < T, T \text{ is a constants} \\ 0, & \text{otherwise} \end{cases}$.

2.2 Timetabling Model

The common adjustment goal is to minimize the total train delay time. When train delays occur, the task is to adjust the timetable in order to sort out the number of delayed trains and the total train delay time.

Assume α is a weight factor, which should be a large enough number to ensure the priority of the first objective, i.e., the number of delayed trains. Variable θ_i (0–1) is used to denote that if the delay time of a train is longer than T, the number of delayed trains will not be taken into account. The aim is to minimizing total delay time of the train instead. The timetabling model can be established as

$$\min Z_1 = \sum_{i=1}^{k} \left\{ \alpha \theta_i y_i + \sum_{s=1}^{m} \left[\left(t_{is}^{\prime a} - t_{is}^{a}\right) + \left(t_{is}^{\prime d} - t_{is}^{d}\right) \right] \right\}$$

s.t.{Timetabling constraints}

2.3 Platforming Model

Taking the volatility of the platform scheme as the first objective and the equalization of the platform scheme as the second objective into consideration, and by using the method of linear weighted summation, the two objectives are transformed into a single objective; suppose the weighting factors β_1 and β_2 ($\beta_1 + \beta_2 = 1$) for each objective, the platforming model can be developed as

$$\min Z_2 = \beta_1 \sum_{i=1}^{k} \left(X_i^{\prime s} - X_i^{s}\right)^2 + \beta_2 \frac{1}{n_s} \sum_{u=1}^{n_s} \left(\sum_{i=1}^{k} x_{iu}^{s} - \frac{k}{n_s} \right)^2$$

s.t.{Platforming constraints}

2.4 Routing Model

The arrival/departure tracks occupied by the trains need to be recognized at each station firstly, then enumerate the routes connecting the track according to the station structure and select the optimal train route scheme. The aim is to balance the use of routes by trains, and the routing model can be established as

$$\min Z_3 = \sum_{u=1}^{n_s} \sum_{h=1}^{g_u} \left\{ \sum_{i=1}^{k} y_{iuh}^{s} - \frac{\sum_{i=1}^{k} x_{iu}^{\prime s}}{g_u} \right\}^2$$

s.t.{Routing constraints}

2.5 A Cooperative Adjustment Model

The occurrence of perturbations in a railway system that require the adjustment of train timetable usually affect the platforming and routing plans. Adjustments of timetabling, platforming, and routing plans in a case of the train delay require adjusting the arrival and departure times of all trains to ensure a conflict-free operation. Therefore, it is necessary to combine timetabling, platforming, and routing models to construct a cooperative adjustment model.

Let $t_{is}^{\prime a}$ and $t_{is}^{\prime d}$ be coupling factors, timetabling model as a decision model (D) of platforming (P) and routing models (F), the cooperative adjustment model can be established as

$$\text{Model(Cooperative Adjustment)} \begin{cases} (D)\text{Model(Timetabling)}: \min Z_1\left(t_{is}^{\prime a}, t_{is}^{\prime d}\right) \\ \quad \downarrow \uparrow \\ (P)\text{Model(Platfroming)}: \min Z_2\left(t_{is}^{\prime a}, t_{is}^{\prime d}, x_{iu}^{\prime s}\right) \\ \quad \downarrow \uparrow \\ (F)\text{Model(Routing)}: \min Z_3\left(t_{is}^{\prime a}, t_{is}^{\prime d}, x_{iu}^{\prime s}, y_{iuh}^{\prime s}\right) \end{cases}$$

3 Methodology

This section describes the algorithms of the proposed cooperative adjustment model. The developed cooperative adjustment model is a three-level programming model. The solution algorithm at each level is developed, and the final algorithm of cooperative adjustment model is designed to realize the real-time and fast adjustment of late trains.

3.1 Timetabling Algorithm

As described in Sect. 2.2, the first objective function of the timetabling model is established to minimize the number of delayed trains and the second objective function is to minimize the total delay time. These two objectives can be achieved by (1) calculating the earliest possible departure time of the delayed train i according to its arrival time $t_{is}^{\prime a}$ and its minimum stopping time τ_{is} at station s. Figure 1 shows relationship between the delayed train and its subsequent trains where a, b, c, d, and e represent the subsequent trains sorted according to the original schedule, (2) comparing the departure times of the adjacent j trains. In the iterative search process, if there is a feasible departure interval existing, the interval will be chosen as the departure interval of the delayed train. For example, if $j = 2$, assuming that the departure time interval of trains b and c is large enough to satisfy the time interval

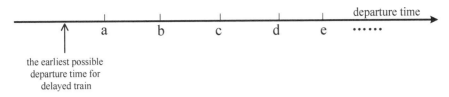

Fig. 1 Sorting relationship between the delayed train and its subsequent trains

requirements after inserting the delayed train i into the interval, then the interval (b, c) is considered as a feasible departure interval of the delayed trains as shown in Fig. 2.

However, if there is no feasible interval after searching all intervals when $j = 2$, then $j = j + 1$, and re-search until finding a solution. It should be noted that since step $j = j + 1$ is based on $j = j$, when $j = j + 1$, there will be no feasible interval in any j or smaller than j adjacent trains combination of any $j + 1$ adjacent train. Therefore, when a $j + 1$ adjacent train has a feasible interval, after inserting the delayed train into this interval, all trains' departure times need to move backward except the first and last train. However, all trains are not allowed to depart in advance. For example, when $j = 3$, assuming that trains c, d, and e constitute a feasible interval, after the delayed train is inserted into this interval, the departure times of the subsequent train d must be shifted backwards to provide the departure space for the delayed train as shown in Fig. 3.

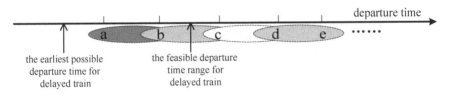

Fig. 2 Search strategy when $j = 2$

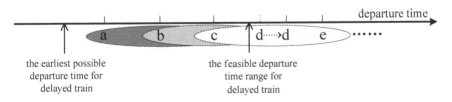

Fig. 3 Search strategy when $j = 3$

As stated above, a MATLAB program is developed for timetabling algorithms as follows:

Timetabling Algorithms

Input: Original timetable, train delay information;

Record the arrival time $t_{is}^{\prime a}$ and minimum stopping time τ_{is} of the delayed train i at station s, and calculate the earliest possible departure time $t_{is}^{\prime a} + \tau_{is}$ of train i.

Calculate the number (k) of all subsequent trains with a departure time greater than $t_{is}^{\prime a} + \tau_{is}$ and sort them from small to large in the original departure time sequence;

for j=2→k
   Search feasible interval sequentially;
   if (Existing feasible interval)
Arrange departure time for delayed train i, and move back the departure time of related follow-up trains.
      if (All trains are not delayed beyond time T)
Get the optimal solution and output the result; turn to Platforming Algorithms;
         break;
      end
      if (Existence of a train delayed beyond time T)
Record the current objective function value, (output the optimal solution recorded after the loop ends)
         continue;
      end
   else if (There is no feasible interval)
      End the algorithm;
   end
end
end

3.2 Platforming and Routing Algorithm

Carey and Crawford [8] and Lusby et al. [9] pointed out that the train platforming problem is an NP-hard problem and no universally valid algorithm was found. It is necessary to develop an algorithm, so that platforming plan due to change in train timetabling can be taken into account in the decision in the case of the train delay. Based on the new train schedule calculated by timetabling algorithms as described in Sect. 3.1, the simulated annealing algorithm is used to find platforming solution. The general framework of the simulated annealing algorithm is not presented in this paper because of page limited. Only the solution structure of the simulated annealing algorithm and the generation of the neighborhood solution are discussed [1]. The structure of the solution: According to the number of trains (k) and the number of the arrival/departure tracks that can be used by the current station, a matrix $A_{k \times 2}$ is used to represent the solution of the algorithm. The first column of the matrix represents the train number, and the second column represents the arrival/departure track number occupied by the trains [2]. Neighborhood solution: Randomly transforming the track number in the current solution matrix, means the neighborhood solution constitutes the current solution.

Therefore, a MATLAB program is developed for platforming algorithm as follows:

Platforming Algorithms

    Input: reference information of platform and the results of Timetabling Algorithms;

    Setting parameters: initial temperature t ; iteration number L ; cooling rate c ; temperature threshold e;

    Initial solution: the initial solution is generated from the new timetable and the original platform scheme;

    while t>=e

        for r=1:L

            Random replacement of a train's stopover track in the current solution to generate a new solution;

            if (The current solution satisfies all constraints)

                Calculate the objective function value of the current solution;

            else

                continue;

            end

            Accept or discard the current solution according to acceptance criteria;

        end

        t=t*c;

    end

    if (The algorithm has a feasible solution under the given parameters)

        Output result, and turn to Routing Algorithms;

    else

        Turn to Timetabling Algorithms;

    end

Once the arrival/departure tracks occupied by trains are determined, the routes connecting the arrival/departure tracks can be enumerated one by one according to the current station structure.

3.3 Cooperative Adjustment Algorithm

The above three sub-algorithms are used to form cooperative adjustment algorithms, in order to coordinate adjustment of delayed trains.

Figure 4 shows cooperative adjustment algorithms' flowchart, and in the event of train delays, it determines the optimal timetable, platform, and route adjustment schemes for each station affected by the delayed train; If there is a second delay

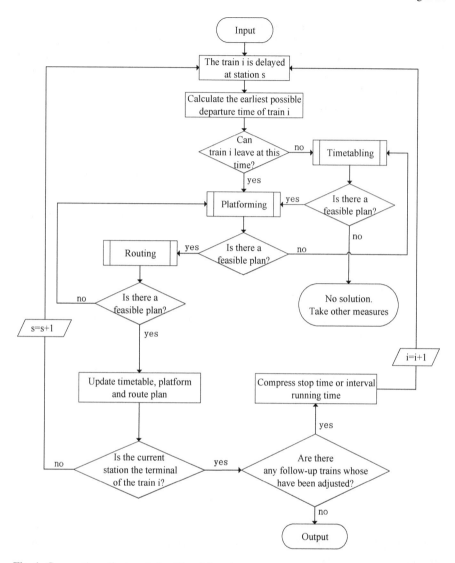

Fig. 4 Cooperative adjustment algorithms' flowchart

in the adjustment process, the same adjustment strategy will be applied again for the trains with subsequent delays. The effect of train delay is gradually absorbed by using the redundancy of train diagram. By using the proposed cooperative adjustment algorithms quickly, adjusting the train operation scheme and restoring the normal train order can be achieved.

4 Case Study

A simulation case study is discussed in this section to verify the effectiveness of the proposed method. 21 pairs of trains from five adjacent stations A, B, C, D, and E between 12:00 and 16:00 are selected in this case. The original train operation scheme is shown in Fig. 5.

As can be seen in Fig. 5, the number of arrival/departure tracks used at stations A, B, C, D, and E is 15, 10, 4, 4, and 12, respectively, with the main line as the segmentation, and the upward and downward trains are stopped in the corresponding receiving/departure yard. The tracking interval of all these five stations in the upward directions is both 5 min, the arrival interval of stations A, B, and E is 5 min, and the continuous non-stop passage interval of stations C and D is 3 min. In all these five stations, $\tau_{us} = 10$ min.

Due to the impact of unexpected events, the upward train ID02 was delayed for 28 min, and the arrival time of station A was delayed from 12:04 to 12:32. On the premise of meeting the above requirements of safe interval and technical operation time of each station, the adjustment scheme obtained by using the proposed method is shown in Fig. 6 (set $T = 60$ min).

According to the adjustment result, in the section between stations A and E, train ID 16 is affected by train ID 02, and train ID 16 is restored to the original planned departure time at station E. The total number of delayed trains is 5, i.e., 2 trains are delayed at station A, 2 trains at station B, and 1 train at station E, and the total delay time is 273 min. (The trains do not stop at D and C, so the delay time and the number of delayed trains are not counted).

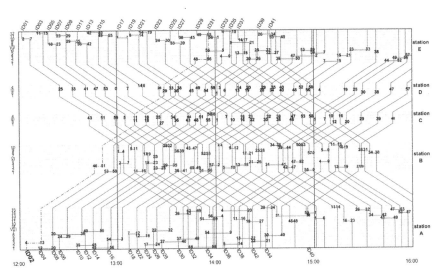

Fig. 5 Original train operation scheme

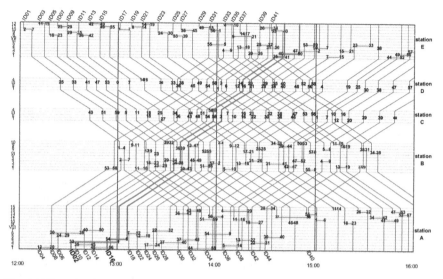

Fig. 6 Adjusted train operation scheme

5 Conclusions

This paper presents a proposed cooperative adjustment algorithm that can be used in the case of train delay to adjust the train operation plan under the condition of train delay, restore the normal train operation order, and reduce the impact of train delay effectively. MATLAB programs have been developed in order to calculate solutions. The case study indicates that proposed algorithms can realize the coordinated adjustment of train timetabling, platforming, and routing schemes quickly and efficiently. Compared with the traditional methods of hierarchical adjustment through multiple single models, the total number of delayed trains and the total delay time of trains can be well controlled by using the proposed method.

Acknowledgements The research is supported by the National Natural Science Foundation of China (Grant No. 71971220), the Natural Science Foundation of Hunan Province, China (Grant No. 2019JJ50829), and the Fundamental Research Funds for the Central Universities of Central South University.

References

1. Dindar S, Kaewunruen S, An M (2019) Rail accident analysis using large-scale investigations of train derailments on switches and crossings: Comparing the performances of a novel stochastic mathematical prediction and various assumptions. Eng Fail Anal 103(2019):203–216

2. Corman F, D'Ariano A, Marra AD, Pacciarelli D, Samà M (2017) Integrating train scheduling and delay management in real-time railway traffic control. Transp Res Part E Logistics Transp Rev 105:213–239
3. Luan X, Wang Y, De Schutter B, Meng L, Lodewijks G, Corman F (2018) Integration of real-time traffic management and train control for rail networks-Part 1: optimization problems and solution approaches. Transp Res Part B Methodological 115:41–71
4. Cacchiani V, Huisman D, Kidd M, Kroon L, Toth P, Veelenturf L, Wagenaar J (2014) An overview of recovery models and algorithms for real-time railway rescheduling. Transp Res Part B Methodological 63:15–37
5. Yang L, Qi J, Li S, Gao Y (2016) Collaborative optimization for train scheduling and train stop planning on high-speed railways. Omega 64:57–76
6. Zhang Y, Lei D, Liu M (2010) Scheduling model and algorithm for track application in railway station. Zhongguo Tiedao Kexue 31(2):96–100 (in Chinese)
7. Samà M, Meloni C, D'Ariano A, Corman F (2015) A multi-criteria decision support methodology for real-time train scheduling. J Rail Transp Plann Manage 5(3):146–162
8. Carey M, Crawford I (2007) Scheduling trains on a network of busy complex stations. Transp Res Part B Methodological 41(2):159–178
9. Lusby R, Larsen J, Ryan D, Ehrgott M (2011) Routing trains through railway junctions: a new set-packing approach. Transp Sci 45(2):228–245

Research on Service Ability Evaluation of Automatic Fare Collection System Based on DEMATEL and VIKOR-Gray Relational Analysis

Yuanyuan Zhou, Jiabing Zhang, Zhihui Zhou and Zongyi Xing

Abstract There are many manufacturers of urban rail automatic fare collection (AFC) systems, and the quality of the products is uneven. There are also no methods of service capability evaluation. So an evaluation model of service ability of AFC system based on DEMATEL and VIKOR-gray correlation analysis is proposed. Firstly, in order to balance expert knowledge and objective reality, the fuzzy DEMATEL is used to determine the weight of index. Secondly, the evaluation method based on VIKOR-gray correlation analysis is constructed to fully exploit the inherent law of sample data. Finally, the model is verified by an example.

Keywords Service ability · Evaluation index · Fuzzy DEMATEL method · VIKOR method · Gray relational analysis

1 Introduction

Along with the development of the rail transit industry, the AFC system [1, 2] is a modern networked toll collection system commonly used in the operation of rail transit in internationalized cities and is an important passenger management tool greatly facilitating passenger travel. As an important part of the automatic fare collection system, service capability evaluation [3] enables operators to keep abreast of the operation of the automatic fare collection system and provides reference for key equipment and system optimization of the automatic fare collection system.

Firstly, it introduces the DEMATEL [4] method to determine the weight of the evaluation index. Secondly, it introduces the VIKOR method and the improved method which will be used if the test conditions are not met. Thirdly, it verifies the example.

Y. Zhou · J. Zhang · Z. Zhou · Z. Xing (✉)
School of Automation, Nanjing University of Science and Technology, Nanjing, China
e-mail: xingzongyi@163.com

2 Determination of Index Weight Based on Fuzzy DEMATEL

DEMATEL is a methodology for the analysis of system elements based on graph theory and matrix tools proposed by the Battelle Association of the University of Geneva [5] in 1971 to solve complex problems in reality. This method makes full use of expert knowledge to deal with complex problems and expresses the logical relationship between elements in a simple and clear way. This paper extends DEMATEL [6] to the fuzzy domain through the semantic transformation of triangular fuzzy and improves the accuracy of DEMATEL [7]. Specific steps are as follows:

- Establishment of indicator system and fuzzy evaluation scale.

According to the establishment principle of the evaluation index system, the evaluation indicators are initially selected, which are numbered as $s_1, s_2, \ldots, s_n$, and the language scale of the expert's evaluation of the relationship between the indicators can be converted into a triangular fuzzy number as shown in Table 1.

- Determination of direct influence matrix.

Experts compare the indicators according to the semantic variables in Table 1 to determine the mutual influence relationship between the indicators. Assuming that the number of experts is m, the evaluation matrix of each expert is $E_1, E_2, \ldots, E_m$, so the direct influence matrix is E and the elements in E are the mean of the evaluation of the corresponding elements by m expert, which is as below $E = (E_1 + E_2 + \cdots + E_m)/m$.

The initial direct influence matrix is $E = \begin{bmatrix} 0 & e_{12} & \ldots & e_{1n} \\ e_{21} & 0 & \ldots & e_{2n} \\ \vdots & \vdots & \ddots & \vdots \\ e_{n1} & e_{n2} & \ldots & 0 \end{bmatrix}$, $e_{ij} = (l_{ij}, m_{ij}, u_{ij})$ is triangular fuzzy number, $e_{ii} = 0, i, j = 1, 2, \ldots, n$.

Table 1 Semantic variables and their corresponding fuzzy evaluation values

Semantic variable	Fuzzy scale
No effect	(0, 0, 0.25)
Little impact	(0, 0.25, 0.5)
Less affected	(0.25, 0.5, 0.7)
Greater impact	(0.5, 0.75, 1)
Great influence	(0.75, 1, 1)

- Determination of the standard influence matrix B

$$B = \begin{bmatrix} 0 & b_{12} & \cdots & b_{1n} \\ b_{21} & 0 & \cdots & b_{2n} \\ \vdots & \vdots & \ddots & \vdots \\ b_{n1} & b_{n2} & \cdots & 0 \end{bmatrix}$$

where $b_{ij} = e_{ij}/p = (l_{ij}/p, m_{ij}/p, u_{ij}/p) = \left(l'_{ij}, m'_{ij}, u'_{ij}\right)$, $p = \max_i \sum_{j=1}^n u_{ij}$.

- Determination of the comprehensive influence matrix Z

$$Z = \begin{bmatrix} z_{11} & z_{12} & \cdots & z_{1n} \\ z_{21} & z_{22} & \cdots & z_{2n} \\ \vdots & \vdots & \ddots & \vdots \\ z_{n1} & z_{n2} & \cdots & z_{nn} \end{bmatrix}$$

where $z_{ij} = \left(l''_{ij}, m''_{ij}, u''_{ij}\right)$, $\left[l'''_{ij}\right] = \sum_{c=1}^{\infty}(B_l)^c = B_l \times (I - B_l)^{-1}$

$$\left[m'''_{ij}\right] = \sum_{c=1}^{\infty}(B_m)^c = B_m \times (I - B_m)^{-1}$$

$$\left[u'''_{ij}\right] = \sum_{c=1}^{\infty}(B_u)^c = B_u \times (I - B_u)^{-1}$$

$$B_l = \begin{bmatrix} 0 & l'_{12} & \cdots & l'_{1n} \\ l'_{21} & 0 & \cdots & l'_{2n} \\ \vdots & \vdots & \ddots & \vdots \\ l'_{n1} & l'_{n2} & \cdots & 0 \end{bmatrix}$$

$$B_m = \begin{bmatrix} 0 & m'_{12} & \cdots & m'_{1n} \\ m'_{21} & 0 & \cdots & m'_{2n} \\ \vdots & \vdots & \ddots & \vdots \\ m'_{n1} & m'_{n2} & \cdots & 0 \end{bmatrix}, \quad B_u = \begin{bmatrix} 0 & u'_{12} & \cdots & u'_{1n} \\ u'_{21} & 0 & \cdots & u'_{2n} \\ \vdots & \vdots & \ddots & \vdots \\ u'_{n1} & u'_{n2} & \cdots & 0 \end{bmatrix}$$

- Determination of centrality and cause

Add Z by row and column to get Z_{ri} and Z_{cj}. The centrality of the indicator s_i is $Z_{cj} + Z_{ri}(j = i)$, which indicates the comprehensive influence value of the indicator s_i on the other indicators. The greater the centrality, the more important the indicator s_i is in the system. The reason is $Z_{ri} - Z_{cj}(j = i)$ indicates that the index s_i is affected by other indicators; if the cause is greater than 0, indicator belongs to the cause indicator. Conversely, it belongs to the result indicator.

$$Z_{ri} = \sum_{j=1}^{n} z_{ij}, \quad i = 1, 2, \ldots, n \quad (1)$$

$$Z_{cj} = \sum_{i=1}^{n} z_{ij}, \quad j = 1, 2, \ldots, n \quad (2)$$

$$\tilde{O}_i = Z_{ri} + Z_{cj} = \left(ol_i''', om_i''', ou_i'''\right), \quad i = j = 1, 2, \ldots, n \quad (3)$$

$$\tilde{R}_i = Z_{ri} - Z_{cj} = \left(rl_i''', rm_i''', ru_i'''\right), \quad i = j = 1, 2, \ldots, n \quad (4)$$

- Defuzzification

Since the CFCS method can effectively distinguish two fuzzy numbers with the same exact value, the CFCS method is used to perform the defuzzification process. Taking $\tilde{O}_i$ as an example, the clear value of $\tilde{R}_i$ can be obtained as follows.

1. Triangular fuzzy standard processing

$$\alpha_{li} = \frac{\left(ol_i''' - G\right)}{\Delta}, \quad \alpha_{mi} = \frac{\left(om_i''' - G\right)}{\Delta}, \quad \alpha_{ui} = \frac{\left(ou_i''' - G\right)}{\Delta} \quad (5)$$

where $H = \max_i ou_i'''$, $G = \min_i ol_i'''$, $\Delta = H - G$.

2. Normalize the left and right values

$$\alpha_i^l = \alpha_{mi}/(1 + \alpha_{mi} - \alpha_{li}), \quad \alpha_i^r = \alpha_{ui}/(1 + \alpha_{ui} - \alpha_{mi}), \quad (6)$$

3. Clear value after deblurring

$$O_i = G + \alpha_i^c \times \Delta \quad (7)$$

where $\alpha_i^c = \{\alpha_i^l \times (1 - \alpha_i^l) + \alpha_i^r \times \alpha_i^r\}/(1 - \alpha_i^l + \alpha_i^r)$.

- Determination of weight Q_i

Geometric mean of centrality and cause determines weight

$$q_i = \sqrt{(O_i)^2 + (R_i)^2} \quad (8)$$

$$Q_i = q_i / \sum_{i=1}^{n} q_i \quad (9)$$

3 Construction of Evaluation Model Based on VIKOR-Gray Correlation Analysis

3.1 Description of the Problem

Suppose that m objects are evaluated by n indicators, and x_{ij} is used to represent the original index value of the jth ($1 \leq j \leq n$) evaluation indicators in the ith ($1 \leq i \leq m$) schemes. The decision matrix is $X = [x_{ij}]_{m \times n}$, and the matrix is normalized by the vector method to obtain the standardized decision matrix $F = [f_{ij}]_{m \times n}$. For the benefit indicator, $f_{ij} = \frac{x_{ij}}{\sqrt{\sum_{i=1}^{m} x_{ij}^2}}$, and for cost indicators, $f_{ij} = \frac{1}{x_{ij}} / \sqrt{\sum_{i=1}^{m} \frac{1}{x_{ij}^2}}$.

3.2 Modeling Process

Calculate positive and negative ideal solutions.

$$f_j^* = \max_i f_{ij}, \quad f_j^- = \min_i f_{ij}, \tag{10}$$

Calculate group benefits and individual regret values.

$$S_i = \sum_{j=1}^{n} w_j d = \sum_{j=1}^{n} w_j \frac{f_j^* - f_{ij}}{f_j^* - f_j^-} \tag{11}$$

$$R_i = \max(w_j d) = \max\left(w_j \frac{f_j^* - f_{ij}}{f_j^* - f_j^-}\right) \tag{12}$$

Calculate the feasible value of the eclectic solution and rank the objects to be evaluated. Feasible value of the eclectic solution is Q_i

$$Q_i = \varepsilon \frac{S_i - S^*}{S^- - S^*} + (1 - \varepsilon) \frac{R_i - R^*}{R^- - R^*} \tag{13}$$

where $S^* = \min S_i$ represents optimal group benefits; $S^- = \max S_i$ denotes the worst group benefit; $R^* = \min R_i$ indicates the minimum individual regret value; $R^- = \max R_i$ indicates the maximum individual regret value; ε represents the proportion of group benefits ($\varepsilon \in (0, 1)$). In the comprehensive evaluation, the value of ε depends on the subjective tendency of the decision maker. In this study, we choose $\varepsilon = 0.5$ in order to simultaneously pursue group benefits and individual regrets. After obtaining the feasible value of the eclectic solution Q of the object to be evaluated, the ranking of the objects to be evaluated may be ranked according to the magnitude of the Q value, and the smaller the Q value is, the better the object to be evaluated is.

Constraint test. The use of the VIKOR method [8] is also limited by specific constraints, and the sorting result according to the Q value may be ranked as the final result between the objects to be evaluated if and only if the Q values of the objects to be evaluated simultaneously satisfy the constraint conditions.

Condition 1: Excellent threshold condition

$$Q_2 - Q_1 \geq 1/(m-1) \tag{14}$$

where Q_1 is the better order of the objects to be evaluated, Q_2 is the worse order of the objects to be evaluated, and m is the number of objects to be evaluated.

Condition 2: Decision reliability condition

The group benefit value and the individual regret value of the object to be evaluated which is the former should be smaller than the later one.

The object to be evaluated that cannot pass the algorithm constraint is filtered out. The index weighted value of the object to be evaluated that fails to pass the constraint condition in the VIKOR method is calculated, which can be obtained by the equation $z_{ij} = w_j f_{ij}$.

Obtain the optimal and worst values of the weighted values of each indicator of the object that failed to pass the constraint.

$$z_{ij}^* = \max_j z_{ij}, \quad z_{ij}^- = \min_j z_{ij} \tag{15}$$

And thus constitute the optimal reference sequence and the worst reference sequence. **Calculate the correlation coefficients r_j^*, and r_j^- of each evaluation index with the optimal reference sequence and the worst reference sequence.**

$$r_j^* = \frac{\min_i \min_j \left| z_{ij}^* - z_{ij} \right| + \xi \max_i \max_j \left| z_{ij}^* - z_{ij} \right|}{\left| z_{ij}^* - z_{ij} \right| + \xi \max_i \max_j \left| z_{ij}^* - z_{ij} \right|},$$

$$r_j^- = \frac{\min_i \min_j \left| z_{ij} - z_{ij}^- \right| + \xi \max_i \max_j \left| z_{ij} - z_{ij}^- \right|}{\left| z_{ij} - z_{ij}^- \right| + \xi \max_i \max_j \left| z_{ij} - z_{ij}^- \right|} \tag{16}$$

where ξ is the resolution coefficient and $\xi \in (0, 1)$. In this study, the optimal reference sequence and the worst reference sequence are equal, so $\xi = 0.5$.

Calculate the degree of association. The degree of association of the most and the worst reference sequence and the comprehensive relevance r_i of the object can be obtained from Eq. 17.

$$r_i^* = \frac{1}{n}\sum_{j=1}^{n} r_j^*, \quad r_i^- = \frac{1}{n}\sum_{j=1}^{n} r_i^-, \quad r_i = \frac{r_i^*}{r_i^* + r_i^-} \tag{17}$$

According to the comprehensive relevance degree of the object to be evaluated, the ranking of the objects to be evaluated is determined. The greater the comprehensive relevance, the higher the ranking of the objects to be evaluated.

4 Case Analysis

According to the basic principles of the evaluation system, the evaluation index system of the AFC system service capacity is determined. It includes ticket vending machine (TVM), automatic gate machine (AGM), and cloud booking office machine (cloud BOM) as shown in Table 2.

This paper selects the second line, the third line, and the fifth line of the Guangzhou Metro with a large flow of people as the verification case.

Table 2 Evaluation index system for service ability of AFC

TVM (A1)	Ticket purchase time (B1)
	Currency first acceptance rate (B2)
	Language suitability (B3)
	Type of payment method (B4)
	Failure rate (B5)
	User-friendliness (B6)
	Feedback effectiveness (B7)
	Change method (B8)
AGM (A2)	Ticket processing time (B9)
	Passing ability (B10)
	Failure rate (B11)
	Checking method diversity (B12)
	Passover comfort (B13)
	Prompt completeness (B14)
	Display content richness (B15)
	Communication rate with SC (B16)
Cloud BOM (A3)	Diversified payment method (B17)
	Query information richness (B18)
	Electronic invoicing capabilities (B19)
	Provide remote video service (B20)
	Type of ticket exception handling (B21)

Table 3 Center degree, reason degree, and weight of index (ticket vending machine)

	B1	B2	B3	B4	B5	B6	B7	B8
W_i	0.098	0.110	0.022	0.122	0.325	0.087	0.100	0.136
M_i	0.067	0.075	0.016	0.090	0.202	0.057	0.074	0.093
N_i	−0.030	−0.031	0	0.015	0.131	−0.030	0	−0.040

Table 4 Center degree, reason degree, and weight of index (automatic gate machine)

	B9	B10	B11	B12	B13	B14	B15	B16
W_i	0.048	0.756	0.066	0.043	0.014	0.024	0.013	0.036
M_i	0.170	0.231	0.203	0.160	0.047	0.076	0.047	0.132
N_i	−0.069	2.871	0.147	0.040	0.022	0.048	−0.02	0.036

Table 5 Weight of index (cloud booking office machine)

	B17	B18	B19	B20	B21
W_i	0.2	0.2	0.2	0.2	0.2

4.1 Determining Weight

See Tables 3, 4, and 5.

According to expert experience, the weight of TVM AGM and cloud BOM is 0.4, 0.4, and 0.2.

4.2 Comprehensive Evaluation of Service Capacity

The initial data of the Guangzhou Metro in June 2018 is shown in Table 6.

Standardize initial data

$$F_A = \begin{bmatrix} 0.551 & 0.584 & 0.640 & 0.616 & 0.611 & 0.625 & 0.530 & 0.616 \\ 0.627 & 0.564 & 0.426 & 0.492 & 0.573 & 0.469 & 0.662 & 0.492 \\ 0.551 & 0.584 & 0.640 & 0.616 & 0.547 & 0.625 & 0.530 & 0.616 \end{bmatrix}$$

$$F_B = \begin{bmatrix} 0.110 & 0.591 & 0.312 & 0.492 & 0.492 & 0.492 & 0.492 & 0.541 \\ 0.095 & 0.550 & 0.937 & 0.616 & 0.616 & 0.616 & 0.616 & 0.595 \\ 0.110 & 0.591 & 0.156 & 0.616 & 0.616 & 0.616 & 0.616 & 0.595 \end{bmatrix}$$

$$F_C = \begin{bmatrix} 0.616 & 0.616 & 0.616 & 0.616 & 0.616 \\ 0.616 & 0.616 & 0.616 & 0.616 & 0.616 \\ 0.492 & 0.492 & 0.492 & 0.492 & 0.492 \end{bmatrix}$$

Table 6 Initial value of each line

	Line 2	Line 3	Line 5
B1	2.2	2.5	2.2
B2 (%)	85	82	85
B3	3	2	3
B4	5	4	5
B5	6	6.4	6.7
B6	4	3	4
B7	4	5	4
B8	5	4	5
B9	1.2	1.4	1.2
B10	43	40	43
B11	0.15	0.05	0.32
B12	4	5	5
B13	4	5	5
B14	4	5	5
B15	4	5	5
B16	100	110	110
B17	5	5	4
B18	5	5	4
B19	5	5	4
B20	5	5	4
B21	5	5	4

Determine positive and negative ideal solutions

$f^* =$(0.627, 0.584, 0.640, 0.616, 0.611, 0.625, 0.662, 0.616, 0.110, 0.591, 0.937, 0.616, 0.616, 0.616, 0.616, 0.595, 0.616, 0.616, 0.616, 0.616, 0.616)

$f^- =$(0.551, 0.564, 0.426, 0.492, 0.547, 0.469, 0.530, 0.492, 0.095, 0.550, 0.156, 0.492, 0.492, 0.492, 0.492, 0.541, 0.492, 0.492, 0.492, 0.492, 0.492)

Overall utility value and individual regret value are given in Table 7.

According to Eq. 13, $Q_1 = 0$, $Q_2 = 1$, and $Q_3 = 0.4954$, which satisfies the test of Eq. 14. Therefore, the service capacity from high to low is Line 2, Line 5, and Line 3.

Table 7 Overall utility value and individual regret value

	S_i	R_i
1	0.1532	0.04
2	0.5896	0.3024
3	0.4356	0.1300
Max	0.1532	0.04
Min	0.5896	0.3024

5 Conclusion

This paper focuses on the service capability evaluation of AFC system based on fuzzy DEMATEL and VIKOR-gray correlation analysis. Firstly, the improved fuzzy DEMATEL method is introduced in detail, and the method of determining the weight of the index is proposed. Secondly, the steps of constructing the evaluation model of VIKOR and the improved algorithm after the test conditions are not satisfied are elaborated. Finally, the feasibility of the evaluation model is verified by an example. Due to the uncertainty of the selection of indicators, there may be some impact on the results of the evaluation. Therefore, the selection and screening of evaluation indicators are also the contents that need to be improved in subsequent research.

Acknowledgements This work is supported by National Key R&D Program of China (2017YFB1201203).

References

1. He J, Wan X (2019) Research on the automatic fare collection system of subway in cloud environment. Sci Technol Innov Appl (15):62–63 (in Chinese)
2. Zhao Y (2017) Research on adaptability evaluation of internal service facilities capacity of Dalian station on fast track line 3. Dalian Maritime University (in Chinese)
3. Ge Y (2015) Research on performance evaluation method for R&D and construction of WX subway automatic fare collection system. Yunnan University (in Chinese)
4. Du Y-W, Zhou W (2019) New improved DEMATEL method based on both subjective experience and objective data. Eng Appl Artif Intell 83:57–71
5. Tseng M-L, Wu K-J, Ma L, Kuo TC, Sai F (2019) A hierarchical framework for assessing corporate sustainability performance using a hybrid fuzzy synthetic method-DEMATEL. Technol Forecast Soc Change 144:524–533
6. Fu X-M, Wang N, Jiang S-S, Yang F, Li J-M, Wang C-Y (2019) A research on influencing factors on the international cooperative exploitation for deep-sea bioresources based on the ternary fuzzy DEMATEL method. Ocean Coast Manag 172:55–63
7. Gul M, Ak MF, Guneri AF (2019) Pythagorean fuzzy VIKOR-based approach for safety risk assessment in mine industry. J Saf Res 69:135–153
8. Li X, He M, Yu X (2019) Comprehensive evaluation of fresh e-commerce based on extended VIKOR method. Jiangsu Agric Sci 47(03):331–337 (in Chinese)

Cloud Computing System of Rail Transit Traction System and Data Collection

Yijun Liang, Ruijun Chen, Jie Chen, Ruichang Qiu and Zhigang Liu

Abstract With the sustained and rapid growth of the national economy, the demand for urban rail transit is increasing, and the requirements for the operation reliability of the traction power supply system are becoming more and more. Therefore, the importance of intelligent operation and maintenance is increasingly prominent. This paper proposes an intelligent cloud computing system for the energy-fed traction power supply system of Hohhot metro line 1. On the basis of the original system, according to the requirements of cloud computing, the network structure scheme of the system is presented, and through the analysis of the types and transmission mode of monitoring data needed in system, to determine the system data acquisition scheme.

Keywords Urban rail transit · Traction power supply · Cloud computing system · Data collection

Y. Liang (✉) · J. Chen · R. Qiu · Z. Liu
School of Electrical Engineering, Beijing Engineering Research Center of Electric Rail Transportation, Beijing Jiaotong University, Beijing 100044, China
e-mail: 18121452@bjtu.edu.cn

J. Chen
e-mail: jiechen@bjtu.edu.cn

R. Qiu
e-mail: rcqiu@bjtu.edu.cn

Z. Liu
e-mail: zhgliu@bjtu.edu.cn

R. Chen
Hohhot Urban Rail Transit Construction Management Co., Ltd., Hohhot, China
e-mail: 864021085@qq.com

1 Introduction

Traction power supply system for urban rail transit is the power source of urban rail vehicle, so the equipment in system will produce large amounts of complex data. Except monitoring some important parameters according to the requirements, how to store the massive amounts of information and use them efficiently, and predicting according to the results of the analysis, is an urgent problem that should be solved in the traction power supply system [1–4]. The intelligent cloud computing system can collect data information from each device of the traction power supply system and store it in the database, so that it can be analyzed with big data and be digitized, standardized, and networked [5].

2 Cloud Computing System

2.1 Introduction of Hohhot Metro Traction System and Construction of Hardware Network

Hohhot, metro line 1, using a new generation of energy feedback device in traction power supply system, includes 2 main substation automation, Hohhot East and Hui District, and 9 traction substation, Xierhuan Road, Hugangdong Road, Ulanhu Memorial, Xinhua Square, Art Institute, Uranjat, Hohhot East, Shdelam Village, and Bayan Village. Medium voltage energy feedback device is applied in the whole line.

The structure of network is shown in Fig. 1. The energy feedback devices at 9 stations of line 1 and the electricity meters in 2 main substations are under control of the system. All the monitored nodes are connected to the main network through the switch, and then connected to the data server through the main switch.

2.2 Cloud Computing System Structure

The cloud computing system consists of 3 parts, including the acquisition layer, the computing layer, and the application layer.

The acquisition layer is mainly used for data acquisition and computing. It collects the key information (voltage, current, temperature, humidity, and switching times) of the key equipment and components of the traction power supply system of the urban rail system and preprocesses the data (data picking, screening, and elimination) to extract the key information data.

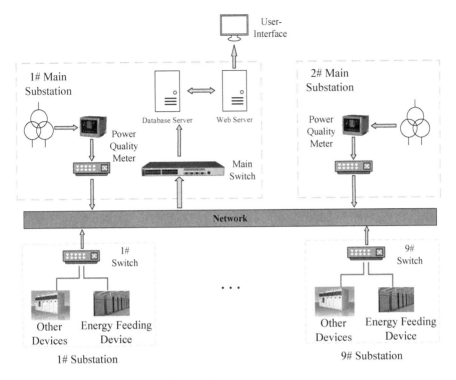

Fig. 1 Structure of intelligent cloud computing system

The computing layer includes the storage, computation, and analysis of data. A large amount of data is exchanged and stored on the online real-time server, and the data is analyzed on the cloud computing platform to realize the four functions of online monitoring, PHM (fault diagnosis and health management), fault analysis, and system evaluation.

The application layer is mainly used for the comprehensive management, information integration, and application of the system. The results of analysis in the computing layer can be displayed on the interface or accessed through the intelligent terminal.

3 Monitoring Data Type

Data of monitoring nodes are mainly from energy feedback devices and electricity meters, which can be divided into two parts according to data sources: the energy feedback device and the watt-hour meter.

3.1 Energy Feedback Device

Figure 2 shows the configuration diagram of energy feeder system, including switch cabinet (821, 80), transformer (NTr), inverter cabinet (NNB), isolation cabinet (85, 86), etc.

The core of the energy feeder device is the inverter. Each control box operates an inverter, and transfers data and instructions. The network of this part is shown in Fig. 3.

Ethernet Communication: The voltage and current waveform data of the two inverters will be uploaded through Ethernet communication, including a total of 16 channels of the two inverters with a sampling rate of 10 K. Ethernet data collection is shown in Table 1.

RS485 Communication: Voltage, current, and other data of the control box other than waveform data are transmitted through RS485 communication, which is set to be collected every 2 s. Table 2 shows serial port data information and Table 3 shows a part of error information table.

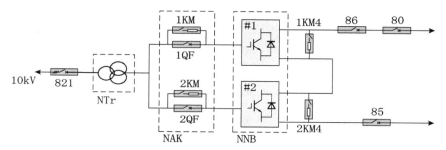

Fig. 2 Topology of energy feedback device

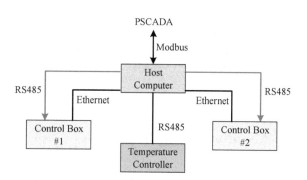

Fig. 3 Network of energy feedback device communication

Table 1 Data from Ethernet

Inverter 1#		Inverter 2#	
U_a	1# line voltage of AB	U_a	2# line voltage of AB
U_b	1# line voltage of BC	U_b	2# line voltage of BC
I_a	1# A-phase line current	I_a	2# A-phase line current
I_b	1# B-phase line current	I_b	2# B-phase line current
I_c	1# C-phase line current	I_c	2# C-phase line current
I_{dc}	1# DC current	I_{dc}	2# DC current
U_{dc}	1# DC voltage	U_{dc}	2# DC voltage

Table 2 Data from serial port

Inverter 1#		Inverter 2#	
I_{a1}	1# A-phase line current	I_{a2}	2# A-phase line current
I_{b1}	1# B-phase line current	I_{b2}	2# B-phase line current
I_{c1}	1# C-phase line current	I_{c2}	2# C-phase line current
I_{dc1}	1# DC current	I_{dc2}	2# DC current
U_{ab1}	1# line voltage of AB	U_{ab2}	2# line voltage of AB
U_{bc1}	1# line voltage of BC	U_{bc2}	2# line voltage of BC
U_{dc1}	1# DC voltage	U_{dc2}	2# DC voltage
P (kV)		Feedback power	
W (KWh)		Feedback energy	
I_{dc}		$I_{dc} = I_{dc1} + I_{dc2}$	
U_{dc}		$U_{dc} = U_{dc1} + U_{dc2}$	

3.2 Watt-Hour Meter

The real-time data collected by the watt-hour meter are the active power P and reactive power Q of the main substation, which are updated every 3 s.

4 Data Collection

For different data types, data acquisition is divided into two transmission modes, Ethernet data acquisition, and serial data acquisition.

Table 3 Part of error information

Inverter 1# and Inverter 2#	
10 kV bus undervoltage	SKA overheated shutdown
10 kV busbar overvoltage	Charging circuit error
Pulse overlap	Parameter reading error
Synchronization error	DC discharge fault
DC undervoltage	Boot failure
SKC error	CAN communication failed
SKC current overflows	Switch 80 trip
SKB error	Transformer overtemperature shutdown
SKB current overflows	Transformer high temperature alarm
SKA error	Three-phase current imbalance
SKA current overflows	…

4.1 Ethernet Data Acquisition

The data acquisition card in the bottom control box can collect data at high speed, real-time waveform monitoring is needed for some important voltage and current, and the sampling rate is high. The upper acquisition program collects data by calling DLL instruction [6, 7]. The flow chart is shown in Fig. 4a. After initializing the DLL file, data is collected, and the sampling rate of the Ethernet communication is up to 10 K.

4.2 Serial Data Acquisition

RS485 communication protocol is also used between the acquisition program and the control box to transmit the monitoring data with low sampling rate. RS485 is a commonly used serial communication standard, which adopts the query-response communication mode, as shown in Fig. 4b.

Query: The important part of the query message is the opcode, which represents which data to send or perform an operation in the device being queried [8]. The data segment contains the actual data information to be transmitted, such as the amount of data to be transmitted. The error detection domain is used by the device to verify that the message content is correct and to prevent an error response.

Response: When the query message received by the device is verified to be error free, the device performs different operation responses according to the opcode. If the query message does not pass validation, no response operation is performed.

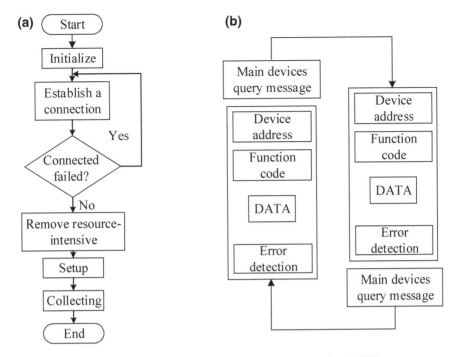

Fig. 4 **a** Flow chart of Ethernet data collection, **b** communication mode of RS485

Frame Format. The RS485 protocol frame used in the communication between the acquisition program and the control box of the energy feeder device adopts the half-duplex mode and is defined as 10 bytes, as shown in Fig. 5.

The starting bit is 0XAA, and the addresses of the two sets of control boxes are 0X71 and 0XF1, respectively. Opcodes are used to uniquely identify the amount of data transmitted. Each monitoring data has a corresponding opcode. The control box determines which operation to be performed by identifying the opcodes in RS485 frame and returns the corresponding data [9]. The data bit has four bytes, which is 0X0X00000000 in the query frame and the actual data in the response frame. When composing a data frame, the actual data needs to be multiplied by 2^{21} and converted into hexadecimal data. The CRC cyclic redundancy check is adopted in the check method. If the check condition is not met, the command of this query frame will not be received. The corresponding relationship between opcodes and parameters is shown in Table 4.

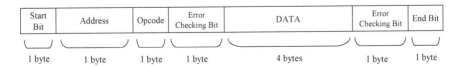

Fig. 5 Frame format of RS485

Table 4 Correspondence of serial communication operation code

Opcode	Parameter
AA	U_{ab}
	U_{bc}
	I_a
	U_{dc}
	I_{dc}
55	U_{ca}
	I_b
	I_c
	SystemState
	SKAtemp
2E	IO state
33	Feedback energy
30	SKBtemp
31	SKCtemp

5 Conclusion

Based on the introduction of the traction power supply system of Hohhot metro line 1, this paper designs the structure of the intelligent cloud computing system, selects data acquisition variables based on data requirements, and introduces the collection methods in detail.

Acknowledgements This work is supported by the National Key Research and Development Program (2017YFB1200802-01).

References

1. Mao y (2015) Design and implementation of massive power data analysis system based on cloud computing. Beijing Jiaotong University (in Chinese)
2. Chen j (2018) Key technologies for traction power supply of new generation intelligent urban rail transit. Electr Age 7:72 (in Chinese)
3. Zhang Y (2014) Design and application of rail transit power monitoring system. East China University of Science and Technology (in Chinese)
4. Luo m (2015) Fault prediction and health management platform design and implementation of high-speed railway traction power supply equipment. Southwest Jiaotong University (in Chinese)
5. Liu k (2017) Design and research of 110 kV smart substation for JIMO power grid. Qingdao University (in Chinese)
6. Wang x (2016) Research and implementation of Ethernet transmission system based on FPGA. Beijing Institute of Technology (in Chinese)

7. Peng g, Du y (2008) Application discussion of Ethernet technology in train communication network. Railw Veh 46(12):25–28 (in Chinese)
8. Compare M, Bellani L, Zio E (2017) Availability model of a PHM-equipped component. IEEE Trans Reliab PP(99):1–15
9. Liu D, Zhou J, Liao H et al (2015) A health indicator extraction and optimization framework for lithium-ion battery degradation modeling and prognostics. IEEE Trans Syst Man Cybern Syst 45(6):915–928

Exploration on Innovation Education of Higher Vocational Automobile Specialty Based on Internet Thinking

Lei Yang, Fanling Zeng and Daoru Yao

Abstract It is necessary for higher vocational automobile professional innovation education with Internet thinking. Will have a profound impact on the employment of automotive graduates and future entrepreneurship. In this paper, through the innovative practice, the automotive cloud service platform is an innovative entity, leading students to actually explore the innovation of automotive industry. The car service platform in this project integrates a large number of merchants and car users. It provides a wide range of car services. Can promote the development of the industry and employment. In this process, let the students truly think and feel the power of innovation. Through the practical application of this platform, students can improve their ability to innovate in their future work.

Keywords Internet · Automobile speciality · Innovative education

1 Introduction

The ubiquitous automobile has played a dominant role in our life. It is the necessary means of transportation for travel. With the development of the market economy, cars have become an indispensable part of life. Cars play a very important role in our lives. At the same time, Internet technology has made rapid progress. While enriching people's lives, the Internet is also growing and growing. The future development trend of Internet technology is time-domain, interactive and numerous users [1].

However, cars often experience various faults while driving. This general phenomenon has seriously threatened personal safety and property security.

L. Yang (✉) · F. Zeng · D. Yao
Anhui Vocational and Technical College, Hefei 230000, China
e-mail: 739849206@qq.com

2 Introduction to Automotive Cloud Service Platform

2.1 Current Problem

The resources of each platform cannot be effectively utilized and integrated. Some service points are in a downturn, the geographical location is remote, the customers cannot concentrate, and the industry is confused, and the professional knowledge of the service points cannot be utilized [2].

As a student of the Automobile Academy, the entrepreneurial team pays attention to the development direction of the automotive industry. And observe the development status of automobile service enterprises.

In the detailed investigation, entrepreneurs gradually found some problems (Table 1).

In order to solve the appeal problem, bring students to do relevant innovative practices. The purpose of creating a platform is to meet the needs of users and businesses. The user directly understands all the services required for car repair and maintenance through the platform client. Choose the type of service based on actual needs, free and convenient online booking or real-time service.

2.2 Platform Introduction

The car helper service platform is based on automotive services and the Internet platform, combining offline business with online Internet services. In this way, the effective integration of each platform is achieved. Provide integrated information services to all industries and the public within the scope. Integrate data into this cloud platform to realize the comprehensive sharing of data resources, information resources, computing resources, storage resources, knowledge resources and expert resources and solve problems for customers. Work on all aspects of the car. Based on the principle of "convenient, fast, comprehensive, meticulous, low-cost, high service," it not only provides customers with all-round services but also brings profits to the merchants who enter the platform [3].

Table 1 List of questions

List of questions	Most car owners do not know the health of their cars
	4S shop car maintenance costs are higher
	The owner does not trust the quality of service of the auto repair shop
	Worried about car maintenance, the oil is not cleaned, replaced with genuine
	4S shop car maintenance is more greasy, there are industry hidden rules
	Trouble with maintenance cannot solve the problem in time

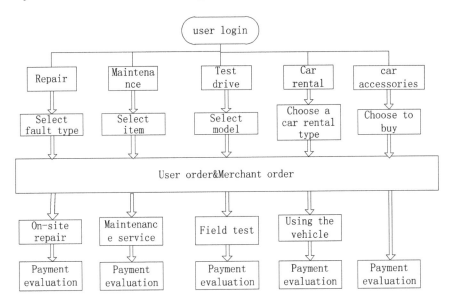

Fig. 1 Automotive project service list

Automotive project service content introduction: Users can obtain these service contents by logging into the platform such as maintenance, car rental services and car supplies.

Specific case: After logging into the platform, the user chooses to repair the project and then selects the type of fault. After the order is placed to the merchant, after the merchant receives the order, the on-site repair is performed according to the fault condition, and after the repair is completed, the user pays and gives an evaluation (Fig. 1).

3 APP Application

3.1 Specific Interface

This product is currently under development, and the specific interface is as follows:

Enter the account to log in, as shown in Fig. 2, you can also use QQ or WeChat, Weibo and other ways to log in. And improve personal information when you log in (nickname, car information, maintenance and repairs, etc.)

Enter the software home page, as shown in Fig. 3, you can clearly choose the services we need: car doctor, annual review, car rental, driving school, vehicle valuation, used car service [4].

This is the car doctor service interface. As shown in Fig. 4, you can clearly see the repair points near our location. When our car fails, you can apply for a service

Fig. 2 Login interface

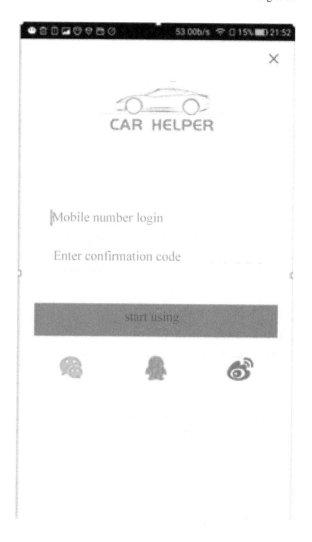

by filling out the fault order to a nearby service point. After the merchant receives the order, they can enjoy the excellent service [5]. Of course, we can also choose the merchants according to the comprehensive star rating of the merchants.

Other interfaces, car maintenance service interface, we can choose according to their own habits or business merchant ratings. The platform will also issue coupons from time to time to improve the user experience.

Other interface, used car service. We can search for the models we want from this interface and filter the used cars according to the brand, price and so on [6].

Fig. 3 Interface options

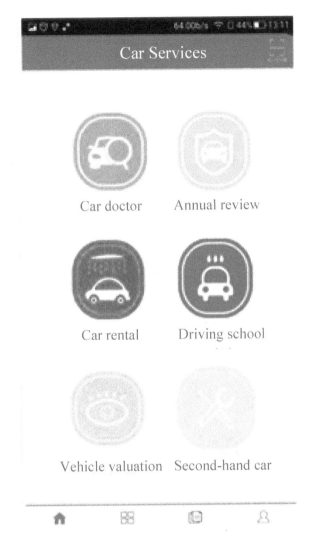

3.2 Product Property Protection

The car helper has applied for national software copyright protection, as shown in Table 2. And rely on professional companies to develop the program. Team members have their own advantages in all aspects and make a lot of effort. In the later period, many car dealers will settle in and provide core services for users of the platform.

Fig. 4 Car repair interface

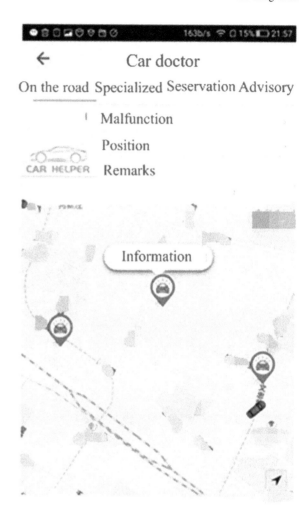

Table 2 Software copyright

Car cloud service mobile phone software copyright certificate	
Software copyright name	Car cloud service mobile APP software V1.0
Registration number	2018SR629328

3.3 Automotive Fault Service Process Introduction

The company's initial development is scheduled to be in the vocational education city of Yaohai District, Hefei City, Anhui Province. Consider the pre-administration funds and risks and the pre-environmental car wash service project. After accumulating

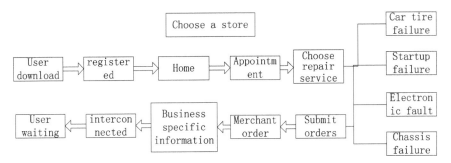

Fig. 5 Car failure service flowchart

customers and funds for half a year, we will gradually promote maintenance, maintenance, car rental and other services [7]. Promote the city one year later. Through the virtuous circle of mutual support of the four major service projects, a good development momentum is formed. Then, gradually, promote our development model [8].

In the process of development, we implemented a scientific management approach, and after the successful trial of small-scale trials, we began to further promote and improve our service projects. The specific service process is as follows (Fig. 5).

4 Summary

This paper completed the research and development of the automotive cloud service platform and the introduction and exploration of application examples through the innovative thinking of the Internet. In this process, let the students truly think and feel the power of innovation.

Inspire students to innovate in the automotive industry. Through the practical application of this platform, students can improve their ability to innovate in their future work.

Acknowledgements Project fund:
1. Application and Research of Compressed Sensing Algorithm Based on Wavelet Transform in Traffic Monitoring. Item Number: KJ2019A0990.
2. Development of a smart toilet with adjustable voice interaction. Item Number: KJ2019A0985.
3. Automotive electrical and auxiliary electronic system wisdom classroom pilot. Item Number: 2017zhkt293.
4. Automotive electrical and auxiliary electronic system technology and maintenance. Item Number: 2015gkk028.

References

1. Li J (2017) In: 2017 international conference on robots & intelligent system (ICRIS)—digital information teaching resources for automobile repair specialty, pp 131–134
2. Ferreira LP, Gómez EA, Peláez Lourido GC et al (2012) Analysis and optimisation of a network of closed-loop automobile assembly line using simulation. Int J Adv Manuf Technol 59(1–4):351–366
3. Lina W, Lin C, Institution WE et al (2018) Exploring the innovative way of education informationization in the new era: a summary of the 16th international forum on educational technology. e-Educ Res
4. Mady E, Ahmed S, Kirsi Y et al (2018) Innovative education and active teaching with the Leidenfrost nanochemistry. J Chem Educ
5. Wells DD (2018) You all made dank memes: using internet memes to promote critical thinking. J Polit Sci Educ 14:240–248
6. Rice R, Amant KS (2018) Introduction. Thinking globally, composing locally: re-thinking online writing in the age of the global Internet
7. Harrison C (2017) Defining and seeking to identify critical Internet literacy: a discourse analysis of fifth-graders' Internet search and evaluation activity. Literacy
8. Raikar MM, Desai P, Naragund JG (2016) Active learning explored in open elective course: Internet of Things (IoT). In: 2016 IEEE eighth international conference on technology for education (T4E). IEEE

A Study on the Redundancy of Ethernet Train Backbone Based on Dual-Link Topology

Xiaodong Sun, Jianyu Jin, Jianbo Zhao and Chenyang Bing

Abstract At present, Ethernet has replaced fieldbus as one of the effective solutions for train communication network. The International Electrotechnical Commission has also issued a standard for train communication network based on Ethernet. Ethernet train backbone (ETB) is responsible for all train-wide communication. The safe operation of trains requires high reliability of the train backbone. In order to improve the reliability of train backbone network, this paper introduces a network redundancy scheme of train backbone network based on dual link. The characteristics of parallel redundancy protocol (PRP) and media redundancy protocol (MRP) protocols proposed by IEC 62439 are analyzed, and a redundancy method suitable for train backbone network is proposed. The network structure of train backbone is optimized, and redundancy is realized through the management of backbone network nodes. During normal operation of the system, data is transmitted through two links, respectively, thus reducing the utilization rate of each link and reducing the transmission delay. In case of network failure, link switching can be carried out through the management of backbone network nodes to avoid communication interruption. The OPNET simulation software is used for simulation analysis. The simulation results verify the high reliability of the scheme and analyze the relationship between network utilization rate and time delay.

Keywords Train backbone · Media redundancy protocol · Parallel redundancy protocol · Network reliability · Dual link · Network simulation · Delay

---

X. Sun (✉) · J. Zhao · C. Bing
CRRC Qingdao Sifang Co., Ltd., 266111 Qingdao, China
e-mail: sunxiaodong.sf@crrcgc.cc

J. Jin
China Academy of Railway Sciences Corporation Limited (Institute of Locomotive and Rolling Stock), 100081 Beijing, China
e-mail: jinjianyu@zemt.cn

1 Introduction

Intelligent rail transit has become one of the important development directions in the field of rail transit. As one of the core train systems, the train communication network is responsible for the transmission of information such as train control, diagnosis, and passenger service. With the increasing demand for intelligence, the amount and types of information transmitted by trains are increasing. The traditional train communication network uses fieldbus technology, and its transmission bandwidth has been difficult to meet the transmission requirements of big data [1].

Ethernet communication protocol has the advantages of high bandwidth, low cost and flexible networking and is widely used in various fields. Through the introduction of Ethernet technology, industrial control field has formed industrial Ethernet technology with high real-time performance and high reliability. Industrial Ethernet technology can solve the problem of low transmission bandwidth when used in train communication network. At the same time, Ethernet provides a variety of implementation methods for complex network environment of trains with its flexible networking mode. At present, the train communication network based on Ethernet has become the main development direction. The International Electrotechnical Commission has also issued the Ethernet train backbone standard IEC61375-2-5 [2] and the Ethernet consist network standard IEC61375-3-4 [3], which have become important reference standards for Ethernet trains.

In the process of high-speed train operation, the operation environment has a great influence on the reliability of the network, and the network failure poses a great threat to the traffic safety. Using redundant network is one of the effective methods to improve network reliability [4, 5]. Common redundant networks include ring networks, mesh networks, parallel networks, etc. Different network topologies also have different redundancy protocols. IEC62439 standard for high availability automation network specifies a variety of network redundancy protocols, including media redundancy protocol (MRP), parallel redundancy protocol (PRP), distributed redundancy protocol (DPR), etc. [6].

The train communication network based on Ethernet is divided into train backbone network and train networking two-layer network architecture [7]. Ring redundant network is adopted for networking, and linear topology network is adopted for backbone network. There is no redundancy in a single train backbone network. Once a fault occurs, communication within the train will be interrupted, which will pose a threat to the safe operation of the train.

In order to improve the reliability of the train backbone network, this paper proposes a network structure using dual links for data transmission. While improving the reliability of the network, it can reduce link congestion and improve the transmission quality.

2 Analysis of Network Redundancy Protocol

Redundant protocols commonly used in industrial Ethernet include parallel redundancy protocol (PRP) and media redundancy protocol (MRP). This section analyzes these two protocols.

2.1 PRP Protocol

PRP protocol requires that the network includes two completely independent networks. Terminal nodes are connected to the two networks. Each network can provide complete communication links for redundant nodes. The terminal node realizes redundant communication by adding link redundant entity (LRE) in the protocol stack. When sending data, LRE will copy the data into two copies, add redundant information in the data frame, and then send the data to the two networks, respectively. After receiving the two copies of data, the receiver will retain one copy of the data in the copied frame and discard the other copy through the discard window algorithm. When one network fails, the other network can communicate normally without data loss [8, 9].

Under normal circumstances, the two networks run in parallel and are not affected. In case of failure, the communication between the two networks can be seamlessly switched and the failure recovery time is 0. Due to the adoption of two independent networks, the cost of the system adopting PRP protocol increases more. The data transmitted by the two networks is exactly the same. During normal operation, half of the data will be discarded, resulting in a waste of network bandwidth.

2.2 MRP Protocol

MRP protocol is a network recovery protocol based on ring network, which realizes network redundancy through switch port state switching. The switches in the network based on MRP protocol are divided into media redundancy manager (MRM) and media redundancy clients (MRC). During normal operation, one end of MRM is blocked, and information sent by terminal equipment is not allowed to pass through. At the same time, MRM sends a detection frame to detect the network status. Once a network failure is detected, the blocking port of the MRM is switched to the forwarding port, allowing the data of the terminal node to pass through, and notifying the MRC to update the network topology [10].

MRP protocol does not change the terminal node but is only applicable to ring network. Moreover, MRM nodes are required to be high. Once MRM fails, reliability cannot be guaranteed.

Both PRP and MRP have improved the network architecture and realized network redundancy by managing terminal equipment or switches. In the next chapter, this paper improves the network architecture of the train backbone network and carries out redundant management on the backbone network nodes.

3 Optimization of Train Network Architecture

At present, the research on train communication network mainly adopts the network architecture shown in Fig. 1. Backbone network adopts linear topology and consists of backbone network nodes (TBN) and links between nodes. Networking adopts ring topology or other types of topology and is composed of networking nodes (CNN) and links between nodes. TBN and CNN have the function of switches and can manage the network. In this architecture, only the redundant network is used for networking, and the backbone network is not redundant. When the backbone network node or the link between nodes fails, the communication within the network is not affected, but the communication between networks will cause interruption and affect the normal operation of vehicles.

In order to realize the redundancy of the backbone network, a ring topology as shown in Fig. 2 can be adopted. The TBN at both ends can be directly connected to form a ring network. When a single-link failure occurs in the backbone network, communication can be recovered through fast link switching. The method of Fig. 2 can realize the redundancy of the backbone network, but the link between TBN and CNN is not within the redundancy range. If the link between TBN or backbone network and the network fails, the network connected with it cannot communicate with other networks, and this method cannot ensure the reliability of all networks.

In order to solve this problem, this paper proposes a network architecture as shown in Fig. 3, in which a backbone network link is added; each backbone network node is connected with the networking node, and the backbone network nodes at both ends are connected to form a ring network. Under the condition that the network structure is not changed, the backbone network is changed from a single link to a double

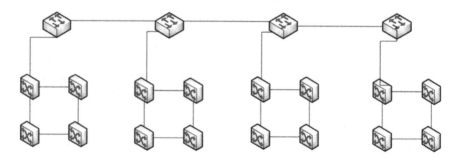

Fig. 1 Traditional train communication network architecture

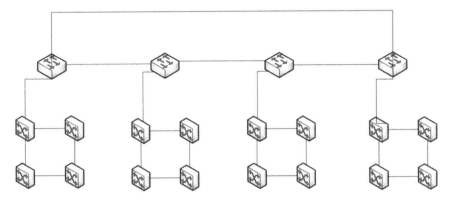

Fig. 2 Network architecture using ring backbone network

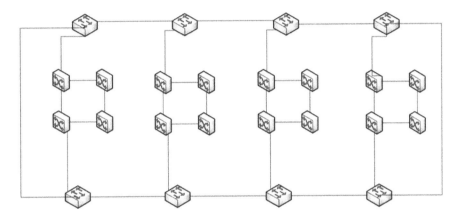

Fig. 3 Dual-link backbone network architecture

link structure. The structure realizes the redundancy of the backbone network, and at the same time, the links between the backbone network and the network are also redundant, thus greatly improving the reliability of the network.

4 A Study on Communication Mechanism

4.1 Transmission Mode

In the network architecture of Fig. 3, two backbone network links are included. When there is no network failure, data is transmitted through one of the links, and the other link can be used as a cold backup or a hot backup. Cold backup means that the backup link does not carry out data transmission and only switches to the backup link for transmission after the working link fails. Hot backup means that the backup link transmits the same data as the working link, but the network does not receive the data. When the working link fails to transmit, the network starts to receive the backup link data.

Both cold backup and hot backup can realize the redundancy of the backbone network, but the backup link does not play a role when the network works normally, and only when the working link fails can the backup link be converted into the working link.

When the network has no fault, both communication links can realize communication within the train range. In order to make more efficient use of the two links, the communication data within the train range can be divided into two parts and transmitted separately through two backbone network connections. This method can reduce the network load, which is equivalent to doubling the original network bandwidth. When transmitting in this way, communication data needs to be classified, and backbone network nodes can identify different data.

In order to realize dual-link transmission of data, it is necessary to distinguish communication data. The data transmitted by link 1 is link data 1 (LD1), and the data transmitted by link 2 is link data 2 (LD2). There are many ways to distinguish communication data: (1) classify according to transmission period. In train communication network, process data is sent periodically and message data is sent aperiodically. Process data can be set as LD1 and message data as LD2; (2) according to the classification of terminal equipment, the data sent by the central control unit can be set as LD1, the data sent by the traction control unit as LD2, and other equipment can be classified in the same way. According to different actual needs, other classification methods can also be adopted.

Backbone network nodes need to distinguish data according to the information of Ethernet data frames. Link identifier LID (link ID) should be set when sending data frames. The data frame format is shown in Fig. 4. LID is located at the end of user data and occupies 2 bytes. LID of LD1 is 0x1001, and LID of LD2 is 0x1002. LID is configured in terminal equipment, which can be determined during equipment initialization or switched according to network transmission quality during operation.

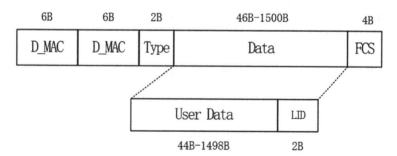

Fig. 4 Data frame format

4.2 Working Principle of Backbone Network Nodes

The backbone network node includes two types of ports, namely a backbone network port (TBP, train backbone port) and a networking port (CNP, consist network port). The backbone network port is connected with other backbone network nodes, and the networking port is connected with the networking nodes. Each port has two states, forwarding state and blocking state. In forwarding state, networking data can be forwarded, while blocking state cannot forward networking data [11].

The backbone network link includes a lead node, which manages the port status of the backbone network node. In order to prevent the failure of the main node, the two backbone network links always each include a main node. When a failure condition is detected, the main node will switch the port status of the backbone network node.

Networking ports will identify LID of data frames and set port status for different LD. Take link 1 as an example. When the network works normally, the networking port pair LD1 is set to the forwarding state and the pair LD2 is set to the blocking state. When the data frame reaches the backbone network node, the networking port identifies LID as 0x1001, forwards the data, identifies LID as 0x1002, and discards the data. For communication within the train range, the networking port is the only entrance for data frames to enter the backbone network. LD2 cannot be propagated in link 1, thus reducing the load rate of the link, thus reducing the communication delay and improving the communication quality MRP.

Backbone network ports can forward all data, but the outer backbone network ports of backbone network nodes located at both ends of the backbone network, i.e., the backbone network ports directly connected to link 1 and link 2, need to be set to a blocking state to prevent the formation of broadcast storms.

4.3 Failure Handling Mechanism

When the backbone network fails, it is necessary to discover the failure and switch links in time to prevent the impact caused by communication interruption. In order to detect faults, neighboring backbone network nodes (including backbone network nodes connected with another link) and between backbone network nodes and networking nodes need to periodically send a heartbeat. Once the signal is interrupted, the node that detects the signal interruption sends fault information to the master node. The master node can locate the fault location by analyzing the fault information, and then, the master node broadcasts the port state information throughout the network. Upon receiving the port state information, the backbone node will immediately switch the state of the networking ports. If link 1 fails, the networking port pairs LD1 and LD2 of the backbone node of link 1 are set to the blocking state, and the backbone node pairs LD1 and LD2 of link 2 are set to the forwarding state. The backbone network is switched from a dual-link mode to a single-link mode, and at the same time, a fault alarm is issued to notify relevant personnel to carry out fault maintenance.

In the single-link operation mode, the node will still send heartbeat signals. Once the recovery of heartbeat signals is detected, the node will send information to the master node. After the master node confirms that the failure is relieved, the port status information will be broadcasted across the network. After the backbone network node receives the port status information, it will switch the networking port mode to the dual-link operation mode before the failure and release the alarm.

4.4 Reliability Analysis

Terminal nodes need to go through two networking links and a backbone network link when communicating within the train range. After adopting a double backbone network link structure, the reliability of the backbone network link is improved. The network block diagram of communication is shown in Fig. 5.

Assuming that the reliability of the networking link is Rc, the reliability of the backbone link is Rb, and the reliability of the entire network transmitted by the single backbone link is

$$Rn = Rb * Rc^2$$

The reliability of the entire network using dual backbone links is

$$Rn = Rb * (2 - Rb) * Rc^2$$

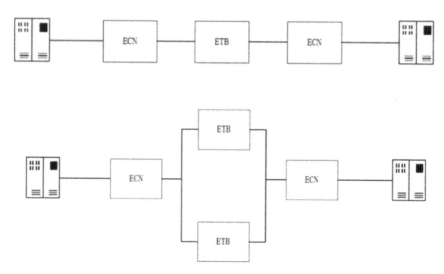

Fig. 5 Communication network block diagram

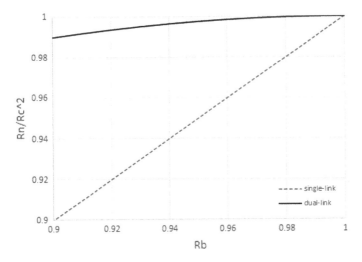

Fig. 6 Network reliability

The reliability of networking is not in the scope of this article. Assuming that the networking reliability of the two architectures is the same, the reliability of the whole network is only related to the reliability of the backbone network. As can be seen from Fig. 6, the reliability of using dual link is much higher than that of single link.

At the same time, under the condition of no fault, the adoption of double links can reduce the network load rate of a single link, and the reduction of the load rate can shorten the queuing delay of data frames in the switch, thus improving the real-time performance of communication.

5 Network Simulation

In order to verify the feasibility of the dual-link scheme, the text uses OPNET Modeler network simulation tool for simulation verification [12]. The network model shown in Fig. 7 is built.

The network model takes the grouping of 8 cars as an example and uses 100 megabytes Ethernet for communication. The backbone network includes two links; each link includes a backbone node in each vehicle section and a network in each vehicle section. Each backbone node is connected with a network, respectively. Networking adopts a ring topology structure; each networking includes 4 networking nodes, and terminal equipment is connected to different networking nodes, as shown in Fig. 8.

As shown in Fig. 9, the backbone network node model includes a plurality of data transceiver ports. TEB_node_manager is responsible for managing each port,

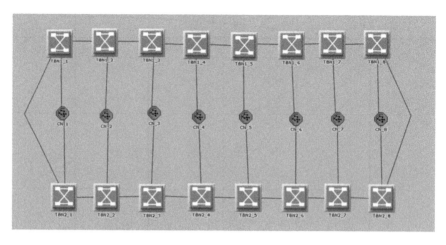

Fig. 7 Backbone network model

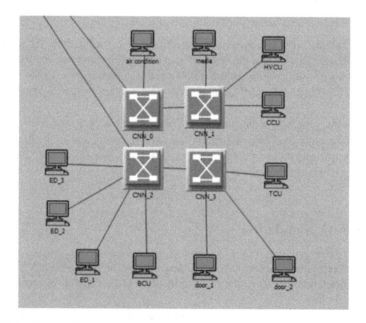

Fig. 8 Networking model

including setting port blocking/forwarding status, sending and detecting heartbeat signals, and sending port status information.

Communication within the train is mainly based on the communication between the central control unit (CCU) and other terminal equipment. Real-time data is transmitted periodically, with a transmission period of 10 ms or 20 ms and a fixed data

A Study on the Redundancy of Ethernet Train Backbone ... 407

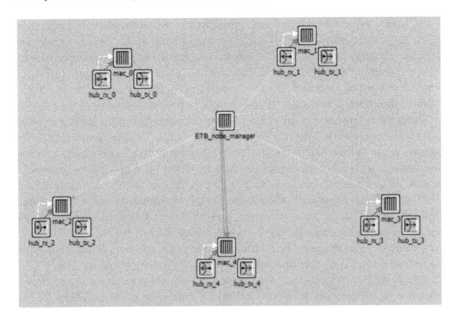

Fig. 9 TBN model

length. The transmission time of non-real-time data is relatively random, and the data length is not fixed.

First of all, the network communication delay under different load conditions is analyzed, and the simulation results are analyzed to obtain the results shown in Fig. 10. As can be seen from the figure, the delay increases with the increase of the utilization rate. When the network utilization rate is below 50%, the network delay

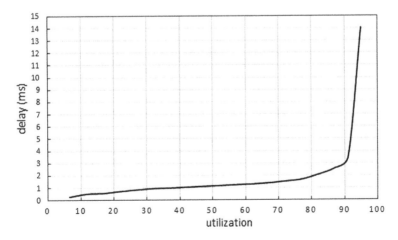

Fig. 10 Network utilization and delay

is less than 1 ms, with high real-time performance. When the network load rate is below 80%, the network delay is basically proportional to the utilization rate, and the network delay is below 2 ms. When the utilization rate of the network exceeds 80%, the delay obviously increases, and when the utilization rate exceeds 97%, the network will not be available. Therefore, reducing the utilization rate of the network is one of the effective methods to improve the real-time performance of the network.

Under the condition of medium load, that is, the sum of the utilization rates of the two backbone network links is 50%, the network failure situation is simulated. The network starts to operate at 100 s; the backbone network link 2 fails at 200 s, returns to normal at 400 s, and the simulation ends at 500 s. By analyzing the simulation results, it can be seen from Figs. 11 and 12 that the bandwidth utilization ratio of link 1 and link 2 is about 25% when the network is in normal operation. When link 2 fails, the data transmission is interrupted, and link 1 undertakes all data transmission tasks. The bandwidth utilization ratio of link 1 is the sum of the utilization ratios of the two links when the network is in normal operation, which is about 50%. When the failure is recovered, the bandwidth of the two links is recovered to 25%.

Analyzing the communication delay, it can be found that the delay in normal operation of the network is about 1 ms, and the delay in failure is about 1.2 ms, as shown in Fig. 13. According to the relationship between the utilization rate and the delay, it can be concluded that the greater the network utilization rate, the more obvious the improvement of network transmission quality by dual-link transmission.

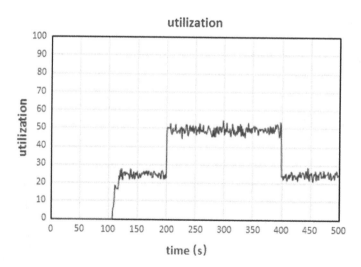

Fig. 11 A network utilization of link 1

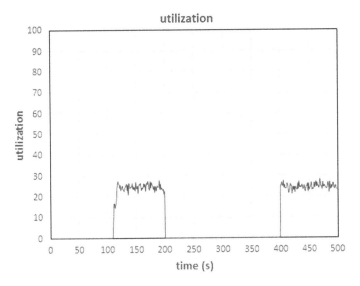

Fig. 12 B network utilization of link 2

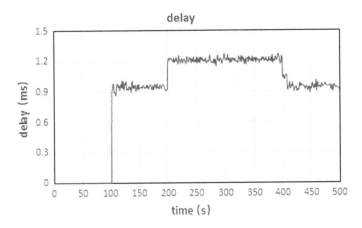

Fig. 13 Network delay

6 Conclusion

The PRP protocol and MRP protocol specified in IEC62439 realize network redundancy through improvement of network architecture and management of network nodes. The train backbone network based on Ethernet adopts linear topology. In order to improve the reliability of the backbone network, this paper refers to the parallel network model of PRP protocol and the switch management mechanism of MRP protocol and proposes a backbone network architecture based on double links. By using the management of backbone network nodes, the network redundancy is realized. Meanwhile, by using two communication links to transmit data simultaneously, the utilization rate of each link can be reduced, thus improving the network communication quality. OPNET simulation results show that the dual-link backbone network architecture can realize fast network switching in case of failure. According to the relationship between network utilization and time delay, it can be concluded that when the network load is large, dual-link communication has a better effect on improving the network communication quality.

References

1. Zhou J, Wang L (2016) Scheduling algorithm of PROFINET IRT based on train communication network. J Beijing Jiaotong Univ 40(3):7–13 (in Chinese)
2. IEC.IEC 61375-2-5 (2014) Electronic railway equipment—Train communication network (TCN)—Part 2-5: Ethernet Train Backbone (ETB)
3. IEC.IEC 61375-3-4 (2014) Electronic railway equipment—Train communication network (TCN)—Part 3-4: Ethernet Consist Network (ECN)
4. Tien NX, Rhee JM (2017) RPE: a seamless redundancy protocol for Ethernet networks. IEICE Trans Commun E100.B(5):711–727
5. Jian J, Wang L, Jin J, Shen P (2018) Reliability analysis of train communication network redundancy based on Ethernet. J Beijing Jiaotong Univ 42(2):76–83 (in Chinese)
6. IEC.IEC 62439-1 (2010) Industrial communication networks—High availability automation networks—Part 1: General concepts and calculation methods
7. Zhang Y-z, Cao Y, Wen Y-h (2015) Modeling and performance analysis of train communication network based on switched Ethernet. J Commun 36(9):181–187 (in Chinese)
8. Araujo JÁ, Lazaro J, Astarloa A et al (2015) PRP and HSR for high availability networks in power utility automation: a method for redundant frames discarding. IEEE Trans Smart Grid 6(5):2325–2332
9. Jin J, Wang L, Jian J, Shen P, Liu B (2017) Research on parallel redundancy method and protocol of train communication network. J China Railw Soc 39(12):76–85 (in Chinese)
10. Giorgetti A, Cugini F, Paolucci F et al (2013) Performance analysis of media redundancy protocol (MRP). IEEE Trans Ind Inf 9(1):218–227
11. Xu L-s, Hu L-s (2015) Research on method of improving industrial Ethernet reliability based on IEC 62439-2 standard. Control Instrum Chem Ind 42(9):1009–1012 (in Chinese)
12. Zhang Z-g (2017) High-speed train communication network simulation based on industrial ethernet. Comput Mod 9:1–6 (in Chinese)

Xiaodong Sun male, born in 1992, majoring in traction control and network.

Jianyu Jin male, born in 1991, focuses on traction control and network.

System Implementation of a Processing Subcontracting Business in SAP ERP Non-standard Subcontracting Scenarios

Chenyang Bing, Guohui Yang and Jinlong Kang

Abstract Based on a non-standard subcontracting business scenario of SAP ERP, the business model, data model and application architecture are established from four angles of business, application, data and system, and the system is finally realized. The project node (WBS), operation number (OPID) and subcontracted materials are taken as the control dimensions of the subcontracting business of the operation, covering the whole process of subcontracting demand, purchase approval and price limit, subcontracting contract signing, subcontracting execution, inventory management and financial accounting. SAP ERP provides two standard subcontracting business scenarios, namely subcontracting and processing subcontracting. Subcontracting scenarios are triggered by inventory demand. Pricing is based on subcontracting service fees. Enterprises provide raw materials, recover processed products, carry out inventory management on issued materials and finished products and settle processing fees. Operation subcontracting is triggered by production operations. Based on operation pricing, it does not manage inventory but only settles processing fees. The difference between the above standard scenario and the business of an enterprise mainly lies in the need to outsource the production process, to carry out inventory management on the process, and to divide the process into work steps for subcontracting price limit. In order to solve the above differences, this paper designs the subcontracting price limit function and the virtual inventory management function and realizes the information management of this kind of processing subcontracting business.

Keywords Processing subcontracting · Processing fee accounting · Price limit approval for subcontracted processes

---

C. Bing (✉) · G. Yang · J. Kang
CRRC Qingdao Sifang Co., Ltd., Qingdao, China
e-mail: 010500018765@crrcgc.cc

G. Yang
e-mail: yangguohui.sf@crrcgc.cc

J. Kang
e-mail: kangjinlong.sf@crrcgc.cc

Classification number of the middle map: TM??

1 Introduction

With the further improvement of the world free trade system and the establishment of the global transportation system and communication network, international economic cooperation and exchanges have become increasingly close. Global industry has entered an important period of structural adjustment. According to relevant data, the eastward shift of large international manufacturing centers has become a trend. Under the favorable international environment, the domestic rail transit equipment manufacturing industry will cooperate with its foreign counterparts at a broader level and will also face severe challenges from global rail transit manufacturing enterprises. According to the 13th Five-Year Plan of informatization, according to the company's new organizational structure, the planning, technology, procurement, logistics, manufacturing, quality and finance under the business chain will be based on product manufacturing + service. At present, it is necessary to improve the informatization of processing subcontracting business management. With the improvement of the company's fine management level, the whole process of processing subcontracting needs to be controlled. It is urgent to solve the outstanding problems such as multi-processing subcontracting, more temporary subcontracting, price management, less fine cost accounting and frequent engineering changes. It is urgent to resort the business processes and data. On the basis of the three-flow integration of ERP, PDM, BPM, MES, MRO, QMS, SGO, supplier coordination platform and other information systems (capital flow, information flow, material flow), the company's collectivization and globalization development strategy is supported. The core of the strategic measures to improve the fine management of business and build an information system supporting enterprise development and intelligent manufacturing is that informatization should "support enterprise development" to greatly promote the reduction of informatization cost and improve enterprise information security. Under this business background, SAP ERP is taken as the core system to carry out processing subcontracting informatization.

SAP ERP system provides two standard subcontracting business scenarios, namely subcontractor and processing subcontracting.

In the subcontract business scenario (as shown in Fig. 1), the subcontract demand comes from the inventory demand, and the purchase object is the subcontracting supplier. The enterprise provides or partially provides raw materials to the subcontracting processing suppliers. After the suppliers finish processing, the enterprise recovers the finished products and pays the processing fees to the suppliers. The business can be initiated from a subcontract purchase application, and then a purchase order can be created, or the order can be initiated. After the subcontract purchase order is created, raw materials are first provided to the subcontract processing supplier, then the raw materials are received from the supplier and deducted from the supplier, and finally, the invoice is verified by receiving the supplier processing fee invoice.

System Implementation of a Processing Subcontracting Business ...

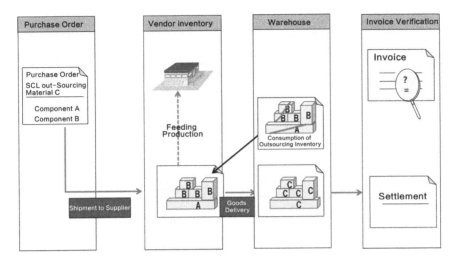

Fig. 1 Subcontract business scenario

The business scenario of processing subcontracting is due to technical or capacity reasons. An enterprise outsources a certain process (such as pipeline cleaning) in the production process to a supplier. After the supplier completes the service, the finished product is delivered according to the process, and then the invoice verification is completed through the process, and the processing fee is settled. The operation can bring components, but cannot carry out subcontract inventory management. When components are consumed, they directly enter the production cost.

The above two business scenarios either do not carry out inventory management or are not subcontracted based on operations, neither meet the subcontracting requirements of enterprise's operations, nor provide the functions of process refinement and price verification.

Starting from the specific needs of enterprises, this paper designs and implements a processing subcontracting management system under non-standard scenario based on the processing subcontracting scenario and absorbing the advantages of inventory management of subcontracted material components and finished products under subcontract scenario.

2 Design Principle

2.1 General Design Principles

The system design follows the design principles of business standardization, data standardization and process automation, of which data standardization is the most

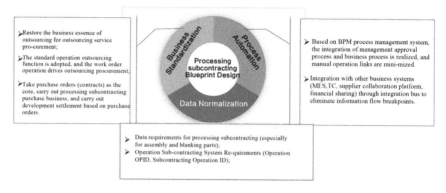

Fig. 2 Design principles of processing subcontracting system

important, especially whether process and BOM data can meet the requirements, is the premise of business standardization and process automation (Fig. 2).

2.2 Principle of Business Capability

A comprehensive review of subcontracting business requirements has been made. From the five aspects of subcontracting demand management, production management, procurement management, inventory management and financial management, 17 core business capabilities that an enterprise should have for processing subcontracting have been determined. These business capabilities reflect the competitive capabilities that an enterprise should have in order to obtain a competitive advantage in the market. The system design takes these business capabilities as its starting point and foothold (Fig. 3).

3 Overall Scheme Design

3.1 Overall Business Plan

There are two major types of processing subcontracting in an enterprise, i.e., processing subcontracting and temporary subcontracting. Processing subcontracting refers to processes explicitly designated as subcontracting from the time the process is formulated. Such subcontracting is mainly due to technical reasons. Temporary subcontracting refers to the subcontracting procurement of processes that need temporary subcontracting selected by the production branch factory from the original self-made processes in the production process. Such subcontracting is mostly due to production capacity.

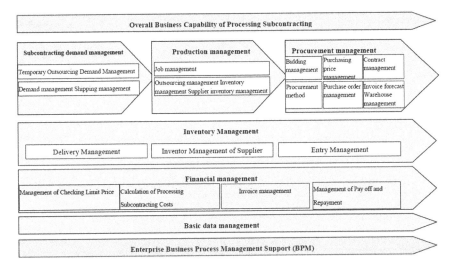

Fig. 3 Processing subcontracting capacity

The demand for processing subcontracting comes from the process route of the process department. The demand for temporary subcontracting comes from the self-made process selected in the workshop production process. The two kinds of subcontracting requirements are distinguished by process control codes (SF15, SF16) and are also distinguished from the self-made process. Temporary subcontracting requires BPM approval process and deduction of self-made working hours.

In order to meet the price verification requirements, the subcontracted processes are further refined into work steps (a subset of processes). The technical subcontracting is refined by the process department, while the temporary subcontracting is refined by the finance department according to the maintained refinement strategy. The refined work step information is subject to the subcontracting price verification by the finance department as the price limit for external bidding.

After the subcontract operation is confirmed, an operation subcontract schedule is generated. Part of this schedule uses detailed operations as the basis for price verification. On this basis, the bidding center will initiate the evaluation of procurement methods, determine the provisional evaluation criteria, and approve and sign the subcontract.

The specific subcontracting execution is driven by the production plan. The production department issues a three-day plan, drives the work order to be issued, triggers the subcontracting purchase application and establishes the purchase order according to the contract.

The branch factory/logistics center delivers the subcontracted parts, and the receipt triggers the work order submission after completion. The issued parts and the recovered finished products are managed through virtual inventory, which does not affect financial bookkeeping.

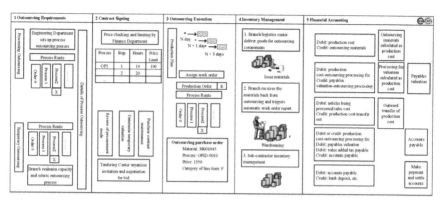

Fig. 4 Overall business design

The final financial settlement involves two financial cycles, namely, cost cycle: processing fee settlement into production cost, and subsequent transfer into sales cost or in-process product according to product sales invoicing. A/P cycle: An A/P estimation is formed for subcontracting receipt. After receipt of invoices, the original estimation is reversed to form an A/P account for supplier processing fees. This financial cycle is subsequently completed through payment settlement (Fig. 4).

3.2 Application Architecture Design

Processing subcontracting mainly involves business process management system (BPM), enterprise resource planning system (ERP), manufacturing execution system (MES), financial sharing system (FSSC) and supplier collaboration platform system and completes information exchange through enterprise bus (ESB). BPM mainly realizes the functions of temporary subcontracting approval, procurement method review, contract review, etc. ERP realizes the functions of subcontracting business demand triggering, subcontracting execution management, inventory management and financial accounting and reporting, MES system is responsible for planning release, financial sharing system realizes processing fee reimbursement and settlement, supplier coordination platform realizes price verification management, supplier picking plan, supplier purchase order query, etc. (Fig. 5).

3.3 Basic Data Specifications and Models

Basic data is the most important thing to realize processing subcontracting informatization, and the standardization of basic data model is the key to smooth operation.

System Implementation of a Processing Subcontracting Business ...

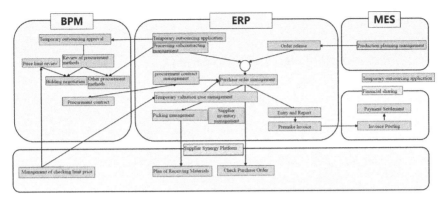

Fig. 5 Processing subcontracting application architecture

The processing subcontracting project determines the BOM, process route, provisional valuation grid, contract price maintenance specifications and principles and various basic data sources.

The overall data model is built based on WBS, operation and material. Blanking parts must maintain material BOM and process route, and the subcontract operation code (OPID) and subcontract materials must be maintained in the operation.

The provisional valuation is maintained with WBS (to the production node), operation code (OPID) and material as the primary key, while the contract price is maintained with WBS (to the production node), operation code (OPID), material and supplier as the primary key (Fig. 6).

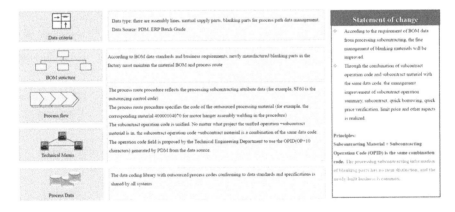

Fig. 6 Data model specification

4 Key Plan

4.1 Subcontracting Demand Generation

In order to clarify the actual demand sources of processing subcontracting and temporary subcontracting, control the subcontracting business from the demand source and achieve the purpose of visualizing the demand of the subcontracting source in the system, ERP system uses process control codes to realize the conversion of the subcontracting business and changes the process data according to certain rules (the work center is provided with the subcontracting identification X). The system automatically generates the subcontracting purchase application according to the change of the control codes and the information such as the procurement organization, procurement group and material group of the completed subcontracting purchase. The purchase application cannot be manually modified, and the change of the subcontracting process path will automatically trigger the change of the corresponding purchase application, such as quantity change and application deletion. (Fig. 7).

The operation start time and completion time specified in the production schedule determine the subcontract production plan. MRP generates subcontracting production plans based on the parameters of operation "preparation time," "processing time" and "purchase cycle" (Fig. 8).

Production Order Display : Operation Overview

Material		Capacity	Operations		
Order	310000003891			Type	ZZ00
Material	40000038560			Plant	1100
Sequence	0	Standard s...			

Operation Overview

Op.	SOp	Start	Work Ce...	Plant	Co...	Std Text	Operation short text	T...		SysStatus	User Stat
0010		2017.11.21	08:00:00	E12ZG011	1100	Z000001	Welding of seat component	☐		REL	
0020		2017.11.24	08:00:00	E50ZX001	1100	Z000002	Welding the hanging seat of motor at end 1	☐		REL	
0030		2017.11.29	08:00:00	E50ZX001	1100	Z000003	Welding the beam I	☐		REL	
0040		2017.12.04	08:00:00	E12ZG011	1100	Z000004	Welding the beam II	☐		PCNF EOP	
0050		2017.12.06	17:00:00	E50ZX001	1100	Z000005	Welding the beam III	☐		REL	
0060		2017.12.06	17:00:00	E50ZX001	1100	Z000006	Welding Grinding	☐		REL	

Fig. 7 Requirements for subcontracting process

Display Routing: Operation Overview

Work center CompAlloc Sequences PRT Inspection Characteristics

Material 40000038560 Grp.Count1

Sequence 0

Operation Overv.

Op...	SOp	Work ce...	Plnt	Cont...	Standard...	Description	L...	Base Q...	Uni...	Preparatio...	Unit	Processing Time	Unit	Labor Expense	Unit	Activit...	Labor Attached
0010		E12KG006	1100	SF10	YC00003	Preassemble Electrical Parts	l		EA	0	H	10	H	10	H	1001	10
0020		E12KG006	1100	SF10	YC00002	Preassemble Electrical Parts	l		EA	0	H	1	H	1	H	1001	1
0030		E12KG006	1100	SF12	CD00001	Preassemble Electrical Parts	l		EA	0	H	2	H	2	H	1001	2

Fig. 8 Subcontracted working hours

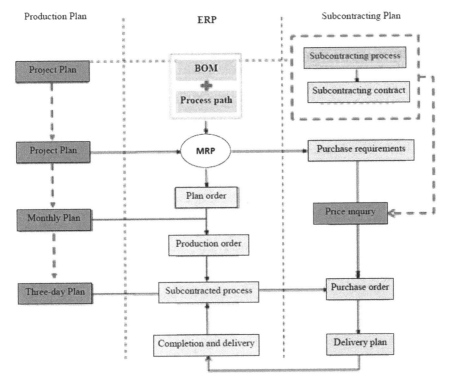

Fig. 9 Planning system

Subcontracting production plan further improved the production planning system consisting of project plan, main production plan, monthly plan and three-day plan (Fig. 9).

4.2 Subcontracting Demand Release

The subcontracting procurement executive department (production department) summarizes the subcontracting requirements automatically generated by the system, organizes the procurement review and bidding according to the price limit set by the finance department and enters the contract price confirmed by the bidding in ERP according to the bidding results after the bidding is completed, forming the subcontracting price database and making the contract in ERP.

4.3 Subcontracting Inventory Management

We will improve the control over the process of issuing and retrieving subcontracted materials. As a bridge for information interaction between an enterprise and suppliers, the collaboration platform can realize information transmission with suppliers.

After the subcontract purchase demand is issued (purchase order), the supplier obtains the processing demand through the collaborative platform and makes the picking plan according to the capacity. The picking plan is transferred to the ERP system. The branch factory issues the materials according to the picking plan. After the supplier completes the processing, the branch factory makes the delivery plan, and the branch factory arranges to receive the materials from the receivable team.

The original extensive management mode of component inventory was changed. The branch factory increased the quantity of components to 1110 virtual factory according to the picking plan based on the original Kanban. The special inventory type O of the virtual factory is the component inventory of the supplier. The supplier inventory increases when the branch factory issues the materials. The component inventory decreases the quantity of components according to the delivery plan when the branch factory receives the materials after processing. The output inventory increases to the next process. The branch factory can query the supplier inventory through ERP1110 factory.

Inter-department Material Movement (Fig. 10).

Data Flow between Systems (Fig. 11).

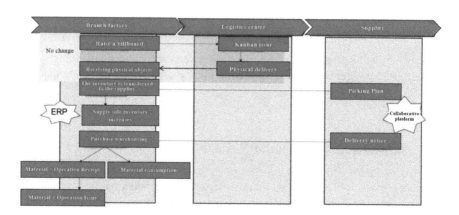

Fig. 10 Material movement

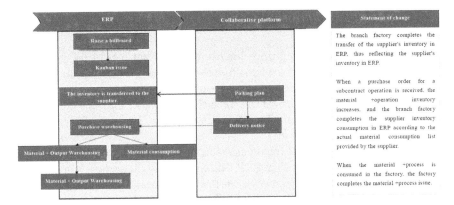

Fig. 11 Data flow

4.4 Blanking Management

Blanking parts have universality and are not managed by project (WBS). The current management mode of blanking parts will be continued after the system is implemented. Basic data has no project attribute. If the process in the process data is subcontracted, then the production of this blanking part will be subcontracted.

After the demand for blanking parts is released, the branch factory can split the production order according to the subcontracting scope. The subcontracting method of blanking parts is based on the production order method (Fig. 12).

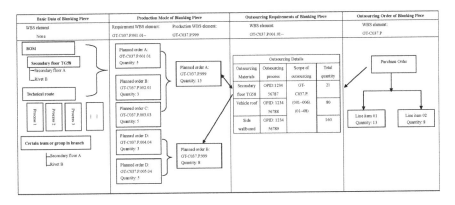

Fig. 12 Blanking piece management process

4.5 Financial Valuation

The processing subcontracting price limit is an important part of the processing subcontracting demand and is the basis for processing subcontracting procurement, temporary valuation maintenance, contract maintenance, subsequent cost accounting and accounts payable management. The financial price verification plan realizes the control of the verification demand, verification price and limit price review process (Fig. 13).

The information of process requirements for processing subcontracting and temporary subcontracting comes from ERP. The engineering department is responsible for refining the subcontracting process. The finance department organizes the temporary subcontracting process to be refined and maintained to the supplier coordination platform, respectively.

The production department imports and releases subcontracting procedures and allocates suppliers, and after the suppliers log into the supplier collaboration platform, they, respectively, specify the subcontracting procedures that need to be targeted and fill in the detailed process information such as working hours and machine hours, and the finance department analyzes the process drawings, refers to historical process data and information filled in by the suppliers and performs subcontracting price verification.

After the completion of the price limit, the review process of the check price limit will be triggered. At the same time, the price limit information will be sent to the production department and merged into the subcontract demand. Contracts will be signed according to different purchase methods, triggering ERP to update the subcontract provisional valuation.

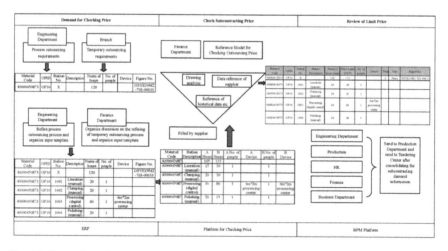

Fig. 13 Processing subcontracting pricing model

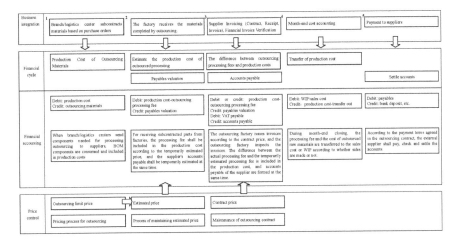

Fig. 14 Financial accounting

4.6 Financial Accounting

Processing subcontracting financial management is mainly divided into three parts: subcontracting price accounting and management, subcontracting cost management and supplier A/P management.

Subcontract price verification management mainly manages the subcontract provisional valuation and subcontract contract price.

In subcontract cost management, the estimated processing fee is formed according to provisional valuation after the purchase order is received, and the estimated subcontract fee is included in the production cost. After the purchase invoice is received, invoice verification is performed, and the final subcontract processing fee formed according to the contract price is included in the production cost. This part of the cost is transferred to the sales cost with the product cost.

In A/P management, supplier A/P estimation based on provisional valuation is formed after purchase receipt. After receipt of invoices, accounts payable are formed according to purchase contract price. At the same time, the original A/P estimation is offset. Subsequently, processing fees of suppliers payable are paid according to the contract accounting period and the accounts are settled (Fig. 14).

5 Conclusion

Based on the analysis of an enterprise's processing subcontracting business requirements and the pre-set standard subcontracting business scenarios in ERP system, this paper makes a comparative difference analysis and establishes a non-standard processing subcontracting procurement mode from the aspects of business scheme, application framework, data model and key schemes, which covers the system design and development of the whole process management of subcontracting demand triggering, contract management, subcontracting execution, inventory management and financial accounting closed loop. It has realized the overall informatization of a certain enterprise's processing subcontracting business and improved the management level and control ability of the processing subcontracting business.

Appendix: Description of Processing Subcontracting Financial Accounting

(1) **Issue of Subcontracting Components**

When the branch factory/logistics center sends out the components needed for processing subcontracting to the supplier, the BOM components will be consumed and included in the production cost.

Debit: Production Cost
Credit: Material Cost.

(2) **Withdrawal of Subcontracted Finished Products**

When the branch factory receives the foreign parts, the processing fee shall be included in the production cost according to the provisional estimate, and the accounts payable of the supplier shall be estimated at the same time.

Debit: Production Cost-Subcontracting Expenses
Credit: Estimated Payable.

(3) **Verification of Subcontract Invoice**

Subcontracting manufacturers make out invoices according to the contract price, and foreign suppliers check the invoices. The difference between the actual processing fee and the estimated processing fee is included in the production cost, and accounts payable of suppliers is formed at the same time.

Borrowing or Loan: Production Cost-Subcontracting Fee
Debit: A/P Estimation
Value added tax payable
Credit: accounts payable.

(4) **Production Cost Carry Forward**

During month-end closing, the processing fee and the cost of subcontracted raw materials are transferred to the sales cost or work in process according to whether sales are made or not.

Debit: WIP/Cost of Sales
Credit: Production Cost-Transfer Out.

(5) **Pay and Settle Accounts**

According to the payment terms agreed in the subcontracting contract, the external supplier shall pay, check and settle the accounts.

Debit: accounts payable
Loan: bank deposit, etc.

References

1. Le L (2013) SAP logistics module implementation strategy. China Machine Press. ISBN: 9787111426615 (in Chinese)
2. Tauseef NAS (2013) SAP ERP finance: configuration and design (trans: Chen C, Lan Y), 2nd edn. Posts & Telecom Press Co., Ltd. ISBN: 9787115307521 (in Chinese)
3. help.sap.com

Chenyang Bing male, born in 1983, holds a master's degree in the field of Railway Transportation Electrification Commissioning Technology and Manufacturing Informatization.

Guohui Yang male, born in 1987, holds a master's degree in Railway Transportation Electrification Construction Process.

Jinlong Kang male, born in 1986, holds a master's degree in Railway Transportation Electrification Design.

Research on Passengers' Choice of Subway Travel Intention Based on Structural Equation Model

Yu ping Xu, Siwei Chen and Wei Xu

Abstract Based on the survey results of Nanchang subway service quality and passenger satisfaction, combined with the theory of planned behavior, the four factors of attitude, subjective norm, perceptual behavior control, and subway selection intention were used to establish the corresponding TPB-based subway selection intention model. The subway selects the intention questionnaire and conducts a formal questionnaire survey based on the pre-survey data. The model is analyzed by AMOS22.0 software. Based on this, the model parameters are estimated and the hypothesis is verified. At the same time, the survey samples were divided into six groups: male, female, student, office worker, low-income, middle-high income group, and the differences in interaction between different groups were explored.

Keywords Subway selection intention · TPB interaction variable

1 Introduction

Behavioral intention is a very important part of behavior research. Fishbein and Ajzen believe that the most direct and accurate means of predicting consumer behavior is by discriminating from behavioral intentions [1]. Boulding et al. believe that consumers' satisfaction with overall service is related to service quality [2]. Parasuraman et al. believe that the individual's behavioral intention is directly affected by the quality of service [3]. In recent years, domestic scholars have applied behavioral intentions to traffic research, and Ge Xuefeng constructed the related variables to measure the intentional behavior of tourism destination selection [4]. Yan et al. explore the influencing factors and factors of the shared bicycle use intention [5]. Tang Wei believes that passenger satisfaction with rail transit services will determine the choice of rail transit [6]. Jing Peng and other constructive hybrid selection models predict the travel mode selection behavior [7]. Yuan Junyi studied the traveler's differentiated traffic travel choice behavior [8].

Y. Xu · S. Chen (✉) · W. Xu
East China Jiaotong University (ECJTU), Nanchang, Jiangxi, China
e-mail: 867569016@qq.com

2 Planned Behavior Theory

The theory of planning behavior (TPB) is a behavioral model obtained by Ajzen (1980, 1991) based on the theory of reasoned action (TRA) and combined with the actual situation. The predictive ability is better than TRA. There has been a significant increase [9] in the theory of rational behavior, and it assumes that the individual's will can completely control the occurrence of behavior; but in fact, the occurrence of behavior is affected by many other factors, which leads to a significant reduction in the interpretation of the individual's specific behavior. According to the theory of planned behavior, behavioral intentions are influenced by attitudes and subjective norms and are also influenced by perceived behavioral control (PBC). The degree of individual control over behavior will determine the individual's ability to perform on behavior. In TPB theory, attitudes, subjective norms, and perceptual behavioral control will affect behavioral intentions in whole or in part, and behavioral intentions will ultimately determine the specific behavior of the human body.

3 TPB-Based Subway Selection Intention Model

3.1 Model Construction

In order to analyze the influence of potential factors in passengers' choice of subway travel, a TPB-based subway selection intention model was established according to the planned behavior theory, and the following assumptions were made.

Hypothesis 1 (H1): There is an interaction between the three psychological factors: traveler attitude, subjective norm, and perceptual behavior control;
Hypothesis 2 (H2): The attitude of psychological factors to the subway positively affects the intention of the subway;
Hypothesis 3 (H3): Subjective norms of psychological factors positively affect the intention of subway selection;
Hypothesis 4 (H4): Psychological factors of perceptual behavior control positively affect the subway selection intention.

The structure of the subway selection intention model consists of four psychological latent variables, namely the attitude of the traveler to the choice of subway travel (AT), subjective norms (SN), perceived behavior control (PBC), and subway selection intention (SCI).

3.2 Model Content

According to the selection results of latent variables and measured variables, combined with the structural equation theory and the model hypothesis of this paper, the measurement model and structural model of the TPB-based metro selection intention model can be obtained, respectively.

(1) Measurement model is used to describe the relationship between the explicit variables X and Y and the latent variables ξ and η, respectively.

$$X = \Lambda_X \xi + \delta \tag{1}$$

$$Y = \Lambda_Y \eta + \varepsilon \tag{2}$$

where X is a vector consisting of the measured values of 19 independent variables; ξ is a vector composed of 3 exogenous latent variables; Λ_X and Λ_Y are the factor-matching matrix of X versus Y and Y versus η; δ, ε are the measurement error of X and Y, respectively.

(2) Structural model is the description of the causal relationship model of the latent variable, reflecting the connection between the latent variables.

$$\eta = B\eta + \Gamma\xi + \zeta \tag{3}$$

where B is the path coefficient, that is, the relationship between the internal latent variables; since there is only one internal latent variable, namely the subway selection intention, the influence between the internal potential latent variables is not considered; the path coefficient, i.e. the structural equation model exogenous The relationship between the latent variable attitude to the subway, the subjective normative $\xi 2$, the perceptual behavior control $\xi 3$, the passenger expectation $\xi 4$, the relationship of the intrinsic variable subway selection intention η; ζ is the error vector of the model.

3.3 Data Collection

The formal survey of the questionnaire was conducted in the form of a combination of online survey and field research. After editing the questionnaire on the questionnaire star, it was widely promoted through social software such as WeChat and QQ. It was also distributed in local forums such as forums and post bars in Nanchang, and in Nanchang subway. Lines 1 and 2 will be distributed on the spot by means of a follow-up survey. 300 questionnaires were distributed online, 265 valid questionnaires were collected, the effective rate was 88%, 200 questionnaires were distributed in the field, 142 valid questionnaires were collected, the effective rate was 71%, 500 questionnaires were distributed in total, and 407 valid questionnaires were collected. The effective rate was 81.4%. Descriptive statistical analysis showed that the proportion of males in sex was 52.58% slightly higher than 47.42% in females; in terms of age, the proportion of 19–35 years old was 84.77%, and that under 18 years old was 5.41%, 36–55 years old, 7.62%, 2.21% over the age of 56; occupationally, the highest proportion of middle school students was 56.51%, followed by civil servants and institutions accounting for 18.92%, and corporate employees were 14.74%, which accounted for 90.17%.

4 Empirical Analysis

This paper takes Nanchang Metro as the research object. As of June 2018, Nanchang Metro has opened two lines: Nanchang Subway Line 1 and Nanchang Subway Line 2. A total of 41 stations and a transfer station were built, with a total length of 48.4 km. At present, there are three lines under construction in Nanchang Metro, with a total length of 108.1 km. In 2017, Nanchang Metro transported a total of 10,976,700 passengers, of which the passenger traffic of Line 1 was 10,136,460 times, the passenger traffic of Line 2 was 8,397,100 times, and the total number of passengers was 149,830. The punctuality rate was 99.95%, and the redemption rate was 99.99%.

4.1 Measurement Model Analysis

(1) Overall fitness analysis

The results of the subway selection intention model adaptation test are shown in Table 1. The specific index of the model is calculated by the chi-square value X^2 of 490.66, the degree of freedom (DF) is 146, $X^2/\text{df} = 3.361$, and the standard is less than 5, RMSEA is close to 0.08 and acceptable; all other indicators have reached the adaptation standard value, in line with the standard, indicating that the model fit degree meets the requirements.

(2) Scale reliability test

In Table 2, the normalized load of 19 measured variables is greater than 0.6 except for AT2 and PBC6 greater than 0.5, indicating that the measured variable has high response capacity to latent traits; T and P values show that p is less than 0.001. In the case, the normalized load of each measured variable reached a significant level; the combined reliability of the four latent variables was greater than 0.7, indicating that the model has higher reliability; the average variance extraction is greater than 0.5, indicating that the measured variable has sufficient explanation and ability with sufficient convergence validity. Overall, the model has good explanatory power.

Table 1 TPB-based subway selection intention model adaptation indicator table

Evaluation index	X^2	df	X^2/df	RMSEA	CFI	NFI	GFI	IFI	AGFI
Standard	–	–	<5	<0.08	>0.9	>0.9	>0.8	>0.9	>0.8
Actual	490.66	146	3.361	0.082	0.928	0.901	0.870	0.928	0.830

Table 2 TPB-based subway selection intention model parameter summary table

Latent variable	Index	Nonstandardized factor load	S.E.	E.R.	P	Standardized factor load	SMC	CR	AVE
Attitude toward the subway	AT7	1.000				0.678	0.460	0.900	0.569
	AT6	0.948	0.069	13.743	***	0.810	0.656		
	AT5	0.976	0.066	14.702	***	0.877	0.769		
	AT4	0.902	0.077	11.732	***	0.679	0.461		
	AT3	0.964	0.072	13.386	***	0.786	0.618		
	AT2	0.917	0.095	9.606	***	0.548	0.300		
	AT1	0.988	0.069	14.281	***	0.847	0.717		
Subjective norm	SN3	1.000				0.671	0.450	0.856	0.668
	SN2	1.399	0.098	14.218	***	0.874	0.764		
	SN1	1.392	0.097	14.383	***	0.888	0.789		
Perceptual behavior control	PBC5	1.000				0.776	0.602	0.885	0.565
	PBC6	0.802	0.070	11.379	***	0.596	0.355		
	PBC4	1.074	0.066	16.337	***	0.810	0.656		
	PBC3	1.132	0.080	14.097	***	0.717	0.514		
	PBC2	1.171	0.079	14.860	***	0.750	0.563		
	PBC1	1.121	0.066	17.007	***	0.837	0.701		
Subway selection intention	SCI1	1.000				0.827	0.684	0.857	0.666
	SCI2	0.938	0.050	18.872	***	0.827	0.684		
	SCI3	1.062	0.060	17.715	***	0.794	0.630		

Table 3 Structural equation model parameter table

Dependent variable/argument	R-square	Path coefficient	T	P
Subway selection intention	0.98			
Attitude toward the subway		0.232	4.247	***
Subjective norm		0.277	4.031	***
Perceptual behavior control		0.547	7.749	***

Note * indicates $p < 0.05$, ** indicates $p < 0.01$, *** indicates $p < 0.001$

4.2 Structural Equation Model Analysis

The main function of the structural equation model is to explain the structural relationship between the latent variable and the latent variable. The relationship between them is mainly reflected by the load coefficient and the path coefficient and analyzed the paths and parameters of the structural equation model.

The R-square value of the subway selection intention is 0.98, indicating that the three latent variables of attitude, subjective norm, and perceptual behavior control of the subway can explain 98% of the intention of the subway. The standardized path coefficient of the attitude of the subway to the subway selection intention is 0.23, the standardized path coefficient of the subjective norm for the subway selection intention is 0.28, and the standardized path coefficient of the perceptual behavior control for the subway selection intention is 0.55, indicating that the three latent variables are, respectively, for the subway and select the effect of the intent. The T values of the three paths reach a significant level in the case of $P < 0.001$, and the sign of each path coefficient also conforms to the influence relationship of the research hypothesis, indicating that the model parameter estimation is valid (Table 3).

In summary, the attitude, subjective norms, and perceptual behavior control of the subway have a significant positive impact on the subway selection intention, among which the perceptual behavior control (0.55) is the most important factor, followed by the subjective norm (0.28) is the weakest. It is the attitude toward the subway (0.23).

4.3 Sample Grouping Study

A summary analysis of the effects of three submersible variables on the attitude, subjective norms, and perceptual behavior control of the subway in the six different groups is shown in Table 4.

It can be seen from the table that for the six groups divided, the positive influence of perceptual behavior control on the subway selection intention is the most significant; the male group (0.650) is the highest, and the female group (0.469) is the lowest; the positive influence of subjective norms on the subway selection intention is significant. The highest is the low-income group (0.360) and the female group (0.352), and the

Table 4 Clustering path analysis results affecting subway selection intention

	Low income	Medium and high income	Student	Office worker	Male	Female
Attitude toward the subway	0.148*	0.403***	0.234**	0.405***	0.207*	0.182*
Subjective norm	0.360***	0.111*	0.230**	0.191*	0.245**	0.352**
Perceptual behavior control	0.533***	0.569***	0.579***	0.500***	0.650***	0.469***

Note * indicates $p < 0.05$, ** indicates $p < 0.01$, *** indicates $p < 0.001$

lowest is the middle-income group (0.111). The positive impact on the subway's intention is also significant. The highest is office workers (0.405) and middle-income groups (0.403), and the lowest is low-income groups (0.148).

In summary, the male group has higher control over their behavior than the female; the most affected group is the low-income group and the female group; the most affected by the surrounding group is the middle-to-high income group; the office workers; and the middle-to-high income group on the subway. Attitude is significantly higher than other groups in choosing the subway.

5 Conclusion

According to the analysis results of the model, combined with the actual situation at the beginning of the opening of Nanchang Metro, the following suggestions are proposed:

(1) Based on the newly opened operation of Nanchang Metro, Nanchang passengers are able to meet the safe and fast travel demand for the subway. Under the premise of ensuring operational safety, reduce the train departure interval and improve the cleanliness inside the vehicle; the requirements for comfort are slightly lower than the former. Since Nanchang Metro has only opened two lines at present, the convenience of transfer has little effect. After the medium-term subway line reaches a certain scale, the convenience of subway transfer is improved in station.

(2) While optimizing the quality of service, it will enhance the passengers' psychological positioning of the subway and their attitude toward the subway. Through the imperceptible influence, the passengers think that the subway should be the most convenient, comfortable, fast, and safe way to travel. In this way, the subway selection intention is enhanced by improving passenger expectations and attitude toward the subway.

(3) On the basis of the government's preferential policies to encourage subway travel, increase the media propaganda and enhance the passengers' choice of the subway. Since the subjective norms of women and low-income groups have the highest impact on subway selection intentions, the focus can be shifted to women and low-income groups.

Acknowledgements Foundation: 2018 Annual Research Base of Humanities and Social Sciences of Colleges and Universities in Jiangxi Province—Tendering Project of Communication and Engineering Application Translation Research Center of East China Jiaotong University (JD18104); Evaluation and Optimization of the Distribution of Comprehensive Passenger Transportation Hub in Nanchang City (JJ201802).

References

1. Fishbein M, Ajzen I (1975) Belief, attitude, intention and behavior: an introduction to theory and research. Addison-Wesley, Reading, MA
2. Boulding W, Kalra A, Staelin R, Zeithaml VA (1993) A dynamic process model of service quality: form expectations to behavior intentions. J Mark Res 30:7–27
3. Parasuraman A, Zeithaml VA, Berry LL (1996) The behavioral consequences of service. Quality 60:31–46
4. Gou X (2012) Study on the influencing factors of tourism destination choice intention. Dalian University of Technology (in Chinese)
5. Yan R, Chu J, Yang X (2017) Urban residents use shared bicycle intention research—taking Hefei as an example. Traffic Inf Saf 35(6) (in Chinese)
6. Wei T (2017) Research on the influence mechanism of urban rail transit passenger satisfaction. Chongqing Jiaotong University (in Chinese)
7. Peng J, Zhicai J, Qifeng Z (2014) Travel mode choice behavior model considering psychological latent variables. China J Highw Transp 27(11) (in Chinese)
8. Junyi Y (2015) Research on structural model of differentiated traffic travel choice behavior. Southeast University (in Chinese)
9. Ajzen I, Fishbein M (1980) Understanding attitudes and predicting social behavior. Prentice-Hall, Englewood Cliffs, NJ

Adaptive Radio Resource Management Based on Soft Frequency Reuse (SFR) in Urban Rail Transit Systems

Ye Zhu and Hailin Jiang

Abstract Due to limited spectrum resources and intercell interference (ICI), the increasingly high throughput requirements cannot be satisfied in urban rail transit system in future. Soft frequency reuse (SFR), as a technology of ICI coordination (ICIC), has been proposed to reduce ICI. A static SFR scheme, whose resource management is fixed prior to system deployment, limits the potential performance of users. Especially in urban rail transit system, trains are in moving states. An adaptive intercell resource-allocation algorithm is proposed in this paper, focusing on the performance from the perspective of individual directly. The algorithm dynamically optimizes resource allocation and transmission power in multi-cell wireless communication network to improve the throughput of every Train Access Unit (TAU). For the given locations of trains, we adopt exhaustive search and greedy descend methods to calculate resource allocation and transmission power and then repeat that in all cells until the pre-defined convergence criterion is satisfied. The simulation shows that the algorithm can effectively improve the performance for individuals in the general scenarios compared to where no joint optimization methods of resource allocation and transmission power are performed.

Keywords Long-term evolution for metro (LTE-M) · Soft frequency reuse (SFR) · Power optimization · Resource allocation · Transit system · Interference coordination (ICIC)

1 Introduction

Communication-Based Train Control (CBTC) systems achieve continuous bi-directional and large-capacity train-wayside wireless information exchange to ensure

---

Y. Zhu (✉) · H. Jiang
State Key Laboratory of Rail Traffic Control & Safety, Beijing Jiaotong University, 100044 Beijing, People's Republic of China
e-mail: 17120302@bjtu.edu.cn

H. Jiang
e-mail: lhjiang@bjtu.edu.cn

safety control and effective operation of trains in urban rail transit systems [1]. In recent years, 3GPP long-term evolution (LTE) has been employed in train-wayside wireless communication system of the CBTC systems, which is called as 3GPP long-term evolution for metro (LTE-M) [2]. Unfortunately, spectrum resources distributed in urban rail transit systems are very limited [3]. Besides, both image monitoring system (IMS) service and passenger information system (PIS) service put forward high requirements of individual's throughput. Thus, it is necessary to reuse the resource in each cell. However, intra-frequency networking inevitably causes intercell interference (ICI). Then, ICI coordination (ICIC) techniques have been proposed [4].

And there are two major methods about ICIC, i.e., soft frequency reuse (SFR) [5] and fractional frequency reuse (FFR) [6]. In both FFR and SFR, the spectrum is divided into several segments and respective groups of subcarriers are allocated to users. Besides, the transmission power should be improved in the cell center while the transmission power should be reduced in the cell edge. It has been proved that SFR achieves higher spectrum efficiency than the FFR [7].

Giambene et al. [8] proposed SFR-3 schemes and presented the iterative methods to solve the problem related to power control parameters and cell association, which is non-convex. Nevertheless, in urban transit systems, the number of trains in a cell is quite small, so there is no need to apply a multi-level method. In static SFR schemes, the resource allocation and transmission power in different regions of cells are fixed beforehand. However, the throughput in each cell significantly fluctuates with time and space domains due to user mobility. Thus, the static resource allocation is not suitable in the urban rail transit systems. Qian et al. [4] determined the number of subcarriers and transmission power for different regions of cells and did not consider the performance from the perspective of the individual.

In our paper, we consider a distributed adaptive-SFR scheme related to power control and resource allocation. As shown in Fig. 1, to consider safe distance between high-speed trains and simplify the resource scheduling method, we only consider the SFR in which the subcarriers are divided into two parts, center and edge, respectively.

The adaptive SFR determines the major and minor subcarriers and ensures that the frequencies used in adjacent cell edge areas must be orthogonal to each other to minimize the ICI among the cells. Besides, the SFR operates by decomposing the multi-cell problems into a sequence of subproblems in the single cell [9]. The

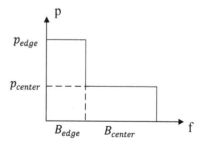

Fig. 1 Frequency segments and power control in soft frequency reuse

decomposition results in a dramatically lower computation than centralized resource allocation schemes.

2 System Model

In this section, we consider the scenario about downlink communication transmission in LTE-M network with a group of C cells. We assume the cells are in liner and zonal distribution, as shown in Fig. 2. Each cell can calculate its resource allocation and transmission power locally based on resource occupation in its adjacent cells. Then, SFR-2 scheme is designed for LTE-M.

Then, we assume that there are K_i trains in cell i in bandwidth B. And $\{k\}_t$ mean that locations of trains in the cells at the time t. The total system spectrum of LTE-M is divided into a set of N physical resource blocks (PRBs). We select one PRB as a unit of resource allocation, because the subcarriers are grouped into PRBs. For each cell $i \in C$ we chooses N_i^{major} and N_i^{minor} PRBs as its major and minor resource groups, and transmission power along with the major and minor PRBs is denoted by P_i^{major} and P_i^{minor}, respectively. Each cell is divided into center and edge. The boundary between the two parts in each cell is defined based on the distance to the serving BS in this paper.

In wireless systems, the throughput of trains in a cell is determined by large-scale fading and the related message received from its adjacent cells. And the channel gain on all the PRBs $j \in N$ between a train and the serving BS are statistically identically independent because the path losses are only determined by the locations of the trains. We define the channel gain for train k operated in cell i (served by BS i) on the PRBs $j \in N$ transmitted by BS m as $g_{i,m,k}$, where $k \in K_i$, $m \in C$.

Assuming that for train k, the interference only comes from adjacent cells. Thus, the SINR on PRB $j \in N$ is calculated as

$$\mathrm{SINR}_k^j = \frac{p_i^j g_{i,i,k}}{\sum_{m \in B_{adj}^i} p_m^j g_{i,m,k} + N_0 B_{sub}} \quad (1)$$

where p_i^j is transmission power on PRB j in cell i. Besides, p_i^{major} and p_i^{minor} denote PRB j deployed as a major PRB and a minor PRB in cell i, respectively; B_{adj}^i represents

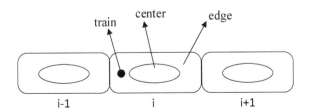

Fig. 2 Cells distribution in urban transit rail system

the cells adjacent to the cell i. N_0 is thermal noise power density. B_{sub} is bandwidth of one PRB.

In the system used SFR-2 scheme, the throughput of trains in edge and in center of cell i using (1) and Shannon capacity formula can be written as

$$R_i^E = N_i^{major} B_{sub} \log_2 \left(1 + \frac{p_i^{major} g_{i,i,k_{i,E}}}{\sum_{m \in B_{adj}^i} p_m^{minor} g_{i,m,k_{i,E}} + N_0 B_{sub}} \right) \quad (2)$$

Similarly,

$$R_i^C = N_i^{minor} B_{sub} \log_2 \left(1 + \frac{p_i^{minor} g_{i,i,k_{i,E}}}{\sum_{m \in B_{adj}^i} p_m^{major} g_{i,m,k_{i,E}} + N_0 B_{sub}} \right) \quad (3)$$

3 Problem Formulation

We adopt the maximization of the minimum throughput among the trains to achieve the objective function. The method is to determine the number of PRBs in different regions and their corresponding transmission power in each cell. The mathematical formulations about the resource allocation problem are written as

$$\max_{q_{i,j}, p_i^{major}, p_i^{minor}} \min_{i \in C} R_i^k \quad (4)$$

$$\text{s.t.} \quad R_i^E \geq r_{min} \quad (5)$$

$$R_i^C \geq r_{min} \quad (6)$$

$$p_i^{total} \leq P_{max} \quad (7)$$

$$N_i^{major} + N_i^{minor} = N \quad (8)$$

$$N_i^{major}, N_i^{minor} \in \{1, 2, \ldots, N-1\} \quad (9)$$

$$N_i^{major} = \sum_{j \in N} q_{i,j} \quad (10)$$

$$q_{i,j} \in \{0, 1\} \quad \forall j \in N, \ m \in B_{adj}^i \quad (11)$$

$$\sum_{m \in B_{adj}^i} q_{k,j} \leq 1 \qquad (12)$$

where $r_{\min}$ is the minimum data rate requirement for trains. Constraints (5) and (6) represent that all trains should satisfy $r_{\min}$; Constraint (7) limits the transmission power in cells, denoted by p_i^{total}. And Constraints (8) and (9) indicate that the number of PRBs should be integers and within total number N. Constraints (10) and (11) ensure that the optimal reuse factor could be achieved when the major PRBs allocated in cell i are orthogonal to adjacent cells, where $q_{i,j}$ is the PRB allocation variable indicating the PRB j is deployed in edge or in center of a cell.

4 Adaptive Soft Frequency Reuse Resource Allocation Algorithm

The algorithm decomposes the total optimization in (5) into C single-cell SFR optimization stages. The variables N_i^{major}, N_i^{minor}, P_i^{major}, P_i^{minor} should be found in each cell. We propose the solving algorithm in a single cell, then it can be applied in the multi-cell scenarios sequentially.

We assume that the information about resource occupation in adjacent cells is known. Based on the locations of trains, we first calculate the minimum transmission power that satisfies minimum rate requirement $r_{\min}$ for each train. Then, we reallocate the remaining power to improve the throughput. And the new throughput of trains will be denoted as $r_{\min}$ in the next iteration. We repeat the process until the single cell throughput does not change anymore.

We determine the optimal number of PRBs allocated in different regions and their corresponding transmission power based on the locations of the trains. The formulations are written as

$$\min \ N_1^{\text{major}} \times P_1^{\text{major}} + P_1^{\text{minor}} \times N_1^{\text{minor}} \qquad (13)$$

$$\text{s.t.} \ N_1^{\text{major}} \leq N_1^{\max} \qquad (14)$$

Constraint (14) indicates that the major PRBs in a given cell are non-overlapping with adjacent cells. And $N_1^{\max}$ denote that the maximum PRBs can be used as major PRBs for the single cell, which can be formulated as

$$N_1^{\max} = N - \max \left\{ \sum_{B_{adj}^i} N_i^{\text{major}}, \ i = 2 \right\} \qquad (15)$$

The formula (15) indicates that the maximum PRBs used as major PRBs in cell i are related to the major PRBs allocation in the adjacent cells. And the minimum transmission power could be achieved when the Constraints (5) and (6) become equality constraints. And P_i^{major}, P_i^{minor} can be formulated as follows:

$$P_1^{\text{major}} = (2^{\frac{R_1^E}{N_1^{\text{major}} \times B_{\text{sub}}}} - 1) \cdot \frac{p_2^{\text{minor}} g_{1,2,k_{1,E}} + N_0 B_{\text{sub}}}{g_{1,1,k_{1,E}}} \quad (16)$$

$$P_1^{\text{minor}} = (2^{\frac{R_1^C}{N_1^{\text{minor}} \times B_{\text{sub}}}} - 1) \cdot \frac{p_2^{\text{major}} g_{1,2,k_{1,C}} + N_0 B_{\text{sub}}}{g_{1,1,k_{1,C}}} \quad (17)$$

By using (16), (17), and (8), we can calculate transmission power P_1 with $N_i^{\text{major}}, N_i^{\text{minor}}, P_i^{\text{major}}, P_i^{\text{minor}}$. First, we propose an exhaustive search among all possible ($N_i^{\text{major}}, N_i^{\text{minor}}$) that yields the lowest transmission power, denoted $P_{\min}$.

If $P_{\min}$ is less than the maximum available power, we reallocate the remaining power to the cell edge by major PRBs. Thus, they can be transmitted with higher power. Therefore, a smaller number of major PRBs are required to be employed for the trains in the cell edge. This modification can bring more considerable PRBs used in the adjacent cells.

To allocate the remaining power, we gradually decrease major PRBs and increase their transmission power using the descent method. And the iterative algorithm stops when the maximum power is achieved. Besides, the optimization objective is nondecreasing as the number of iterations increases.

Single-cell algorithm is illustrated as follows:

1. Input the locations of trains.
2. For each ($N_1^{\text{major}}, N_1^{\text{minor}}$) that satisfies (8) and (9), find the corresponding ($P_1^{\text{major}}, P_1^{\text{minor}}$) according to (16) and (17).
3. Calculate the minimum transmission power

    (3.1) If $P_{\min}(k) \leq P_{\max}$, go to step 4;
    (3.2) If $P_{\min}(k) > P_{\max}$, terminate the algorithm.

4. If $P_{\min}(k) \leq P_{\max}$, repeat:

    (4.1) Fixed P_1^{minor}, $N_1^{\text{minor}} \leftarrow N_1^{\text{minor}} + 1$. $R_1^C(k+1) > R_1^C(k)$.
    (4.2) Update the number of major PRBs $N_1^{\text{major}} \leftarrow N - N_1^{\text{minor}}$; Calculate the major transmission power P_1^{major}. $R_1^E(k+1) > R_1^E(k)$.
    (4.3) Calculate p_1^{total}; If $p_1^{\text{total}} < p^{\max}$, use the newly obtained throughput as the target value for the next iteration.

5. Set $k = k+1$ and repeat steps 2–4 until the objective does not change any more.

5 Simulation

We consider that the system bandwidth is 10 MHz (50 PRBs). Besides, the simulation is built on the scenario that there exist two trains in the edge and the region of a cell.

We record the results on how resource allocation and transmission power to each train change during the iterative process when the trains' locations are given. From Fig. 3, we can see that the resource allocation satisfying the minimum requirement for each train keeps monotony and remains stable after 25 iterations. Note that after the power is reduced, more resources are needed to ensure the minimum rate requirement.

And the power distributed to the train in the central area is gradually increasing because more resources could be allocated to the user with better conditions of the communication channel.

As the number of iterations increases, the throughput difference gradually decreases, as Fig. 4 shows. The simulation validates the expectation that the resource should not be allocated to the users with good conditions as much as possible.

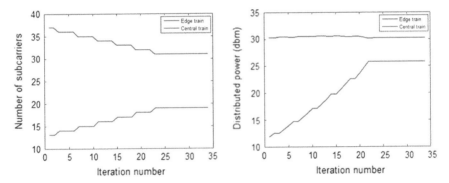

Fig. 3 Number of PRBs and transmission power versus iterations

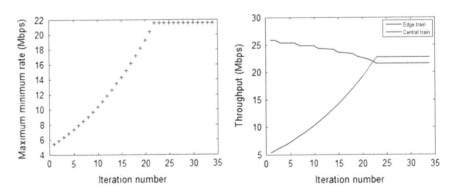

Fig. 4 Maximum minimum rate and individual throughput versus iterations

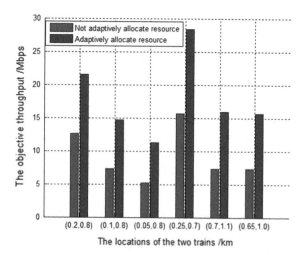

Fig. 5 Maximum minimum throughput of two trains under different scenarios

Then, we compare the maximum value of the minimum throughput of two trains according to the different locations. Note that the adaptive resource allocation algorithm can dramatically improve the throughput of the train with the smallest throughput in the cell, as Fig. 5 shows.

6 Conclusion

In this paper, we have studied a joint optimization of transmission power and resource allocation based on SFR that aims to maximize the minimum value among the train's throughput, provided the locations of trains. We have presented an algorithm jointly optimizing resource allocation and transmission power. The proposed algorithm effectively improves the performance for individuals in the general compared with no joint optimization case and exhibits a low computation complexity with reliable convergent property.

Acknowledgements The research work was supported by the research project (No. I17H100021) from Beijing Municipal Science and Technology Commission, Beijing Laboratory of Urban Rail Transit, Beijing Key Laboratory of Urban Rail Transit Automation, and Control.

References

1. Xu TH, Tang T, Gao CH et al (2009) Dependability analysis of the data communication system in train control system. Sci China Ser E: Technol Sci 52(9):2605–2618

2. Zhao H, Jiang H (2017) LTE-M system performance of integrated services based on field test results. In: Advanced information management, communicates, electronic & automation control conference. IEEE
3. Zhao H, Xu W, Cao Y et al (2016) Integrated train ground radio communication system based TD-LTE. Chin J Electron 25(4):740–745
4. Qian M, Hardjawana E, Li Y et al (2015) Adaptive soft frequency reuse scheme for wireless cellular networks. IEEE Trans Veh Technol 64(1):118–131
5. Huawei (2005) Soft frequency reuse scheme for UTRAN LTE. 3GPP TSG RAN WG1 Meeting #41 Project Document R1-050507, Athens, Greece, May 2005 (online). Available at: http://www.3gpp.org
6. Ali SH, Leung VCM (2009) Dynamic frequency allocation in fractional frequency reused OFDMA networks. IEEE Trans Wirel Commun 8(8):4286–4295
7. Venturino L, Prasad N, Wang X (2009) Coordinated scheduling and power allocation in downlink multicell OFDMA networks. IEEE Trans Veh Technol 58(6):2835–2848
8. Giambene G, Le VA, Bourgeau T et al (2017) Iterative multi-level soft frequency reuse with load balancing for heterogeneous LTE-A systems. IEEE Trans Wirel Commun PP(99):1-1
9. Yu Y, Dutkiewicz E, Huang X et al (2012) Adaptive power allocation for soft frequency reuse in multi-cell LTE networks. In: International symposium on communications & information technologies. IEEE

Research on Passenger Flow Distribution in Urban Rail Transit Hub Platform

Long Gao, Lixin Miao, Zhongping Xu, Liang Chen, Zhichao Guan, Hualian Zhai, Limin Jia, Yizhou Chen and Jie Zhuang

Abstract Urban rail transit has become the main mode to ease traffic congestion. Researches on the dynamic variation regularity of passenger flow distribution and passenger flow volume of urban rail transit hub platform have important implication in hub capacity design, operation organization and risk prevention. Simulation designs are built with Java and Anylogic, which contain simulation system functional framework, implementation path, simulation processes and simulation models. By case study, the performance comparison between two simulation methods indicates that simulation designs are scientific and accordant with the reality scene.

Keywords Passenger flow distribution · Urban rail transit hub platform · Simulation and application · Performance evaluation

L. Gao · L. Miao (✉)
Graduate School at Shenzhen, Tsinghua University, Shenzhen, China
e-mail: 11114234@bjtu.edu.cn

L. Gao · Z. Xu (✉) · L. Chen · Z. Guan · H. Zhai
Transport Bureau of Shenzhen Municipality, Shenzhen, China
e-mail: 11114234@bjtu.edu.cn

L. Gao · Z. Xu · L. Chen · Z. Guan
Shenzhen Transportation Operation Command Center, Shenzhen, China

L. Jia
State Key Laboratory of Rail Traffic Control and Safety, Beijing Jiaotong University, Beijing, China

Y. Chen
Shenzhen Rongheng Industrial Group Co. Ltd, Shenzhen, China

J. Zhuang
Shenzhen Linrun Industrial Group Co. Ltd, Shenzhen, China

© Springer Nature Singapore Pte Ltd. 2020
B. Liu et al. (eds.), *Proceedings of the 4th International Conference on Electrical and Information Technologies for Rail Transportation (EITRT) 2019*, Lecture Notes in Electrical Engineering 640, https://doi.org/10.1007/978-981-15-2914-6_42

1 Introduction

In recent years, with the rapid development of urban rail transit system, passenger traffic demand has increased quickly and sustainably. Urban rail transit has become the main mode to ease traffic congestion. As we know, the urban rail transit hub is the core node of urban rail transit system. For an urban rail transit hub, hub platform is the most important area for passenger flow distribution. Under the viewpoint of traffic safety and service level, researches on the dynamic variation regularity of passenger flow distribution and passenger flow volume of hub platform are very necessary and meaningful. The prospective research findings could provide the reference for train scheduling, hub capacity design and operation management.

Aiming at the researches on modeling, simulation and application of passenger flow distribution in urban rail transit field, domestic and foreign scholars have done mountains of work and achieved abundant findings. Generally, there were two sides to the related researches, one side was the research on modeling and simulation of passenger flow distribution; the other side was the research on application of passenger flow distribution. The purposes of these researches were to improve passenger traffic service quality and reliability.

According to the hierarchy of modeling and simulation, passenger flow models consist of macroscopic model, mesoscopic model and microscopic model. Early researches of passenger traffic mainly focused on the relations among traffic flow, speed and density; research findings represented by Fruin [1] were adopted as an effective macroscopic analysis approach for passenger traffic in the US Highway Capacity Manual [2]. As a representative of the mesoscopic model, lattice gas model was firstly proposed to passenger traffic simulation by Muramatsu et al. [3]. With the rapid development of computer technology, microscopic model was proposed and applied to passenger traffic simulation, which mainly included magnetic model [4], benefit cost cellular model [5], queuing network model [6], social force model [7] and cellular automata model [8, 9]. On the other hand, application of passenger flow distribution mainly concerned with passenger traffic system planning, operation organization design, facility performance evaluation, emergency evacuation and so on. Due to the different development stages, major cities of the western developed countries had reached an advanced level, such as New York, Berlin, Copenhagen and London [10, 11].

With respect to transit hub distribution efficiency and passenger behavior analysis, passenger individual behavior model was firstly set up by Gipps and Marksjo [5], and they supposed that passenger movement obeyed the short-circuit law and put forward a simple route choice model. Helbing [12] illustrated the complex characteristics of passenger flow and built social force model. Xiong et al. [13] proposed a continuous-time random walk model for pedestrian flow walking behavior simulation. Lu et al. [14] explored the effects of different walking strategies on bi-directional pedestrian flow in the channel with cellular automata formulation. Daamen et al. [15, 16] summed up that passenger flow crowd degree in hub interlayer facility as the key factor directly affected route choice behavior by analyzing the relations between interlayer facility layout and passenger path-finding behavior. Lin and Trani [17] analyzed passenger flow characteristics inside of the hub and identified passenger flow distribution bottleneck. Similarly, scholars researched on the passenger flow through corridor bottlenecks, experiments results showed that the bottleneck capacity was almost linearly increased with the width, and jamming occurred below the maximum capacity [18–20]. In recent researches, Duive et al. [21] proposed the state-of-the-art crowd motion simulation models to explain different phenomena of crowd motion such as lane formation, stop-and-go waves, faster-is-slower effect, and turbulence and zipper effect. Guillermo [22] built a mathematical model of the formation of lanes in crowds of pedestrians moving in opposite directions. Bandini et al. [23] improved the traditional floor field cellular automata model to simulate the negative interaction among pedestrians of high density. Xie and Jia [24] and Wang et al. [25] proposed the forecasting methods of passenger flow based on hybrid temporal-spatio forecasting model and modular neural network.

In view of the above researches, it could be found that kinds of models and methods were widely used to explore passenger flow characteristics and behavior in transit hub. Their findings motivate the present work. However, there were few researches related with passenger flow distribution in hub platform. Therefore, the objective of this paper is to modeling and simulation of passenger flow distribution in urban rail transit hub platform and explores the practical application.

2 Basic Analysis of Passenger Flow Distribution

2.1 Passenger Flow Distribution Process

In urban rail transit hub platform, passenger flow distribution process refers to traffic behaviors of passenger group dynamics influenced by platform facilities and train scheduling. Passenger flow distribution morphology is regularly changed with train departure interval.

What is more, passenger flow distribution process presents obvious phases during one departure interval. Especially after the train door opening, the process of passenger getting on or off the train can be divided into three phases.

2.2 Influence Factors of Passenger Flow Distribution

Based on the relationship between interaction agents, influence factors of passenger flow distribution are derived from three parts, which are platform, train and passenger flow.

Platform: available area of platform, traffic capacity of passage, traffic capacity of stairs and traffic capacity of the escalator.

Train: train type, train formation, train departure interval and train dwell time.
Passenger flow: temporal distribution of passenger flow and spatial distribution of passenger flow.

3 Simulation Design of Passenger Flow Distribution

In this chapter, based on the passenger flow distribution models of hub platform, functional framework and simulation process of simulation system are designed by Java 8.0. On the other hand, passenger flow distribution simulation environment of hub platform is built by using Anylogic 7.2.0. Simulation design research in this chapter could provide technical approaches for the case study and model performance verification following.

By using the event marching method, simulation process of passenger flow distribution models in hub platform is built, as shown in Fig. 1. Meanwhile, event is defined as the variation of passenger flow volume in hub platform, which pushes

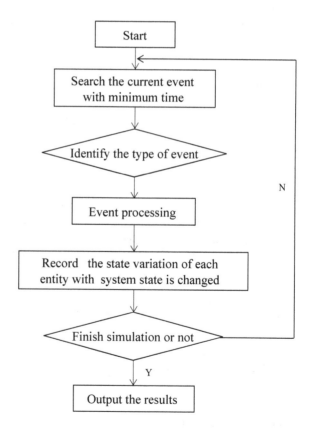

Fig. 1 Simulation process of simulation system

Research on Passenger Flow Distribution in Urban Rail Transit ...

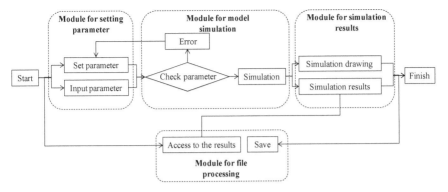

Fig. 2 Functional framework and implementation path of simulation system

the simulation clock forward. Functional framework and implementation path of simulation system are designed by Java 8.0, as shown in Fig. 2.

4 Case Study

As an urban rail transit hub station, Beijing South Subway Station is located in the first floor underground of Beijing South Railway Station, which contains Beijing subway line 4 and line 14. Passenger could transfer among the different traffic modes such as high-speed railway, urban rail transit and urban public bus. In this paper, the platform of subway line 4 is selected as the hub platform for research, meanwhile, the passenger flow in the platform of subway line 4 is selected as the research object.

According to the above research and practical investigation data, the value of passenger flow volume in the hub platform is calculated by Java 8.0 and Anylogic 7.2.0. The schematic diagrams of simulation interface display are shown in Fig. 3. Statistical curves of passenger flow volume in the hub platform are shown in Figs. 4 and 5. Then, by using contrastive analysis based on the calculation results, more results could be obtained, as shown in Table 1.

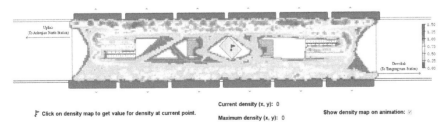

Fig. 3 Schematic diagram of simulation interface display with density map showing

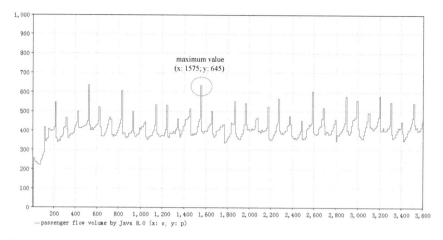

Fig. 4 Statistical curve of passenger flow volume in the hub platform by Java

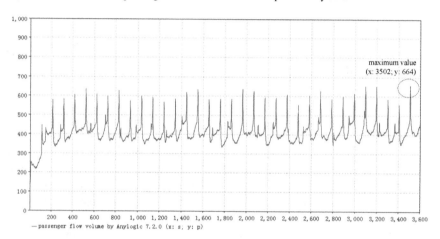

Fig. 5 Statistical curve of passenger flow volume in the hub platform by Anylogic

Table 1 Results of contrastive analysis between two statistical curves

Result	Calculating by Java	Calculating by Anylogic
Maximum value of passenger flow volume (p)	645	664
Average density (p/m^2)	0.39	0.40
Average error	2.86%	

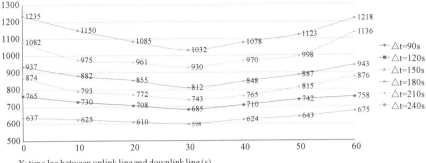

Fig. 6 Statistical curve of maximum passenger flow volume with different departure intervals

More experiments are taken with different departure intervals of subway line 4. Calculation results of passenger flow volume in the hub platform are shown in Fig. 6. It is shown that the minimum value among the calculation results is 598, as $\Delta t = 90$ and $\left|\Delta t^+ - \Delta t^-\right| = 30$.

By the model performance experiment, some conclusions are summarized as follows.

(1) It is proved that passenger flow distribution models which are built in Chap. 2 have a good performance on describing passenger flow distribution process of the hub platform.
(2) From the above calculation results, it is shown that simulation designs by Java and Anylogic for passenger flow distribution models which are built in Chap. 3 are scientific and accordant with the reality scene.
(3) The average error between two simulation methods is 2.86%, and it is indicated that the simulation precision is comparatively ideal.
(4) The average density of passenger flow in the hub platform is less than $0.40\,p/m^2$, and it is indicated that the service level of hub platform is reliable.

5 Conclusion

For an urban rail transit hub, hub platform is the most important area of passenger flow distribution. Passenger flow volume of hub platform has become a fundamental and crucial data for hub capacity design, operation organization and management. In this paper, in order to research the dynamic variation regularity of passenger flow distribution and calculate the passenger flow volume of hub platform, some studies are carried out and some conclusions are obtained.

In addition, this research is not free of mistakes and inconveniences. Future researches could be conducted to further improve the performance and practicality of passenger flow distribution models for more scientific and reasonable.

Acknowledgements This work is supported by China Postdoctoral Science Foundation funded project (No. 2018M641380).

References

1. Fruin JJ (1971) Designing for pedestrians: a level of service concept. Highw Res Rec 355(12):1–15
2. Transportation Research Board (2000) Highway capacity manual. National Research Council Press, Washington, DC, pp 635–670
3. Muramatsu M, Irie T, Nagatani T (1999) Jamming transition in pedestrian counter flow. Phys A 267(3):487–498
4. Okazaki S (1979) A study of pedestrian movement in architectural space, part 1: pedestrian movement by the application on of magnetic models. Trans Archit Inst Jpn 283(3):111–119
5. Gipps PG, Marksjo B (1985) A micro-simulation model for pedestrian flows. Math Comput Simul 27(2):95–105
6. Lovas GG (1994) Modeling and simulation of pedestrian traffic flow. Transp Res Part B 28(3):429–443
7. Helbing D, Molnar P (1995) Social force model for pedestrian dynamics. Phys Rev E 51:4282–4286
8. Blue VJ, Adler JL (2001) Cellular automata microsimulation for modeling bi-directional pedestrian walkways. Transp Res Part B 35:293–312
9. Li XM, Yan XD, Li XG (2012) Using cellular automata to investigate pedestrian conflicts with vehicles in crosswalk at signalized intersection. Discrete Dyn Nat Soc 3:1555–1565
10. Edward LF, Gabe K (2010) Pedestrian and bicyclist safety and mobility in Europe. U. S. Department of Transportation Federal Highway Administration, Washington, DC, p 2
11. Barbara M, Edward S (2010) Public policies for pedestrian and bicyclist safety and mobility. U. S. Department of Transportation Federal Highway Administration, Washington, DC, p 8
12. Helbing D (2004) Collective phenomena and states in traffic and self-driven many-particle systems. Comput Mater Sci 30:180–187
13. Xiong HW, Yao LY, Wang WH (2012) Pedestrian walking behavior revealed through a random walk model. Discrete Dyn Nat Soc 12:1951–1965
14. Lu LL, Ren G, Wang W (2013) Exploring the effects of different walking strategies on bi-directional pedestrian flow. Discrete Dyn Nat Soc 2:1–9
15. Daamen W, Bovy PHL, Hoogendoorn SP (2005) Influence of changes in level on passenger route choice in railway stations. In: 84th annual meeting of the Transportation Research Board. National Research Council Press, Washington, DC, pp 12–20
16. Daamen W, Bovy PHL, Hoogendoorn SP (2006) Choices between stairs, escalators and ramps in stations. In: 10th international conference on computer system design and operation in the railway and other transit systems, pp 3–12
17. Lin YD, Trani AA (2000) Airport automated people mover systems: analysis with a hybrid computer simulation model. Transp Res Rec 1703:45–57
18. Seyfried A, Rupprecht T (2007) New insights into pedestrian flow through bottlenecks. Transp Sci 43(3):395–406
19. Schadschneider A, Klingsch W (2009) Evacuation dynamics: empirical results, modeling and applications. Encycl Complex Syst Sci 5:3142–3176

20. Seyfried A, Boltes M (2010) Enhanced empirical data for the fundamental diagram and the flow through bottlenecks. Pedestrian and evacuation dynamics, pp 145–156
21. Duive DC, Daamen W, Hoogendoorn SP (2013) State-of-the-art crowd motion simulation models. Transp Res Part C 37(3):193–209
22. Guillermo HG (2015) A mathematical model of the formation of lanes in crowds of pedestrians moving in opposite directions. Discrete Dyn Nat Soc 2:56–63
23. Bandini S, Mondini M, Vizzari G (2014) Modelling negative interaction among pedestrians in high density situations. Transp Res Part C 40(1):251–270
24. Xie ZY, Jia LM (2013) A hybrid temporal-spatio forecasting approach for passenger flow status in chinese high-speed railway transport hub. Discrete Dyn Nat Soc 5:248–259
25. Wang SW, Zhou RG, Zhao L (2015) Forecasting Beijing transportation hub areas's pedestrian flow using modular neural network. Discrete Dyn Nat Soc 2013:1–6

Simulation Analysis of Urban Rail Transit Hub Based on Complex Network

Long Gao, Lixin Miao, Zhongping Xu, Liang Chen, Zhichao Guan, Hualian Zhai, Limin Jia, Yizhou Chen and Jie Zhuang

Abstract Aiming at the complex environment for passenger flow distribution and considering the related research achievements in URTH, based on the theory and method of complex network, topology structure attribute, the measurement indexes sets and the reliability analysis method are developed; simulation method for network reliability-oriented is built; further, the typical URTH is selected for case study. In conclusion, the theory and method of the modeling and simulation of URTH are established.

Keywords Urban rail transit hub · Simulation analysis · Complex network

1 Introduction

With the rapid development of urban rail transit system in our country, network operating pattern of urban rail transit has been formed gradually. Urban rail transit hub (URTH) is considered as the core node in urban rail transit network, and the functional structure for passenger flow distribution in URTH has evolved into a

---

L. Gao · L. Miao (✉)
Graduate School at Shenzhen, Tsinghua University, Shenzhen, China
e-mail: 11114234@bjtu.edu.cn

L. Gao · Z. Xu · L. Chen · Z. Guan · H. Zhai
Transport Bureau of Shenzhen Municipality, Shenzhen, China

L. Gao · Z. Xu · L. Chen · Z. Guan
Shenzhen Transportation Operation Command Center, Shenzhen, China

L. Jia
State Key Laboratory of Rail Traffic Control and Safety, Beijing Jiaotong University, Beijing, China

Y. Chen
Shenzhen Rongheng Industrial Group Co., Ltd, Shenzhen, China

J. Zhuang
Shenzhen Linrun Industrial Group Co., Ltd, Shenzhen, China

rather complex service network system. Due to the impacts by many factors such as economy, culture and policy in different cities, there are significant differences in structure and function among different hubs. And then, passenger traffic behavior and passenger flow characteristics among different hubs are diverse.

The theory and method of complex networks provide a new perspective for the study of real complex systems and a theoretical basis for the design of networks with good performance. Domestic and foreign scholars' research on the application of complex network in urban rail transit is mainly divided into two aspects. First, in terms of network topological structure characteristics analysis, Derrible [1] empirically analyzed the topological characteristics of subway network in 33 major cities around the world, and the results showed that most subway networks were scale-free. Latora [2], Seaton and Lee [3] put forward node degree distribution, clustering coefficient, network efficiency, mean shortest distance and node average degree as the measurement indexes, and empirically analyzed the topological structure characteristics of Boston metro network, Vienna metro network and Seoul metro network, respectively. The results proved that the above metro network had small-world characteristics and scale-free characteristics. Liu et al. [4] put forward the node degrees, average shortest distance and clustering coefficient for metrics, topological statistical characteristics of Guangzhou urban rail transit network were analyzed, and the road network global efficiency and local efficiency and maximal connected subgraph relative size of reliability index, the influence degree of the failure of urban rail transit network performance were analyzed.

On the other hand, in terms of network performance, the research mainly focuses on robustness, invulnerability and connectivity reliability. Li [5] calculated the robustness of Beijing subway network with connectivity, average shortest distance and network diameter as the measurement indexes, and the results showed that the network had better robustness when it was subjected to random attack, but there was poor destructiveness when it was subjected to deliberate attack. Zhang [6] analyzed the vulnerability of the undirected complex network system from the structure and function. Based on the theory of reliability and network traffic distribution, Gao [7] established the anti-damage reliability evaluation model of urban rail transit network system, further adopted weighted average method to build five reliability indexes for the network anti-destroying ability.

Aiming at the complex environment for passenger flow distribution and considering the related research achievements in URTH, based on the theory and method of complex network, topology structure attribute, the measurement indexes sets and the reliability analysis method are developed; simulation method for network reliability-oriented is built; further, the typical URTH is selected for case study. In conclusion, the theory and method of the modeling and simulation of URTH are established.

2 Network Model

2.1 Basic Description

The inner space environment of urban rail transit hub is composed of various pedestrian facilities and other building entities. The interaction between the above two could realize the structural division of the hub space. Cognitive schema theory [8] reveals that passenger's cognition of the spatial structure relationship among the internal facilities of the hub is usually expressed in the form of network in real life, as shown in Fig. 1. In which, an example of the internal facilities network diagram of the urban rail transit hub is shown.

Therefore, this section begins with a basic description of the basic attributes of the internal facilities of the hub and their spatial structural relations, and its form could be defined as a set of quaternions:

$$F = (N, A, S_N, S_A) \qquad (1)$$

where N refers to the facility set. A refers to the facility adjacency set. S_N refers to the basic attributes of facilities. S_A refers to the adjacency basic property. What is more, the basic attributes of facilities include classification attributes, physical properties and spatial attribute. The adjacency basic property is the adjacency relation between facilities.

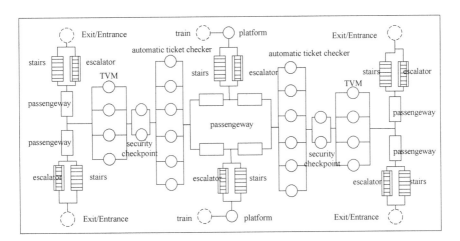

Fig. 1 Internal facilities network diagram of the urban rail transit hub

2.2 Network Model

At present, there are three topological mapping methods for traffic network structure: L space method (geospatial) [9], P space method (transfer relation) [10] and R space method (line relation) [11]. In this paper, L space method is adopted to build the hub topological network model. Namely, hub facilities are taken as nodes; if and only if two nodes are on the same passenger line and adjacent to each other, there is an undirected adjacency (edge) between the two nodes. Hub topological network is shown as follows.

$$G = (V, E, L, P) \qquad (2)$$

where V refers to the facility nodes set, $V = \{v_i\}$, $i = 1, 2, ..., N$. In which, v_i refers to the node i, and N refers to the number of nodes. E refers to the edges set, $E = \{e_{(vi, vj)}\}$, $i, j = 1, 2, ..., N, i \neq j$. In which, $e_{(vi, vj)} = 1$, if v_i is connected to v_j. Otherwise, $e_{(vi, vj)} = 0$. E refers to the passenger flow lines set, $L = \{l_1, l_2, ..., l_m\}$, m refers to the number of passenger flow lines. P refers to the nodes paths set, $P = \{p_{(vi, vj)}\}$, $i, j = 1, 2, ..., N, i \neq j$. In which, $p_{(vi, vj)}$ refers to the nodes paths between v_i and v_j.

2.3 Network Reliability

It is of great practical significance to study the reliability of complex networks. At present, study on the reliability of complex networks mainly focuses on network self-adaptability and network survivability, which mainly measure the fault-tolerant ability or attack resistance of network systems against random failures or deliberate attacks. Albert and Barabasi [12] pioneered in the research on the attack modes of complex network, including random fault and intentional attack. Random fault refers to the network node or edge fails randomly with a certain probability. Intentional attack refers to network node or edge is selectively damaged by a specific attack strategy.

(1) Network Self-adaptability

The self-adaptability of hub topological network reflects the uneven distribution of node or edge structure, its index is heterogeneity. The establishment of "information entropy" theory [13] provides a technical means for study the state evolution of complex systems. On this basis, the proposal of "graph entropy" [14] provides a concrete method for measuring the complexity of hub topological network. Therefore, entropy is adopted in this paper to analyze the heterogeneity of hub topology network structure, which includes distribution entropy of node structure degree, distribution entropy of node structure betweenness and distribution entropy of edge structure betweenness.

1. Distribution entropy of node structure degree

$$H(\text{Nsd}) = - \sum_{v_i, v_j \in V, i \neq j} [p(\text{Nsd}_{v_i}) \cdot \ln p(\text{Nsd}_{v_i})] \qquad (3)$$

$$p(\text{Nsd}_{v_i}) = \frac{n(\text{Nsd}_{v_i})}{N} \qquad (4)$$

where $H(\text{Nsd})$ refers to the distribution entropy of node structure degree. $p(\text{Nsd}_{v_i})$ refers to the probability as node structure degree is Nsd_{v_i}.

2. Distribution entropy of node structure betweenness

$$H(\text{Nsb}) = - \sum_{v_i, v_j \in V, i \neq j} [p(\text{Nsb}_{v_i}) \cdot \ln p(\text{Nsb}_{v_i})] \qquad (5)$$

$$p(\text{Nsb}_{v_i}) = \frac{n(\text{Nsb}_{v_i})}{N} \qquad (6)$$

where $H(\text{Nsb})$ refers to the distribution entropy of node structure betweenness. $p(\text{Nsb}_{v_i})$ refers to the probability as node structure betweenness is Nsb_{v_i}.

3. Distribution entropy of edge structure betweenness

$$H(\text{Esb}) = - \sum_{v_i, v_j \in V, i \neq j} [p(\text{Esb}_{e_{(v_i, v_j)}}) \cdot \ln p(\text{Esb}_{e_{(v_i, v_j)}})] \qquad (7)$$

$$p(\text{Esb}_{e_{(v_i, v_j)}}) = \frac{n(\text{Esb}_{e_{(v_i, v_j)}})}{M} \qquad (8)$$

where $H(\text{Esb})$ refers to the distribution entropy of edge structure betweenness. $p(\text{Esb}_{e_{(v_i, v_j)}})$ refers to the probability as edge structure betweenness is $\text{Esb}_{e_{(v_i, v_j)}}$.

In conclusion, the higher the structure entropy, the more uneven the distribution of node or edge structure in hub topology network, and the higher the degree of structure heterogeneity and the lower the self-adaptability. On the contrary, the smaller the structure entropy, the more uniform the distribution of the node or edge structure in the hub topology network, and the lower the degree of structure heterogeneity and the stronger the self-adaptability.

(2) Network survivability

In this section, structural connectivity reliability [15] is used to represent the structure connectivity reliability of hub topology network after partial nodes or edges failure. On this basis, network survivability is characterized by the variance of structure connectivity reliability of topology network.

1. Structure connectivity reliability of topology network

$$\text{SCR}_{TN} = \frac{1}{N(N-1)} \sum_{v_i, v_j \in V, i \neq j} \left[\text{SCR}_{(v_i, v_j)}\right] \quad (9)$$

$$\text{SCR}_{(v_i, v_j)} = \frac{r'_{(v_i, v_j)}}{r_{(v_i, v_j)}} \quad (10)$$

where SCR_{TN} refers to structure connectivity reliability of topology network after partial nodes or edges failure. $\text{SCR}_{(v_i, v_j)}$ refers to structure connectivity reliability between node v_i and v_j after partial nodes or edges failure. $r_{(v_i, v_j)}$ refers to the number of effective paths between node v_i and v_j in normal situation; $r'_{(v_i, v_j)}$ refers to the number of effective paths between node v_i and v_j after partial nodes or edges failure.

2. Variance of structure connectivity reliability of topology network

$$\sigma^2_{\text{SCR}_{TN}} = \frac{1}{K} \sum_{v_i, v_j \in V, i \neq j} \left[\Delta(\text{SCR}_{TN}) - \bar{\Delta}(\text{SCR}_{TN})\right]^2 \quad (11)$$

$$\Delta(\text{SCR}_{TN}) = 1 - \text{SCR}_{TN} \quad (12)$$

$$\bar{\Delta}(\text{SCR}_{TN}) = \frac{1}{K} \sum_{v_i, v_j \in V, i \neq j} \left[\Delta(\text{SCR}_{TN})\right] \quad (13)$$

where $\sigma^2_{\text{SCR}_{TN}}$ refers to the variance of structure connectivity reliability of topology network. $\Delta(\text{SCR}_{TN})$ refers to the change value of SCR_{TN} after partial nodes or edges failure. $\bar{\Delta}(\text{SCR}_{TN})$ refers to the average value of $\Delta(\text{SCR}_{TN})$. K refers to the number of associated failures for node or edge failure strategy.

In conclusion, the larger the $\sigma^2_{\text{SCR}_{TN}}$, the higher the dispersion degree of the connectivity reliability change, which indicates that network survivability is weaker. On the contrary, the smaller the $\sigma^2_{\text{SCR}_{TN}}$, the lower the dispersion degree of the connectivity reliability change, which indicates that network survivability or destruction resistance is stronger.

3 Case Study

In this research, Beijing South Subway Station (BSSS) is selected for case study. Through investigation and drawing, topology network diagram of BSSS is shown in Fig. 2.

Fig. 2 Topology network diagram of BSSS

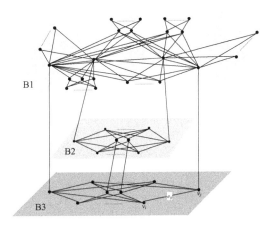

(1) Network Self-adaptability

It is calculated that if and only if all nodes in the topology network are completely unevenly distributed, the maximum value of structure entropy exists. The heterogeneity calculation results of topology network structure shown in Table 1.

(2) Network survivability

In this section, the topology network is deliberately attacked according to the descending order of the node structure betweenness and the edge structure betweenness. The calculation results of damage resistance of topological network of BSSS under different failure strategies are as follows.

Table 1 Structure entropy of topology network of BSSS

Index	$H(\text{Nsd})$	$H(\text{Nsb})$	$H(\text{Esb})$
Structure entropy	1.296	1.475	1.857
Nonuniform coefficient (%)	24.41	27.79	34.98

Fig. 3 Variation curve of structure connectivity reliability of topology network under node failure in turn

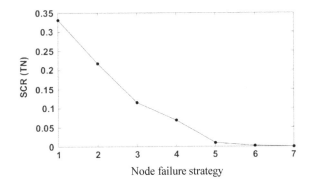

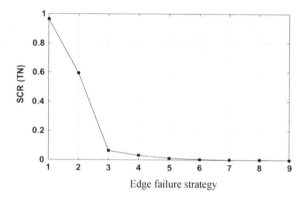

Fig. 4 Variation curve of structure connectivity reliability of topology network under edge failure in turn

Table 2 Variance of structure connectivity reliability of topology network of BSSS

Failure strategies	$\sigma^2_{SCR_{TN}}$
Node failure	0.0129
Edge failure	0.0735

From Fig. 3, it could be calculated that when the node is deliberately attacked, the structure connectivity reliability of the topology network gradually decreases as the number of failed nodes increases. When the number of attacks reaches 7, the structure connectivity reliability of the topology network drops to 0 after the number is greater than or equal to 0.0404 fail one after another, the topology network is in a collapse state. At this time, the number of failure nodes is 33, accounting for 16.34%.

From Fig. 4, it could be calculated that when the edge is deliberately attacked, the structure connectivity reliability of the topology network gradually decreases as the number of failure edges increases, and the reliability decrease in this case is larger than that in the case of node failure. When the number of attacks reaches 9, the structure connectivity reliability of the topology network decreases to 0 after successive failures of edges with the number greater than or equal to 0.0028, the topology network is in a collapse state. At this time, the number of failure edges is 112, accounting for 7.53%.

Based on the above results, the variance of structure connectivity reliability of topology network of BSSS could be calculated as shown in Table 2. In which, $\sigma^2_{SCR_{TN}}$ as node failure is smaller than $\sigma^2_{SCR_{TN}}$ as edge failure, indicating that the damage resistance of topology network in the case of node failure is stronger than that in the case of edge failure.

4 Conclusion

In this paper, aiming at the complex environment for passenger flow distribution and considering the related research achievements in URTH, based on the theory and method of complex network, topology structure attribute, the measurement indexes sets and the reliability analysis method are developed; simulation method for network reliability-oriented is built; further, the typical URTH is selected for case study. In conclusion, the theory and method of the modeling and simulation of URTH are established. Findings of this research are of practical significance in the performance evaluation of passenger flow distribution.

Acknowledgements This work is supported by China Postdoctoral Science Foundation funded project (No. 2018M641380).

References

1. Derrible S, Kennedy C (2010) The complexity and robustness of metro networks. Physica A 389(17):3678–3691
2. Latora V, Marchiori M (2002) Is the Boston subway a small-world network? Physica A 314(1–4):109–113
3. Lee K, Jung W, Park JS (2008) Statistical analysis of the Metropolitan Seoul subway system: network structure and passenger flows. Physica A 387(24):6231–6234
4. Liu ZQ, Song R (2010) Reliability analysis of Guangzhou rail transit with complex network theory. J Transp Syst Eng Inf Technol 1(5):194–200
5. Li J, Ma JH (2009) Research of complexity of urban subway network. J Xidian Univ (Social Sciences Edition) 19(2):51–54
6. Zhang JH, Xu XM, Hong L (2011) Networked analysis of the Shanghai subway network in China. Physica A 390(23–24):4562–4570
7. Gao J, Shi QZ (2007) Definition and evaluation modeling of metro network invulnerability. J China Railw Soc 29(3):29–33
8. Smith A (1990) The body in the mind: the bodily basis of meaning, imagination, and reason, 1 edn. University Of Chicago Press, Chicago
9. Sienkiewicz J, Holyst JA (2005) Statistical analysis of 22 public transport networks in Poland [J]. Phys Rev E: Stat, Nonlin, Soft Matter Phys 72(4):1–10
10. Kurant M, Thiran P (2006) Extraction and analysis of traffic and topologies of transportation networks. Phys Rev E 74(3):036114
11. Xu X, Hu J, Liu F (2007) Scaling and correlations in three bus-transport networks of China. Physica A 374(1):441–448
12. Albert R, Jeong H, Barabási AL (2000) Error and attack tolerance of complex networks. Nature 406(6794):378–382
13. Shannon CE (1948) A mathematical theory of communication. Bell Syst Tech J (27):379–423, 623–656
14. Rashevsky N (1955) Life, information theory, and topology. Bull Math Biol 17(3):229–235
15. Asakura Y (1999) Evaluation of network reliability using stochastic user equilibrium. J Adv Transp 33(2):147–158

Regret Minimization and Utility Maximization Mixed Route Choice Model

Chenran Sun, Shiwei He and Yubin Li

Abstract Based on regret theory and utility theory, this paper studies the passenger travel route choice behavior in urban rail transit. Aiming at the problem of neglecting passenger heterogeneity and the scale effect of OD in existing passenger route choice models of urban rail transit, a semi-compensated route choice model considering passenger heterogeneity is constructed, which combines regret minimization and utility maximization. In the route choice of semi-compensatory behavior, the generation of effective route set and the selection of optimal route from effective route set are considered. The passenger's individual parameters and regret coefficient were regarded as random variables to express the heterogeneity of passengers. Based on the surveyed data of Chengdu Metro, the transfer volume of Chengdu Metro after the opening of Line 10 is predicted.

Keywords Route choice · Regret theory · Utility theory · Semi-compensatory

1 Introduction

With the continuous opening and operation of new lines, which have formed the network operation situation of urban rail transit. Under the condition of urban rail transit network, there are many effective routes between an OD pair, so it is necessary to dig deeply into the rules of passenger route choice. Passenger route choice behavior analysis can provide basic data such as transfer capacity, line flow and cross-section flow for urban rail transit operation management department. This data can support customization of train operation plan, passenger flow organization and regulation, ticket clearance, line network optimization, impact evaluation of new line access

C. Sun · S. He (✉) · Y. Li
MOE Key Laboratory for Urban Transportation Complex Systems Theory and Technology, Beijing Jiaotong University, Beijing, People's Republic of China
e-mail: shwhe@bjtu.edu.cn

C. Sun
e-mail: 17120876@bjtu.edu.cn

on existing line network, etc. Therefore, the research on the behavior analysis of passenger route choice in urban rail transit has important practical significance.

The traditional research on urban rail transit route choice behavior is based on expected utility theory (EUT) [1] and random utility maximization (RUM) [2]. It assumes that passengers are completely rational and always choose the route with the greatest expected utility in the effective routes. However, in actual choice, people will be in the bounded rationality between complete rationality and irrationality. So, Bell [3] and Loomes [4] put forward the regret theory. Regret theory uses regret function to describe the influence of people's regret and pleasure factors on route choice. Passenger's choice behavior is not only related to the inherent utility of the route, but also related to regret or pleasure after comparison with other routes in the effective route set. After that, Chorus et al. [5] established a random regret minimization model (RRM) based on regret theory. Then, Chorus [6] proposes a mixed RRM-RUM model, which considers that for all attributes of a selection scheme, some attributes obey the utility maximization criterion and some attributes obey the regret minimization criterion. Boeri et al. [7] also studied the heterogeneity of two decision criteria. When passengers are not familiar with the decision-making environment, it is more likely to lead to avoid regrets.

The route choice model based on discrete selection theory is mainly based on multinomial logit model (MNL) [8] and derivation models. However, the random term of MNL model obeys the identical and independent) Gumbel distribution (IID), and there are OD scale effect problems and route overlapping problems. The traditional RUM model achieves a comprehensive consideration of the factors affecting different dimensions by establishing an alternative relationship between the influencing factors. This substitution relationship is a compensatory alternative relationship, that is, when the intention factor is at a disadvantage, the purpose of enhancing utility can be achieved by improving other factors. Non-compensatory means that even if all factors of a route are in excellent condition, as long as one factor exceeds the threshold, the route will not be considered. Martinez and Aguila [9] put forward the constrained multinomial logit model (CMNL), which combines the MNL model with a non-compensating factor, in which the non-compensating factor is a binomial logit model that characterizes a smooth transition from compensation to non-compensation.

In summary, this paper constructs a mixed route choice model of regret minimization and utility maximization. Combining the two decision criteria, this paper uses bounded rationality to better describe passengers' route choice behavior. The semi-compensation model accurately reflects the differences between different passengers, including the difference in the compensation selection process and the difference in the passenger tolerance threshold during the non-compensated effective route set generation process. Based on the surveyed data of Chengdu Metro, the model parameters were calibrated, and the fitting of the mixed decision criteria model and the random utility maximization model are compared. At the same time, the prediction performance of the mixed decision criteria model under the condition of new line access is analyzed.

2 Route Choice Modeling

2.1 Semi-compensation and Heterogeneity of Urban Rail Passenger Route Choice

Figures 1 and 2 show the statistical results (12,807 samples) based on the surveyed data of Chengdu Metro in December 2017. Figure 1 depicts the longest travel time that passengers can tolerate if the minimum travel time is 30 min; Fig. 2 shows the result of the passengers' choice of number of transfer tolerance threshold when the minimum number of transfers is 0. The above results show that different passengers have different tolerance limits on the travel time and the number of transfers of the route. Different threshold parameters will form different effective route sets. Therefore, this paper set the threshold parameters to a random variable to reflect the passengers' heterogeneity, combined the two processes of compensatory and non-compensatory, and endogenously estimated parameters.

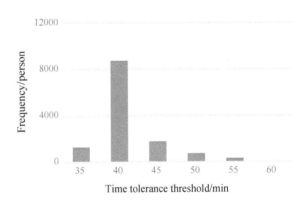

Fig. 1 Time tolerance threshold distribution (relative to 30 min)

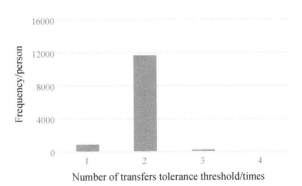

Fig. 2 Number of transfers tolerance threshold distribution (relative to 0)

2.2 Non-compensatory Choice Part

The generation of effective route set is the direct reflection of non-compensatory behavior in the process of passenger route choice. Non-compensatory means that passengers have tolerance limits on certain factors affecting route choice. If a certain factor of a route exceeds the tolerance limit, the route will not be considered as an alternative. In actual travel, the route with too long travel time and too many transfer times will not be considered by passengers. Therefore, the tolerance thresholds of the route are constructed based on these two factors.

Travel time can reflect the constraints of route choice from the time dimension and is a continuous value. According to the Chengdu Metro survey data, the different longest travel times that passengers can tolerate in the case of different minimum travel times are obtained. It is obvious that as the shortest travel time increases, the difference between the longest travel time and the shortest travel time that can be tolerated increases, and the increase trend is similar to the logarithmic function. The function of travel time threshold is as follows:

$$b_{nt}^{rs} = \omega_{nt} \cdot \ln(|X_{\min 1}^{rs}| + 1) + X_{\min 1}^{rs} \tag{1}$$

where b_{nc}^{rs} is travel time threshold of OD pair between rs, $X_{\min 1}^{rs}$ is minimum travel time of OD pair, ω_{nt} is the undetermined coefficient.

The number of transfers can reflect the constraints of route choice from the spatial dimension and is discrete. Therefore, the function of the threshold of number of transfers is as follows:

$$b_{nc}^{rs} = \omega_{nc} + X_{\min 2}^{rs} \tag{2}$$

where b_{nt}^{rs} is number of transfers threshold of OD pair between rs, $X_{\min 2}^{rs}$ is minimum number of transfers of OD pair, ω_{nc} is the undetermined coefficient.

By fitting, when the travel threshold parameter $\omega_{nt} = 0.389$, the average absolute percentage error is 6.12%. In the investigation of transfer times, the threshold of number of transfers selected by passengers was averaged, and the parameter $\omega_{nc} = 1.45$ is obtained.

Screening out the routes with the travel time or the number of transfers exceeding the threshold greatly reduces the calculation time and greatly reduces the number of routes in the set and the number of effective route sets.

2.3 Compensatory Choice Part

Regret theory considers the influence of psychological factors such as regret and joy on route choice. It assumes that passengers' choice of a route is not only related to the physical utility of the route itself, but also related to regret or joy obtained by comparing with other alternative routes. In this paper, we only consider the influence of regret on route choice and construct a new cost function, which takes into account both regret minimization and utility maximization. The function is as follows:

$$H_{nk}^{rs} = V_{nk}^{rs} - \sum_{h=1}^{H} R[\alpha_h (X_{\min h}^{rs} - X_{kh}^{rs})] \tag{3}$$

where X_{kh}^{rs} is the hth influencing factor variable of the route compensation item, H is the number of influencing factors, V_{nk}^{rs} is the utility determinant (compensation item), regret coefficient α_h is the level of regret of the passengers to the hth influencing factor and $\alpha_h = 0$ means no regret. The larger α_h is the more regret the passengers will have. The function $R(x)$ is as follows:

$$R(x) = 1 - \exp(-x), x \in [0, +\infty) \tag{4}$$

For passenger n, the perceived cost of the kth route from starting point r to ending point s is:

$$U_{nk}^{rs} = H_{nk}^{rs} + \varepsilon_{nk}^{rs} = V_{nk}^{rs} - \sum_{h=1}^{H} R[\alpha_h (X_{\min h}^{rs} - X_{kh}^{rs})] \tag{5}$$

where ε_{nk}^{rs} is random error term of cost function.

The traditional MNL model neglects the passenger's tolerance limit to some factors, such as passengers who do not consider the route of travel time being too long or too many transfer times. This indicated that passengers' route choice needs to consider the passengers' tolerances for certain influencing factors. The correction term of the effect of route tolerance on route utility was added to the function.

$$U_{nk}^{rs} = H_{nk}^{rs} + \frac{1}{\mu} C_{nk}^{rs} + \varepsilon_{nk}^{rs} = V_{nk}^{rs} - \sum_{h=1}^{H} R(X_{\min h}^{rs} - X_{kh}^{rs}) + \frac{1}{\mu} C_{nk}^{rs} + \varepsilon_{nk}^{rs} \tag{6}$$

where C_{nk}^{rs} is the correction term of the effect of route tolerance on route utility, also known as non-compensation term, ε_{nk}^{rs} is the random error term, which obeys the Gumbel distribution, μ is the scale coefficient of Gumbel distribution which the random error term obeys, and generally takes 1. $\frac{1}{\mu} C_{nk}^{rs}$ indicates that the variance of passenger's perception errors of route utility increases with the increase of OD's scale (i.e., the decrease of scale coefficient), and the penalty of this correction term for route utility increases.

The correction term of the effect of route tolerance on route utility need to satisfy the following relationships: when the route tolerance is 100%, the value should be 0, i.e., there is no need to modify the route utility; when the route tolerance is 0, the route can be excluded from the route set; when the route tolerance is between 0 and 100%, the amendment term should be a continuous function; and the logarithmic function can realize the above relationship. The function of the correction of the route tolerance's effect on the route utility is as follows:

$$C_{nk}^{rs} = \ln \phi_{nk}^{rs} \tag{7}$$

where ϕ_{nk}^{rs} is passenger n's tolerance of route k. When the tolerance is 0, the route is excluded from the effective route set.

However, in the above models, the random error term is still assumed to obey the independent and unrelated agreed Gumbel distribution. In order to further solve the scale effect of OD and the problem of route overlapping, the following models are established:

$$U_{nk}^{rs} = V_{nk}^{rs} - R(X_{\min 1}^{rs} - X_{k1}^{rs}) + \theta_{CPS} \cdot CPS_{nk}^{rs} + \frac{1}{\mu} C_{nk}^{rs} + \varepsilon_{nk}^{rs} \tag{8}$$

where V_{nk}^{rs} is the compensation term. With the increase of OD scale (i.e., the decrease of OD scale coefficient and the increase of the shortest travel time between OD pairs), the influence of factors other than travel time (such as number of transfers and comfort degree) on the route utility increases gradually. The coefficients of these factors can be expressed as the reciprocal function of the scale coefficient, that is:

$$V_{nk}^{rs} = \theta_{n1} X_{k1}^{rs} + \frac{1}{\mu_n^{rs}} \sum_{h=2}^{H} \theta_{nh} X_{kh}^{rs} \tag{9}$$

For the route k, X_{k1}^{rs} is the travel time and X_{kh}^{rs} ($h = 2 \sim H$) is the factor besides the travel time. CPS_{nk}^{rs} is a correction term for characterizing the route overlapping problem under semi-compensated behavior. The formula is as follows:

$$\text{CPS}_{nk}^{rs} = \ln \sum_{a \in \Gamma_k} \frac{L_a}{L_k} \frac{\phi_{nk}^{rs}}{\sum_{j \in A_n} (\delta_{aj} \cdot \phi_{nj}^{rs})} \tag{10}$$

where L_a is the length of section a, Γ_k is the set of sections constituting route k; if route k passes through section a, then $\delta_{aj} = 1$, otherwise $\delta_{aj} = 0$. θ_{CPS} is the parameter to be estimated, C_{nk}^{rs} is the non-compensating term, μ_n^{rs} is the scale coefficient of Gumbel distribution of random error term, $\mu_n^{rs} = \frac{\mu'}{|X_{\min 1}^{rs}|}$ and μ' is set to 1. The formula of route choice probability can be deduced as follows:

$$P_{nk}^{rs} = \frac{\exp\{\mu_n^{rs} H_{nk}^{rs} + \mu_n^{rs} \theta_{CPS} \cdot CPS_{nk}^{rs} + C_{nk}^{rs}\}}{\sum_{j \in A_n} \exp\{\mu_n^{rs} H_{nj}^{rs} + \mu_n^{rs} \theta_{CPS} \cdot CPS_{nj}^{rs} + C_{nj}^{rs}\}} \tag{11}$$

In this paper, the parameters (travel time threshold and regret coefficient) in the cost function obey a certain probability distribution, in order to show the randomness and difference of passenger route choice preferences. Regret coefficient α_h indicates the level of repentance aversion of passengers to the hth influencing factor. α_h is set to obey a probability distribution. It mainly deals with the heterogeneity of the regret degree of travel time and transfer times. Comparing the degree of passenger regret and travel time threshold according to the survey, it is found that the distribution of regret coefficient and travel time threshold parameters conforms to the normal distribution. The mixed route choice model is established as follows:

$$\begin{cases} P_{nk}^{rs} = \int P_{nk}^{rs}(\alpha) f(\alpha) d\alpha \\ P_{nk}^{rs}(\alpha) = \dfrac{\exp\{\mu_n^{rs} H_{nk}^{rs}(\alpha) + \mu_n^{rs} \theta_{CPS} \cdot \text{CPS}_{nk}^{rs} + C_{nk}^{rs}(\alpha)\}}{\sum_{j \in A_n} \exp\{\mu_n^{rs} H_{nj}^{rs}(\alpha) + \mu_n^{rs} \theta_{CPS} \cdot \text{CPS}_{nj}^{rs} + C_{nj}^{rs}(\alpha)\}} \\ H_{nk}^{rs}(\alpha) = V_{nk}^{rs} - \sum_{h=1}^{H} R[\alpha_h (X_{\min h}^{rs} - X_{kh}^{rs})] \\ V_{nk}^{rs} = \theta_{n1} X_{k1}^{rs} + \dfrac{1}{\mu_n^{rs}} \sum_{h=2}^{H} \theta_{nh} X_{kh}^{rs} \\ \text{CPS}_{nk}^{rs} = \ln \sum_{a \in \Gamma_k} \dfrac{L_a}{L_k} \dfrac{\phi_{nk}^{rs}}{\sum_{j \in A_n} (\delta_{aj} \cdot \phi_{nj}^{rs})} \\ C_{nk}^{rs} = \ln \phi_{nk}^{rs}(\alpha) \end{cases} \quad (12)$$

where P_{nk}^{rs} is the probability that passenger n chooses route k from effective route set A_n^{rs}; α is the unknown parameter matrix corresponding to each variable (including regret aversion level parameter and route tolerance parameters); $P_{nk}^{rs}(\alpha)$ is the probability that passenger n chooses route k according to the model when parameter α is taken; $\alpha = [\alpha_1, \alpha_2, \cdots]^T$ is the parameter to be estimated and obeys normal distribution; $f(\alpha)$ is the joint distribution density function; X_{kh}^{rs} is the hth influencing factor variable of the compensation term of route k; H is the number of influencing factors; $H = 3$ in this paper. The factors considered are as follows: Travel time is X_{k1}^{rs}, including inbound time, waiting time at the starting platform, train running time, transfer time, platform parking time and outbound time. The number of transfers is X_{k2}^{rs}, which indicates the total number of transfers of passengers from incoming station to outgoing station in a trip. Comfort degree is X_{k3}^{rs}, which is divided into five levels. In the Chengdu Metro survey, passengers were asked to rate on a five-point scale how the comfort of travel.

3 Algorithm

When estimating parameters, the method of combining Markov chain Monte Carlo algorithm with data expansion algorithm is adopted. Let the parameter to be estimated be $\Phi = \{\alpha, \mu, \sigma^2\}$, where α is a random parameter and (μ, σ^2) is the mean and variance of the random parameter.

Step 1: Random generation of parameter Φ_l. When $l = 0$, $\Phi_l = 0$; when $l \geq 1$, the random parameter $\Phi_l = \{\alpha_l, \mu_l, \sigma_l^2\}$ is obtained based on $\Phi_l \sim N(\Phi_{l-1}, \xi^2)$ (where ξ^2 sets constants according to the range of random parameters).

Step 2: Calculate the probability P_{nk}^{rs} of passenger n choosing route k. The route choice probability P_{nk}^{rs} is calculated according to Formula 12, where $P_{nk}^{rs}(\alpha)$ is calculated by Formula 12. At the same time, the probability $P(\alpha_l|(\mu_l, \sigma_l^2))$ of α_l is obtained by calculating the distribution parameters of characteristic variables in Φ_l.

Step 3: Update the likelihood function. The likelihood function is changed to $L(\Phi_l)_l = \prod_{n=1}^{N} (P_{nk}^{rs})^{\delta_{nk}^{rs}} P(\alpha_l|(\mu_l, \sigma_l^2))$. In the formula, N is the total number of samples. When passenger n chooses route k in OD pair, δ_{nk}^{rs} is 1, otherwise 0.

Step 4: Let $\lambda = \ln(L_l) - \ln(L_{l-1})$ and make the following judgment: when $\lambda \geq 0$, $\Phi_l = \Phi_l$, otherwise $\Phi_l = \Phi_{l-1}, L_l = L_{l-1}$.

Step 5: Make $l = l + 1$, repeat the above process until Q-times are repeated and stop. Remove the previous Q_l-times data and then get the final parameter estimate $\hat{\Phi} = \frac{(\sum_{l=Q_l+1}^{Q} \Phi_l)}{Q - Q_l}$.

Parameter estimation is mainly based on survey data. Survey data includes passengers' tolerance thresholds for various influencing factors and AFC data of the subway company, etc. Passengers' preferences and regret parameters are estimated by investigating passenger route choice. Algorithm implementation mainly used MATLAB.

4 Model Comparison and Application

Based on the RP data, the maximum likelihood estimation method is used to calibrate the constrained multinomial logit model (CMNL), and the mixed route choice model of regret minimization and utility minimization is calibrated by Markov chain Monte Carlo algorithm and data expansion algorithm. The calibration results are shown in Table 1.

Analyzing the parameters in the two tables above, the absolute value of t value is greater than 1.96, which meets the requirements of statistical significance test, ρ^2 is greater than 0.2. It shows that both models have better fitting effect, and the ρ^2 of the mixed model is larger than that of the CMNL model, which proves that the fitting effect and prediction effect of the mixed model are better.

In the mixed model, the negative value of travel time coefficients obeys lognormal distribution, and its expected value can be expressed as $-\exp(\mu_1 + \frac{\sigma_1^2}{2})$. The calculated result is -4.8155, which is close to the estimated value of travel time coefficients in CMNL model. There is no comparative item in the regret coefficient of each factor, but according to the fitting effect, it shows that the regret factor can explain the part of the utility that cannot be covered, so that the fitting degree of the model is higher, and it is more in line with the passengers' route choice behavior in

Table 1 Estimation results

Variable	CMNL	Mixed route choice model	
	Distribution parameter value (t value)	Parametric distribution form	Distribution parameter value (t value)
Travel time/h	-5.2269 (-27.172)	Lognormal distribution $-\alpha_1 \sim \log N(\mu_1, \sigma_1)$	$\mu_1 = 1.5496$ (17.287) $\sigma_1 = 0.2109$ (12.112)
Transfer times/times	-1.4773 (-23.295)	Fixed value	-1.354 (-12.833)
Comfort degree	0.0050 (5.798)	Fixed value	0.0056 (3.987)
Travel time regret coefficient	–	Lognormal distribution $\alpha_4 \sim \log N(\mu_4, \sigma_4)$	$\mu_4 = 0.2198$ (11.206) $\sigma_4 = 0.0392$ (7.536)
Transfer times regret coefficient	–	Lognormal distribution $\alpha_5 \sim \log N(\mu_5, \sigma_5)$	$\mu_5 = 1.0196$ (8.317) $\sigma_5 = 0.1879$ (5.502)
Comfort degree regret coefficient	–	Lognormal distribution $\alpha_6 \sim \log N(\mu_6, \sigma_6)$	$\mu_6 = 0.0196$ (10.397) $\sigma_6 = 0.0031$ (3.929)
CPS	–	Fixed value	0.238 (4.912)
ρ^2	0.4386	0.6719	
Sample size	12,807	12,807	

urban rail transit. In summary, the mixed model is better than the CMNL model, and the prediction is more accurate.

Based on the 15-min OD statistics of Chengdu Metro on December 27, 2017, and the effective route set generated, stochastic network passenger flow loading was carried out to calculate the all-day transfer volume of different transfer directions of each transfer station. The results obtained are compared with those provided by Chengdu Metro, as shown in Fig. 3.

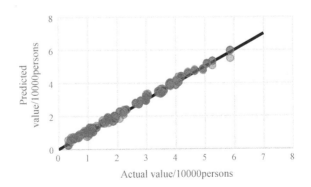

Fig. 3 Predictive check-up diagram of all-day transfer volume

As shown in Fig. 3, the vertical axis is the predicted value, the horizontal axis is the check value (ticket clearing value provided by Chengdu Metro), the red dots represent the mixed model, and the blue dots represent the CMNL model. It is found that the red dots are closer to the 45° line than the blue dots, indicating that the mixed model had less error than the CMNL model. Through calculation, the relative error of the mixed model was 4.32%, and the relative error of the CMNL model was 5.16%. It is indicated that the mixed model had better fitting effect and prediction effect than the CMNL model.

5 Conclusion

The route choice process includes two processes: effective route set generation and route choice. This paper proposed a mixed model of regret minimization and utility maximization. It considered that passengers will take regret and utility into account when choosing routes. The cost function is divided into compensation item and non-compensation item, which further reflects the heterogeneity of passengers. Based on RP survey data, the model was calibrated and the following conclusions were drawn: The mixed route choice model is superior to CMNL model in fitting effect and forecasting effect. Regret coefficient is logarithmic normal distribution, which basically reflects the regret situation that passengers feel after comparing the optimal route. The thresholds of travel time and transfer times obeyed lognormal distribution, which were more similar to the survey results, and confirms the correctness of the calibration results.

This paper discusses the application effect of the model. In view of the opening of the new Chengdu Metro Line 10, the transfer volume of each transfer station is predicted and compared. It is proved that the mixed model can be well applied to the passenger flow prediction of urban rail transit.

Acknowledgements This work was supported by the Fundamental Research Funds for the Central Universities under Grant 2017JBZ106.

References

1. Morgenstern O, Von Neumann J (1953) Theory of games and economic behavior. Princeton University Press
2. Mc Fadden D (1974) Conditional logit analysis of qualitative choice behavior. Front Econom (1):105–142
3. Bell DE (1982) Regret in decision making under uncertainty. Oper Res 30
4. Loomes G, Sugden R (1982) Regret theory: an alternative theory of rational choice under uncertainty. Econ J 92(368):805–824
5. Chorus CG, Arentze TA, Timmermans HJP (2008) A random regret-minimization model of travel choice. Transp Res Part B 42(1):0–18

6. Chorus C, Rose JM, Hensher DA (2013) Hybrid models of random utility maximization and random regret minimization: results from two empirical studies. Transportation Research Board Meeting
7. Boeri M, Scarpa R, Chorus CG (2014) Stated choices and benefit estimates in the context of traffic calming schemes: Utility maximization, regret minimization, or both? Transp Res Part A Policy Pract 61:121–135
8. Mc Fadden D (1975) The revealed preferences of a Government Bureaucracy: theory. Bell J Econ 6(2):401–416
9. Martínez F, Aguila F, Hurtubia R (2009) The constrained multinomial logit: a semi-compensatory choice model. Transp Res Part B Methodol 43(3):0–377

Urban Rail Transit Passenger Flow Monitoring Method Based on Call Detail Record Data

Lejing Zheng, Liguang Su and Honghui Dong

Abstract The interior of urban rail transit system has always been the blind area of traditional passenger flow monitoring methods. The call detail record (CDR) data is a kind of data containing passenger travel information with the advantages of wide coverage and no dead angle. It can well complement the shortcoming that traditional passenger flow monitoring method has blind areas. This paper proposes an approach of urban rail passenger flow monitoring based on CDR data which can directly monitor passenger flow in metro station and passenger flow on different routes. It makes up the defect that previous research can only estimate passenger flow by models. Experiment with actual data, the one-week passenger flow of Beijing Metro Line 2 is monitored. It shows that the characteristics of passenger flow are closely related to the nature of land use and passengers are more likely to be affected by transfer times when transferring. The experimental results match reality and verify the feasibility of the algorithm.

Keyword CDR data · Passenger flow monitoring · Characteristic analysis · Transfer route selection

1 Introduction

Urban rail transit has unique advantages such as high efficiency, energy saving, safety, and punctuality, so it has gradually become the first choice for people to travel and commute. With the growth of passenger flow, the situation of operation, organization, and guarantee is increasingly severe. As a basic factor in the traffic field, the distribution and status of passenger flow have important guiding significance for the operation and organization.

L. Zheng · L. Su · H. Dong (✉)
State Key Laboratory of Rail Traffic Control and Safety, Beijing Jiaotong University, Shang Yuan Cun, Hai Dian District No. 3, Beijing, China
e-mail: hhdong@bjtu.edu.cn

At present, the passenger flow studies of urban rail transit are mainly based on AFC data which has the advantages of complete statistical, clear direction of passenger flow and is suitable for the research work of passenger flow distribution and prediction. This kind of data is collected when passengers swipe their cards, only including the location information and time stamp of entry and exit. Scholars have conducted in-depth research on passenger flow assignment based on AFC data and continuously optimized and improved models [1–4]. Although the accuracy of estimation improves a lot, there is still a big gap with the actual situation due to the limitations of the data itself.

With the development of modern technologies, mobile phone detection technology has gradually matured, and the quality of CDR data has continuously improved. Scholars found that the data characteristics are very suitable for the transportation field. They have successfully obtained information from CDR data such as travel speed [5, 6] and travel time [7] and carried out the analysis of passenger flow characteristics [8–10]. It is easy to find that CDR data can collect the traffic information during the whole journey of passengers, which can make up for the limitation of AFC data.

Based on the limitation of passenger flow research based on AFC data and the advantage of CDR data, this paper proposes a passenger flow monitoring approach based on CDR data. The rest of the paper is organized as follows: Sect. 2 briefly introduces CDR data; Sect. 3 describes the algorithm in detail; in Sect. 4, it performs an experiment and analyzes the results. Finally, the main results of this paper are summarized.

2 Call Detail Record Data

The CDR data is the location data that is generated by the communication interaction between base stations and mobile devices carried by mobile users. It can reflect the users' movement in the mobile communication network. The general format of CDR data is shown in Table 1. IMSI is a unique identification code associated with the network for each user; LAC and CI can identify the base station where the user is located; TIME_IN and TIME_OUT represent the time when the user enters and leaves the coverage of the base station.

Table 1 CDR data format

IMSI	LAC	CI	TIME_IN	TIME_OUT
053e29ed956f3c2b21d266545a84e2cf	4259	11,033	20160928090000	20160928090000
ed572af4b2b6dea85e031a75197e114f	4233	10,941	20160928090000	20160928090000
f4c483c3272bb22ee620a0d796b34f67	4589	51,778	20160928090112	20160928090112

In addition, the traffic research also needs the base station attribute data, which includes LAC, CI, longitude, latitude, station name, and so on. The attribute data correlates to the real position of base station, so the movement under the communication network can correspond to the real traffic network. For the sake of confidentiality, this paper does not show the examples of base station attribute data.

Urban rail transit system has a special communication coverage network independent of the ground. It is located in the strong signal area covered by mobile signals all day and there is no signal blind zone, so it can completely detect the mobile phone user groups in all stations. Besides, China's mobile phone penetration rate exceeds 96/100 people, and the use frequency is high which invisibly increases the density and quantity of CDR data. Therefore, it is reliable to use the mobile phone user group monitored by CDR data to represent the passenger group.

The accuracy of position monitoring is determined by the coverage of the base station. Most of the rail transit systems are located in densely populated areas. The density of base stations is large, and the coverage radius of base stations in urban areas is generally less than 300 m. In some special areas, the coverage radius of base stations is below 100 m. Passenger flow allocation is relatively macroscopic traffic parameter. The existing monitoring accuracy fully meets the monitoring requirements and can ensure the accuracy of the data. Based on the integrity of the monitored object and the accuracy of the location monitoring, the CDR data is very suitable for the monitoring of rail transit passenger flow.

3 Approach

The passenger flow monitoring approach proposed in this paper is divided into two parts: (1) passenger flow monitoring in the subway station, monitor the total passenger flow in the subway station according to the base station identification in the CDR data, extract the rate of entry and exit of each station; (2) transfer behavior monitoring, track single-passenger travel trajectories, and monitor the distribution of passenger flow in different transfer paths.

3.1 Metro Station Passenger Flow Monitoring

Passenger flow monitoring method: In this paper, taking 10 min as the minimum analysis period, the passengers monitored by the base stations are taken as the passenger of the subway station including passengers in the station and passengers in the train. In combination with the previous analysis, the specific monitoring method for passengers in subway stations is as follows:

$$C_i = \frac{\sum_{j=1}^{n} C_i^j}{\alpha} \qquad (1)$$

$$C_i^j = \text{count}\left\{\text{pos}_p | \text{pos}_p = \text{pos}_i^j\right\} \tag{2}$$

where c_i is the passenger flow in the ith subway station; c_i^j is the passenger flow monitored by the jth base station within the ith subway station; n is the base station number of the ith subway station; α is the proportion of monitored passengers in actual passengers; pos_i^j is position information of the jth base station in the ith subway station; and pos_p is the base station position information in user p's CDR data.

Since the passenger flow is monitored on a 10-min basis. If a passenger has not got off the train, it may pass through three stations in 10 min. Assumed that the passenger always has communication interaction with the base station, it will produce at least three pieces of data which causes the monitored results containing redundant data. In order to avoid the noise interference caused by passengers' duplicate position information, this paper improves the monitoring method. For passengers who have multiple occurrences of position information within a 10-min period, only the last appearing position can be saved. The removal of redundant data greatly improves the accuracy of the algorithm.

Inbound and outbound traffic rate monitoring algorithm: Due to the huge amount of CDR data and privacy requirements, it is difficult to continuously monitor users' position and perform statistical work. This paper uses a sampling method and sets the time window to monitor the inbound and outbound traffic rate at the sampling time point. Taking the inbound rate of station A as an example, the monitoring method flowchart is as shown in Fig. 1.

The algorithm is described in detail as follows:

Step 1: Set the parameters. Set the time window length to 1 min, set the sampling period T to 10 min, set the initial sampling time point to t, establish two sets O_j ($j = 1, 2, \ldots, n$) and DA, and the initial sets are empty.

Step 2: According to the time axis, when the time reaches $t - 1$, start to continuously monitor the n base stations adjacent to the subway station A, and put the mobile users within the coverage of these base stations into the set O_j.

Step 3: According to the time axis, when the time reaches the sampling time point t, the accumulation of the set O_j is stopped, and the base stations in the range of the subway station A start to be continuously monitored. Update the set DA, put the mobile users within the coverage of these base stations into the set DA.

Step 4: According to the time axis, when the time reaches $t + 1$ min, the updating of the set DA is stopped, and then filter the elements in the DA. If the user in the set DA appeared in the set O_j, keep the data; Otherwise, delete the data. Calculate and output the size of the set DA that is the inbound rate of station A at the sampling time point t.

Step 5: Clear the collection DA and O_j, update the sampling point to $t + T$, return to Step 2, calculate the next sampling period.

Since the statistical process does not take the inbound passenger flow beyond the sampling period into account, there is a difference between the monitoring results and the actual passenger flow rate, but the trend is the same. In the same way, the method of monitoring the outbound traffic rate is similar.

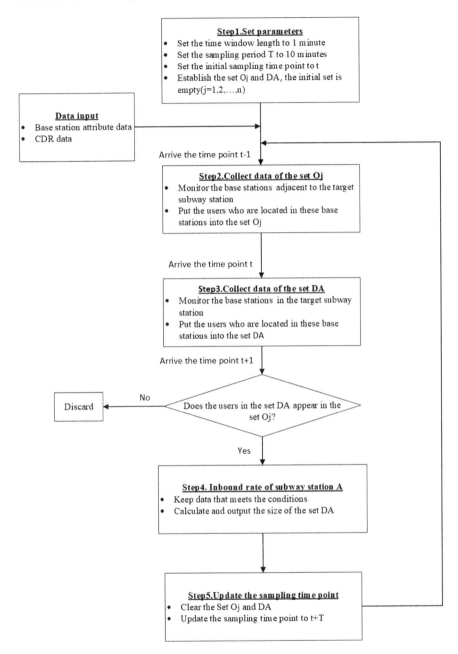

Fig. 1 Inbound traffic rate monitoring algorithm flowchart

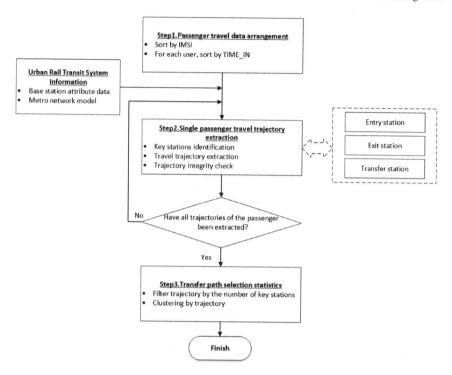

Fig. 2 Passenger flow transfer monitoring algorithm flowchart

3.2 Passenger Flow Transfer Monitoring

The passenger flow transfer monitoring algorithm solves the problem of passenger transfer path selection behavior and passenger flow transfer path allocation. The core of this algorithm is the passenger travel trajectory extraction. The algorithm flowchart is shown in Fig. 2.

The CDR data is defined as R_i (I, L_i, C_i, RN_i, SN_i, T_i), where I represents IMSI; L_i, C_i represents LAC and CI; RN_i, SN_i represents the metro line number and the name of the station, set the value to null if it is not a metro base station; and T_i represents the time when passengers enter the service range of the base station. Passengers travel trajectory is defined as P_j (I, S_1, S_2, S_3, S_4, S_5), where S_1, S_2, S_3, S_4, S_5 represents the key stations during passengers' travel, only records the entry, transfer, and exit stations. This paper assumes that passengers' metro travel generally does not exceed three transfers.

Entry station identification: Iterative judge to single-passenger CDR data sorted by time. When R_i appears and satisfies $R_i(RN_i, SN_i) \neq$ null, $R_{i-1}(RN_{i-1}, SN_{i-1})$ = null, this passenger is considered having entrance behavior. The corresponding station of R_i is an entry station, so that $P_j(I) = R_j(I)$, $P_j(S_1) = R_j(SN_i)$.

Transfer station identification: When R_i appears and satisfies $R_i(RN_i, SN_i) \neq$ null, $R_{i-1}(RN_{i-1}, SN_{i-1}) \neq$ null, $R_i(RN_i, SN_i) \neq R_{i-1}(RN_{i-1}, SN_{i-1})$, this passenger is considered having transfer behavior and the corresponding station of R_i is a transfer station.

Verify the corresponding station of R_i, judging whether the station is a real transfer station according to the metro network model. If it is a real transfer station, the corresponding station of R_i is considered to be an effective transfer station, so that $P_j(S_2) = R_i(RN_i)$. If it is not a transfer station, it is considered to be invalid and needs to be corrected. Search for the transfer station with the shortest distance to the corresponding station of R_i on the line which the last key station belongs to and assign value to $P_j(S_2)$. Continue to judge iteratively, assign the following transfer stations to $P_j(S_3, S_4)$. If there are no subsequent transfer stations, record them as null value.

Exit station identification: When R_i appears and satisfies $R_i(RN_i, SN_i) \neq$ null, $R_{i+1}(RN_{i+1}, SN_{i+1}) =$ null, this passenger is considered having exit behavior. The corresponding station of R_i is an exit station, so that $P_j(S_5) = R_j(SN_i)$. Let $j = j + 1$, continue the iterative analysis until all the data is analyzed.

Path Check: Determine whether adjacent track points belong to the same metro line, and if the conditions are satisfied, the path is considered complete. If they do not belong to the same line, take these two points as the endpoints, use the Dijkstra algorithm to find the shortest path, then add the finding transfer stations to the passenger travel trajectory. Redetermine whether the adjacent track points meet the conditions, and repeat the steps until the path is complete.

Passenger flow transfer monitoring: If a passenger's travel trajectory contains 3 or more trajectory points, it indicates that this passenger has transfer behavior. Clustering by trajectory, different paths, and the passenger flow distribution can be obtained.

4 Experiment

This paper takes Beijing Metro Line 2 as the research object and uses the CDR data of mobile phone users who have traveled on the target line from September 28, 2016, to October 4, 2016, as the experimental data. According to the above algorithm, the instantaneous passenger flow, inbound, and outbound traffic rate of each station are obtained, and the passenger transfer behavior and passenger flow assignment of transfer paths are monitored. Some experimental results are shown in Tables 2, 3, and 4.

The experimental results can be analyzed in many ways.

Based on the experimental results of passenger flow in stations, this paper analyzes the spatial and temporal distribution characteristics of passenger flow.

Figure 3 shows the distribution of passenger traffic for all stations on Line 2 for one week. It can be seen from the figure that different passenger flow characteristics

Table 2 Passenger flow in station

TIME	DSST	DZM	YHG	ADM	GLDJ	JST	XZM	CGZ	FCM	……
201609280800	917	2091	1963	903	1979	1832	3542	1720	1593	……
201609280810	1087	2385	2180	1110	2238	2233	4067	2018	1810	……
201609280820	1352	2600	2424	1205	2352	2683	4551	2360	2174	……
201609280830	1592	2705	2531	1191	2298	2684	4725	2377	2353	……
201609280840	1744	2961	2519	1203	2293	2501	4618	2338	2398	……
……	……	……	……	……	……	……	……	……	……	……

Table 3 Inbound and outbound passenger flow

TIME	HPM		QM		CWM		BJZ		.....
	IN	OUT	IN	OUT	IN	OUT	IN	OUT	
20160928080	113	58	307	140	381	195	222	127	.....
201609280810	161	58	244	145	389	206	191	138	.....
201609280820	143	87	234	169	413	235	209	159	.....
201609280830	137	77	290	206	426	286	218	175	.....
201609280840	106	90	299	205	354	255	204	209	.....
.....	.....	.....	.....	.....	.....	.....	.....	.....	.....

Table 4 Passenger transfer distribution

OD	Route	Percentage
BJNZ–BJZ	BJNZ-XWM-BJZ	86.5
	BJNZ-CSK-CQK-CWM-BJZ	0.4
	BJNZ-PHY-CWM-BJZ	7.3
JSBWG–BJZ	JSBWG-FXM-BJZ	59.2
	JSBWG-XD-XWM-BJZ	6.5
	JSBWG-DD-CWM-BJZ	4.5
	JSBWG-JGM-BJZ	28.5
……	……	……

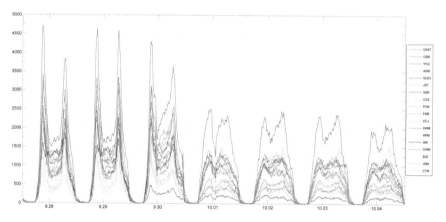

Fig. 3 Spatial and temporal distribution of weekly passenger flow

are presented according to the different land use of the station. Qianmen (QM) Station, as a tourist attraction station, has a relatively balanced passenger flow during the opening hours of the working day. During the National Day, due to the restriction of tourists in the Tiananmen area, the vehicles at the QM Station will not stop, so the passenger flow throughout the day is at a very low level. As a connecting hub for intercity traffic and urban traffic, Beijing (BJZ) Station's passenger flow has a jagged fluctuation. It is closely related to the train station arrival schedule and basically unaffected by residents' commuting. The remaining stations have similar characteristics: The passenger flow on working day has significant commuting characteristics with morning and evening peak; the travel demand of the residents is less in holiday, and the fluctuation throughout the day is relatively stable.

According to the experimental results of inbound and outbound traffic, the occurrence and attraction characteristics of different sites are analyzed.

Figure 4 shows the daily inbound and outbound traffic at QM Station, XZM Station, and BJZ Station. The blue line represents XZM Station, the red line represents QM Station, and the green line represents BJZ Station. The solid line indicates the inbound passenger flow, and the dotted line indicates the outbound passenger flow.

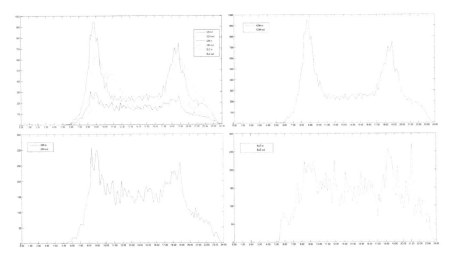

Fig. 4 Daily inbound and outbound traffic at three stations

As is shown, the occurrence and attraction characteristics of XZM Station, QM Station, and BJZ Station are gradually weakened according to the different land use of the corresponding regions. XZM Station, as a commercial center and a large hub station for the transfer of three lines, is active in the region. Its passenger flow volume is much larger than the others. According to the passenger flow situation at the XZM Station, the value of the inbound passenger flow is greater than the outbound passenger flow before the arrival of the morning peak, which reflects the occurrence characteristics of the XZM Station are stronger than the attraction characteristics during the morning peak period. It remains balanced during the mean hours. Then, the passenger attraction characteristics of the XZM Station become stronger than the occurrence characteristics during the evening peak period. QM Station is a site connecting tourist attractions. In the morning, the attraction characteristics of passenger flow are obviously stronger than the characteristics of occurrence when the occurrence characteristics of passenger flow are stronger than the attraction characteristics in the evening. It reflects the travel mode that passengers enter the scenic spot during the day and leave the scenic spot at night. The passenger flow at BJZ Station is mainly affected by the large railway. Therefore, the whole-day passenger flow situation is determined by the arrival and departure trains, and the occurrence characteristics are roughly the same as the attraction characteristics.

The experimental results of passenger flow assignment on transfer paths show that: From the perspective of overall situation, 91.97% of passengers transfer once, 7.96% of passengers transfer twice and 0.07% of passengers transfer three times. This shows that Beijing Metro Network has good connectivity and most passengers can achieve their travel only by one transfer. Taking the two pairs of ODs of BJNZ Station–BJZ Station, JSBWG Station–BJZ Station as an example, the passenger route selection analysis is carried out. The distribution of passenger flow on different paths is shown in Table 4. It can be seen that passengers have obvious preferences in path selection. They are more sensitive to the transfer times than travel distances and prefer to select paths with fewer transfer times.

5 Conclusion

The independence of the rail transit communication network and the data characteristics of CDR data make it possible to apply the data to solve internal area perception problem and monitor passenger flow. In addition, the integrality of the monitored object and the accuracy of the monitoring position ensure the accuracy of passenger flow monitoring. Based on this, this paper has done the following work:

(1) A method of passenger flow monitoring in station is proposed. According to the change of time stamp and position information in CDR data, the instantaneous passenger flow and the inbound and outbound traffic rate at stations are extracted.
(2) An algorithm of passenger flow transfer monitoring is proposed. Taking passenger trajectory extraction as the key, passenger flow path allocation and passenger transfer path selection behavior are counted.
(3) Conduct experiment. It is shown that regional land use has a greater impact on the traffic characteristics. Most subway stations have remarkable commuting characteristics with the morning and evening peak in working days, and the passenger flow is relatively stable during holidays. Passengers are more sensitive to transfer times when choosing transfer routes.

Future research will focus on how to extract traffic information from real-time data and achieve real-time monitoring of urban rail transit passenger flow. In addition, the analysis of metro passenger travel pattern can also be carried out in depth.

Acknowledgements This work was supported by Science and Technology Support Plan (Grant No.2016YFB1200104).

References

1. Su J (2009) Research on flow assignment of urban rail transit, master dissertation. Beijing Jiaotong University, Beijing (in Chinese)
2. Liu JF (2012) Transfer-based modeling flow assignment with empirical analysis for urban rail transit network [Ph.D.]. Beijing Jiaotong University, Beijing (in Chinese)
3. Yan Y (2015) Traffic assignment problem of urban railway transit based on seamless transfer. Master Dissertation. Southwest Jiaotong University, Sichuan (in Chinese)
4. Cai XC (2011) Research on passenger flow assignment model and algorithm of urban rail transit. Master Dissertation. Beijing Jiaotong University, Beijing (in Chinese)
5. Smith BL, Pack ML, Lovell DJ, Sermons MW (2001) Transportation management applications of anonymous mobile call sampling. In: ITS America 11th annual meeting and exposition, ITS: Connecting the Americas, Florida, America

6. Wilson E, Whittaker RJ (2005) Real-time traffic monitoring using mobile phone data. Mol Cell Biol 13(3):1315–1322
7. Gundlegard D, Karlsson JM (2006) Generating road traffic information from cellular networks—new possibilities in UMTS. In: International conference on ITS telecommunications IEEE, pp 1128–1133
8. Hu YK (2017) Urban rail transit passenger travel behavior analysis methods based on cellular data. Ph.D, Southeast University, Jiangsu (in Chinese)
9. Guan ZC, Hu B, Zhang X, Qiu WY (2012) Research on urban transport planning, de-signing, construction, management decision-making support based on mobile data. In: China intelligent transportation annual conference
10. Lai JH (2014) Research on data mining and analysis in transportation based on mobile communication location, Ph.D. Beijing University of Technology, Beijing (in Chinese)

Research on Dynamic Route Generation Algorithms for Passenger Evacuation in Metro Station

Weiming Dong and Zhengyu Xie

Abstract The subway station has dense passenger flow and complicated facilities. The traditional path planning model only takes the geometric shortest path as the optimal path, and rarely considers the influence of the real-time density of passenger flow and the difficulty of path access on the selection of passenger evacuation path. The application is in the actual safe evacuation path planning, which is not conducive to rapid evacuation. During the movement of passengers inside the metro station, the popularity of the passengers has a certain similarity with the ant colony system. This paper aims to study the improved ant colony algorithm to generate the passenger flow dynamic evacuation route of the subway station.

Keywords Safe evacuation · Real-time passenger flow · Ant colony algorithm · Dynamic evacuate

1 Introduction

At present, the main problem in the study of evacuation routes for subway stations gathered by passenger flow is that the evacuation routes generated for passenger flow congestion are pre-set emergency plans, which are static and single. Swarm intelligence algorithm is widely used in evacuation route research. Pelechano N [1] proposed a multi-grid evacuation model; Nagatani [2] proposed a mean-field model. But these algorithms are not ideal in providing path planning for global optimization. Therefore, a path planning model based on ant colony algorithm is proposed to find the optimal path of safe evacuation, and fully consider the different information of the passenger flow state in the evacuation process [3–6]. According to the real-time passenger flow density state, a real-time evacuation path is generated.

W. Dong · Z. Xie (✉)
School of Transport and Transportation, Beijing Jiaotong University, No.3 Shangyuancun, Haidian District, Beijing, People's Republic of China
e-mail: xiezhengyu@bjtu.edu.cn

2 Basic Principle of Algorithm

Studies have shown that ants tend to choose foraging routes with high pheromone concentrations. In the same time, shorter paths will accumulate more pheromones, and high concentrations of pheromones will attract a large number of ants to choose this route to guide ants. Groups look for the shortest path for foraging.

In the path planning problem based on ant colony algorithm, the following settings need to be made: m is number of evacuated personnel; d_{ij} is the distance between nodes i and j; $\tau_{ij}(t)$ is pheromone function; $\eta_{ij}(t)$ is heuristic function; $Tabu_k$ ($k = 1, 2, 3, \ldots, m$) is taboo table; $p_{ij}^k(t)$ is path selection strategy, and the formula is:

$$p_{ij}^k(t) = \begin{cases} \dfrac{\tau_{ij}^\alpha(t)\eta_{ij}^\beta(t)}{\sum_{x \in Allowed_k} \tau_{ij}^\alpha(t)\eta_{ij}^\beta(t)}, & \text{if } j \in Allowed_k \\ 0, & \text{otherwise} \end{cases} \quad (1.1)$$

where $Allowed_k$ represents the set of nodes currently selected by evacuation personnel k; α is the pheromone heuristic factor; β is the expectation heuristic factor.

After the evacuation personnel complete each iteration, the pheromone concentration remaining on the path is updated. The pheromone update formula is as follows:

$$\tau_{ij}(t+n) = (1-\rho) \times \tau_{ij}(t) + \Delta\tau_{ij}(t) \quad (1.2)$$

$$\Delta\tau_{ij}(t) = \sum_{k=1}^{m} \Delta\tau_{ij}^k(t) \quad (1.3)$$

$$\Delta\tau_{ij}^k(t) = \begin{cases} Q/L_{ij}, & \text{If the } k\text{th ant passes the path } (i, j) \text{ in this loop} \\ 0, & \text{otherwise} \end{cases} \quad (1.4)$$

where

ρ—Pheromone volatility;
Q—Pheromone intensity.

2.1 Algorithm Improvement

In the classic application of ant colony algorithm for path planning, the heuristic function $\eta_{ij}(t) = 1/L_{ij}$ (L_{ij} indicates the geometric distance between nodes i and j) indicates that when the evacuation personnel is transferred from the current node to the next node, the shorter the geometric distance, the greater the probability of being selected.

However, in the actual dynamic evacuation process, since the evacuation efficiency of pedestrians is affected by environmental factors and passenger flow density factors, the setting of the heuristic function depends not only on the geometric distance between the two nodes, but also on the path between the two nodes. Factors such as difficulty level and passenger flow density. Therefore, we introduce the path access difficulty λ_{ij}, passenger flow density influence factor $f_{ij}(\theta)$, equivalent length D_{ij}.

1. Path access difficulty λ_{ij}

In the dynamic evacuation process of subway stations, the speed of evacuation personnel will be affected by the ease of evacuation of passages. So we introduce λ_{ij} to indicate the difficulty of path travel and the correction of the length of the geometric path. The formula is:

$$\lambda_{ij} = \frac{v_0}{v_t'} = \begin{cases} 1, & \text{Straight paths such as station hall, platform} \\ 1.4, & \text{Complex paths such as stairs and passages} \end{cases} \quad (1.5)$$

2. Passenger flow density influence factor $f_{ij}(\theta)$

According to Nelson H E's research in the SFPE Fire Engineer's Handbook [7], we obtain the coefficient of influence of personnel density on passenger travel speed $M_{ij}(\theta)$. In order to more intuitively reflect the influence of passenger flow density on the optimal evacuation path generation, the reciprocal of $M_{ij}(\theta)$ is taken here to indicate the correction of the geometric path length by the passenger flow density. The formula is:

$$f_{ij}(\theta) = \begin{cases} 1, & \theta \leq 1.87 \\ \frac{1.2}{1.4\theta - 0.3724\theta}, & 1.87 < \theta < 3.75 \\ \infty, & \theta \geq 3.75 \end{cases} \quad (1.6)$$

3. Equivalent length D_{ij}

The equivalent length is a correction to the geometric length, which reflects the influence of various factors on the evacuation path during the evacuation process, including the real-time passenger flow density and the difficulty of different road conditions. The formula is:

$$D_{ij} = L_{ij} \times \lambda_{ij} \times f_{ij}(\theta) \quad (1.7)$$

where

D_{ij}—Path equivalent length;
L_{ij}—Path geometry length;
λ_{ij}—Path access difficulty;
$f_{ij}(\theta)$—Passenger flow density influence factor.

2.2 Algorithm Implementation Flow

According to the mathematical model of the dynamic evacuation ant colony algorithm, the flowchart of the evacuation model is shown in Fig. 1.

3 Subway Evacuation Simulation Experiment

The environment is initialized according to the characteristics of the subway site environment, and a raster map of the evacuation environment is obtained. According to the difficulty of the path, the equivalent length of each outlet can be obtained. Then, by setting the parameters of the ant colony algorithm, combined with the passenger flow density information of the subway station from 18:00 to 18:20 on a certain day, the real-time evacuation path at different moments is generated.

3.1 Environment Initialization

Create a 20 * 20 raster map to simulate an evaculation environment simulation map (Fig. 2).

3.2 Algorithm Simulation

In the MATLAB simulation environment, the parameters are set as follows: $\alpha = 1$, $\beta = 7$, $\rho = 0.3$, $m = 100$, $Q = 1$, $K = 50$. In the process of evacuation, the evacuation rate of each evacuation environment is as shown in Table 1, and the density of each passenger flow at each time is as shown in Table 2.

According to the algorithm flow, the optimal evacuation route from the evacuation starting point to the exit at each moment is shown in Fig. 3.

3.3 Experimental Results

According to the five rounds of experiments, it can be seen that in the evacuation process, the evacuation route with the shortest geometric distance is not the optimal evacuation route under actual conditions. From the time change of 18:00 → 18:05 → 18:10 → 18:15 → 18:20, the passenger flow density changes at any time, so the optimal evacuation path changes continuously, from B → A→D → C→A. According to the experiment, it can be concluded that in the evacuation of passenger

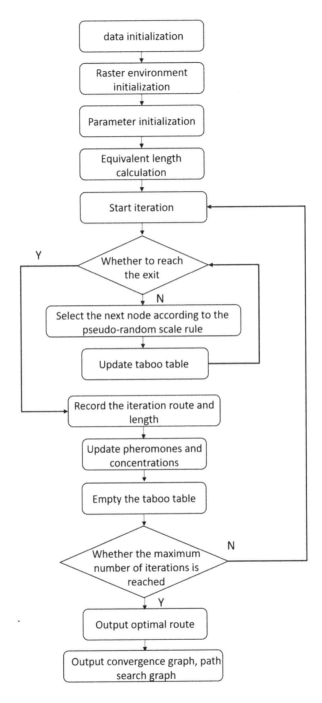

Fig. 1 Flowchart of the evacuation model

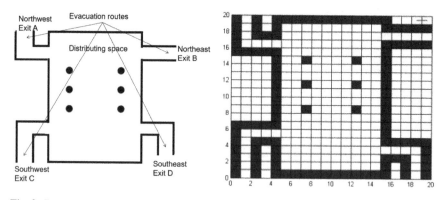

Fig. 2 Evacuation environment simulation map and raster map

Table 1 Evacuation speed

Evacuation environment	Evacuation speed (m/s)
Distributing space	1.2
Evacuation aisles	0.85

Table 2 Density of each passenger flow

Parameter	Numerical value				
m	100	100	100	100	100
T	18:00	18:05	18:10	18:15	18:20
θ_a	<1.87	2	2.8	3	2.6
θ_b	<1.87	3	3.2	3.8	3.6
θ_c	<1.87	<1.87	2	2.3	2.7
θ_d	<1.87	<1.87	2.2	2.7	2.9

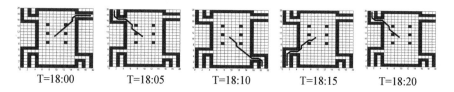

T=18:00   T=18:05   T=18:10   T=18:15   T=18:20

Fig. 3 Real-time optimal evacuation route

flow, real-time dynamic passenger flow data needs to be combined, and the optimal path is selected according to the passenger flow density in different regions.

3.4 Evaluation Index

The optimal evacuation route effect generated by the algorithm can be evaluated by the equivalent length of each route. By calculating the equivalent length of the evacuation route from the evacuation starting point to each exit at each moment, the optimal evacuation route results are the same as the experimental simulation results.

4 Case Analysis

Based on the passenger flow distribution of Yangji Station at a certain time, conduct a safe evacuation simulation of Line 5 of Yangji Station. Station hall, station passenger flow distribution and grid map are shown in Fig. 4.

4.1 Algorithm Simulation

The paper discusses the two cases separately and finds the shortest evacuation route of each exit under the passenger flow distribution in the MATLAB simulation environment. The passenger flow density of each node in the subway environment at this moment is shown in Table 3.

According to the algorithm, the following results are obtained as shown in Figs. 5 and 6.

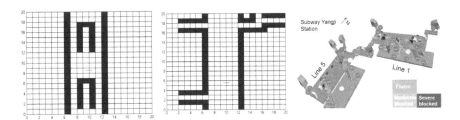

Fig. 4 No. 5 line platform, station hall raster map

Table 3 Passenger flow density at each node

Experimental variable	Numerical value	Experimental variable	Numerical value
m	100	θ_E	<1.87
θ_{F1}	3.2	θ_F	<1.87
θ_{F2}	<1.87	θ_{HC}	2.7
θ_A	<1.87		

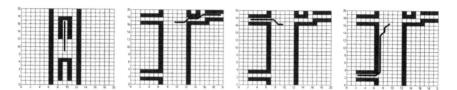

Fig. 5 Starting point is to evacuate the shortest route to each exit via the F1 stairs

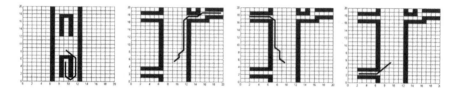

Fig. 6 Starting point is to evacuate the shortest route to each exit via the F2 stairs

Fig. 7 Equivalent length of each exit evacuation route

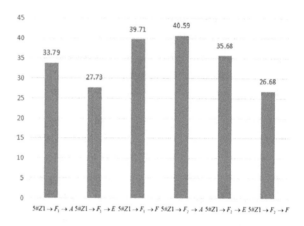

After comparison, as shown in Fig. 7, the shortest evacuation route from the starting point to each exit at this moment can be obtained.

4.2 Experimental Results

Through experimental analysis, it can be seen that under the passenger flow distribution at this moment, the optimal evacuation route of Line 5 of Yangji Station is not the shortest 5#Z1 → F1 → E, but 5#Z1 → F2 → F.

4.3 Evaluation Index

After monitoring the data in real time and using the evacuation route generated by the algorithm to carry out the evacuation work, the congestion of each node of the subway is effectively alleviated. In the subsequent evaluation of the dynamic evacuation route, the effectiveness of the algorithm can be verified by analyzing the evacuation of the evacuated passenger flow.

5 Conclusion

The subway station has dense passenger flow and complex building space structure. The safety evacuation of passengers is affected by many complicated factors such as the construction environment and personnel psychology. Based on the ant colony algorithm, the dynamic route generation of passenger flow evacuation in metro station is studied. Based on the classical path planning model, the real-time dynamic optimal evacuation is generated by introducing the heuristic function of the ant colony algorithm with improved path access difficulty and passenger flow density influence factor.

References

1. Pelechano N, O'Brien K, Silverman B et al (2013) Crowd simulation incorporating agent psychological models, roles and communication. In: First international workshop on crowd simulation, 21–30
2. Ghoseiri K, Nadjari B (2010) An ant colony optimization algorithm for the bi-objective shortest path problem. Appl Soft Comput J 10(4):1237–1246
3. Xu R, Ming D, Qiu S, Wang X, Qi H, Zhang L, Wan B (2013) Wireless walker dynamometer design and static calibration based on ant colony system. Wirel Sensor Syst IET 3(3):233–238
4. Jayaprakash A, KeziSelvaVijila C (2019) Feature selection using ant colony optimization (ACO) and road sign detection and recognition (RSDR) system. Cognit Syst Res
5. Wu G, Bo N, Wu H, Yang Y, Hassan N (2018) Fuzzy scheduling optimization system for multi-objective transportation path based on ant colony algorithm. J Intell Fuzzy Syst 35(4)
6. Liu M, Zhang F, Ma Y, Roy Pota H, Shen W (2016) Evacuation path optimization based on quantum ant colony algorithm. Adv Eng Inf 30(3)
7. Nelson HE, Mowrer FW (2002) Emergency movement. In: Di Nenno PJ, Walton WD (eds) The SFPE handbook of fire protection. Society of Fire Protection Engineers, Bethesda, MD 3-367-363-380

Single Image Dehazing of Railway Images via Multi-scale Residual Networks

Zhi wei Cao, Yong Qin and Zheng yu Xie

Abstract In this paper, an end-to-end single image defogging algorithm is proposed which can achieve significant defogging effect under railway conditions. We present a method of fusing multi-scale feature information based on the residual network, which can extract more effective information at different scales. The method uses multi-scale convolution kernels to obtain different scale feature information, but it will increase computational cost and reduce the network depth. By increasing the bottleneck layer, we can deepen the residual network and ensure that the network can achieve better dehazing results. The proposed method can achieve better results than the state-of-the-art algorithms based on synthetic datasets and railway images.

Keywords Dehaze · Railway images · Multi-scale · Residual networks

1 Introduction

In the foggy weather, it is necessary to perform image dehazing to improve the accuracy of algorithms such as image recognition. Currently, railway surveillance video data is rising exponentially. However, image quality is degraded due to fog and haze interference in bad weather. Both traditional algorithms and deep learning algorithms cannot achieve favorable results in image detection and recognition. It is an urgent problem to be solved for restoring a clear image from a hazy image. In the image defogging method, single image dehazing is more practical and difficult. The

---

Z. Cao · Y. Qin
State Key Laboratory of Rail Traffic Control and Safety, Beijing Jiaotong University, No. 3 Shangyuancun, Haidian District, Beijing, People's Republic of China
e-mail: caozhiwei@bjtu.edu.cn

Y. Qin
e-mail: yqin@bjtu.edu.cn

Z. Xie (✉)
School of Traffic and Transportation, Beijing Jiaotong University, No. 3 Shangyuancun, Haidian District, Beijing, People's Republic of China
e-mail: xiezhengyu@bjtu.edu.cn

purpose of a single image dehazing is to recover a clear image from a foggy image caused by fog and dust. The atmospheric scattering model can be formulated as

$$I(x) = J(x)t(x) + A(1 - t(x)) \tag{1}$$

where $I(x)$ and $J(x)$ represent the hazy image and the true scene radiance, A is the global atmospheric light, and $t(x)$ is the transmission map. When the haze is homogeneous, the transmission map is formulated as $t(x) = e^{-\beta d(x)}$ where β is the medium extinction coefficient, $d(x)$ is the scene depth, and x is the pixel in an image. In (1), only the hazy image $I(x)$ is given, and it is necessary to estimate the transmission map $t(x)$ and the global atmospheric light A for restoration of the scene radiance $J(x)$.

It is a classic ill-posed problem to estimate the transmission map and the global atmospheric light. Most scholars focus on improving the accurate estimation of transmission map and further estimating the global atmospheric light. However, if the accuracy of transmission map is poorly estimated, the accuracy of global atmospheric light estimation will decrease. In order to solve the above problems, we propose an end-to-end image dehazing neural network.

We add multi-scale convolution kernel on the basis of the residual network, which can extract more useful information by fusing the features of the three sets of convolution kernels. Multi-scale convolution fusion increases computational cost and reduces network depth. It is well known that the deeper the residual network, the better. We solve this problem by increasing the bottleneck layer which is to add a 1 × 1 convolution kernel to the neural network. It can reduce the computational cost and increase the depth of the network.

2 Related Work

This section presents the state of the art of the single image dehazing method and the application of the residual network in image processing.

2.1 Single Image Dehazing

There are two main solutions to the dehazing of single image: handcrafted prior-based methods and data-driven methods.

Handcrafted Prior-Based. Fattal [1] assumed that transmission and surface color are locally uncorrelated, thereby estimating the transmission in a single image fog map. Tan [2] proposed to improve the visibility of scenes in bad weather based on Markov random fields. In [3], it was found that at least one of the color channels in the outdoor clear image has a very low-intensity pixel based on the key observations, and a dark channel prior estimated transmission map is proposed. Fattal [4] proposed a color line-based method in which pixels of small image blocks typically exhibit a 1D distribution in the RGB color space. Similarly, Berman et al. [5] observed a haze line where each color cluster in a sharp image became a line in RGB space. They used these haze lines to restore the deep map and the clear image.

Data-driven Methods. Cai et al. [6] proposed a dehazenet based on CNN and designed a new nonlinear activation function, which achieved fine results. Ren et al. [7] proposed a multi-scale deep neural network for single image dehazing by learning the mapping between hazy images and their corresponding transmission maps. Different from the above method, [8] proposed an all-in-one dehazing network to generate clear images directly through a lightweight CNN network without estimating the transmission map and atmospheric light. Li et al. [9] presented a encoder and decoder architecture- based conditional generative adversarial network to estimate clear images through end-to-end trainable neurons network. Zhang et al. [10] proposed a densely connected pyramid dehazing network that maximizes the flow of information from different levels of features. Ren et al. [11] proposed a gated fusion network to solve the single image dehazing problem by learning the input confidence map.

2.2 Residual Network

In [12], He et al. proposed using a deep residual network to solve image recognition problems. The result of deep residual network won the first places on the tasks of ImageNet detection, ImageNet localization, COCO detection, and COCO segmentation. Since then, the residual network has been used in various fields of image processing and has achieved remarkable results. In addition, the networks such as the dense network and Mask R-CNN learn the idea of residual network and improve the effect of the network.

3 Proposed Method

In this section, we introduce the multi-scale residual network proposed. First, we introduce the overall architecture of the proposed network, then present the structure of the convolution kernels of three different scales, and finally fuse multi-scale information.

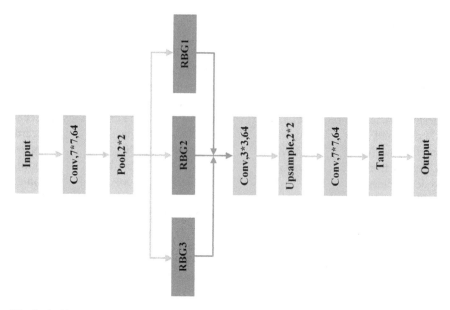

Fig. 1 Architecture of multi-scale residual networks

3.1 Multi-scale Residual Network

Our proposed multi-scale residual network consists of an encoder and a decoder as shown in Fig. 1. The encoder includes shallow feature extraction, residual block groups (RBG), and multi-scale fusion.

We use convolution operations and average pooling to extract shallow features. The residual block group includes three sets of deep residual networks, and the features at different scales can be obtained by changing the size of the convolution kernel. We fuse features obtained at different scales and get more effective information in terms of multi-scale fusion. The decoding process consists mainly of upsampling operations and nonlinear spatial transfer. The parameters of the network architecture are shown in Table 1.

3.2 Residual Block Group

The residual block group includes three sets of deep residual networks: RBG1, RBG2, and RBG3, and the corresponding residual block convolution kernels have sizes of 1×1, 3×3, and 5×5.

The residual block group is shown in Fig. 2 with RBG2 as an example. Multi-scale convolution fusion will increase calculation to limit the network depth, but it is well known that the deeper the residual network, the better the result. We join the

Table 1 Parameters of multi-scale residual networks

Formulation	Type	N	Channels	Filter	Pad
Feature extraction	Convolution	1	3	7×7	3
	Average pooling	1	–	2×2	0
Multi-scale context aggregation	Convolution	1	64	1×1	0
	Convolution	24 24 24	64	1×1 3×3 5×5	0 1 2
	Convolution	1	192	3×3	0
Nonlinear regression	Upsampling	1	64	2×2	0
	Convolution	1	64	7×7	3
	Tanh	1	–	–	–

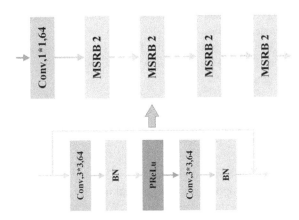

Fig. 2 Multi-scale residual block 2

bottleneck layer. It turns out that this operation greatly reduces computational cost and ensures that the multi-scale residual network can guarantee a deeper network. Each residual block includes two convolutional layers, two batch normalizations and one activation function.

3.3 Multi-scale Feature Fusion

We make full use of the convolutional features of RBG2 in multi-scale fusion. The 3×3 convolution kernel is widely used and can achieve better results. According to our experiments, RBG1 and RBG2 are fused, and the same is true for RBG3 and RBG2. The fusion results are convoluted and then fused with RBG2.

4 Experimental Results

In this section, we compare the proposed algorithm with the state-of-the-art algorithms based on an outdoor synthetic dataset and visually observe the dehazing results of real-world images.

4.1 Synthetic Dataset

We only consider outdoor image dehazing and synthesize an outdoor dataset. We randomly selected 800 training images and 100 testing images from Make3D datasets [13–15]. Before the synthesis, the clear image and the corresponding scene depth are adjusted to a specification size of 512×512 pixels. We select four atmospheric light A, where $A \in [0.8,1]$, and the scattering coefficient $\beta \in [0.7,1.6]$ is randomly sampled. We synthesize the hazy image $I(x)$ according to (1) by giving a clear image $J(x)$ and the corresponding depth $d(x)$. Therefore, there are a total of 3200 training images and 400 testing images. We make sure there are no testing images in the training set.

4.2 Training Details

During the training, mean squared error (MSE) is used as the loss function. We use the Adam optimization algorithm to train our network and set the learning rate of 0.0001 and the batch size of 1. We implemented the proposed method using the PyTorch and trained the network for 400 iterations. It took 8 days to train on the NVIDIA Titan XP GPU.

4.3 Quantitative Evaluation

We evaluate the proposed algorithm with the state-of-the-art algorithms using peak signal-to-noise ratio (PSNR) and structural similarity index (SSIM). The method achieves higher average value of PSNR and SSIM than other algorithms based on our synthesized test dataset. The results are shown in Table 2.

Table 2 Quantitative evaluation on the synthetic testing dataset

Metric	He [3]	Zhu [16]	Berman [5]	Ren [7]	Cai [6]	Ours
PSNR	18.40	18.37	17.48	18.98	19.91	25.01
SSIM	0.8925	0.8725	0.8606	0.8946	0.9017	0.9079

Single Image Dehazing of Railway Images ...

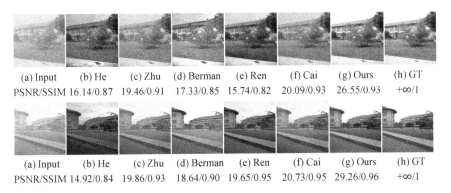

(a) Input	(b) He	(c) Zhu	(d) Berman	(e) Ren	(f) Cai	(g) Ours	(h) GT
PSNR/SSIM 16.14/0.87	19.46/0.91	17.33/0.85	15.74/0.82	20.09/0.93		26.55/0.93	+∞/1

(a) Input	(b) He	(c) Zhu	(d) Berman	(e) Ren	(f) Cai	(g) Ours	(h) GT
PSNR/SSIM 14.92/0.84	19.86/0.93	18.64/0.90	19.65/0.95	20.73/0.95		29.26/0.96	+∞/1

Fig. 3 Dehazing results on the testing datasets using several state-of-the-art dehazing methods

We select two examples from the testing images to show the dehazing results in Fig. 3. The results by He et al. [3] show obvious artifacts when dealing with sky scenes, which is due to the inability to accurately estimate the transmission map. The inaccurate transmission map leads to more hazy residuals in dehazing results of Zhu et al. [16] There are hazy residuals in results by method of Berman et al. [5], especially in distant scenes. The dehazing results from Ren et al. [7] have color distortions and a number of hazy residuals. The method-based CNN proposed by Cai et al. [6] is not working well. The proposed method achieves the best dehazing results in both PSNR and SSIM, and the visual effect is also excellent. In the first row of Fig. 3, our dehazing results are well restored to the details on the ground and in the sky. In the second row, our hazy residuals are much less than other dehazing results.

4.4 Real Image Haze Removal

In order to demonstrate the dehazing effect of the proposed algorithm, we evaluate the dehazing results of the proposed algorithm on some real-world images. The four real-world images are dehazed by the state-of-the-art methods and our proposed method, and the corresponding dehazing results are obtained as shown in Fig. 4.

The dark channel dehazing proposed by He et al. [3] can remove the haze of the real-world images, but there are obvious artifacts. The dehazing results by Zhu et al. [16] have more hazy residuals. After we used the method of Berman et al. [5] to remove haze of the real-world images, there are certain artifacts and hazy residuals. The methods of Ren et al. [7] and Cai et al. [6] rely on the CNN model to estimate atmospheric light and transmission maps for dehazing, but there are residual dehazing. The proposed method achieves better dehazing results, not only without artifacts and colors distortion but also with less haze residuals than other methods.

Fig. 4 Dehazing results on natural hazy images from railway using several state-of-the-art dehazing methods

5 Conclusions

We propose a dehazing model of an end-to-end multi-scale residual network, which is effective for defogging. Based on the residual network, we combine multi-scale features to get more effective information. Bottleneck layer is added to further extend the depth to solve the problem of multi-scale network depth, which can significantly improve network performance. After extensive evaluation, the proposed method can achieve better results than the state-of-the-art algorithms based on synthetic datasets and railway images.

Acknowledgements The study is sponsored by multi-source information fusion and big data analysis technology for air-conditioned vehicles-02(T17B500040).

References

1. Fattal R (2008) Single image dehazing. ACM Trans Graph 27(3):72
2. Tan RT (2008) Visibility in bad weather from a single image. In: 26th IEEE conference on computer vision and pattern recognition, IEEE, Anchorage
3. He K, Sun J, Tang X (2011) Single image haze removal using dark channel prior. IEEE Trans Pattern Anal Mach Intell 33(12):2341–2353
4. Fattal Raanan (2014) Dehazing using color-lines. ACM Trans Graph 34(1):1–14
5. Berman D, Treibitz T, Avidan S (2016) Non-local image dehazing. In: Proceedings—29th IEEE conference on computer vision and pattern recognition, pp 1674–1682. IEEE, Las Vegas
6. Cai B, Xu X, Jia K, Qing C, Tao D (2016) Dehazenet: an end-to-end system for single image haze removal. IEEE Trans Image Process, 5187–5198
7. Ren W, Liu S, Zhang H, Pan J, Cao X, Yang MH (2016) Single image dehazing via multi-scale convolutional neural networks. In: Computer vision—ECCV 2016, pp 154–169. Springer, Amsterdam
8. Li B, Peng X, Wang Z, Xu J, Feng D (2017) AOD-Net: all-in-one dehazing network. In: 2017 IEEE international conference on computer vision, pp 4780–4788. IEEE, Venice
9. Li R, Pan J, Li Z, Tang J (2018) Single image dehazing via conditional generative adversarial network. In: 31st meeting of the IEEE/CVF conference on computer vision and pattern recognition, pp 8202–8211. IEEE, Salt Lake City
10. Zhang H, Patel VM (2018) Densely connected pyramid dehazing network. In: 31st meeting of the IEEE/CVF conference on computer vision and pattern recognition, pp 3194–3203. IEEE, Salt Lake City
11. Ren W, Ma L, Zhang J, Pan J, Cao X, Liu W et al (2018) Gated fusion network for single image dehazing. In: 31st meeting of the IEEE/CVF conference on computer vision and pattern recognition, pp 3253–3261. IEEE, Salt Lake City
12. He K, Zhang X, Ren S, Sun J (2016) Deep residual learning for image recognition. In: Proceedings—29th IEEE conference on computer vision and pattern recognition, pp 770–778. IEEE, Las Vegas
13. Saxena A, Chung SH, Ng AY (2006) Learning depth from single monocular images. In: 2005 annual conference on neural information processing systems, pp 1161–1168. Neural information processing systems foundation, Vancouver
14. Saxena A, Chung SH, Ng AY (2008) 3-d depth reconstruction from a single still image. Int J Comput Vision 76(1):53–69

15. Saxena A, Sun M, Ng AY (2008) Make3d: learning 3d scene structure from a single still image. IEEE Trans Pattern Anal Mach Intell 31(5):824–840
16. Zhu Q, Mai J, Shao L (2015) A fast single image haze removal algorithm using color attenuation prior. IEEE Trans Image Process 24(11):3522–3533

System Dynamic Model and Algorithm of Railway Station Passengers Distribution

Zhe Zhang, Limin Jia and Sai Li

Abstract The precondition for both reasonable use of facilities and scientific passenger flow organization scheme of rail passenger station is mastering the distribution characteristics of passenger flow. Through the analysis of railway passenger station's passenger distribution service network structure, taking railway passenger station as mutual integration of passenger, facilities, environment, and management, the passenger distribution system dynamic model with double-level structure has been proposed and established. Considering macroscopic properties including the traffic mode, facility capacity, and microscopic properties including passenger speed and route choice, a passenger distribution model based on the system dynamics has been presented. The proposed model can express the passenger distribution process interdependency clearly and adapt to different layout and operational characteristic of railway station. The proposed model has been applied in a multi-terminal railway station using the real data and produces some useful results.

Keywords Railway station · Passenger distribution · Model · System dynamic

1 Introduction

As the demand for railway services is increasing, railway stations have shown issues incompatible with such a growth, there is urgent need for providing more efficient and safe railway station service, which can be evaluated by the simulation of pedestrian distribution of railway station [1].

In the operation of railway station, some variability and stochastic events may happen to a process, such as random arrivals, delays, variations in train schedule, capacity, and level change of service; every process has its own service goal and tries to perform as well as possible, which resulted in local improvements but poor performance of other related process. For example, if the number of security scan

---

Z. Zhang (✉) · L. Jia · S. Li
State Key Laboratory of Rail Traffic Control and Safety, Beijing Jiaotong University, Beijing 100044, China
e-mail: zhangzhe@bjtu.edu.cn

is shorter, passengers can arrive at security check point in less time, and then offer higher level of walking service to passengers. However, this would result in higher passenger aggregation for the sub-sequent process of security check, reducing its level of service [2].

Therefore, the passenger distribution of railway passenger station is a system engineering problem rather than individual process and facility optimization problem is in an urgent need of using the system thinking approach to understand dynamic behavior of passenger distribution in time and space and the complicated relationships among various processes.

Passenger distribution characteristics of railway passenger station are determined by traffic sharing rate, passenger flow organization procedure, channel and service capacity, and train time table [3]. Researchers have studied the passenger distribution process inside railway station from the macroscopic and microscopic perspective.

Macroscopic models are designed to deal with more policy analysis and strategy development. Davis studied the level of service (LOS) of walking facilities in transfer station and gets the service standard of channel facilities for passenger walking [4]; Lu Sheng and Li Xuhong put forward network utility model to evaluate and optimize the facilities layout of transfer station [5]; a network optimization-based methodology was proposed to seek the optimal reconfiguration of layout to support efficient crowd management in large public gatherings held in transportation station. A number of simulation tools (e.g., PEDROUTE [6], PAXPORT [7], and SimPed [8]) have macroscopic rules describing the behavior of the pedestrian flow [9], however, these tools are designed for specific purpose frequently and their application area is limited.

2 Model Formulation

2.1 Passenger Generation Model and Algorithm

Passengers including boarding and alighting passengers are generated by timetable; the following data should be prepared to transform the train timetable into input data for the proposed model (passenger arrival and departure patterns).

The number of boarding passengers expected for each departure train, which is calculated by multiply seating capacity of train based on train type and occupancy rate of train. The number of alighting passengers expected for each arrival train, which is calculated by multiply seating capacity of train based on train type and transfer rate of train. Passenger arrival distribution describes proportion of passengers arriving at the railway station in interval of time ahead of the scheduled departure time of their train. The parameters of arrival distribution and the time interval of arrival are different based on the trip time of a day, trip habits, and the lead time of ticket check.

The algorithm steps to calculate the total number of passengers arriving at railway station per unit time can be depicted in Fig. 1.

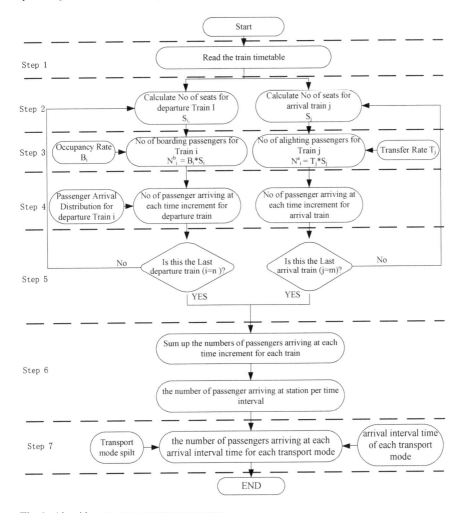

Fig. 1 Algorithm steps to generate passengers

2.2 SD Model of Service Facilities

There are a series of service facilities in unpaid and paid areas, the associated process of each service facility in the railway station is depicted as stock-and-flow structure, which is based on four entities: (1) Stocks (▢) represent the accumulation level of a process (e.g., number of passengers in route); (2) flows (⌼) represent the rate of change of this level, increase the preceding stock and decrease the behind stock (e.g., number of passengers served per unit time); (3) converters(○) are the variables (e.g., number of service channel) determining flows, and calculates

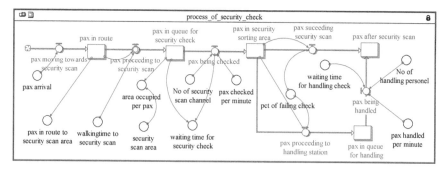

Fig. 2 Example of a processing facility—security check facility

algebraic relationships (space occupied by per person), and converts inputs into outputs (e.g., waiting time of service) (4)connectors ( ) connect above entities.

Figure 2 shows the SD model of security check facility, passenger's arrivals need security check ("pax moving toward security scan") pass the route ("pax in route") and take time ("walking time to security scan") to proceed to security scan(pax proceeding to security scan), queue for the service process ("pax in queue for security scan"), and leave the security scan at the rate of check ("pax being checked"), which depends on the number of security scan channel ("No of security scan channel") and the number of passengers inspected per minute ("pax checked per minute"), passengers leaving security scan ("pax in security sorting area") are divided into two parts: (1) passengers success to pass the process can leave the security scan(pax succeeding security scan); (2) passengers failing security scan proceed to handling station ("pax proceeding to handling station"); All passengers can leave the facility ("pax after security scan") after performing the corresponding process.

The performance of security check process can be expressed through two indicators: waiting time ("waiting time for security check"), the space occupied per passenger ("area occupied per pax") which will be used to access the level of service of service facility.

2.3 SD Model of Passenger Flow

Passengers flow model describes the passengers transfer behavior between the various facilities of the station, considering the general relationship among speed, density, and flow; the walking speed of passengers depends on the density of route (area occupied by per passenger) and has shown randomness as a result of varied characteristic of passengers.

However, railway station provides a special case of pedestrian platoon flow, which is different from highway in terms of traffic patterns, directional characteristics, walkway geometries, and pedestrian spatial preferences. Davis and Braaksma [4] gives

Table 1 Speed distribution used in model

LOS	Density (m²/p)	Flow (p/min/m)	Speed interval (m/min)	Normal distribution
A+	≤2.3	[0, 37]	[84, 93]	N (88.5, 1.125)
A	[1.7, 2.3]	[37, 46]	[78–84]	N (81.0, 0.75)
B	[1.3, 1.7]	[46, 57]	[72,78]	N (75, 0.75)
C	[1.0, 1.3]	[57, 68]	[66,72]	N (79, 0.75)
D	[0.8, 1.0]	[68, 75]	[60, 66]	N (63, 0.75)
E	[0.7, 0.8]	[57, 75]	[42.0, 60]	N (51, 225)
F	[0, 0.7]	≤57	[0, 42]	N (21, 5.25)

the pedestrian speed interval for different Level of Service (LOS) threshold for platoon flow in transportation terminals [4], as presented in Table 1, and by implication, the transportation terminals refers to locations where platoon flow is prevalent along pedestrian walkways, for example, railway station and airport. Consequently, the LOS thresholds for platoon flow in transportation terminals can be used in the passenger flow model of railway station, and the normal distribution of walking speed is designed for each LOS threshold to capture the stochastic of walking speed.

In order to express the normal distribution, two parameters including mean value (μ) and standard deviation (σ) serve as a measure of speed randomness and should be calculated. We assume that the speed interval is $[a, b]$, according to the characteristic of normal distribution, the mean value is the center of interval range, and four standard deviations account for 99.99% of the sample population being studied. Therefore, the following equations can be used to estimate μ and σ

$$\mu = (a+b)/2$$
$$u + 4\sigma = b$$
$$u - 4\sigma = a \tag{1}$$

And then, we can get

$$\sigma = (b-a)/8 \tag{2}$$

So the common normal distribution $N((a+b)/2, (b-a)/8)$ can be used to represent the probability of walking speed, in this respect, there is only a small percentage 0.01% outside the speed interval $[a, b]$.

In addition, the normal distribution of walking speed on stairway is depicted as Table 2.

Table 2 Speed distribution of stairway used in model

Direction	Speed interval (m/min)	Normal distribution
Ascending	[30.6, 34.2]	N (32.4, 0.45)
Descending	[40.2, 46.2]	N (43.2, 0.75)

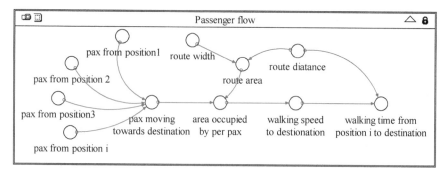

Fig. 3 Passenger flow model

Figure 3 shows the SD model structure of passenger flow, the density is decided by the number of passenger in route ("passenger moving toward destination") and the route area (route area), the walking speed ("walking speed to destination") is worked out by the logical "if-else" based on the density, and then the walking time can be obtained based on the route distance and the walking speed.

3 Model Application: Beijing South Station

In order to validate the proposed model, the proposed model has been used to model and analyze the passenger flow distribution of Beijing south station which is a distribution center of four transportation modes including railway, metro, bus and car. The distribution service network of boarding passenger is depicted as Fig. 4. By using

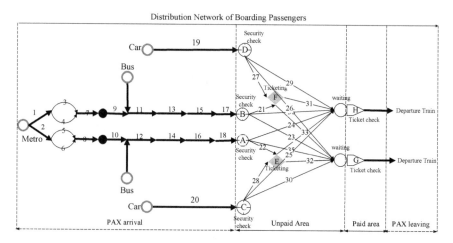

Fig. 4 Distribution service network of boarding passengers at Beijing south station

the real data, the SD model of passenger distribution network is established.

In terms of passenger arrival pattern, there are 76 pairs of trains including 40 pairs of intercity high-speed train (seating capacity: 610), 36 pairs of high-speed trains (seating capacity 1230) in Beijing south station. This application is based on the assumption that 50% of Beijing south station users would use the metro, 20% the public buses, 28% cars, and 2% other buses. The arrival interval time is 3 min for metro, 10 min for buses, 12 min for cars, and 20 min for others. The lead time of ticket check is 15 min for city train and 20 min for others. Considering the trip habits, the arrival distribution is lognormal (3.5769, 0.34256) for the departure train in morning (6:30–9:00), lognormal (4.2227, 0.4553) for other time, and the occupancy rate is set 100%. The simulation starts from 5:00 a.m. to 23:00 p.m. Based on the algorithm steps of transferring timetable to input data, the passenger arrival pattern is depicted as Fig. 5. As it can be observed, one severe peak occurs in the early hours of the day (i.e., 7:25–7:50 a.m., and the other two lower peaks occur in afternoon (i.e., 13:35–13:55, 15:30–16:05).

As far as service facilities are concerned, the acceptable maximum number of passengers in queue for security check is 35. Figure 6 shows the results of the simulation referring to the security scan facility. Based on the real scenario analysis performed, the average number of passengers in queue amounts to 7 in A and B and 8 in C and D, whereas the maximum number reaches 19 in A and B and 22 in C and D, given that the number of security scan is 12 in A and B and 4 in C and D. Figure 7 shows the LOS of security check (where LOS $A = 1, B = 2, C = 3, D = 4, E = 5, F = 6$) over time.

In order to evaluate the response of passenger flow to different operational strategies, the operators of railway station can use the proposed system dynamics model to design a set of scene hypothesis experiments, and further make the most efficient decision to meet the required level of service. For example, as illustrated in Fig. 8,

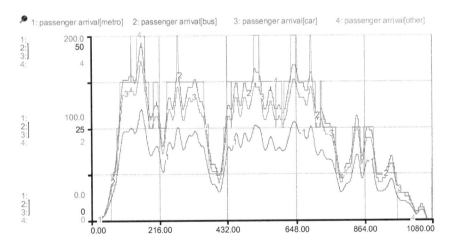

Fig. 5 Passenger arrival pattern of Beijing south station

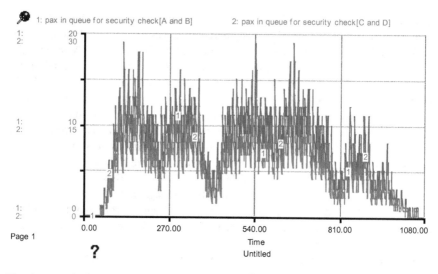

Fig. 6 Number of passengers in queue for every security scan

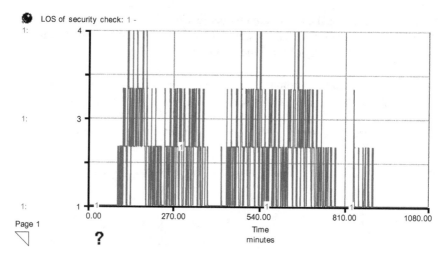

Fig. 7 LOS of security check

in order to analyze the reaction from the number of passengers in queue under different number of security scan, as depicted in figure, when the number of security scan is reduced from 16 (curve 1 in Fig. 8) to 14 (curve 2 in Fig. 8), and then 12 (curve 3 in Fig. 8), the average number increases from 7.2 to 8.3, and then 10, while the maximum number of passengers in queue increases from 20 to 23, and then 35, which in any case is unacceptable level of service and would exceed the maximum number of passengers in queue for security check if without capacity restriction.

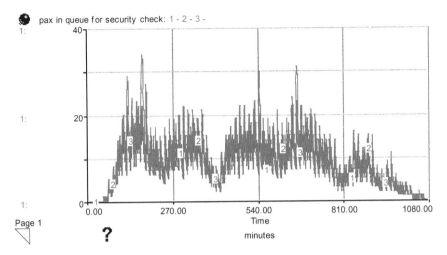

Fig. 8 Average number of passengers in queue for security check with different number of security scan

However, the operators not only have the desire to improve the level of service, but also expect to enhance the level of facility utilization; a balancing scheme can be obtained based on the scene hypothesis experiments. For example, based on the above experiments, more security scans can be used in severe peak period to meet the required level of service and less can be used in lower peak period to enhance the level of facility utilization. In this respect, as shown in Fig. 9, different number of security scan is used in different period (curve 1), which results in the relative

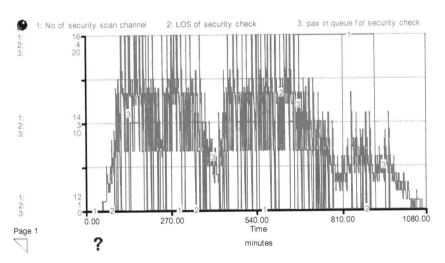

Fig. 9 Number of passengers in queue and LOS under proposed balancing scheme

stable aggregation of passengers before facility (curve 3) and acceptable LOS (curve 2), through the statistical analysis, the average number of passengers in queue is 8.6 compared to 7.2 in the real scenario, while the maximum does not change and the LOS is still above level D, while the percent of the time when the number in queue is larger than 10 46% compared to 33% in the real scenario. So no obvious peak occurs and the facility is busy most of the time after adopting the proposed balancing scheme.

4 Conclusion

This paper has presented a system dynamic model of passenger distribution in railway station, considering both the microscopic and macroscopic characteristic of passenger flow. The proposed model has some advantages compared to other models:

(1) It is easy to express the passenger distribution process interdependency clearly;
(2) It is easy to change the model structure and parameters to adapt to different layout or service process of railway station.

The theoretical and practical value has been reflected by the model demonstration at Beijing south station, and the results of the model demonstration suggested that:

(1) The proposed model can be used efficiently to model the railway station.
(2) It takes little time to present the corresponding passenger flow response to different operational strategy, so it is convenient for station operators to choose the best strategy timely in different period of time by taking the scene hypothesis experiments.
(3) It produces some useful results to reflect the level of service of facility and the level of facility utilization.
(4) It can be used to conduct sensitive analysis among different service processes.

In further study, the proposed model needs to be validated by the comparison of the results from the proposed model against both actual data and the output of other simulation tools such as Legion. The proposed model can also be used to optimization of facility capacity, layout, and train timetable.

Acknowledgements This work was financially supported by the National Key Research and Development Program of China (No. 2018YFB1201403), the National Natural Science Foundation of China (No. 71801009) and the State Key Laboratory of Rail Traffic Control and Safety (Contract No. RCS2018ZT008).

References

1. Jia H, Yang L, Tang M (2009) Pedestrian flow characteristics analysis and model parameter calibration in comprehensive transport terminal. J Trans Sys Eng IT 9(5):117–123

2. Manataki IE, Zografos KG (2010) Assessing airport terminal performance using a system dynamics model. J Air Transp Manag 16(2):86–93
3. Aili W, Baotian D, Chunxia G (2013) Assembling model and algorithm of railway passengers distribution. J Transp Syst Eng Inf Technol 13(1):142–148
4. Davis DG, Braaksma JP (1987) Level of service standards for platooning pedestrians in transportation terminals. Inst Transp Eng 57(4):31–35
5. Lu S, Li X (2006) Utility analysis of facilities layout alternatives of urban inter modal passenger transfer points. 36(6):1024–1026
6. Halcrow Group Ltd. PEDROUTE. 2002. www.halcrow.com/pdf/urban_reg/pedrt_broch.pdf.2
7. Halcrow Group Ltd. PAXPORT Passenger Simulation Model. 2002.www.halcrow.com/pdf/urban_reg/pasport_broch.pdf
8. Daamen W, Hoogendoorn SP (2002) Modeling pedestrians in transfer stations. Presented at 81st annual meeting of the Transportation Research Board, Washington, DC
9. Serge PH, Miklos H, Nuno R (2007) Applying microscopic pedestrian flow simulation to railway station design evaluation in Lisbon, Portugal. Transportation Research Board annual meeting 2007. National Academy Press, Washington DC, 83–94

Estimation of Line Capacity for Railway

Xu Wei and Jie Xu

Abstract To estimate the line capacity of a railway, the influencing factors on railway line capacity are investigated. First, various states of the China Train Control System (CTCS) under severe ice and snow conditions are discussed. Second, descriptive and analysis models of train operation under adverse situations are given based on a stochastic Petri network. Since the SPN and Markov chain model (MCM) are isomorphic, the transition probability of train operation conditions is analyzed in depth by MCM. Then dynamic line capacity transition probability and fuzzy comprehensive evaluation matrixes are constructed for analyzing the capacity variation of a railway line.

Keywords Estimation · Railway capacity · Adverse condition · SPN · MCM

1 Introduction

Railway capacity is a concept that is not easily defined or quantified [1]. In China, railway line capacity is the maximal number of trains that can go through a specified line or section during a certain period, such as 24 h. Currently, railway managers and researchers usually calculate line capacity by using empirical or analytical methods tested over a long period [2–5]. Numerous capacity studies have been performed to provide as many as possible route slots for railway operators and to know how much railway traffic can be supported by the current railway network [1, 3–6]. The goal of the capacity analysis is to determine the maximum number of trains that can be operated over a given railway infrastructure during a specific phase [4]. To analyze and enhance capacity utilization of an existing timetable, UIC 406 was developed for capacity evaluation, in which macro- and micro-capacity utilizations were described

---

X. Wei
Safety Supervision and Management, Office of the Commissioner of China Railway Corp., Shanghai, China

J. Xu (✉)
State Key Laboratory of Rail Control and Safety, Beijing Jiaotong University, Beijing, China
e-mail: jxu1@bjtu.edu.cn

based on the discrete nature of capacity utilization [7]. In accordance with well-defined rules, train operations were simulated to obtain the capacity [5]. Several international companies are also working on similar computer-based systems, e.g., DEMIURGE, which can evaluate a network's capacity to absorb additional traffic and locate bottlenecks [8]. RAILCAP can measure how much of the available capacity is utilized by a given program and analyze bottlenecks. However, previous studies have focused on daily operation [4]. As such, it is also necessary to present an analysis approach that provides capacity planning under abnormal weather conditions. In this paper, the negative effects of abnormal weather conditions on train operation facilities and equipment will be discussed, and the main factors that can affect capacity will be analyzed. Next, an analytical method based on Petri nets will be presented to analyze train operation within railway sections that are experiencing severe weather conditions to reach conclusions about the line capacity of railway sections. After the model is decomposed further, the performance of the train operation model is calculated by using an isomorphic Markov chain.

The remainder of the paper continues as follows. In Sect. 2, the process of train operations based on speed limits is described with the given various facilities and equipment function impairments, and train operation rules from railway authorities of China are depicted. Then, the influencing factors of line capacity and capacity variation under severe weather conditions are analyzed. Train operation process under severe weather condition is modeled based on a Petri net model in Sect. 3. Then the model is decomposed, and the performance of the train operation model is calculated by using a Markov chain method. To estimate capacity variation during a period of severe weather conditions, a fuzzy comprehensive evaluation method is discussed.

2 Process of Train Operation

Numerous factors include train speed, signal format of the train control system, the length of block section, headway time, train stop plan [5], influence the capacity. At present, the utilization of a moving-like block system has been implemented for controlling train operations in China's railway network. Distance-to-go control mode is adopted to manage speed and the headway of trains in this system. To avoid potential collisions, headway intervals are divided into three or four blocks. Each block is permitted to be occupied by each train at the same time. Signals are usually placed at the start point of the blocks to indicate whether the approaching train is permitted to access the block. Trains follow each other at operational speeds according to the operation rules and are separated by a headway interval, which is composed of different numbers of blocks due to the type of signaling system [9]. In China, four-aspect signaling arrangements are the most common systems for train control, which is shown in Fig. 1.

When the train passes by signal H, it will be changed to red to indicate that the block section protected by this signal is occupied. A green light implies that three block sections are unoccupied, green plus yellow means that two block sections are

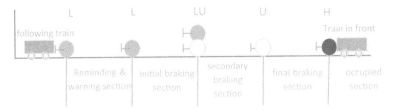

Fig. 1 Interval between trains in four-aspect block section

unoccupied, while a yellow light means only one further block is clear, with a speed limit of zero at the end. Generally, the braking distance of the train should be less than the length of three block sections. The braking distance includes four block sections. Therefore, the minimum headway time is calculated as follows:

$$t_I = [(4L_s + L_t)/v] \times 3.6 \quad (1)$$

where L_s is the length of the block (m), L_t is length of train (m) and v is the speed of the train (km/h).

2.1 Function Impairment of Operation Facilities

Diverse operational scenarios can cause function reduction in train operation facilities, such as with slippery track conditions, power loss of traction power supply systems, equipment damage to CTCS, which prevents train operations at a normal speed. To ensure safe operation, trains have to slow down during severe conditions. In China, the speeds of trains are controlled by CTCS. Both CTCS and line capacity can be affected by severe conditions; therefore, some train operation scenarios are defined in CTCS to deal with the failure of facilities, and nine main train operation scenarios are designed to deal with facility failures. The different scenarios correspond to various facility failures, such as track circuit failure and balise transmission module (BTM) failure, specific transmission module (STM) failure, driver machine interface (DMI) failure, train speed recorder failure, and automatic train protection (ATP) failure. To evaluate failure levels, the speed limits can be classified into two grades, i.e., facility failures and control modes. Speed limits are chosen in the range of 160, 120, 40, and 20 km/h.

2.2 Capacity Variation of Railway Line Under Severe Weather Conditions

The functions of train operation facilities and equipment can be hampered by severe weather; once an adverse condition occurs, the railway line capacity will drop. At the beginning of a severe condition, the speed of a train can be constrained and line capacity will keep falling. According to operation rules, train speeds will be increased, and line capacity will recover after the emergent situation. Variation in available line capacity with severe weather is shown in Fig. 2, where C_1 is the current line capacity under normal condition, and C_2, C_3, C_4, C_{m-1} denote the beginning, the middle, and the end of the whole process of a severe weather situation, respectively.

According to the rule, trains are required to run at speeds within certain sections. Since the speed of trains has to decrease, this will lead to a drop-in line capacity. The maximal numbers of trains that can go through the affected section will be diminished at a unit time. Just like the stage t_1 shown in Fig. 2, the line capacity C_1 is more than any capacity under normal operation conditions. When severe conditions become increasingly serious, train speed limits have to be reduced further according to operation rule, and line capacity drops to C_2. When the condition becomes overly severe and there is no time to implement emergency responses, the trains have to come to a halt and line capacity drops down to 0. Otherwise, the capacity will keep constant. As condition improves and railway maintenance staff undertake effective measures, both the speed limit and line capacity will gradually recover, as stage t_3 and t_m as shown in Fig. 2.

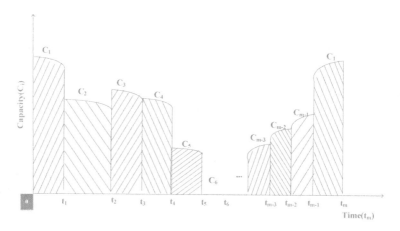

Fig. 2 Line capacity variations under severe conditions

3 Model

3.1 Petri Net Model

Petri net is a graphical and mathematical modeling paradigm applicable to communication, synchronization, and resource sharing systems, which is similar to that of lines shared by trains [10]. The Petri net is formulated as 3-tuple $N = (P, T, F)$, where T and F are finite, nonempty and disjoint sets. P is a set of places, and T is a set of transitions. $F \subseteq (P \times T) \cup (T \times P)$ is the flow relation. Given a net, $N = (P, T, F)$ and a node $X \in P \cup T$, $*x = \{y \in P \cup T | (y, x)\}$ is the prior set of x, while $x* = \{y \in P \cup T | (y, x) \in F$ is the post-set of x. Let $N = (P, T, F)$ be a Petri net. A marking x of N is a mapping from P to η, where $\eta = \{0, 1, 2, \ldots\}$. In general, we use multi-set notation $\sum_{p \in P} m(p) p$ to denote vector m, where $m(p)$ indicates the number of tokens in P at m. The Petri net is used to represent the operation process of a train between stations under ice and snow conditions and to model the relation between headway and speeds of related trains, as shown in Fig. 3. In the Petri net model, a railway section is defined as places that permit occupation by only one train, and a train is considered as a token, which can move from one place to another. The movements by which tokens move from one place to another mean that one train moves from one section (or block) to another. The entering track section is occupied, and the leaving one is free. Under different train operation scenarios, the maximum speed of a train changes according to train operation rules and the capacity of the railway line changes as well.

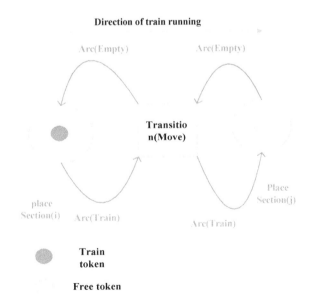

Fig. 3 Basic Petri net model of train operation

The process of a train running from one place to another is defined as a transition, which needs to trigger two conditions.

3.2 Train Operation Model Based on Petri Nets

According to the three principles of limited speeds, the train operation model is built with various operation scenarios based on a Petri net model as shown in Fig. 4. When approaching a given disaster-effect section, a train will suffer from one type of negative weather factor; meanwhile, operation facilities (F_i) will be impaired by the poor weather. Dispatchers will then give specific instructions and the trains will run at various control modes (M_i) corresponding to facility fault; thus, the speed will be decreased to $\min(V_i)$.

Figure 4 illustrates a Petri net model under CTCS failure scenarios, where F_i, M_i, and V_i are facility failure, state of train operation, and train speed under an adverse scenario, respectively; the model can be simplified as Fig. 5, where $F = \{F_i | i = 1, 2, \ldots, 8\}$ is the set of various failures of CTCS, $M = \{M_i | i = 1, 2, \ldots, 8\}$ is the set of control modes of CTCS corresponding to M_i, $V = \{V_i | i = 1, 2, \ldots, 8\}$ is the set of speed limits under the various control modes. Furthermore, M_0 refers to a daily train operation condition with a speed limit of 300 km/h. Otherwise, M_9 refers to judgment operation mode, which means the train will run into the next section at a speed of V_9, i.e., 40 km/h.

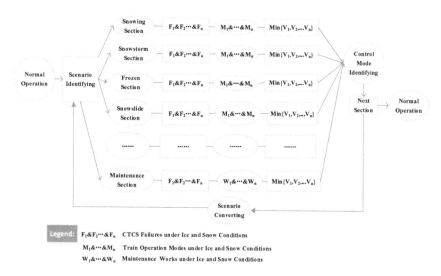

Fig. 4 Petri net model under CTCS failure scenarios

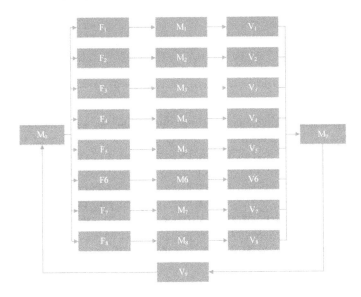

Fig. 5 Model of train operation under CTCS failure situations

To illustrate the changing process of train operations with various scenarios, a specified route of the Petri net model was selected and analyzed. Suppose that the process changing of scenarios is as follows:

$$M_0 \rightarrow F_5 \rightarrow M_5 \rightarrow V_5 \rightarrow M_9 \rightarrow V_9 \rightarrow M_0 \tag{1}$$

At the beginning of scenario, the train is operated under scenario 1 (M_0) with a speed of 300 km/h. When the condition gets worse, the BTM malfunctions during train operation and failure F_5 begins to emerge, and then the train control mode has to be changed to isolation mode M_5. Correspondingly, trains need to run at a speed of 120 km/h (V_5) in this section. When the train is approaching the next section (or block), the control mode is changed to determine mode (M_5) when the condition gets worse, and the train would run at a speed of 40 km/h (V_9) into the next section (or block). After the emergency response actions are put in use and conditions improve, the train begins to resume normal operations and speed gets recovered (M_0). The model describes modes of train operation under a CTCS failure situation.

3.3 Markov Chain Model

Since the speeds of trains under an adverse situation can be divided into several grades. To illustrate capacity variation, the speed set of trains can be defined as

Fig. 6 Markov processes of various capacities between scenario states

$\{N_1, N_2, N_3, N_4\}$. A simple transiting procedure among scenarios based on stochastic Petri nets (SPN) is shown in Fig. 6. The transition probability of scenarios is depicted with SPN model. And the transiting processes are known as stochastic. The transition probability of train operation scenarios will be depicted as Markov chain model (MCM) [10]. The calculation procedure of transition probability for different scenarios will be considered as the transiting probability of state of Markov process. Markov chain has the character of no aftereffects, and the expected transition probability of railway section capacity at time t_{n+1} is calculated based on the scenario state of time t_n. Transition probability (P^{n+1}) of the state of time t_{n+1} will be denoted in the equation below:

$$P^{n+1} = P_{n+1} \times P^n \tag{2}$$

where P_{n+1} is the transition probability of scenario from state $P^{(n)}$ to $P^{(n+1)}$. If P_{n+1} is constant, the transition probability can be calculated as follows:

$$P^{(n+1)} = P^{n+1} \tag{3}$$

The model, shown in Fig. 6, can be simplified into Fig. 6 without the cause of transition probability. In the model of Fig. 6, C_1 means the capacity of a railway section under normal speed (300 km/h), C_2 means capacity under a limited speed of 200 km/h, C_3 means the capacity under the limited speed of 120 km/h, C_4 means the traffic disruption under a limited speed of 0–40 km/h, transition deterioration means the carrying capacity of a railway section reduces, and transition recovery means the carrying capacity increases due to effectiveness of emergency responses. This occurs because the Petri net model (Fig. 6) is isomorphic to a Cardiff homogeneous Markov process [10]. The transition processes of scenarios for various control modes of CTCS are depicted as a Markov chain, as shown in Fig. 6. Because it is no aftereffect in Markov chain, changing the probability of section capacity is only determined by the previous process. The transition probability of the moment n is denoted as follows:

$$P\{\zeta(n+1) = j / \zeta(n) = i\} \tag{4}$$

Similarly, m-step transition probability is defined as follows:

$$P_{ij}(m)(n) = P\{\zeta(n+m) = j / \zeta(n) = i\} \tag{5}$$

Because the changing process has been proven to be a Cardiff homogeneous Markov process, whose state transition can be expressed by C-K equation, i.e., $P_{ij}^{m+r}(n) = \sum P_{ik}^{(m)}(n) \times P_{kj}^{(r)}(n+m)$ $(i, k, j \in I; i < k < j)$, $P^{(m)}$ is m-step transition probability matrix of the train operation scenario. Simply, the C-K equation can be written as:

$$P_{ij}^{m+r} = P_{jk}^{m} \times P_{kj}^{r}, \tag{6}$$

or

$$P^{m+r} = P^m \times P^r \tag{7}$$

Because Markov chain has finite state, m is a positive integer. Assume any stage of the state space $i, j, \exists P_{ij}^m \succ 0$, then $\forall \lim_{m \to \infty} P^m = \pi$ and π have two basic properties:

$$\sum_{i \in I} \pi_i P_{ij} = \pi_j \quad (\pi_i \succ 0, P = \pi) \text{ and } \sum_{i \in I} \pi_i = 1 \tag{8}$$

when $P_{ij}^m \succ 0$ (i, j for any state), the process will reach steady state, which is known as the ergodicity of a Markov chain.

3.4 Fuzzy Comprehensive Evaluation Model

Uncertainty is an important feature of line capacity under severe weather conditions [4]. At first, all parameters of train operation, such as speed of train, dwell time, headway and departure and arrival interval of train, are uncertain [2]. Second, both maintenance capacity (emergency response capacity) of railway authorities and weather severity have stochastic and fuzzy characters. Therefore, the best method for capacity estimation for a railway line is to apply a fuzzy method [2]. Fuzzy Markov chain will be used to estimate the capacity of railway lines. Given two finite sets $U = [u_1, u_2, \ldots, u_m]$ and $V = [v_1, v_2, \ldots, v_m]$. U denotes the set composed of all the evaluation factors of railway line, and V means the set composed of all the remark grades; p_{ij} is the result of the assessment of an effective factor u_i to remark v_j, so that the assessment decision-making matrix p of the evaluation factors is as follows:

$$P = \begin{bmatrix} P_1 \\ P_2 \\ \ldots \\ P_m \end{bmatrix} = \begin{bmatrix} p_{11} & p_{12} & \cdots & p_{1n} \\ p_{21} & p_{22} & \cdots & p_{2n} \\ \cdots & \cdots & \cdots & \cdots \\ p_{m1} & p_{m2} & p_{m3} & p_{mn} \end{bmatrix} \tag{9}$$

If the weight of each evaluation factor is one fuzzy subclass of set V, then V can be figured out by applying operation of a fuzzy transform, which provides comprehensive evaluation result:

$$B = A \times P = [b_1, b_2, \ldots, b_4] \quad (10)$$

where B stands for a fuzzy set of V, and the weights fuzzy set for each factor is expressed by $A = a_1, a_2, \ldots, a_m$.

4 Model Analysis

If r_{i1} represents the element of the first i factor in factors set U to first element in evaluation sets V, and fuzzy sets $R_i = (r_{i1}, r_{i2}, \ldots, r_{in})$ express the first r element single factor evaluation result. The matrix R is composed of m single factor evaluation set $R_1, R_2, \ldots, R_m$ and called a fuzzy comprehensive evaluation matrix. To solve the Markov chain model, which depicts line capacity under a severe weather condition, we use a fuzzy comprehensive evaluation method to calculate the carrying capacity of each state by this matrix:

$$R = \begin{Bmatrix} 0.0 & 0.3 & 0.5 & 0.2 \\ 0.0 & 0.1 & 0.2 & 0.7 \\ 0.0 & 0.5 & 0.2 & 0.3 \\ 0.7 & 0.1 & 0.2 & 0.0 \\ 0.5 & 0.3 & 0.2 & 0.0 \end{Bmatrix} \quad (11)$$

And
$A_1 = (0.7, 0.1, 0.1, 0.05, 0.05)$, $A_2 = (0.1, 0.0.0.1, 0.5, 0.3)$, $A_3 = (0.2, 0.6, 0.2, 0.0, 0.0)$, $A_4 = (0.3, 0.0, 0.5, 0.0, 0.2)$.

The results of the transition probability matrix are calculated as follows:

$$\begin{bmatrix} 0.06 & 0.29 & 0.41 & 0.24 \\ 0.50 & 0.22 & 0.23 & 0.05 \\ 0.00 & 0.22 & 0.26 & 0.52 \\ 0.10 & 0.40 & 0.29 & 0.21 \end{bmatrix} \quad (12)$$

Initially, the train operates at a regular speed of 300 km/h. Meanwhile, railway line capacity is denoted as N_1. The decision-making vectors of the scenarios can be expressed as follows:

$$P(0) = [1000] \quad (13)$$

Estimation of Line Capacity for Railway

Table 1 Calculation of each step

$P(i)$	$C_i - C_1$	$C_i - C_2$	$C_i - C_3$	$C_i - C_4$	$P(max)$
$P(1)$	0.060	0.29	0.410	0.240	C_3
$P(2)$	0.173	0.267	0.267	0.293	C_4
$P(3)$	0.173	0.285	0.286	0.255	C_3
$P(4)$	0.178	0.278	0.285	0.258	C_3
$P(5)$	0.175	0.279	0.286	0.259	C_3

Table 2 Values of the steady-state probabilities under various conditions

π	π_1	π_2	π_3	π_4
Value	0.176	0.279	0.286	0.259

Then, we will figure out the kth stage of the expected carrying capacity, and equation C-K can be written as:

$$P(k) = P(0) \times P^k \quad k = 1, 2, 3, \ldots \tag{14}$$

Therefore, the probabilities of the transition probability of line capacity between the scenarios are calculated and provided in Table 1.

A Markov chain was applied to predict the maximum probability of railway line capacity variation. The result reflects a trend from the change in carrying capacity and the ergodicity of a Markov model. As is known, after a sufficient long time period has passed, the railway line carrying capacity of each state will reach steady-state probabilities. According to $\sum_{i \in I} \pi_i P_{ij} = \pi_j$, we can figure out the steady-state probabilities π (see Table 2).

The maximum steady-state probability is $\pi_3(0.286)$, which means C_3 marked the longest period of snow and ice and train operation with a restricted speed of 120 km/h. Eventually, the computational method is given as:

$$N_i = \min\left\{\sum_{i=1}^{4} P(i)_i \times C_i, \sum_{i=1}^{4} \pi_i \times C_i\right\} \tag{15}$$

During severe snow and ice conditions, N_i reflects the comprehensive flexible capacity. With the continuous improvement of Petri nets and railway line carrying capacity transition probability matrix, the results will be more accordant with actual conditions.

5 Conclusions

This paper identified the influencing factors of railway line capacity variations under severe weather conditions. Based on the variations, the functional deterioration and fluctuation of facilities and equipment for train operations were investigated. By constructing a transiting model of railway line capacity with Petri nets, we applied a Markov chain and a fuzzy comprehensive evaluation method to predict trends and reach a steady state under various negative conditions. The conclusions further our understandings on how to effectively manage train flow variation and reasonably arrange dispatching operations during severe weather conditions.

Acknowledgements The authors gratefully acknowledge the support provided by China National "13th Five-Year" key research project "safety assurance technology of urban rail system" (Grant No. 2016YFB1200402).

References

1. Mussone L, Wolfler Calvo R (2013) An analytical approach to calculate the capacity of a railway system. Eur J Oper Res 228(1):11–23
2. Wang L, Qin Y, Xu J (2014) Capacity determination approach of railway section in speed restriction conditions. J Tongji Univ (Natural Science) 42(6):880–886
3. Burdett RL, Kozan E (2006) Techniques for absolute capacity determination in railways. Transp Res Part B Methodol 40(8):616–632
4. Abril M, Barber F, Ingolotti L (2008) An assessment of railway capacity. Transp Res Part E Logist Transp Rev 44(5):774–806
5. Harrod S (2009) Capacity factors of a mixed speed railway network. Transp Res Part E Logist Transp Rev 45(5):830–841
6. D'Ariano A, Pacciarelli D, Pranzo M (2008) Assessment of flexible timetables in real-time traffic management of a railway bottleneck. Transp Res Part C Emerg Technol 16(2):232–245
7. Sameni MK, Landex A, Preston J (2005) Developing the UIC 406 method for capacity analysis, 1–19
8. Kim K (2003) Allocation of rail line capacity between Ktx and conventional trains under different policy goals with mathematical programming. J East Asia Soc Transp Stud 5(10):236–251
9. Wang H, Schmid F, Chen L (2013) A topology-based model for railway train control systems. IEEE Trans Intell Transp Syst 14(2):819–827
10. Murata T (1989) Petri nets: properties, analysis and applications. Proc IEEE 77(4):541–580

Research on the Process Control for the Acceptance of Urban Rail Transit Equipment and Facilities

Cheng xin Du, Jun hua Zhao, Ming Zhang, Zhi fei Wang and Chao Zhou

Abstract Taking the standardization and informatization for the acceptance management of urban rail transit equipment and facilities as the goal, the standardized processes of pre-acceptance and completion acceptance of rail transit equipment and facilities are put forward to optimize the acceptance management system of equipment and facilities, aiming at the non-standard acceptance process of equipment and facilities, unclear explanation of problems and incomplete rectification measures. In order to improve the level of intelligent acceptance control, using advanced workflow and Activiti parallel routing process technologies, a unified equipment and facilities multi-specialty management platform is implemented, which combines online status detection and offline resource caching. The platform realizes modularization of functions, humanization of operation, standardization of data, real-time information and visualization of approval. Many problems such as separation of multi majors, serious isolation of data and asynchronous information of multiple units in acceptance are solved.

Keywords Urban rail transit · Equipment and facilities · Multi-specialty · Standardized process · Informatization

1 Introduction

Urban rail transit, as an integral part of the urban comprehensive transportation system, presents the trend of network operation in the rapid development of the 12th five-year plan and the grand plan of the 13th five-year plan. Urban rail transit plays an increasingly important role in guiding and supporting urban development, satisfying people's travel needs, easing traffic congestion, reducing air pollution and

---

C. Du · M. Zhang · Z. Wang · C. Zhou
Institute of Computing Technology, China Academy of Railway Sciences Corporation Limited, Beijing 100081, China

J. Zhao (✉)
Beijing Jingwei Information Technology Co.Ltd., Beijing 100081, China
e-mail: zhjhky@126.com

other aspects and has become the preferred mode of public transportation for urban people's daily travel. Meanwhile, with the rapid growth of operating mileage and passenger flow, the pressure and challenge of safe operation of urban rail transit are increasing gradually. The acceptance of equipment and facilities before the initial operation is the first safety pass for the new line, which plays an important role in ensuring the safe operation of urban rail transit.

The acceptance work of urban rail transit involves the participation of transportation commission, construction management unit, operation unit, property right unit, supervision unit, construction unit, design unit and other units to complete acceptance of AFC, PIS and other majors [1–5]. In the process, each participating institution shall timely understand the process and status of equipment and facilities, manage relevant documents generated in the course and archive image data and problems such as key parts and concealed works. At the same time, the acceptance management of equipment and facilities of new urban rail transit lines is faced with some difficulties. So far, there is no online system to realize information-based office, which makes the monitoring data granularity not fine enough and frequency not high. It is difficult to understand the acceptance process in a timely and transparent manner due to the lack of unified platform and isolated data. Multiple units participate in the acceptance, and it is difficult to coordinate unified management and realize macro control. And it lacks intelligent application of data analysis and evaluation based on big data.

In order to further standardize the pre-acceptance and completion acceptance processes for the operation equipment and facilities of new lines, guide the standardization of the acceptance work of multiple units and avoid non-standard acceptance processes, unclear problem description and incomplete rectification measures, it is urgent to study the standardized acceptance processes and realize unified control. In the meantime, in order to strengthen the supervision of the acceptance work [6–8], the traditional manual and empirical acceptance work is changed to the system and data in an information and intelligent way. An information management platform is established to optimize the acceptance management system of equipment and facilities, so as to achieve precision and data standardization for the acceptance of each major of equipment and facilities [9], providing powerful data support for the "looking back" of rail transit acceptance and the acceptance of new line equipment and facilities. Meanwhile, the application of the acceptance management platform can reduce the loss of life and property caused by unqualified products and hidden problems, ensuring the smooth progress of the acceptance work and providing an early safety basis for urban rail transit operation.

2 Three-Level Management of the Acceptance

The acceptance management of new line operation equipment and facilities aims to realize the three-level management mode of "line management, professional management and process management", and it includes every line for urban rail transit equipment and facilities of pre-acceptance and completion acceptance management.

The equipment majors include vehicles, power supply system, communication system, signal system, comprehensive monitoring system, automatic ticket selling and checking system, passenger guidance and identification system, platform door system and depot maintenance equipment. Facilities majors include railway lines, guardrails and evacuation platforms.

3 Process Control

In order to realize the management of the project process of rail transit equipment and facilities, as well as the management of documents and images in the acceptance process, the pre-acceptance process and the completion acceptance process are designed to streamline and standardize the acceptance work so as to ensure the safety, efficiency and orderly progress of the acceptance organization.

According to different roles in the acceptance process, the units participating in the acceptance of equipment and facilities are generally divided into seven categories. Table 1 gives the operation contents of each type of units.

3.1 Pre-acceptance Process

According to the acceptance mode of equipment and facilities, the pre-acceptance logic procedure (as shown in Fig. 1) is presented. The pre-acceptance process (as shown in Fig. 2) is designed according to the pre-acceptance logic procedure. Each node of the process is in a forward direction. The data submitted by the nodes should

Table 1 Units' operation

Units	Multi-specialty acceptance	Single professional acceptance	Submit documents	Check documents
Transportation commission	✓			✓
Construction management unit	✓		✓	✓
Operation unit	✓		✓	✓
Property right unit	✓			✓
Supervision unit		✓	✓	✓
Construction unit		✓	✓	
Design unit		✓	✓	

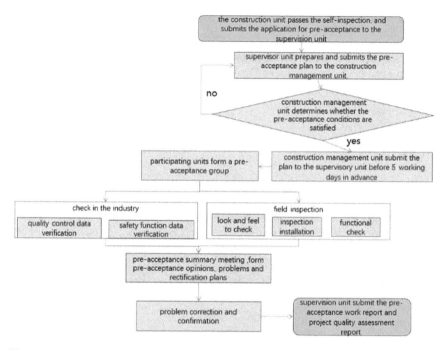

Fig. 1 Pre-acceptance logic procedure

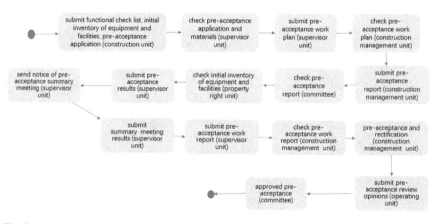

Fig. 2 Pre-acceptance process

be completed, and the process nodes should be smoothly handed over. It cannot proceed to the next node operation until it has completed the current node operation.

Step 1: The construction unit shall submit pre-acceptance application, functional inspection list, initial inventory of equipment and facilities and other relevant inspection materials to the supervision unit.

Step 2: The supervision unit shall check the materials submitted by the construction unit and submit the pre-acceptance work plan to the construction management unit for review.

Step 3: The construction management unit shall check the pre-acceptance work plan and submit the pre-acceptance report to the committee for approval.

Step 4: The property right unit shall review the format and standardization of the initial inventory of equipment and facilities submitted by the construction unit. If the file does not meet the requirements, repeat step 1–step 4.

Step 5: For the approved scheme, the supervision unit can submit the pre-acceptance results in real time and carry out the cycle step 1–step 5 for the pre-acceptance majors (except vehicle major) in batches. After all batches pass the acceptance, a notice of pre-acceptance summary meeting will be issued, and the results of the meeting and the pre-acceptance work report will be submitted to the construction management unit for review.

Step 6: The construction management unit shall submit the pre-acceptance status and rectification status to the operation unit, and the operation unit shall check the relevant documents and submit the pre-acceptance review opinions to the transportation commission.

Step 7: Pre-acceptance shall be completed when all documents submitted by the transport commission are approved. The pre-acceptance of the vehicle in batches shall be carried out step 1–step 7 in a circular manner until the completion of the pre-acceptance of all batches, and the completion acceptance is entered.

3.2 Completion Acceptance Process

Similar to the pre-acceptance process, this paper proposes the completion acceptance logic procedure (as shown in Fig. 3) and designs the completion acceptance process. After the pre-acceptance is completed, the node flows smoothly to the completion acceptance process, as shown in Fig. 4.

4 Design of Acceptance Management Platform

4.1 Modular Design

In order to promote the standardization and information process of equipment and facilities acceptance management of lines, a perfect and unified platform has been established from the original paper office to the efficient transformation of online approval.

According to the overall business process for the acceptance of urban rail transit equipment and facilities, Table 2 shows the overall function of the platform.

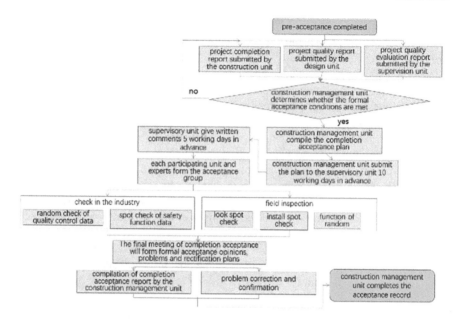

Fig. 3 Completion acceptance logic procedure

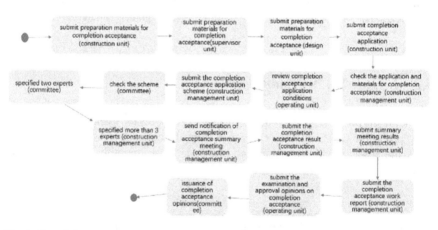

Fig. 4 Completion acceptance process

1. "Fining" of the acceptance management: document data, image data and acceptance problems in the whole acceptance process shall be recorded; image data shall be uploaded on-site; and problems arising shall be recorded in real time, making management more convenient and efficient. Acceptance criteria guide site acceptance in real time, providing information basis for site acceptance. The information input module can record process data, record items such as rejection and delayed inspection, so as to facilitate future acceptance.

Table 2 Functional structure of equipment and facility acceptance management platform

Equipment and facilities acceptance management platform	Material management	File data
		Image data
		Expert database
	Records management	Acceptance list
		Problem record
		Operating history
		Information input
	Standard	Inspection of various standards
	Business management	Acceptance project management
		Acceptance process management
	Management centre	Statistical information
		Alerts
		Data-level permissions configuration

2. "Status" of the acceptance process: dynamically display the process and status and regularly push reminders to the next operation node, so as to ensure timely office operation of multiple inspection units.
3. "Visualization" of the acceptance control: the examination and approval process of each link are visualized; the control right of each participating unit is clearly divided; and group management is implemented.
4. "Customization" of the acceptance process: the platform can realize the customization process, and different process templates can be applied for each major.

In order to make the operation more convenient and efficient for on-site personnel and realize the real-time uploading of on-site acceptance data, the supporting mobile terminal function can be added to realize mobile office and improve the comprehensive control level. Table 3 shows the supporting application functional structure.

1. Message push: persistent connection is adopted to realize message push function through XMPP protocol.
2. Offline storage: offline storage functions are realized through offline resource cache, online state detection and local data storage in HTML5 offline functions.
3. Online submission: on-site acceptance data can be uploaded in real time.
4. Field acceptance module functions:

    (1) Acceptance list templates: download or view the acceptance list template online to guide the scope of the current acceptance.
    (2) Information input: input the name of equipment, quantity, status and other information, which can be stored locally and uploaded to remind users when entering the network. Record the problems found in time, check the list of

Table 3 Supporting application functional structure of the platform

Mobile terminal	Project management	
	Material management	
	Image data	
	Expert database	
	Record management	
	Statistical information	
	On-site acceptance management	Acceptance list templates
		Information input
		Upload of image data
		Message push

historical problems and modify the status of the solved problems found in this acceptance.

(3) Upload of image data: it has the function of calling and uploading local image data and can directly shoot video and upload photographs through mobile devices.

Mobile terminal can meet the requirements of on-site personnel inspection and online mobile office, accelerate the approval process and realize online and offline integrated management.

4.2 Platform Technical Solution

The platform adopts a layered architecture, as shown in Fig. 5. The client realizes the separation of business and presentation logic. The business logic layer adopts the spring framework and realizes the business integration and data transmission of external systems through the unified interface platform. The data access layer realizes the mapping of Java classes to data tables. The data layer stores information such as system business data, configuration data and basic dictionary data.

1. Adopting workflow Activiti technology: adopting advanced workflow technology among nodes to implement process control of acceptance [10]. The introduction of the Activiti concurrent sub-process addresses the need for multiple users to process tasks simultaneously. It put developers focus on writing the business, tie the business to the workflow, render the workflow clear and reliable and make it easier for developers to respond to changes in the process.
2. Using micro-service, loosely coupled, service component technology: supporting for multiple events, multi-process, multi-user multiple concurrent state of data recovery and preservation technology. It supports structured and unstructured

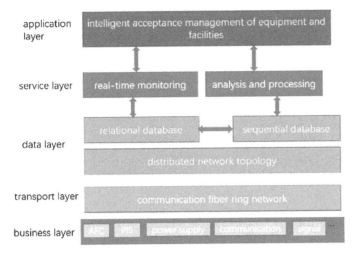

Fig. 5 Platform technical framework

data synchronization process, media flow and data flow, so as to realize the data synchronization and information release of multiple units.

4.3 Network Security

MD5 encryption method is adopted for the platform. User account, password and other information are encrypted and saved to the database. The encryption method is irreversible.

Using two-level heterogeneous firewall, network gate, IDS, security audit and other technical means, it monitors and solves network security problems in the process of network operation to eliminate network security risks. Meanwhile, the hot standby scheme based on storage sharing is used regularly for data backup.

5 Conclusion

In view of the characteristics of unclear acceptance process, multiple units involved and complicated document management in traditional equipment and facilities acceptance work; a standardized acceptance process has put forward; and a corresponding unified management information platform has present by means of information technology to realize pre-acceptance and completion acceptance of enterprise equipment and facilities. The platform has improved the management mechanism, guaranteed the quality of acceptance information, achieved fine management and get real-time statistical analysis and follow-up of acceptance data. Information sharing has realized among different users, and we can look forward to the development prospects of intelligent acceptance of equipment and facilities.

Acknowledgements This work is supported by the R&D Programme of China Railway (No. J2019X005), the Research Programme of China Railway Network Co. Ltd (DFYF19-12), and Foundation of China Academy of Railway Sciences (1852DZ3801).

References

1. Zheng M, Chen W (2017) Emphases in environmental acceptance inspection for completion of rail transit project. J Shanghai Ship Shipp Res Inst 40(2):57–62 (in Chinese)
2. Jin N (2017) A brief discussion on the inspection and acceptance process of urban rail transit maintenance process equipment. Gansu Sci Technol 33(1):82–83 (in Chinese)
3. Wang D (2016) Acceptance testing for urban rail transit CBTC system. Urban Mass Transit 19(3):21–27 (in Chinese)
4. Pan Y (2012) Discussion on the conditional acceptance system on key point of construction in the project of urban rail transit. Qual Manag 30(10):16–18 (in Chinese)
5. Fan J, Jiang X (2018) Key points in technical management and control of urban rail transit signal construction enterprises in the early engineering stage. Chongqing Archit 17(4):56–58 (in Chinese)
6. Zheng D, Ma L, Yan F (2018) Research and prospects of intelligent maintenance management of metro equipment. Urban Rapid Rail Transit 31(5):122–127 (in Chinese)
7. Yang G, Jia J, Liang Q (2006) A new acceptance procedure for urban rail transit engineering. Urban Rapid Rail Transit 19(2):17–19 (in Chinese)
8. Wang Y, Guo Z, Yi X (2015) Study on evaluation system of safety acceptance of urban rail transit engineering. Railw Transp Econ 37(10):91–96 (in Chinese)
9. Jia H, Tang Z (2014) Application of information technology in metro operation. Urban Rapid Rail Transit 27(5):21–24 (in Chinese)
10. Yong S (2016) Design and implementation of attendance workflow system based on Activiti. Comput Era 2:75–78 (in Chinese)

A Model-Based Quantitative Analysis Methodology for Automatic Train Operation System via Conformance Relation

Ruijun You, Kaicheng Li, Yu Liu and Zhen Xu

Abstract V&V processes specified in international standards focus on systematic functional testing (black-box) for safety-critical systems in order to achieve the high trustworthiness of the system based on specification. However, there are different understandings of the specification which can lead to different or even error implementations. It is inefficient and not always accurate to attach most attention on manual analysis for complex Chinese Train Control System (CTCS) with automatic train operation (ATO) function. In order to verify implementations after performing a large number of repeated tests in the laboratory, we propose a methodology to analyze the device using input-output conformance relation (named CSPIO) based on communication sequence process (CSP). We model specification and implementation, respectively, and achieve quantitative results. The verification using PAT for refinement checking is automatically performed. The novelty of our approach is that it handles state spaces in a fully automated manner.

Keywords ATO · CSP · Conformance · Analysis

1 Introduction

The automatic train operation (ATO) system is based on the train control system that the vehicle side adds ATO unit and related ancillary equipment. The ground part adds functions on the server, centralized traffic control (CTC), train control center (TCC) and others [1]. Because of the complexity of the specification published in natural language, different manufacturers may have different specifications in the process of development which may lead to different or even wrong implementation.

R. You (✉) · K. Li · Y. Liu
National Engineering Laboratory for Urban Rail Transit Communication and Operation Control, Beijing Jiaotong University, Beijing 100044, China
e-mail: 17120292@bjtu.edu.cn

Z. Xu
Signal & Communication Research Institute, China Academy of Railway Sciences, Beijing 100044, China

Functional testing and verification are needed to check the correctness of the system by comparing observations with expectations [2] in the safety-critical system's life cycle.

Moreover, it is data analysis rather than the phenomenon that is easier to be ignored, especially some sporadic and implicit problems [3]. At present, there are methods based on image recognition of DMI [4, 5] or knowledge-based analysis for train control system [6]. But in most cases, some implicit and sporadic problems usually require detailed analysis of the data. Formal methods can traverse the internal state space and are used for the trustworthiness using suitable conformance relation. The literature [7] proposed a method based on time automata (TA) model in order to verify the design meets the specified safety requirements and applies to rail-road crossing protection application. The component-based modeling can prevent the state explosions. And the article [8] proposed a strategy using TA to analyze the compatibility of the equipment not quantitatively using search algorithms. The paper [9, 10] proposed CSPIO as a formal relation to generate test cases. It takes only seconds to generate test cases with a coverage of 94.8% which tells us the superiority of the communication sequence process (CSP) mode in concurrency. Through the practice in the automotive industry, nuclear industry and other fields, the CSP model has a strong superiority in describing the system [11] contrasting with other operational methods. The paper [12] automated a previously known, but completely manual technique, namely Josephss relational approach, to prove CSP refinements [13] and use the tool FDR to handle infinite state spaces. This paper presents a strategy of conformance verification and quantitative analysis for the ATO system by using PAT for refinement checking and SMT solving without complex algorithms. Formal methods are used to describe the specification and implementation to traverse the state space and achieve an accurate answer of conformance automatic. The function of core onboard module is analyzed in different scenes to avoid state explosion.

2 Test System and Operation Scenario

The laboratory testing of the ATO system is mainly around the ATO module onboard. The ATO module onboard cooperates with the server to reach ATO functions such as operation plan transmission, station door linkage, start-up and logout. A supporting environment, the ATP and ATO modules, is built to generate real signals of each interface according to the requirements to test the device and record the test data of each interface. It is meaningful that the information through the interfaces, such as wireless transmission interface between the train and the ground, is real information. The ATO module is used as the system under test (SUT), and the excitation signals from the test environment (TE) are acquired through these interfaces, and the response of the device under test is recorded as the output. Our work is to analyze the recorded data of the system (Fig. 1).

As shown in Fig. 2, when the train runs in the interval of AB, it reports the train position and status information to the server. The server reports the position and

A Model-Based Quantitative Analysis Methodology ...

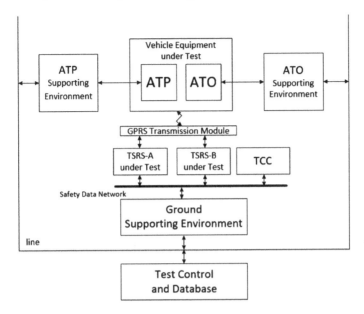

Fig. 1 Schematic diagram of the test platform

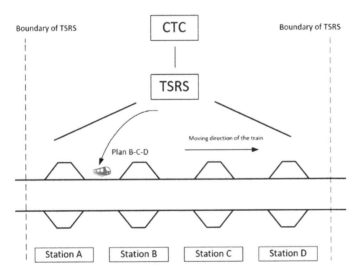

Fig. 2 Schematic diagram of advanced plan sending

status of the registered train to the CTC. Afterward, the CTC sends the train plan to the server, and the server sends the plan to the train. When the train receives the B-C-D plan in this section which means sending the plan in advance, it should be judged that the plan is invalid within 6 s and the plan is not used. When the plan is received in the interval B-C, it should be processed as valid.

3 Modeling the System with CSP

Both the specification and the implementation can be modeled as data-flow reactive systems (DFRS) based on the communication sequence process (CSP). We transform the description and reaction of sending plan in advance to model without considering the internal implementation logic of each device.

3.1 Data-Flow Reactive System Based on CSP Theory

CSP is a formal theory based on process algebra. It is designed specifically for the verification of the reaction system which can express the behavior of the system intuitively. CSP has overwhelming advantages to analyze the concurrent systems. The CSP uses the process as the basic unit to describe the behavior of the system through the events and processes. Although CSP is an expressive process algebra to model the system, there are no semantic differences between the input and output events. The SUT and specification can be seen as DFRS, so there is a separation between the inputs and outputs.

The natural language analysis of specification follows the principles of relative correctness and finite range. We focus on the relation between the ATO module onboard and the server. The ground equipment associated server is abstracted as the resource, and the ATO module is abstracted as the client, and the model is constructed to describe the behavior of the concurrent system (Fig. 3).

Each verb in the description can be associated with case framework (CF) in the Case Grammar linguistic theory. Different components in the sentence represent different thematic roles (TRs). The TR Condition Modifier is defined by us, whereas the others are defined in the related literature. For the conformance between the specification and the model, Table 1 shows some of the CFs that the plan is sent in advance. Using Table 1, we can describe all the requirements defined by the specification and then map them to internal models. A corresponding formal model is established by building a relationship between verbs and state transitions.

Fig. 3 Simplification of the ATO system

Table 1 Examples of case frames

Condition#1-Main Verb (CAC): is							
CPT	Locate	CFV	–	CMD	Between	CTV	AB
Condition#2-Main Verb (CAC): is							
CPT	Plan	CFV	–	CMD	Equal	CTV	BC
Condition#3-Main Verb (CAC): is							
CPT	Plan	CFV	–	CMD	Received	CTV	BC
Action #1-Main Verb (ACT): send							
AGT	plan	TOV	Invalid	PAT	Server	–	–
Action #2-Main Verb (ACT): send							
AGT	Plan	TOV	Invalid	PAT	Resource	–	–

In fact, client components include processes sending and receiving data cyclically, sending invalid data, receiving advance planning, deferred planning, regular planning and none planning. So

$$\text{Client} = df \begin{Bmatrix} ClientRecvAdv || Clientinit || \\ ClientRcy || ClientRcvNor || ClientRecvNone \end{Bmatrix} \quad (1)$$

The description is about the use of CSP for system components without time constraints, focusing on the distribution of components and interaction with the environment. The model is built and converted to the CSP language supported by PAT with good visual interface. Due to space limitations, only part of the trace is shown in Fig. 4. The trace can be exported for further research.

The initial state of the sending plan process can represent as three system inputs (isato = true, position = 0, plan = 0), one system output (isvalid = false) and a counter for the position without time constraints.

3.2 Verification of the Model

Compared to other model checkers, PAT is a user-friendly model checker for all users. It is evolved to be a self-contained framework to support reachability analysis, deadlock-free analysis, full linear temporal logic (LTL) model checking, refinement checking as well as a powerful simulator. Model verification is the formal verification of the canonical CSP model. According to whether the attribute to be verified requires domain knowledge, the verification content is divided into domain-independent and domain-related feature verification. The former covers the basic characteristics that security critical systems need to meet. These include deadlock, reachability, certainty, nondeterminism and divergence. The latter gives the special characteristics that need to be met in the field of train control systems, namely safety features. In order to

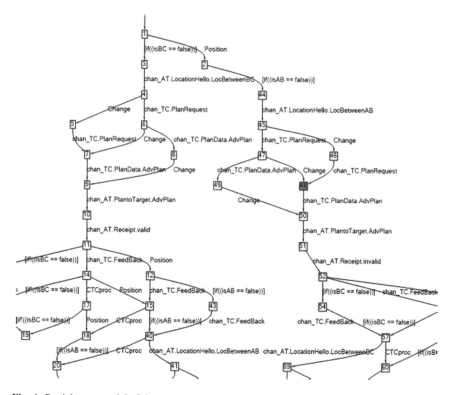

Fig. 4 Partial trace model of the process

prove that the model has a good description ability for the specification, the PAT reserved words and the LTL grammar set should be applied to describe the nature of the verification. Given a model and a property, we can obtain the traces that satisfy the property. If there is a counterexample, check the trace and find the reason. It is known that PAT has good checking attributes (Table 2).

According to the verification of features and the existing case to search the model state space, the obtained model has strong compatibility with the specification.

Table 2 Feature verification and results of plan sending in advance

No.	Feature verified	Verification language	Result
1	Deadlock	Assert SYS deadlock-free	OK
2	Reachability	Assert SYS reaches goal	OK
3	Deterministic	Assert SYS deterministic	OK
4	Nonterminating	Assert SYS nonterminating	OK
5	Divergence	Assert SYS divergencefree	OK
6	Function	Assert SYS	= [] <> FeedBack	OK

4 Conformance Verification and Empirical Analysis

As the implementation is a data-flow reactive system, it could be modeled using the same theory. The tool PAT could help us in the conformance verification of the specification and the implementation to find counter examples. Then, the conformance relation score (CRS), average conformance relation score (ACRS) and weighted conformance relation score (WCRS) are used to the result. Finally, two classical cases in ATO system (send plan in advance and start-up) should be tested and analyzed to evaluate the ability of verification.

4.1 Input and Output Conformance Relation

The CSP input-output relation (CSPIO) is an extension of ioco, and both use input and output events to define conformance. Both the specification and the implementation can be regarded as data-flow reactive systems. The specification is based on semantics and the implementation relation based on input-output LTS (IOLTS) [10]. Let SYSTEM be a CSP process for the specification of a data-flow reactive system S, A the alphabet of SYSTEM, the input/output alphabet of SYSTEM, the 3-tuple represents the CSP I/O process corresponding to S. The CSPIO relation is defined as:

$$A \; cspio \; M \; iff \forall \sigma \in L(A) : out(M \; after \; \sigma) \subseteq out(A \; after \; \sigma) \qquad (2)$$

4.2 Empirical Analysis

According to the specification, considering the plan sending example, we create implementation and get the data of SUT.

$\sigma 1 = pos? \cdot posnor$, $\sigma 1 \in L(Spe1)$, $OUT(Spe \; after \; \sigma 1) = \{posnor, plannor, valid\} \cup \{posnor, planadv, invalid\} \cup \{out!\}$. The channels of the others in the specification are hidden. $OUT(SUT1 after \sigma 1) = \{posnor, plannor, valid\}$; $OUT(SUT2 after \sigma 1) = \{posnor, planadv, invalid\}$; $OUT(SUT3 after \sigma 1) = \{posnor, plannor, valid\}$; $OUT(SUT4 after \sigma 1) = \{posnor, plannon, invalid\} \cup \{out!\}$. $\forall \sigma \in L(Spe) : out(SUT1 after \sigma) \in out(Spe after \sigma)$, so $SUT1 \in Spe$. $\forall \sigma \in L(Spe) : out(SUT2 after \sigma) \in out(Spe after \sigma)$, so $SUT2 \in Spe$. $\forall \sigma \in L(Spe) : out(SUT3 after \sigma) \notin out(Spe after \sigma)$, so $SUT3 \notin Spe$. $\forall \sigma \in L(Spe) : out(SUT4 after \sigma) \notin out(Spe after \sigma)$, so $SUT4 \notin Spe$.

Table 3 Statistics and results of the two cases

Case	CRS	ACRS	WCRS	μDeadlock (ms)	μDivergence (ms)	μDeterminsim (ms)	μcspio, PAT (s)	Accuracy (%)
SendAdvplan	0.98	0.97	0.96	61	15	18	1.21	98.9
Start-up	0.92			101	56	78	2.75	100

Even though the SUT2 is satisfied with the condition, the system cannot work. The error occurs because of interconnection. As we analyzed SUT3 by refinement, it means that when the position is unknown and the plan is normal; the plan is used by the train which is allowed by no means. It is suggested that the implicit problems that could be found out which are not easy for manual analysis. It is of great significance to verify the accuracy of the method. Together with the start-up case, the analysis results are displayed in Table 3. Here, we define the conformance relation score, conformance relation score and weighted conformance relation score which are used to evaluate the ability of verification average. It can be inferred from Table 3 that the method is of great value.

5 Conclusion

This paper proposed a kind of conformance verification method based on successful techniques complex algorithms for ATO system. Then, we come to a conclusion that when the model of specification and implementation is correct, the verification result is of great value. The quantitative analysis results can be achieved with the improvement of accuracy. However, the time of modeling cannot be calculated because of human behavior. Moreover, the formal method is not appropriate for someone that has no formal knowledge. As future work, we intend to: (1) build the ATO model using the strategy defined in [10] from natural languages, (2) analyze connectivity of ATO system and improve the method.

References

1. Du Y, Wang J (2011) The research and application of GSM network coverage based on high-speed railway. In: International conference on electrical & control engineering
2. Tretmans J (1999) Testing concurrent systems: a formal approach
3. Shao L (2013) Research and application of Bluetooth application specification consistency and interoperability test. Ph.D. thesis, Fudan University. (in Chinese)
4. Wu J (2015) Research on automatic identification method and tool of CTCS-3 level vehicle DMI interface information. Ph.D. thesis, Beijing Jiaotong University. (in Chinese)
5. Mutation TAIO-based test and evaluation of safety function for new train control system. Ph.D. thesis, 2018. (in Chinese)
6. Josephs MB (1988) A state-based approach to communicating processes. Distrib Comput 3(1):9–18
7. Sango M, Duchien L, Gransart C (2014) Component-based modeling and observer-based verification for railway safety-critical application s. In: International conference on formal aspects of component software, Springer, pp 248–266
8. Research on automatic analysis method of interconnected test of CTCS-3 train control system based on model. Ph.D. thesis, Beijing Jiaotong University, 2015
9. Nogueira S, Sampaio A, Mota A (2014) Test generation from state based use case models. Formal Aspects Comput 26(3):441–490
10. Carvalho G, Barros F, Lapschies F, Schulze U, Peleska J (2013) Model-based testing from controlled natural language requirements. Commun Comput Inf Sci 419:19–35

11. Carvalho G, Carvalho A, Rocha E, Cavalcanti A, Sampaio A (2014) A formal model for natural-language timed requirements of reactive systems. In: International conference on formal engineering methods
12. Kundu S, Lerner S, Gupta R (2007) Automated refinement checking of concurrent systems. In: 2007 IEEE/ACM international conference on computer-aided design, IEEE, pp 318–325
13. Aichernig BK, Jobstl E, Tiran S (2015) Model-based mutation testing via symbolic refinement checking. Sci Comput Program 97:383–404

Research on the Framework of New Generation National Traffic Control Network System and Its Application

Qian Li, Rusi Chu and Haiyang Wang

Abstract Based on the strategic background of transportation powerful country, China has opened up a research and construction task for a new generation of national traffic control network integrating heaven and earth, three-dimensional and intelligent. This paper is based on combing and summarizing the development status of typical national ITS and its system framework, according to existing theories, combines cutting-edge emerging technologies such as big data technology, Internet of Things, space awareness, cloud computing, and 5G communications, comprehensive use of modern scientific methods such as transportation science, systems, and management science, creatively constructed a new generation national transportation control network system framework model. Finally, this paper takes the driverless train in the field of rail transit as an example of application scenario, designs the structure of the driverless train operation control system based on the new generation national traffic control network, and describes and introduces the structure of the system to show the advancement and feasibility of the new generation of national traffic control network technology.

Keywords Traffic control network · System framework · Driverless train · Operation control

1 Introduction

In recent years, with the rapid development of cutting-edge emerging technologies, it has brought important opportunities for China's intelligent transportation innovation. ITS is a traffic-oriented service system based on data communication, electronic information, and so on [1]; the new generation of national traffic control network

---

Q. Li · H. Wang
Integrated Transport Research Center, China Academy of Transportation Sciences, Beijing 100029, China

R. Chu (✉)
School of Traffic and Transportation, Beijing Jiaotong University, Beijing 100044, China
e-mail: m13137729771@163.com

is from the national level, carries out technological innovation, upgrades improvement based on the existing intelligent transportation system, and it is a border-open complex giant system composed of vehicle, infrastructure, support operations, and service systems by combining a new generation of information technology such as Internet of Things, big data, cloud computing, and other technologies such as space-sensing, space-sensing, 5G communication, and Bei-Dou navigation system with the functions of fast and accurate traffic information perception, collection, processing, decision making, and control. And support real-time data exchange between various parts and subsystems; support system, subsystem and vehicle system to make specific adjustments (control strategy, induction and restriction measures, operation mode, service coordination and so on) according to specific conditions (real-time traffic status, passenger flow trend, meteorological conditions and so on) so that the transportation system is in the process of dynamic adjustment and optimization based on real-time data. Due to its high strain, reliability, and safety, it also supports vehicles with in-vehicle control functions to achieve autonomous operation under controlled conditions [2].

At the National Transportation Work Conference in January 2019, Li Xiaopeng, Minister of Transport, proposed to "develop vigorously to promote smart transportation and continue to promote the pilot of a new generation of national traffic control network" [3]. The system framework of the new generation national traffic control network is an integral part of the transport framework system and also an important basis for serving the strategic framework system of transportation power. Studying the system framework constructed in the context of China's national conditions, it can be used to guide the planning, design, and construction of the traffic control network system and related technical products produced by it. Based on these, on the basis of previous studies, relying on the driving force of new frontier technologies and scientific methods, this paper creatively constructs a new generation of national traffic control network system framework and demonstrates the powerful functions of the new generation of national traffic control network technology through the application scenario of driverless trains in the field of rail transit.

2 Overview of ITS in Typical Foreign Countries

Throughout the development of foreign ITS, ITS has experienced three stages as shown in Fig. 1.

Following is a brief overview of the three leading research camps in the field of ITS in the world.

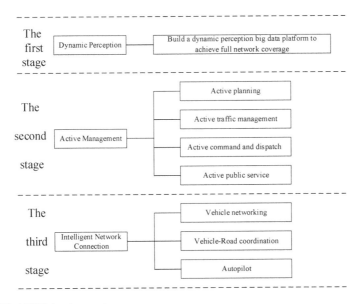

Fig. 1 World ITS development process

2.1 The USA

Currently, the USA is one of the most successful countries in the world to study the application of ITS [4]. The U.S. Department of Transportation launched the research on ITS in the 1990s and adopted the process-oriented research method to construct the national ITS system framework [5].

2.2 The European Union

At the beginning of 2000, EU constructed a national ITS framework through process-oriented research method. Its purpose is to service the ITS to be constructed and developed by the methodology adopted in the system framework and to emphasize the expansion and development of the corresponding system framework from the perspective of users' needs [6].

The EU ITS system framework pays more attention to the detailed analysis and research of key subsystems related to the long-term development of the system, which also reflects the guiding role of the ITS system framework on actual construction and planning [7].

2.3 Japan

At the end of the twentieth century, Japan adopted the object-oriented research method to construct its national ITS system framework. Its system framework draws on the characteristics of the American and European ITS system framework. The overall framework is more similar to the American ITS system framework, but more attention is paid to the synergy between the various components and basic structures of the research system.

3 National Traffic Control Network System Framework Construction Model

The national traffic control network system framework is macroscopic, constructing the top-level design and main structure of the system from the macro level, and setting a clear and universal system structure, so that the interdependence, information interaction, and transmission relationship exist between the subsystem elements are clearly visible, which improves the compatibility, integration and integration of the system.

From the development experience of ITS in typical foreign countries, such as the USA, EU, and Japan, and the content of their ITS framework (Table 1), we can see that the formulation steps of the framework of national traffic control network system are as follows: Plan, Service Architecture, Requirement Analysis, Functional Framework, Physical Architecture, Economic and Technological Assessment. This method has become a general method for establishing the model of transportation system. Its comprehensiveness, completeness, and rationality have been tested by many countries and regions. Therefore, this method can be used for reference in constructing the framework of national traffic control network system.

In the study of the national traffic control network system framework, the service framework of national traffic control network is defined firstly, then the demand analysis of user service and main service is carried out, and the demand system is constructed, after the functional framework and physical framework are formulated to determine the relationship of information architecture flow between subsystems

Table 1 ITS system framework of the USA, EU, and Japan

Name of country	ITS system framework content description
USA	User services, logical architecture, physical architecture, market package, standard system
EU	User requirements, functional architecture, physical architecture, communication architecture, organizational architecture
Japan	User services, logical architecture, physical architecture, standard architecture

and elements, emphasizing the synergy between elements. As shown in Fig. 2, the construction steps of the national traffic control network system framework are as follows:

Step1: Planning the development of the national traffic control network, including the general definition, objectives and characteristics of the national traffic control network, the development principles, and long-term planning of the national traffic control network;

Step2: User service is the starting point of establishing the framework of the national traffic control network system. Therefore, first of all, user subject and main service are defined, and both sides of the service are defined;

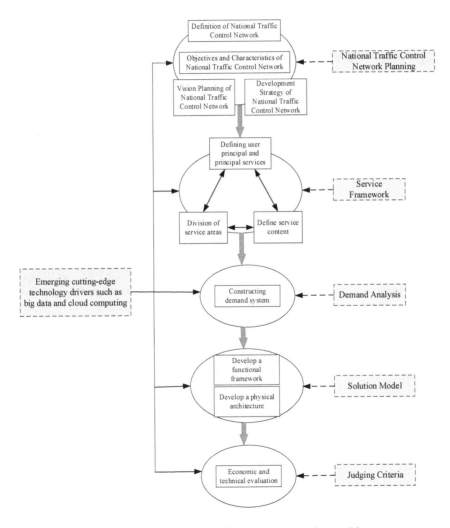

Fig. 2 National traffic control network system framework construction model

Table 2 Traffic control network system framework

Name of framework component	Content description
User subject	Served object
Service subject	Providers of services
Demand system	Identify what services users need the system to provide
Functional framework	Functional decomposition of services and organization of logical functions
Physical architecture	Put forward physical entities to implement logical functions and make clear how the system should provide specific services

Step3: Secondly, starting from the characteristics of China's transportation, we study user's needs, build a demand system, divide service areas, and define service content;

Step4: Classify and merge the functions of the system, establish the functional framework of the national traffic control network, divide the subsystems according to the functions of the system, and determine the physical structure of the national traffic control network;

Step5: Conduct economic and technological evaluation to provide relevant theoretical methods and technical support for the establishment and implementation of the new generation of national traffic control network-related projects.

The framework of national traffic control network system is mainly composed of user service, demand system, functional framework, and physical framework. The corresponding contents include user subject and service subject, traffic control network standard, traffic control network evaluation, etc. Since the system framework is based on user services, the content meaning of each part is explained from the perspective of user services and their corresponding constituent relationships. The relationship description can be seen in Table 2.

Through describing and analyzing the content of the framework of the new generation national traffic control network system, we can see that the formulation of the framework of the system of traffic control network has realized the sublimation from theoretical logic to practical application on the basis of fully considering the needs of users.

4 A New Generation of National Traffic Control Network Application Research—Driverless Train

Facing the future development direction of rail transit, according to the framework of the new generation of national traffic control network system to guide the planning and construction of the space earth integrated traffic control network system, applies the new generation of national traffic control network technology to the operation control of driverless train [8].

Table 3 Train automation level

Automation level	Train operation mode	Train adjustment during operation	Parking	Close the door	Running under emergency
GoA1	ATP	Driver	Driver	Driver	Driver
GoA2	ATO	Automatic	Automatic	Driver	Driver
GoA3	DTO	Automatic	Automatic	Automatic	Train attendant
GoA4	UTO	Automatic	Automatic	Automatic	Automatic

Train control center monitors the whole line network through a large computer, so that the signal system, inter-station links, vehicle scheduling, train operation, and other fully automated trains [9] are called driverless trains. At present, the domestic driverless trains except the Chengdu Metro Line 9 are under the unattended column automatic operation (UTO) pilot mode [10], and the rest are mostly manned to the following vehicle self-operating (DTO) mode [11]. The International Public Transport Association classifies train automation levels [12] as shown in Table 3.

The operation control system of driverless train based on the new generation national traffic control network consists of three parts: infrastructure, train and support operation and service system, and supports real-time data exchange among various parts and subsystems; support system, subsystem, and on-board system are subject to specific adjustments (control strategy, command and dispatch plan, operation mode, service coordination, etc.) according to specific conditions (real-time traffic status, passenger flow trends, meteorological conditions, etc.), so that the train system is in the process of dynamic adjustment and optimization based on real-time data, and finally realizes the fully automatic operation of the train. And its structural relationship is shown in Fig. 3.

5 Conclusion

The digitization and intellectualization of traffic, which is based on the new generation of national traffic control network, are the inevitable choice for the transformation and upgrading of China's traffic field and the realization of traffic modernization. Facing the development process of the new generation of national traffic control network, this paper combs and summarizes the development status of ITS and its system framework in typical foreign countries. Based on the existing theory, combined with the driving force of big data technology, cloud computing, 5G communication, and other frontier emerging technologies, comprehensively use modern scientific methods to build a new generation of national traffic control network system framework, and introduce its application by taking the driverless train operation control system as an example. This research is a holistic description of the complex system of traffic

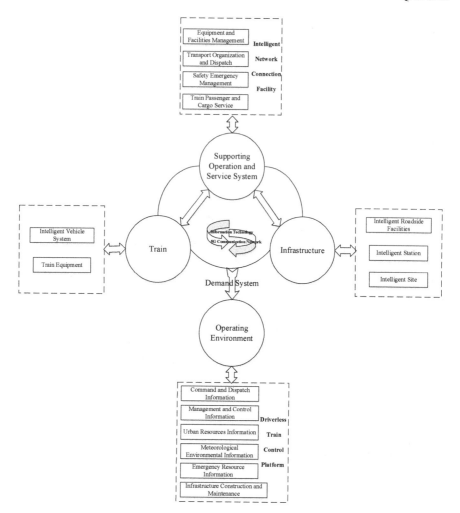

Fig. 3 Structure diagram of driverless train operation control system based on new generation national traffic control network

control network. It can clearly understand the physical composition, mutual logic, and transfer relationship between the subsystems and their sub-modules in the system. It can be used to guide the planning, design, and construction of the traffic control network system and its related products, and it will provide important support for the future construction and development of smart cities and the realization of strong transportation countries.

Acknowledgements This paper is supported by the basic research business project of the central public welfare research institutes—Research on Framework of New Generation National Traffic Control Network System (20186106).

References

1. Wu Y (2013) Discussion on intelligent transportation system and related technology. Build Saf 28(12):63–65 (in Chinese)
2. Anonymous (2015) New generation national traffic control network will be launched. China Log Purchas (20):55–55 (in Chinese)
3. Anonymous (2019) Li Xiaopeng's speech at the 2019 national transportation work conference. East China Highw (01):10–21 (in Chinese)
4. Wei Wu (2012) American road collaboration system and intelligent transportation. World Technol Econ Outlook 27(11):19–21 (in Chinese)
5. Cui W (2015) Research on the structure of intelligent transportation system in underdeveloped areas in the west. Lanzhou Jiaotong University (in Chinese)
6. Bossom R (2000) Developing architectures and systems from the European ITS framework architecture. In: IEE seminar its system architecture
7. Richard APB, Jesty PH (2005) Different types of its architectures and their uses. In: The 12th word congress on intelligent transport systems and services. The 12th its congress. San Francisco, USA
8. Qin Y, Sun X, Ma X, Wang M, Jia L, Tang T (2019) Intelligent railway 2.0 system framework and application research. J Beijing Jiaotong Univ 43(01):138–145 (in Chinese)
9. Xiao P, Liu W (2017) Discussion on the design of subway unmanned driving system. China Equip Eng (21):100–101 (in Chinese)
10. Anonymous (2019) Chengdu metro line 9 fully automatic driverless train appeared. Tunnel Constr (Chinese and English) 39(02):239 (in Chinese)
11. Zheng W (2017) System function and scene analysis in fully automatic unmanned mode. Urban Rail Transit Res 20(11):107–109 + 136 (in Chinese)
12. Zhao B (2018) Preliminary study on dispatching and commanding mode of driverless metro. China's Hi-tech (12):55–57 (in Chinese)

A Fuzzy and Bayesian Network CREAM Model for Human Error Probability Quantification of the ATO System

Jianqiang Jin, Kaicheng Li and Lei Yuan

Abstract Automatic train operation (ATO) system, whose function is to drive the train automatically, can improve the efficiency of operation. With the automation of the system, its failure depends more on human factors. How to evaluate the reliability of people becomes a problem to be solved. In order to solve the inherent uncertainty of cognitive reliability and error analysis method (CREAM), the fuzzy mathematical principle was used to realize the fuzzy CPC evaluation, and the Bayesian network was established based on CREAM's analysis process to complete the reasoning process of the control mode obtained by CPC, and it was proved that the modeling efficiency of Bayesian network was higher than that of fuzzy model. The membership degree of the control mode is output by Bayesian network, and then the human error probability (HEP) is calculated by defuzzification. Finally, the HEPs of the five drivers are calculated and the rationality is proved; the improved CREAM can provide reliable HEP of the driver in ATO system of the railway.

Keywords ATO · Fuzzy · Bayesian network · Human error probability

1 Introduction

Automatic train operation (ATO) system is one of the key technologies of intelligent railway, which can automatically control train operation under the supervision of ATP and take into account the objectives of punctuality, comfort, energy saving, and

---

J. Jin · K. Li
National Engineering Research Center for Rail Transportation Operation and Control System, Beijing Jiaotong University, Beijing 100044, China
e-mail: 17120222@bjtu.edu.cn

K. Li
e-mail: kchli@bjtu.edu.cn

L. Yuan (✉)
State Key Laboratory of Rail Traffic Control and Safety, Beijing Jiaotong University, Beijing 100044, China
e-mail: lyuan@bjtu.edu.cn

parking accuracy. With the introduction of the system, the reliability of equipment has always been the focus of research. However, the driver's errors often lead to the failure of the system in actual operation. Railway system itself is a complex man-machine system, although the new ATO functions reduce the involvement of the driver in angle of manipulation, monitoring and troubleshooting still depend on the driver. With the improvement of automation of this man-machine system, the invalidity of the system will depend more and more on human factors [1]. Therefore, human reliability analysis (HRA) is of great significance to improve the safety and efficiency of the system.

The first generation of HRA method mainly studies human output behavior errors. Its representative method is technology for human error-rate prediction method (THERP) [2]. The second generation method regards environment as the most critical factor affecting human behavior failure. So it focuses on the relationship between environment and the HEP of related people and represented by cognitive reliability and error analysis method (CREAM) [3, 4].

Fuzzy reasoning model can well deal with uncertain problems. He [5] established a simplified calculation model of HEP based on the relationship between control mode and error probability in CREAM. But based on the original structure of CREAM, the number of reasoning rules to be set will reach 8748, so the modeling work is very large and the model is not flexible enough to be adjusted easily. Bayesian network can effectively reduce the computational burden of the model. Kim [6] applied Bayesian network to CREAM and determined the control mode through probability method. Martin [7] constructed a human-induced failure Bayesian network for every basic event in collision accidents of ocean freight ships and finally obtained the HEP of each basic event.

To solve the above problems, this paper, based on the basic principle of CREAM method, comprehensively utilizes the advantages of fuzzy mathematics and Bayesian network reasoning to complete the quantitative analysis of driver's factor reliability and applies it to the quantification of HEP in ATO system.

2 Cognitive Reliability and Error Analysis Method

The basic CREAM classifies the influence of human reliability into nine factors [8] called common performance conditions (CPCs): (i) adequacy of organization, (ii) working conditions, (iii) adequacy of man-machine, (iv) availability of procedures/plans, (v) number of simultaneous goals, (vi) available time, (vii) time of the day, (viii) training and experience, and (ix) crew collaboration quality. The level of CPCs is described by different language variables. There are four characteristic control modes which are determined by the CPCs, namely scrambled, opportunistic, tactical and strategic.

As shown in Fig. 1, the graph shows all possible combinations of CPCs, adjusting the relationship between CPCs and control modes. These CPCs have a negative impact on the reliability of people on the X-axis and a positive impact on the reliability

Fig. 1 Connections between CPCs and control modes

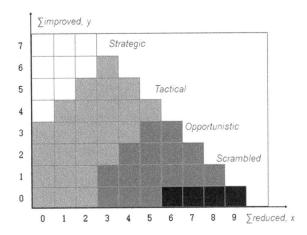

Table 1 Control modes and action failure intervals

Control mode	Probably of action failure
Strategic	(0.000005, 0.01)
Tactical	(0.001, 0.1)
Opportunistic	(0.01, 0.5)
Scrambled	(0.1, 1.0)

of people on the Y-axis. If the negative effect number of CPC is i ($i = 0, 1, 2, \ldots, 9$) and if the positive effect number of CPC is j ($j = 0, 1, 2, \ldots, 7$), we could get $x = i$, $y = j$. Among them, the relationship of x and y has 52 cases. For a given teams, individuals, or events, the level of performance of CPCs will correspond to different control modes. The interval of HEP under different control modes is shown in Table 1.

3 Methodology

The original CREAM provided a method to convert environmental assessment into human error rate, which was applied in the new system or the situation without direct data. Although the original CREAM approach is simple and clear, there are two defects: the original CREAM did not consider the uncertainty of the language assessment. It is also not good at the accuracy of the results. By using Bayesian networks, the problem of uncertain input and output can be solved. After the probability distribution of the control mode is obtained, the region center (COA) is used for defuzzing analysis to obtain the HEP.

3.1 Fuzzy Model

The fuzzy logic can be characterized as the many-valued logic with special properties aiming at modeling of the vagueness phenomenon and some parts of the meaning of natural language by graded approach [9]. It is the result of the graded approach to the formal logical systems. Mamdani's method was proposed in 1975 by Ebrahim Mamdani as an attempt to control a steam engine and boiler combination by synthesizing a set of linguistic control rules obtained from experienced human operators. Recently published works on fuzzy modeling include classification of aged wine distillates, fire crisis management, accident rate estimation in road transport of hazardous materials, analysis of system reliability, assessment of operator response time for off-normal operations in nuclear power plants and system failure engineering [10].

Define discussion scope E, Any mapping from E is in interval $[0, 1]$, $\mu_A: E \rightarrow [0, 1]$, $e \rightarrow \mu_A(e)$. It determines a fuzzy subset of E, which is denoted as A. $\mu_A(e)$ is the membership function of the fuzzy set. $\mu_A(e)$ is the degree to which element e belongs to A, called as membership degree. In order to express the degree to which eigenvalue x belongs to set A, triangular function is commonly used to express the fuzzy membership function. The eigenvalue of any triangular fuzzy number is between three points of $X[l, m, u]$, which is defined as follows.

$$\mu_A(x) = \begin{cases} x/(m-l) - l/(m-l) & x \in [l, m] \\ x/(m-u) - u/(m-u) & x \in [m, u] \\ 0 & \text{others} \end{cases} \quad (1)$$

where $l \leq m \leq u, \{x \in R, l < x < u\}$. It has continuity and convexity.

Fuzzification: The inputs for the proposed model in this research are nine CPCs, and the linguistic terms for the CPCs' level are (i) inadequate, (ii) acceptable and (iii) adequate. Usually, the level of CPCs is not a definite linguistic term. Therefore, it is necessary to fuzzify the evaluation of CPCs. The starting point of CREAM is the effects of CPCs on human reliability. The universes of discourse of the fuzzy sets for CPCs which have three effect terms are [0, 50], [10, 90] and [50, 100], and CPCs which have two effect terms are [0, 90] and [50,100], as shown in Fig. 2.

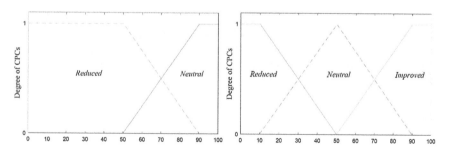

Fig. 2 Membership effects for the CPC

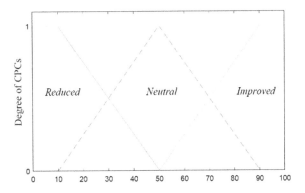

Fig. 3 Membership degree of control mode

The discourse ranges of discourse of the fuzzy sets for the control modes are $[-5.3, -2]$, $[-3, -1]$, $[-2, -0.3]$ and $[-1, 0]$ as shown in Fig. 3. The data are obtained by logarithm of the failure probability interval corresponding to the control mode.

Defuzzification: Defuzzification is to transform the uncertainty conclusion of fuzzy reasoning into accurate value. In this study, the probability distribution of the control mode can be obtained by fuzzy inference. Combined with the failure probability interval of the control mode, HEP can be obtained by logarithmic operation. The defuzzification methods include center of area (COA), maxima, mean of maxima (MOM), weighted mean of maximums (WMOM) and central average weighting (CAW) methods [11].

The COA method is a basic general defuzzification operator for ordered universes. It can be calculated by using two running sums: one coming from the left and one coming from the right [12]. The method determines the central of the area of the combined membership functions by the following equation.

$$CV = \frac{\int_X x f(x) dx}{\int_X f(x) dx} \quad (2)$$

where CV is the crisp value, $f(x)$ is the aggregated membership function for control modes, which can be calculated by $f(x) = f_{Str}(x) + f_{Tac}(x) + f_{Opp}(x) + f_{Scr}(x)$. Integral method is applied to calculate the CV. Therefore, the HEP can be obtained based on the following equation, $HEP = 10^{cv}$.

3.2 Bayesian Network

Bayesian network with N nodes can be represented by $N = \langle \langle V, E \rangle, P \rangle$. Directed graphs $\langle V, E \rangle$ imply the conditional independence hypothesis, that is, given the parent set $Pa(V_i)$. The parent set Vi is conditionally independent from the non-descendant set $A(V_i)$: $P(V_i|pa(V_i), A(V_i)) = P(V_i|pa(V_i))$. When evidence e is given, there

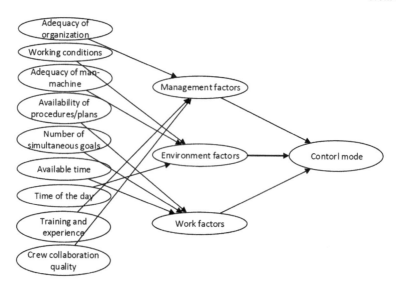

Fig. 4 Bayesian network for determining the distribution of the control modes

are $P(V|e) = \frac{p(V,e)}{P(e)} = \frac{p(V,e)}{\sum_v P(V|e)}$. As shown in Fig. 4, the first layer corresponding to the CPCs is the input layer, which can be calculated by fuzzy logic theory for the CPCs. There are nine CPCs, thus there are $3^7 \times 2^2 = 8748$ possible states. So the CPCs are divided into three groups as shown in the second layer for reducing the calculation load which can effectively reduce the amount of computation. The third layer in the network is the control mode, which is the output of the network.

4 Case Study

4.1 Data Input

The input of Bayesian network is divided into two parts: the prior probability of root node variables and the conditional probability of intermediate nodes. We worked with the drivers and managers at the phase of ATO system test, got data through questionnaires and discussions. The conditional probability of intermediate nodes is directly defined according to the operating environment characteristics of railway industry.

Five drivers were invited to self-evaluate their feelings about the nine CPCs. Each CPC score ranged from 0 (inappropriate) to 100 (appropriate). According to the membership function of 9 CPCs, the input of Bayesian network–membership can be calculated according to Fig. 2, as shown in Table 2.

Table 2 Fuzzy survey results of five drivers

No.	Effects	CPC 1	CPC2	CPC3	CPC4	CPC 5	CPC6	CPC7	CPC8	CPC9
1	Improved	0.75	0.55	0.75	1	0	0	1	1	0.75
	Neutral	0.25	0.45	0.25	0	0.65	0.75	0	0	0.25
	Reduced	0	0	0	0	0.35	0.25	0	0	0
2	Improved	0.5	0.375	0.375	1	0	0	0.75	0.75	0.9
	Neutral	0.5	0.625	0.625	0	0.4	0.625	0.25	0.25	0.1
	Reduced	0	0	0	0	0.6	0.375	0	0	0
3	Improved	0.875	0.45	0.5	0.95	0	0	0.775	0.775	0.95
	Neutral	0.125	0.55	0.5	0.05	0.5	0.65	0.225	0.225	0.05
	Reduced	0	0	0	0	0.5	0.35	0	0	0
4	Improved	0	0.4	0.625	0.875	0	0	0.625	0.625	0.65
	Neutral	1	0.6	0.375	0.125	0.25	0.5	0.375	0.375	0.35
	Reduced	0	0	0	0	0.75	0.5	0	0	0
5	Improved	0.95	0.5	0.7	1	0	0	0.25	0.25	0.75
	Neutral	0.05	0.5	0.3	0	0	0.25	0.75	0.75	0.25
	Reduced	0	0	0	0	1	0.75	0	0	0

4.2 Bayesian Reasoning and Failure Probability Calculation

As shown in Fig. 5, the probability distributions of the control modes for driver No. 1 are 54% (strategic), 22% (tactical), 20% (opportunistic) and 4% (scramble). The next step is to defuze to get the exact value.

The probability of the control mode is input into Fig. 6. In the calculation, the upper and lower limits of the integral are the intersection of the graph with the X-axis or the intensity of the control mode.

Therefore, according to Formula 2, the HEP for driver No. 1 is as follows: $HEP_{d1} = 10^{(-7.8376/2.0336)} = 1.40 \times 10^{-4}$. The results of probability distribution for the control modes and human error probability for the five drivers can be seen in Table 3.

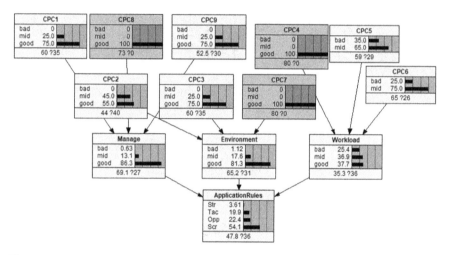

Fig. 5 Bayesian network reasoning results for driver No. 1

Fig. 6 Control mode results for driver No. 1

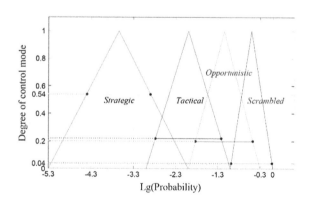

Table 3 Control mode and HEP of five drivers

No.	Str (%)	Tac (%)	Opp (%)	Scr (%)	CV	HEP
1	54	22	20	4	−3.854	1.40e−4
2	38	29	27	6	−2.537	2.90e−3
3	43	27	25	5	−2.806	1.56e−3
4	28	31	30	11	−1.945	1.14e−3
5	33	32	30	5	−2.251	5.61e−3

In this research, all drivers have experience with CTCS systems other than ATO systems. For the other drivers, especially because of his seniority and hierarchy, No. 1 plays the role of the teacher. Nos. 4 and 5 are new drivers of ATO system. According to Table 3, most of the drivers are associated with the dominant control mode of strategic, and if we only get this solution, we will not see the difference between each person. But as the exact value is obtained, we can recognize new drivers who are significantly more likely to make mistakes than older drivers which are also with common sense.

5 Conclusion

This paper studies the quantification of driver's HEP in ATO system of the railway. The fuzzy mathematics principle is used to realize the fuzzy CPC evaluation, and the Bayesian network is established based on CREAM's analysis process to complete the reasoning process of the control mode obtained by CPC, effectively solving the inherent uncertainty problem in CREAM. Only the conditional probability in the reasoning was defined according to industry characteristics, and the corresponding mode of basic CREAM was restored to the greatest extent. Finally, the HEP was obtained by defuzzification calculation of the output of Bayesian network. This method can not only reduce the amount of calculation, but also ensure the accuracy of analysis results. Therefore, as a rapid prediction method of HEP, it has application value in quantitative risk analysis. The HEP results can be used for reference by managers, such as targeted training, which is also the following research of this paper.

Acknowledgements This work was supported by the National Natural Science Foundation of China (No. U1734210) and by National Engineering Laboratory for Urban Rail Transit Communication and Operation Control.

References

1. He X, Tong J, Huang X (2005) Application of Bayes method in estimation of human error probability. J Tsinghua Univ 45(6):843–845
2. Marseguerra M, Zio E, Librizzi M (2010) Human reliability analysis by fuzzy cream. Risk Anal 27(1):137–154
3. Hollnagel E (1998) Cognitive reliability and error analysis method (CREAM)
4. Proctor RW, Van Zandt T (2018) Human factors in simple and complex systems. CRC press
5. He X, Wang Y, Shen Z, Huang X (2008) A simplified cream prospective quantification process and its application. Reliab Eng Syst Saf 93(2):298–306
6. Kim MC, Seong PH, Hollnagel E (2006) A probabilistic approach for determining the control mode in cream. Reliab Eng Syst Saf 91(2):191–199
7. Martins MR, Maturana MC (2013) Application of bayesian belief networks to the human reliability analysis of an oil tanker operation focusing on collision accidents. Reliab Eng Syst Saf 110:89–109
8. Zhou Q, Wong YD, Loh HS, Yuen KF (2018) A fuzzy and bayesian network cream model for human reliability analysis—the case of tanker shipping. Saf Sci 105:149–157
9. Novák V, Perfilieva I, Mockor J (2012) Mathematical principles of fuzzy logic, vol. 517. Springer Science & Business Media
10. Konstandinidou M et al (2006) A fuzzy modeling application of CREAM methodology for human reliability analysis. Reliab Eng Syst Saf 91(6):706–716
11. Runkler TA (1996) Extended defuzzification methods and their properties. In: Proceedings of IEEE 5th international fuzzy systems, vol. 1. IEEE, pp 694–700
12. Van Leekwijck W, Kerre EE (1999) Defuzzification: criteria and classification. Fuzzy Sets Syst 108(2):159–178

Operation Organization Methods for Automated People Mover Systems at Airports

Wenqian Liu, Xiaoning Zhu and Li Wang

Abstract In recent years, China's air passenger throughput has grown rapidly; some new airports being built or planned for expansion are designed with more and larger terminals and facilities that may be spread over large areas. Accompanied with the increasing scale and complexity of airports, the automated people mover (APM) systems are in high demand which provide safe, reliable and efficient passenger transport services on relatively closed, independent lines. Based on the brief description of current situation of APM systems, this paper takes Beijing Capital International Airport as an example to analyze the passenger flow. Subsequently, we propose the organization methods of passengers and the operation planning for APM systems in terms of influence factors so as to ensure the efficiency of APM systems at airports. Finally, the adjustment methods of operation planning are addressed considering the fluctuation of daily passenger flow in the actual operation.

Keywords Automated people mover system · Airport · Passengers organization · Operation planning

1 Introduction

An automated people mover (APM) at an airport is a transportation system with fully automated operations to allow passengers to move more quickly over longer distances between dispersed terminals providing passengers with full-time service.

APM system originated in the USA and is widely used in large airports in Europe and the USA[1–4]. Comparatively speaking, the development of China's airport

---

W. Liu · X. Zhu (✉) · L. Wang
School of Traffic and Transportation, Beijing Jiaotong University, Beijing 100044, China
e-mail: xnzhu@bjtu.edu.cn

W. Liu
e-mail: 18120839@bjtu.edu.cn

L. Wang
e-mail: liwang@bjtu.edu.cn

APM system is still in its infancy; only Beijing Capital International Airport has implemented APM systems, while there are 44 APM systems operating at airports worldwide as of 2010 [5, 6].

Under the background of the new economic situation and the major strategic layout of "The Belt and Road", air transport plays a very important role in promoting the development of "The Belt and Road" because of its unique advantages such as high degree of internationalization, fast speed and being able to meet the demand for long-distance access. According to the latest civil aviation statistics, China has 37 airports with an annual passenger throughput of more than 10 million by 2018. Large hub airports, especially those with multiple terminal units, have the characteristics of large terminal size and a long distance from the check-in counter to the boarding gates [7], therefore auxiliary facilities for transporting passengers must be used. Compared with automatic footpath system and ferry vehicles, APM system is an important means of passenger transportation which has the advantages of fast, safe and convenient.

Recognizing the importance of mobility to passengers as well as employees, some airports such as Shanghai Hongqiao International Airport, Shenzhen Bao'an International Airport, Chongqing Jiangbei International Airport are planning to operate APM systems [8, 9]. With the growth of air passenger volume and the increasing scale and complexity of the airport, it is urgent to build APM systems to provide reliable and efficient passenger transport services on relatively closed and independent lines.

APM systems are critical to the operation of many airports because they provide the fastest and sometimes the only means of movements of passengers within the airport. Serious problems can occur when the APM systems at airports fail to operate properly or stop entirely. In view of the current development status of APM systems at airport, this paper mainly studies the operation and organization methods of APM systems to provide guidance for the planning and design of APM systems of large hub airports.

2 Passenger Organization for APM Systems

Operation organization methods for APM systems at airports are closely related to the passenger organization at airports [10], which depend on the changing rules of passenger volumes and their flow directions at different times of the day.

2.1 Passenger Flow

In the organization of passenger flow, it is necessary to grasp and understand the different needs of passengers in different conditions, so as to provide corresponding services and realize the scientific and reasonable operation of APM systems at airports.

According to the different characteristics of flights and APM stations, it can be further divided into domestic departing passengers, international departing passengers, domestic arriving passengers, international arriving passengers, domestic transit passengers and international transit passengers. This section takes the APM system of Beijing capital international airport as an example to analyze the types and flow of passengers taking APM.

The APM system of Beijing Capital International Airport was completed and used in 2008; the total length of the APM line is 2090 m, and three stations are put into use along the line as shown in Fig. 1.

Passengers at T3C station. Departing passengers who have checked in and gone through security check procedures can go to the T3C APM station according to the guiding signs and get on the APM vehicles from the middle platform to the next station. Arriving passengers who arrive at T3C station by APM get off from the broadside platforms and proceed to the luggage claim to pick up their luggage and finally leave the airport.

Passengers at T3E station. International arriving passengers go to the T3E APM platform by special escalator and get on the vehicles from the middle platform to next stations. International departing passengers arriving at T3E station can get off from the broadside platforms and then go to the boarding gates after passing frontier inspection, quarantine and customs declaration successively.

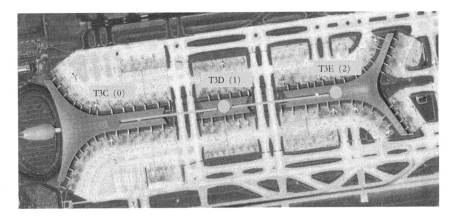

Fig. 1 APM system of Beijing capital international airport

Passengers at T3D station. T3D station mainly provides services for domestic departing and arriving passengers. Passengers here get on the APM vehicles from the middle platform and join the international passengers to the next station.

2.2 Passenger Organization

Passenger flow organization refers to the process of organizing passengers by reasonably distributing related equipment and facilities of passenger transport and taking effective diversion or guiding measures for passenger flow so as to ensure the safety of passenger, avoid congestion and minimize passenger travel time. Specifically speaking, the passenger transport plan that conforms to the actual situation of APM system is developed according to the layout of station facilities and the demand of passengers in different time periods.

For the configuration and operation of each channel and related facilities of the APM system, the baggage carried by passengers and passengers walking speed should be fully taken into consideration in view of the capacity of each facility to provide corresponding services flexibly in order to meet passenger expectations.

The organization of passengers at the platform. First of all, it is necessary to strengthen the guidance for passengers waiting for APM vehicles. By placing railings, arranging guiding signs and adding staff, passengers can be as evenly distributed as possible while waiting vehicles to arrive. The second is to organize the passengers to line up in order to prevent passengers from crowding each other. Finally, passengers are ought to leave quickly and avoid long delays on the platform.

The organization of passengers in the channels. The organization of passengers in the channels should reduce the intersection and convection of the passenger lines as much as possible, which can be controlled by setting isolation belts and arranging staff guidance, so as to ensure the smooth passage of passengers.

3 Operation Planning for APM Systems

In order to reasonably organize the passengers of APM systems at large hub airports, the passenger flow plan of the APM systems should be compiled in terms of passenger flow statistics and flight regularity; on this basis, fleet sizing, operation zone and stop schedule plan should be determined. Only by properly organizing the operation of APM systems, can the passengers be transported safely, quickly and conveniently, and at the same time, the APM vehicles and various equipment can be used economically and reasonably.

3.1 Compositions

The operation planning for APM systems consists of operation zone, fleet sizing and stop schedule plan.

Operation zone. The operation zone of APM system is composed of the starting and ending stations of the trains and their route alignment. In most cases, the best system is the simplest system that will fulfill the planning criteria with only one route. Therefore, it is only necessary to determine the starting and ending stations of the APM system on this route.

Fleet sizing. Fleet sizing refers to the number of APM vehicles determined in a certain direction or in a certain zone in order to meet the demand of passenger flow, with taking the fluctuation of passenger flow into consideration. This is a crucial aspect of the overall planning process since it will dictate the efficiency of operation of the APM system and greatly influenced the APM's cost.

Stop schedule plan. The stop schedule plan mainly focuses on station dwell time and the stop modes which is to decide whether the APM trains will stop at some stations and how long it will stay. For the APM system of large hub airports, the mode of stopping at each station is generally adopted, and the station dwell time is mainly used for passenger boarding and alighting, which is generally a fixed value.

3.2 Influence Factors

The operation planning is affected by the passenger flow, maximum train length, line capacity as well as train performance, among which the passenger flow has the greatest impact.

One of the foundations for determining the APM operation planning is the estimation of ridership demand. Passenger flow in one day, one week, one month, one quarter and one year may be significantly different affected by various factors such as time, weather, holidays; the actual passenger flow is changing all the time, so there is a large imbalance in the passenger flow demand. Information including a flight schedule for one day and aircraft board/deboard rates, walk distances, walk speeds, flight crew factors, airport employee factors and so on is applied to the ridership analysis to determine APM system ridership throughout the design day.

3.3 Operation Planning Method

The operation planning can be determined by the following steps:

Determine the design capacity of each vehicle α_{max} and the length of trains n_α which is the number of vehicles.

Determine the passenger flow ρ_i in terms of per hour per direction.

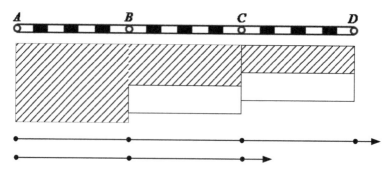

Fig. 2 Schematic diagram of passenger flow in operation zones

Determine the number of trains n_i of each operation zone in different time periods to generate sufficient hourly capacity compared to the surged peak hour demand, which can be calculated using the following formula:

$$n_i = \rho_i / (\alpha_{max} * n_\alpha * \beta) \qquad (1)$$

β represents the level of service artificially set.

The passenger flow in operation zones is shown in Fig. 2, in which A, B, C, D represent the stations along the APM route, while each square represents a different passenger flow between stations. And the specific stop schedule plan can be indicated by arrow lines.

Due to the fluctuation of daily passenger flow at the airport, the operation planning can be adjusted to increase or decrease the transportation capacity through the following methods:

1. Change the fleet size, which means to increase or decrease the number of vehicles of each train. When the number of passengers reaches its daily peak, all vehicles can be utilized to the maximum extent.
2. Change the interval time, which is to shorten or extend the stopping time of the APM trains at stations.
3. Change the stop schedule plan. The APM trains can drive directly through some stations of small number of passengers instead of stopping at each station.

It plays an important role in improving transportation capacity of APM system and equipment utilization to adjust the operation planning flexibly in the actual operation.

4 Conclusion

The rapid growth of air passenger volume as well as the increasing scale and complexity of airports call for high requirements for APM systems. This paper mainly studies the methods of passenger organization and operation planning for APM systems at airports. First, the common types of passengers in the airport are divided, and the flow lines of each type of passengers are analyzed based on the Beijing capital international airport. Moreover, we put forward scientific and reasonable operation organization methods to determine the fleet sizing, operation zone and stop schedule plan of APM systems. Finally, the adjustment methods of operation planning are proposed in view of different amounts of passengers in different time.

The development level of modern air transport industry indicates the degree of political stability and economic prosperity of a country, which is an important field worthy of concern. Therefore, the construction and reasonable operation of APM systems at airports will be a valuable research direction.

Acknowledgements This paper was sponsored by Civil Aviation Administration of China Major projects (No: 201501).

References

1. Gregori G, Moore HL (1997) Design, procurement, and construction of the Leonardo Da Vinci international airport automated people mover. In: International conference on automated people movers
2. Bondada MVA (1997) Florida: the world's showpiece for automated people movers. In: International conference on automated people movers
3. Graham JL, Woodend DL (1997) Los Angeles international airport master plan integration of automated people mover concepts. In: International conference on automated people movers
4. Shen LD (1992) Implications of automated people movers for airport terminal configurations. Ite J 62(2):24–28
5. Elliott L, Incorporated (2012) Guidebook for measuring performance of automated people mover systems at airports
6. Council N (2010) Guidebook for planning and implementing automated people mover systems at airports. ACRP Rep
7. Yu L (2017) Study on automated people mover system of shanghai Pudong international airport. Super Sci 8:142–143
8. Ding Z (2018) Research on the operation management model of airport passenger rapid transit system. Sci Technol Vision 29:269–270 (in Chinese)
9. Wang J (2019) Selection of passenger rapid transit systems for Xianyang international airport in Xi'an city. Urban Mass Transit 22(1):115–119. https://doi.org/10.16037/j.1007-869x.2019.01.025 (in Chinese)
10. Lin W (2018) The overall design of airport landside RTS. Archit Eng Technol Des (30):762–764, 1234 (in Chinese)

Real-Time Train Detection Based on Improved Single Shot MultiBox Detector

Yuchuan Xu, Chunhai Gao, Lei Yuan and Boyang Li

Abstract Autonomous environmental perception can be used in urban rail transit to extend the driver's view. By installing cameras in the front of the train, the running line ahead can be detected. With the help of deep learning algorithm, drivers can identify trains at a long distance. However, complex environment makes the algorithm less robust to the identification of small targets. Therefore, the real-time and accuracy of front-train detection needs to be improved. In this paper, an improved Single Shot MultiBox Detector (SSD) algorithm is proposed. Compared with traditional image recognition method and original SSD, this method has higher accuracy in detection of the train, especially the small one. Moreover, it processes the images faster than traditional method. Experiments show that our method is very robust for the train detection in various illumination environment such as shadow, reflection, glare and high noise, and it reaches 95.38% mean average precision (mAP) and 26.3 frames per second (FPS) on our self-made dataset.

Keyword Real-time · Train detection · Deep learning

1 Introduction

When the train control system breaks down, the driver needs to drive the train visually. However, varied illumination environment and complex circuit structure make the frontal train which is at a long distance hard to distinguish. Therefore, relying on drivers to run the train visually has potential hazards. Hence, computer-assisted

---

Y. Xu
School of Electronic and Information Engineering, Beijing Jiaotong University, Beijing, China

C. Gao · L. Yuan (✉)
State Key Laboratory of Rail Traffic Control and Safety, Beijing Jiaotong University, Beijing, China
e-mail: lyuan@bjtu.edu.cn; 16120266@bjtu.edu.cn

B. Li
Beijing Aiforail Technology Co., Ltd, Beijing, China

train operation has become the mainstream research direction. In this paper, we constructed a train detection algorithm based on deep learning which can provide warning information to drivers.

Currently, many researchers have proposed different frontal train detection schemes, firstly the traditional image processing method, histogram of oriented gradient (HOG features) [1] and SVM classifiers [2] and secondly object detection with lidar [3–5]. However, the processing speed of traditional method is relatively slow which cannot meet the real-time demand, and the accuracy is low because manually designed features can only extract shallow semantic information. The lidar can only get the distance of the object in front but cannot obtain what the object is.

With the advancement of computer vision and deep learning algorithms in recent years, a large amount of image data can be processed in a very short time. Through the video captured by cameras mounted on the train, we can use the deep learning algorithm to identify the train in images. However, original SSD [6] still leaves much to be desired, especially feature extraction ability as well as the detection accuracy. Therefore, the algorithm is improved in this paper to increase accuracy and speed.

2 Real-Time Train Detection Based on SSD

In order to improve the robustness of neural network to detect small target and the ability to extract features in multivariable illumination, we designed an improved SSD network structure, which is shown in Fig. 1. It is mainly divided into three parts; the first part is feature extraction network. We established a 50-layer architecture based on skip connection [7]; the second part is feature fusion structure. Main improvement of this part is deconvolution layers [8], which are used to fuse the high-level and low-level features. The third part is prediction part; a modified structure is established to further improve the detection accuracy. The detailed layers architecture of the proposed network is shown in Table 1.

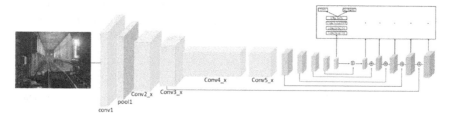

Fig. 1 Network architecture

Table 1 Detailed layers of the proposed network

Layer	Layer type	Size of feature map	Number of feature maps
1	Conv1	161 × 161	64
2–10	Max pooling	80 × 80	64
	Conv2_x	80 × 80	256
11–22	Conv3_x	40 × 40	512
23–40	Conv4_x	20 × 20	1024
41–49	Conv5_x	20 × 20	2048
50	Conv6	20 × 20	2048
51	Conv7	10 × 10	2048
52	Conv8	5 × 5	1024
53	Conv9	3 × 3	1024
54	Conv10	1 × 1	1024
55	Deconv1	3 × 3	512
56	Deconv2	5 × 5	512
57	Deconv3	10 × 10	512
58	Deconv4	20 × 20	512
59	Deconv5	40 × 40	512

2.1 Main Architecture Based on ResNet-50

In the first part of the network, we designed a 50-layer residual neural network based on the skip connection idea, which is the main structure of the residual block. Then, we added five convolution layers on the last of ResNet-50 to compress the feature map resolution. At the end of residual block, the feature map is resized to 1/16 of the original input size. And through the added convolution layers, the resolution is compressed to 1 × 1, which is highly extracted. Figure 2 shows detailed structure of the residual block.

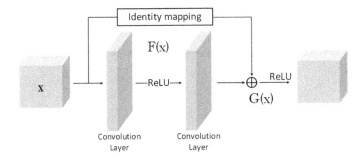

Fig. 2 Skip connection of the residual block

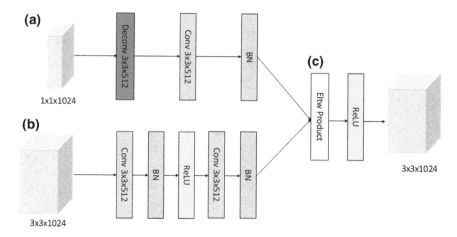

Fig. 3 Detail of the deconvolution layer

2.2 Deconvolution Layer

High-level network has a larger receptive field than low-level; thus, the semantic information can be represented strongly, while the geometric information will be expressed weakly. Therefore, we designed an improved structure inspired by the deconvolution idea. The detail of the deconvolution part is shown in Fig. 3. The A part is a higher feature map and the B part is a lower one. By fusing features from high and low layers, the feature can be extracted effectively under variable light. In addition, the detection accuracy of the train which has a long distance can be significantly improved.

2.3 Improved Prediction Part

Original SSD selected multiple feature maps to create the bounding boxes and sent them to non-maximum suppression [9]. MS-CNN [10] points out that improving the sub-network of each branch can improve accuracy. Based on this, we constructed a residual architecture which is added after each prediction layer. The network is shown in Fig. 4.

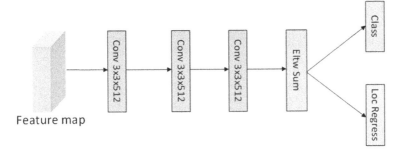

Fig. 4 Detail of the improved prediction part

3 Experiments and Results

3.1 Dataset and Model Training

In order to verify the robustness and processing speed of our algorithm, we have collected video from the high-definition camera which is installed on the train heads. The dataset is composed of 4740 images which are gathered from, Hong Kong, Mass Transit Railway. Every image has a resolution of 1280 * 720. The dataset contains a variety of different lighting and weather. We divided it into two parts, 80% for training and 20% for testing. Then, we labeled each train in the images manually. The dataset example is shown in Fig. 5.

Our experiments are implemented on Caffe [11] framework. After times of training and comparison, we have tested the best group of parameters which are shown in Table 2. Besides, our training loss and testing accuracy is shown in Fig. 6.

Fig. 5 Dataset example

Table 2 A group of optimal parameters

Momentum	Number of iterations	Minibatch size	Learning rate
0.9	50,000	32	0.00025

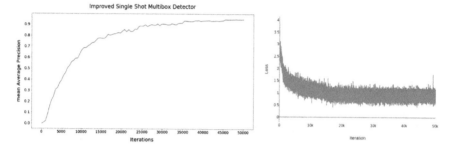

Fig. 6 Training loss and testing accuracy

3.2 Results and Analysis

To understand the effectiveness of the additions to original SSD, multiple sets of comparison experiments were carried out with different input resolutions and different additive structures. The results are shown in Table 3. PP represents the prediction part, and DL means the deconvolution layers. The original SSD is based on VGG16, and it shows 81.2% mAP on the dataset. By replacing the main network with ResNet-50 and ResNet-101, the accuracy reaches 89.2 and 92.1%, which means deeper main network has the advantages in feature extraction. In consideration of the processing speed, ResNet-50 is chosen as the main network. With the addition of prediction part and deconvolution layers, we can see from Table 3 that the accuracy reaches 95.4%, which proves that our addition structure is valid.

In this set of experiments, we compared the accuracy and speed of the algorithm by changing different main networks and input resolutions. It can be seen that selecting a deeper residual network can achieve higher accuracy, but the speed is only a little faster than the traditional method (HOG feature), which cannot meet the real-time

Table 3 Effects of various parts on the self-made dataset

Network architecture	mAP
SSD 321	81.2
SSD 513	85.6
ResNet-101 SSD 321	92.1
ResNet50-SSD 321	89.2
ResNet50-SSD 321 + PP	90.6
ResNet50-SSD 321 + PP + DL (Eltw-sum)	93.5
ResNet50-SSD 321 + PP + DL (Eltw-prod)	95.4

Table 4 Different sets of testing structure

Architecture	Main network	mAP (%)	FPS	GPU	Input resolution
HOG + SVM	\	56	8	GTX1080Ti	1080 × 720
YOLO	\	74	50	GTX1080Ti	360 × 360
SSD 300	VGG16	81.96	45.3	GTX1080Ti	300 × 300
SSD 512	VGG16	83.21	41.5	GTX1080Ti	512 × 512
SSD 321	ResNet-50	85.44	36.8	GTX1080Ti	321 × 321
SSD 321	ResNet-101	89.60	28.3	GTX1080Ti	321 × 321
Deconv-SSD 321	ResNet-50	95.38	26.4	GTX1080Ti	321 × 321
Deconv-SSD 321	ResNet-101	96.77	16.5	GTX1080Ti	321 × 321
SSD 513	ResNet-50	87.63	23	GTX1080Ti	513 × 513
SSD 513	ResNet-101	89.31	25.3	GTX1080Ti	513 × 513
Deconv-SSD 513	ResNet-50	96.37	15.2	GTX1080Ti	513 × 513
Deconv-SSD 513	ResNet-101	97.49	12.9	GTX1080Ti	513 × 513

requirements in our daily operation. In the last group of test, the accuracy reaches 97.46%, which is higher than our algorithm (95.38% mAP). But the processing speed is only 12.9 FPS, which is only a little higher than the traditional method (8 FPS). Thus, this method is not suitable for our study. After comparing multiple sets of testing structure, we found that the proposed algorithm achieves the best balance between the accuracy and processing speed. The details of different sets of testing structure are shown in Table 4. The partial testing results using the proposed method are shown in Fig. 7.

4 Experiments and Results

In this paper, we proposed a real-time and high-accuracy train detection method based on improved Single Shot MultiBox Detector. The proposed method can detect trains in various illuminations such as shadow, low contrast and high noise. In addition, it is highly robust to the detection of small targets.

Compared with traditional object detection method such as HOG feature and deep learning algorithms like YOLO and original SSD, our method achieves higher detection accuracy (95.38% mAP) and faster processing speed (26.4 FPS) on our self-made dataset, which satisfied our daily operating detection requirements.

Fig. 7 Detection example of the proposed method

Acknowledgements This work was supported by National Engineering Laboratory for Urban Rail Transit Communication and Operation Control and by Beijing Laboratory of Urban Rail Transit and National Engineering Laboratory for Urban Rail Transit Communication and Operation Control.

References

1. Dalal N, Triggs B (2005) Histograms of oriented gradients for human detection. In: IEEE computer society conference on computer vision & pattern recognition 2005
2. Tong L et al (2012) Railway obstacle detection using onboard forward-viewing camera. J Transport Syst Eng Inf Technol 12(4):79–83 + 134
3. Stein D, Spindler M, Lauer M (2016) Model-based rail detection in mobile laser scanning data. In: IEEE intelligent vehicles symposium

4. Hmida HB et al (2013) Knowledge base approach for 3D objects detection in point clouds using 3D processing and specialists knowledge. Eprint Arxiv 5.1 and 2(2013):1–14
5. Mockel S, Scherer F, Schuster PF (2003) Multi-sensor obstacle detection on railway tracks. In: IEEE IV2003 intelligent vehicles symposium. Proceedings, IEEE
6. Liu W et al (2015) SSD: single shot multibox detector
7. He K et al (2015) Deep residual learning for image recognition
8. Zeiler MD et al (2010) Deconvolutional networks. In: 2010 IEEE computer society conference on computer vision and pattern recognition, IEEE
9. Neubeck A, Van Gool LJ (2006) Efficient non-maximum suppression. In: International conference on pattern recognition
10. Cai Z et al (2016) A unified multi-scale deep convolutional neural network for fast object detection
11. Jia Y et al (2014) Caffe: convolutional architecture for fast feature embedding

Application of Intelligent Video Surveillance Technology for Passenger Flow Detection in Urban Rail

Shi yao Yu, Zi teng Wang, Yuan Zhao, Xi li Sun, Ai li Wang, Xiang ling Yan and Zheng yu Xie

Abstract With the development of the society, the intelligent video surveillance technology, which is an important part of the safety precaution measures, is more and more used in various industries. At the same time, it has been widely used in the rail transit industry because of the high-speed development of the rail transit industry. This paper introduces the development status of passenger flow detection technology based on intelligent video surveillance in various industries and analyzes the present situation and necessity of its application in rail transit industry, and then a rail transit passenger flow monitoring system based on intelligent video surveillance is proposed, and its advantages are analyzed.

Keywords Intelligent video surveillance · Rail transit · Passenger flow detection

---

S. Yu (✉) · Z. Wang · Y. Zhao · X. Sun · A. Wang
SinoRail Network Technology Research Institute, No. 22 Shouti Nanlu, Haidian District, Beijing, People's Republic of China
e-mail: yushiyao@sinorail.com

Z. Wang
e-mail: wangziteng@sinorail.com

Y. Zhao
e-mail: zhaoyuan@sinorail.com

X. Sun
e-mail: sunxili@sinorail.com

A. Wang
e-mail: wangaili@sinorail.com

X. Yan · Z. Xie
School of Traffic and Transportation, Beijing Jiaotong University, No. 3 Shangyuancun, Haidian District, Beijing, People's Republic of China
e-mail: yanxiangling@bjtu.edu.cn

Z. Xie
e-mail: xiezhengyu@bjtu.edu.cn

1 Introduction

Passenger flow detection technology based on intelligent video surveillance has been applied and developed in various industries. With the increasing demand for people's travel, the large-scale construction of road networks and the increasing traffic volume, the rail transit industry has increased demand for passenger flow information management and control. However, many products and application technologies need to be further improved in accuracy, real-time, environmental personalized adaptability and so on. The effectiveness of the existing manufacturers' technical schemes is lack of verification. The personalized environment in the station leads to great risks in large-scale engineering construction. The popularity of passenger flow intelligent detection and control technology is the "industry pain point". At the same time, due to the large difference between foreign passenger flow and China, its mature experience in passenger flow information collection and safety status assessment cannot be directly applied to China's rail transit passenger flow management. The application and landing of passenger flow monitoring technology based on intelligent video in the rail transit industry also needs to be combined with the needs of individualized scenes. Through technology integration and optimization, we will continue to adapt to the needs of the industry, improve the safety of rail transit passenger flow and further promote the healthy development of China's rail transit industry.

2 Development Status of Passenger Flow Detection Technology Based on Intelligent Video Surveillance in Various Industries

2.1 Large Shopping Malls

Passenger flow statistical analysis, which is based on video, is an important market research method. According to statistics, more than 90% of large-scale shopping malls and chain commercial outlets in developed countries such as Singapore, Japan, South Korea, Europe and America have been widely used in real-time passenger flow analysis systems, such as Dubai City, Plaza Shopping Center.

Passenger flow analysis is an important means of market research, which must be carried out before all large chain business halls, shopping malls; shopping malls and chain commercial outlets in foreign countries make market decisions and management decisions. In the application of large shopping malls, the analysis of customer flow statistics provides an effective scientific basis for management [1]. By studying passenger flow statistics, it is possible to increase customer sales opportunities, turn viewers into commodity consumers, and maximize the potential of consumers to enhance the competitiveness of shopping malls. The traffic flow system based on intelligent video surveillance not only needs to feedback statistical information in real time, but also requires the passenger flow statistics system to be stable, which brings new challenges to passenger traffic statistics.

In China, due to the relative implementation cost, the development of the enterprise scale, and the limitations of management and retailers' own use capabilities, the application is basically in its infancy. However, there are still many business management managers who use passenger flow monitoring methods based on video analytics technology, such as Tianjin Jinhui Plaza, Dalian Wanda, Shanghai Central Plaza. However, at present, it only stays in simple passenger flow statistics and does not realize personalized statistics and deep data analysis mining.

2.2 Public Safety Field

According to the "Guiding Opinions on Further Strengthening the Construction of Social Security Prevention and Control System" issued by the Ministry of Public Security, the Ping An City project has entered a new normal state, and about 30% of the investment is for video surveillance [2]. When an emergency occurs in a public place, video surveillance technology can be used to grasp the situation and development situation of the site at the first time, and it can accurately and intuitively record the impact data of the site, which can help the safety management department accurately grasp the scene of the accident. For large passenger flow public places such as stations and shopping malls, the intelligent video detection technology is used to effectively monitor the crowds and individuals, detect, identify and alarm pedestrian behaviors, assist security personnel to protect personal safety, reduce the probability of public safety incidents and improve the intelligence of security measures.

2.3 Aviation Industry

Passenger flow monitoring based on video analytics technology also has certain applications in the aviation field, such as San Francisco International Airport, Helsinki Airport, Finland. At present, domestic airports can basically achieve full-area and all-round video coverage, but some airports and airport buildings mainly use analog or encoder products, and the resolution is not high. Due to the large area of the terminal building, some camera parameters are insufficient to meet the higher requirements of users. At the same time, the current CCTV monitoring system and other access control, alarm and other systems are not relevant enough, which will cause inconvenience to users affecting operational efficiency [3].

Shanghai Pudong and other airports installed a passenger flow information statistics system to monitor information such as customer retention and density in the waiting area and provide accurate data on the passenger flow distribution capacity of each area of the airport, and the part due to emergencies. Regional passengers are highly gathered to provide early warning, prompting management personnel to

Fig. 1 Qingdao railway station

quickly guide passengers and avoid security incidents. At the same time, it analyzes and excavates the up and down passenger status and distribution balance in various waiting rooms and gates to optimize the usage rate of each gate and improve the passenger experience quality.

2.4 Railway Industry

Qingdao Railway Station and Beijing South Railway Station have also established intelligent video analysis systems. Professional intelligent analysis cameras have been installed in more than 50 crowded locations to accurately collect information such as human traffic and transmit them to the Safe Production Command Center. The super-intelligent computing processing analysis of the background server replaces the human monitoring, which greatly improves the security guarantee capability (Fig. 1).

2.5 Subway Industry

Japan's Tokyo Metro, which is based on video to achieve passenger flow detection, is relatively mature. Video coverage is achieved in both the station hall and the train compartment. Passengers waiting on the platform can directly obtain the congestion of each train that will be reached through PIS, which greatly enhances the passenger experience [4].

The north section of Beijing Metro Line 16 is equipped with 2000 full HD 1080P HD IP cameras to realize security monitoring in the station. The real-time analysis of video images is used to locate, identify and track the targets in the dynamic scene, achieving early detection and active prevention, which helps security personnel to effectively avoid or efficiently deal with threats or sudden incidents. At the same time, the Beijing Metro launches congestion query, which not only improves the passenger experience, but also relieves the traffic pressure of some stations. The construction of Beijing Metro Line 16 network video monitoring center is based on the division of video surveillance area, effectively utilizing the existing rich passenger data and analytical assessment method to achieve a series of alarm responses such as video analysis and passenger flow warning. It escorts the safe travel of passengers and supports the smooth operation and rapid development of rail transit [5].

Hangzhou Metro Line 1 has a total of more than 1500 high-definition cameras, which basically achieves comprehensive coverage. It realizes intelligent analysis functions such as passenger flow statistics and face recognition [6]. It can achieve early warning and its intelligent monitoring test station is under construction, which is used for deepening research on more mature technologies such as video intelligence analysis and promoting the application of related new technologies in engineering. The cameras used in the Hangzhou Metro are mainly variable lenses, which are convenient for accurate and clear monitoring of people or objects within a certain range.

The Shanghai Metro has also introduced intelligent passenger flow monitoring technology to intelligently detect passenger flow speed, density, congestion index, etc., combined with information such as subway train operation and external weather, to predict future traffic changes and help subway staff to conduct passenger flow, emergency dispatch, danger prevention and so on [7]. The Shenzhen Metro Safety Supervision Department and the Chengdu Metro Quality Management Department have established the "Large Passenger Flow Security Early Warning and Monitoring System" as a key project and actively introduced technologies such as intelligent video surveillance and mobile terminal positioning to realize intelligent detection and control of passenger flow information.

The subway line environment is relatively closed, personnel are intensive, and the flow is large. The evacuation of personnel is greatly limited, the ability to resist risks is weak, and the security work is difficult. Once an emergency occurs, it may cause heavy casualties and property losses. The Guangzhou Metro has cooperated with Beijing Jiaotong University to build a passenger flow intelligent detection test station, which installs intelligent monitoring equipment in key areas of the station and realizes intelligent detection, analysis and prediction of passenger flow information (Fig. 2).

Fig. 2 Guangzhou metro station

3 The Necessity of Application of Passenger Flow Detection Technology Based on Intelligent Video Surveillance in Rail Transit Industry

3.1 The Important Part and Stone of Service of Rail Transit Safety

Passenger flow safety is an important part of rail transit safety and security. It is the core of operational safety. The ultimate goal of security guarantees, such as passenger transportation, facilities and equipment, environment and network, is to serve passengers' safety. Obtaining accurate passenger flow information is the service "stone" of depth mining and sharing of passenger data value, and the basic support for accurate analysis, prediction, early warning, assessment and control of passenger flow operation status.

3.2 Urgent Demand for the Development of Rail Transit Passenger Flow Safety Technology

China's rail transit develops rapidly, but passenger flow information detection mainly relies on AFC clearing technology and CCTV system to complete manually. Passenger flow safety detection core technology is lacking of system, standards and specifications are insufficient, evaluation system has not been fully established, and basic research and technical support need to be strengthened. The above problems restrict the further development of the rail transit safety operation level. It is urgent to break through the bottleneck of the fine passenger flow information detection from the aspects of technological innovation and application integration, provide basic data for passenger flow analysis, prediction and early warning, support the upper-level application of safe operation and gradually establish passenger flow. The technical support system for safety and security will improve the innovation capability and level of safety and security and promote the development of China's rail transit industry.

3.3 The Important Support to Improve Passenger Service Capacity and User Experience

Public transportation is people-oriented and being able to get accurate real-time traffic conditions is an important guarantee for a quality travel experience. Real-time monitoring of passenger flow, accurate collection and analysis of passenger flow information can provide passengers with important information such as train to delivery station, passenger load rate, passenger distribution of passengers and congestion of passengers in the station. It is beneficial for travel individuals to optimize their own travel plans, effectively reduce regional passenger flow congestion, improve delivery efficiency and improve overall passenger travel experience.

4 Advantages of Rail Transit Passenger Flow Monitoring System Based on Intelligent Video Surveillance

In this paper, the intelligent monitoring and management system for passenger flow safety status of rail transit proposed uses the real-time monitoring and collection technology of passenger flow to develop a comprehensive and intelligent urban rail transit passenger flow operation decision support platform for real-time monitoring and evaluation of passenger flow in the station. The real-time monitoring of passenger flow at key stations can be realized, and the analysis and evaluation of passenger flow safety status based on multiple scenarios can be carried out to support the optimization of network operation management measures, thereby realizing the refined, intelligent and digital control of urban rail transit network operations (Fig. 3).

4.1 Various Functions

According to statistics, visual information accounts for 60% of the information accepted by human beings [8]. Video image information is the most direct and real response to the objective world. Passenger streaming video acquired by video surveillance system contains important information such as the number and density of passenger flow, the location of personnel distribution, the flow direction of passenger group and the regional environment where passenger flow is located. The above

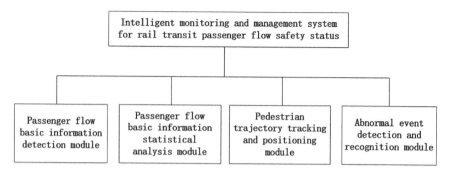

Fig. 3 Intelligent monitoring and management system for rail transit passenger flow safety status

information can be used to obtain the passenger flow characteristic data and provide the basis for passenger flow guidance and evacuation under emergencies.

At the same time, intelligent video analysis technology can realize real-time passenger flow acquisition. It can accurately detect and track the individual trajectory of passenger flow, achieve coverage of a wide range of areas and data acquisition. Through data statistics and analysis, intelligent video analysis technology can get volume and density of passengers and other data indicators which rail transit managers need.

4.2 High Real Time

The smart video surveillance system can get dynamic data and reports with a time interval of 5 s at least. The information contained in the video image is real time and dynamic. By using reasonable and efficient algorithms and high-performance computing and transmission hardware, the passenger flow data information of any scene such as stations, lines and the whole traffic network can be obtained in real time and dynamically. This real-time and dynamic information acquisition capability is conducive to the scientific and rational dynamic adjustment of the disposal measures according to the development of the situation.

4.3 High Flexibility

Passenger flow analysis system can be superimposed on the original video surveillance system or build a new system. The presentation of analysis data can be improved and customized according to the special requirements of customers.

4.4 Alarm Accurately

Intelligent video surveillance system integrates powerful image processing capabilities and runs advanced intelligent algorithms. It has the functions of intelligent analysis and early warning, so that surveillance personnel can more accurately judge the characteristics and nature of security threats, thus effectively reducing false alarm and missed alarm, and reducing the amount of useless and false data and improve work efficiency.

4.5 Automatic 24/7 Monitoring

Intelligent video surveillance system supports 24*7 h continuous surveillance, so whenever an intrusion occurs, it can respond quickly, appropriately and effectively, completely changing the situation in which the monitoring personnel continuously monitor and analyze the monitoring screen. Instead, the monitored video is continuously analyzed and warned by an intelligent video analysis module embedded in the front-end device (network camera) or display terminal.

5 Conclusions

Intelligent video surveillance has been widely used in various industries because of its intelligent, real-time and flexible characteristics. In future applications, we should focus on upgrading the passenger flow detection and analysis technology based on intelligent video surveillance, constantly improve the functions of intelligent video surveillance system, so that the passenger flow detection technology based on intelligent video surveillance can provide greater protection for the operation safety of rail transit and passengers' personal safety.

Acknowledgements The study is sponsored by National Engineering Laboratory project for the Safety Technology of Urban Rail Transit System (Development and Reform Office High Technology [2016] No. 583); 2018, China Railway Information Technology Co., Ltd. Science and Technology Research and Development Program Major Project JGZG-CKY-2018013 (2018A01).

References

1. Zhang L (2013) Passenger flow analysis intelligent video monitoring technology and application. Telecommun Technol 02:12–15 (in Chinese)
2. Shi Z (2017) Research on real-time abnormal behavior detection under video surveillance. Nanjing University of Posts and Telecommunications (in Chinese)

3. Liu B (2016) Application of video surveillance system in airports. China Public Secur (16):141(in Chinese)
4. Jiang Y, Zhang W (2018) Feasibility study and analysis of Xi'an metro passenger flow monitoring and warning information system. Urban Rapid Rail Transit 31(02):10–11 (in Chinese)
5. Zhang Y (2017) Study on the pertinence and accuracy of the alarm threshold of Beijing subway video surveillance. China Railw (05):81 (in Chinese)
6. Xu Q (2013) Intelligent video system guards Hangzhou rail traffic safety. China Public Secur (11):80 (in Chinese)
7. Metro S, Alibaba (2017) Ant financial service reached a strategic cooperation. (2017-12-5) [2019-7-15]. http://www.cnr.cn/shanghai/tt/20171205/t20171205_524051745.shtml (in Chinese)
8. Han Y (2005) Video image stabilization based on gray coded bit plane matching. Harbin Engineering University (in Chinese)

Planning Research for Automated People Mover System at Airports

Shuai Wang, Xiaoning Zhu and Li Wang

Abstract In order to reduce the passenger's walking time and improve the service level of the airport, the airport usually needs the equipment to assist passengers to travel. The automatic people mover system (APM system) is a reliable transportation device. The application of the APM system at the airport can effectively solve the traffic problem, and the APM system has a great impact on the airport. Therefore, this paper analyzes the relationship between the airport master plan and the APM system. The APM system is part of the airport master plan. When planning the APM system, the overall plan of the airport must be considered. At the same time, the APM system plan is divided into site setting, station design and line conditions. Each part has different classifications according to different requirements. Based on this, the APM system can be planned according to different actual conditions to ensure the effective and orderly operation of the airport traffic.

Keywords Automated people mover system · Site setting · Station design · Line conditions

1 Introduction

Automated people mover system (APM system) is mainly used for automatic transportation of short-distance passengers [1]. For example, it undertakes the passenger transportation task within the airport terminal building or between the terminal buildings. It adopts the world's advanced and mature rail transit automatic control technology, and it is also a modern mode of transportation.

S. Wang (✉) · X. Zhu · L. Wang
School of Traffic and Transportation, Beijing Jiaotong University, Beijing 100044, China
e-mail: 18120901@bjtu.edu.cn

X. Zhu
e-mail: xnzhu@bjtu.edu.cn

L. Wang
e-mail: liwang@bjtu.edu.cn

Compared with other passenger transportation methods, the APM system has the main advantages of environmental protection, high reliability and short waiting time for passengers [2–6]. The main disadvantage of this system is that it is very expensive and the maintenance costs are also high.

At some airports, the terminal structure is relatively large due to various factors such as terrain and passenger flow. This results in a long distance if there is no mechanical assistance from the check-in counter to the boarding gate. When the traffic distance is long, the passengers feel tired and the passengers are liable to cause delays in the aircraft, which leads to a decrease in airport service levels. Therefore, it is necessary for some airports to use the automated people mover system to improve service levels.

At present, more than 40 airports have opened and operated the APM system at home and abroad. Among them, the APM system of the three airports in Beijing Capital Airport, Hong Kong Airport and Taipei Taoyuan Airport has already started operations [7]. With the continuous development of the domestic aviation industry, there will be more and more air passenger flow, so the APM system will usher in a development period.

Since the planning of the APM system is a complex process involving a large amount of various expertise, it is necessary to analyze the planning process of the APM system to ensure that the entire airport can operate at the lowest cost while still providing a high level of service. This paper mainly studies the relationship between airport master plan and APM system and the layout planning of APM system to provide guidance for the planning and design of APM system in airport [8].

2 Relationship Between Airport Master Plan and APM Systems

The APM system is an integral part of the airport. Therefore, to plan the APM system at an airport, planners must first understand the overall airport plan. The airport master plan is a graphical representation of the layout of existing and proposed airport facilities and land use planning. The airport master plan includes: important airport facilities, important natural and artificial features, areas reserved for existing or future aviation and non-aeronautical developments and services, existing ground contours and phase-out facilities

The airport master plan can reflect the planning layout of the airport functional area, including the flight zone, terminal zone, ground transportation system, cargo zone and subsidiary zone. At the same time, the APM system is a means of transport for connecting the various functional areas of the airport. Site planning, line conditions, maintenance base configuration and other related plans of the APM system are affected by the airport master plan [9].

The APM system usually undertakes the connection problem among the three functional areas of the terminal area, the flight area and the ground transportation system. It may also be a connection between the interior of the terminal building, between the terminal building and the terminal building, between the terminal building and the satellite hall, between the satellite hall and the satellite hall and between the terminal building and the roadside traffic center.

3 Planning Layout of the APM Systems

This section examines the APM system planning process to analyze and implement better transportation plans. In order to provide a high level of service and ensure the safe operation of the airport, there are three main aspects related to site setting, station design and line conditions [10].

3.1 Site Setting

The airport APM site provides a physical connection between the APM train and the airport facilities it serves, and the site plays a vital role in the efficient operation of the APM system.

When the venue is perfectly integrated with the airport and terminal facilities, the APM system is the most efficient and the passengers travel more easily. This requires planners to establish APM sites scientifically, conveniently and efficiently.

The setting of the airport APM site requires a comprehensive consideration of the characteristics of the airport passenger flow, the spatial layout of the site and the land characteristics.

The platform layout of the APM system includes the single-sided platform, the double-sided platform, the island platform and the island-side platform, as shown in Fig. 1.

The single-sided platform is the simplest type of platform used in monorail and low-traffic stations.

The double-sided platform mainly uses monorail and dual-track systems, and its platform can withstand higher flow rates than the single-sided platform. When using this platform in a single-line system, one station cannot operate due to a failure, while another station can provide a continuous service with half of the passenger flow.

The island platform is a central platform between the two lines that will create a vertical passenger flow. It therefore requires equipment such as escalators, elevators and stairs, which will increase costs and lower service levels.

The layout of the island-side platform is typically used for terminals in huge capacity passenger flow systems. This type can be used in the same level of airport facilities to achieve the highest system capacity.

The design scale of the platform should be reasonable, not only to meet passenger evacuation requirements, but also to reduce resource waste. The length of the platform

Fig. 1 Platform layout

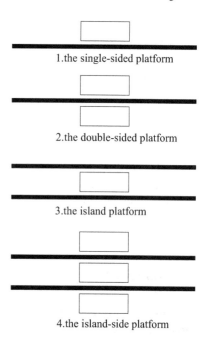

1. the single-sided platform
2. the double-sided platform
3. the island platform
4. the island-side platform

should be able to meet the train length of the largest vehicle group. The terminal limit must be reserved at the end of the terminal so that vehicles outside the parking lot can be buffered to ensure safe operation.

3.2 Station Design

The APM station plays a vital role in the effective operation of the APM system. And the APM station must be appropriately sized, configured and equipped to effectively ensure passenger service levels. Therefore, when planning an APM station, planners must consider some factors such as passenger sorting requirements, passenger baggage characteristics, vertical circulation requirements and queuing areas.

The main facilities of the general rail transit station include station halls, platforms, passages (elevators, stairs, corridors, etc.) and fare collection facilities. The APM system is different from the general rail transit station. Generally, there is no special station floor, but it is integrated into the terminal building or the satellite hall building. The station of the APM system is equipped with the following requirements:

1. Passengers should be guided to the APM platform by clear signs after check-in in the terminal;
2. An appropriate number of escalators should be installed between the station hall and the APM station platform. Because the APM train route and the pedestrian access of the APM service facility are generally different in vertical direction;

3. The design of the station hall should meet the needs of long-term design passenger flow, reasonably organize the passenger flow of the station, and reduce the intersection and interference of the passenger flow lines of the station to ensure the separation of different passenger flow lines.

3.3 Line Conditions

The APM system selection line shall be based on the development plans of the airport in the near-term, medium-term and long-term phases. And it determines the direction of the route, the distribution of the route, the location of the site and the intersection of the lines.

The line planning analysis of APM system is complex and affected by many factors. Therefore, in the process of developing a production line plan, many factors should be considered, such as the impact on existing and planned facilities, passenger flow and line complexity.

The methods of line selection are as follows:

- Analyze the type and quantity of passenger processes

One of the biggest differences between the APM system and urban rail transit is the complexity of its passenger flow. The types and quantities of different airport processes are diverse. Therefore, when planning the configuration, the APM system needs to carefully analyze the types and quantities of passenger processes.

- Passenger capacity analysis of airport terminal and satellite hall

According to the influencing factors such as the processing capacity of the airport, the processing power of the terminal system and the allocation scheme between the route and the terminal, the passenger flow processing capability of each terminal and the satellite hall can be calculated.

- Analysis of passenger flow distribution at airports

When designing the line selection, the APM system should analyze the breakdown of airport passenger volume and calculate the peak hour passenger flow between each terminal and satellite hall.

- Passenger flow allocation at airports

According to the passenger flow analysis between each terminal building and the satellite hall, the passenger flow channel of the airport passenger transportation system was initially determined. At the same time, the passenger flow during the peak hours of the channel was calculated and corrected according to other factors.

- Determine line configuration

After several rounds of adjustments in the above-mentioned work, it is necessary to determine the route configuration of the APM system based on the OD traffic data on the premise of meeting the service transportation purpose.

4 Conclusion

This paper mainly studies the planning and layout methods of the APM system at airports. On the one hand, it analyzes the relationship between the airport master plan and the APM system. The APM system is part of the airport master plan. If the APM system is planned at the airport, planners must first understand the airport master plan. On the other hand, it also analyzes airport site setting, station design and line conditions. In site setting, there are four platform layouts for the APM system; in station design, the station design of the APM system has three requirements; in line conditions, there are four ways to select a line. In determining the relationship between the airport master plan and the APM system, we can use different planning methods to determine the layout of the site, station hall and route of the APM system.

Through the research in this paper, it provides reference and help for the planning and operation of airport planners and APM operators.

Acknowledgements This paper was sponsored by Civil Aviation Administration of China Major projects (No: 201501).

References

1. Shi Y (2019) Design of water supply and drainage and fire protection system of APM rapid transit system in Chengdu Tianfu international airport 55(01):71–76. (in Chinese)
2. Li N (2013) Analysis of the use of passenger transport system in capital international airport. China Sci Technol Inf 20:135–136 (in Chinese)
3. Chen L, Zhang M (2013) Capital airport passenger rapid transit system operation data management (20):86–88. (in Chinese)
4. Xu H (2014) Design of fire trainer control system for airport passenger rapid transit system. China Civil Aviation Univ. (in Chinese)
5. Chen L (2015) Research on transportation organization of Pudong airport rapid transit system 7
6. Wang J (2019) Xi'an Xianyang international airport passenger MRT system standard comparison. Urban rail transit research. pp 115–119
7. Zeng H (2019) Research on transportation organization scheme of airport passenger MRT system. Integrated Transport 41(05):51–56. (in Chinese)
8. Elliott DM, Norton J (1999) An introduction to airport APM systems. J Adv Transport 33(1):35–50
9. Elliott DM (1985) Applications of automated people movers at airports. In Automated people movers, engineering and management in major activity centers, pp 384–398
10. Mckelvey FX (1989) A review of airport terminal system simulation models. Volpe National Transportation Systems Center, Cambridge, Mass, pp 101–110

Design of Vibration Test Bench for Electric Locomotive Bogie

Yunxiao Fu and Boyang Li

Abstract The real vehicle simulation platform is provided for the active vibration reduction control, and the adaptive adjustment of the running stability of the rail vehicle is realized. Based on the HXD2 electric freight locomotive as the prototype, the electric locomotive simulation vibration damping test bench is designed. The test bench can be used to complete the test of vehicle vibration control test. The platform provides vibration excitation to the vehicle through the bottom four actuators. By constructing the sensor monitoring system to sense the vibration information of the locomotive model bogie, based on the perceptual data, judge the vehicle vibration situation, and the active control strategy is formed by the upper computer, and the vibration damper control command is given to realize the vibration damping control of the vehicle model. Firstly, the technical characteristics and main structure of the HXD2 electric locomotive were analyzed. Then, according to the analysis results, the structure of each part was modeled in the SolidWorks software environment. The structure design and assembly were carried out according to the modeling results. Finally, the sensor monitoring system was built and verified by experiments. The bogie vibration test rig can effectively simulate and sense the vibration of the HXD2 locomotive under different road conditions, and provides a more accurate and reliable theoretical reference for the practical design of the HXD2 locomotive bogie vibration reduction control system.

Keywords HXD2 electric locomotive · Vertical vibration · Bogie test bench · Structural design

---

Y. Fu (✉)
CRRC Academy, Beijing 100070, China
e-mail: yunxiaof2012@163.com

B. Li
Beijing Key Laboratory of Performance Guarantee on Urban Rail Transit Vehicles, Beijing University of Civil Engineering Architecture, Beijing 100044, China

Fig. 1 HXD₂ electric locomotive

1 Instruction

The HXD2 is a high-power AC-driven freight electric locomotive designed for heavy-duty transportation with a maximum test speed of 132 km/h, a maximum operating speed of 120 km/h and a total power of 10 MW. The locomotive adopts the international advanced technical standard system and uses the maintenance-oriented life cycle design method. It combines AC drive system, computer network control system, low-power bogie and high-strength body and hooking device. Advanced technology has formed a brand product with obvious technical advantages and international competitiveness, and provided a guarantee for solving the problem of tight transportation capacity of China Railway Coal Transportation Line (Fig. 1).

2 Technical Analysis of HXD$_2$ Electric Locomotive

2.1 Technical Parameters

The HXD2 electric locomotive has powerful power and traction, and can pump 7000 tons of heavy-duty trains in a single machine; the locomotive has multi-machine wireless reconnection remote synchronization control function, and the three-machine reconnection meets the traction requirements of heavy-duty trains of more than 20,000 tons. The locomotive is designed with a weight of 25 tons, and the weight of the car can be removed to achieve a shaft weight of 23 tons. The locomotive can be stored normally under the environment of −40 °C and can be used normally after heating and cold-proof measures.

The HXD2 locomotive bogie adopts 2 (B0-B0) shaft type, which consists of two-section locomotive. The locomotive adopts AC-DC electric drive system and 25 kV/50 Hz voltage system. The locomotive has a weight of 184 tons and a corresponding axle load of 23 tons. It can realize the single-machine traction of 10,000 tons of heavy-duty trains, and the remote re-linking can realize the operation mode of the two-machine traction 20,000-ton heavy-duty train (Table 1).

Table 1 Main technical parameters of HXD$_2$ locomotive

Name	Parameter	Name	Parameter
Curb quality	100 t * 2	Coupler center distance	19,025 * 2 mm
Power supply	25 kv 50 Hz	Gauge	1435 mm
Calibration power	10,000 kw	Axle weight	25/23 t
Maximum speed	120 km/h	Starting traction	760 kN

3 Critical Structure

3.1 Wheel Pair

The wheel pair is composed of an axle, a wheel, and an axle box. The wheel size is designed in accordance with EN13104, EN13260, EN13261, and EN13262. The axle material is A1N of UIC 811-1 "Technical Specifications for the Supply of Locomotives and Vehicle Axles" (A1 is the material grade, N stands for normalizing treatment), and the material purity is higher than Class 1 axles requirement in Product Requirement of EN 13261 "Railway Application—Wheelset and Bogie-Axle—Requirements".

The steel is smelted by electric furnace or converter [1], refining outside the furnace and vacuum degassing to ensure the purity of the axle material. The wheels are made of R7T monolithic steel wheels and meet the requirements of UIC 812-3 and EN 13262. The axle box bearings are fully sealed tapered roller bearings. The bearings meet the requirements of EN 12080. The dimensions of the bearings are in accordance with ISO281/1 [2].

3.2 Architecture

The frame is made of S500MC steel plate and steel casting. The main structure adopts the "Japanese" type steel plate welded structure, which consists of two symmetrical side beams, one beam and two end beams. The side beam and the end beam are connected by welding, the middle beam is fixed on the side beam by bolts, the side beam is a fish-belt box-shaped structure welded by the plate, the beam and the end beam are plate welded box girder, and there are several inside the beam. Ribs—the upper and lower covers of the side frame of the frame [3], respectively, have an integrally formed axle box tie rod seat, a central second spring support seat, a vertical damper seat, and a brake sling. The lower part of the beam is welded with a motor hoisting the seat, the lateral stop and the lateral damper seat are welded with a traction seat on one end of the beam.

3.3 Drive Unit

The driving device is mainly composed of a gear box, an AC traction motor, a driving gear, a driven gear, a bearing housing assembly, a sealing component, and a motor suspension component, and adopts a first-class cylindrical spur gear transmission.

The driving device adopts a rolling hoop semi-suspension method, one side is connected to the axle through a bearing box and two sets of tapered roller bearings, and the other side is suspended on the frame by a motor suspension rod. A rubber elastic element is mounted on the motor hanger to achieve relative displacement between the wheel set and the frame. In order to prevent the motor suspension system from failing, each set of drive unit is equipped with a set of motor anti-fall system to ensure the reliability of the motor suspension system.

The gearbox is made of ductile iron material [4], and the upper and lower boxes are bolted. The lubricating oil in the gearbox is used for the lubrication of the gears, and also for the lubrication of the bearing on the flank and the bearing at the output of the motor.

3.4 Suspension Device

The HXD2 electric locomotive suspension spring is a cylindrical compression coil spring with a fixed pitch and is made of round steel [5]. The size parameter is 38.6 mm spring diameter, 191 mm medium diameter, 4 turns of effective turns, 325 mm free height, and independent axle box spring suspension structure. Each axle box has two coil springs, and each side of the axle box is equipped with one axle box tie rod with different stiffness in three directions. A series of vertical dampers are provided for damping vibration and absorbing vibration energy. The second-type suspension device consists of two sets of two series suspended spiral steel springs and rubber mats connected in series as the main body. At the same time, two series of transverse dampers and two series of vertical dampers are arranged between the frame and the vehicle body. Its design takes into account all the movement of the bogie under the body. The primary and secondary suspension springs are Class B springs and Class A springs in accordance with European Standard EN 13298. The end of the spring is defined as the D type in the ISO 2162-2 standard [6].

4 Technical Route

According to the vibration response analysis of the steering frame of the HXD_2 electric locomotive, considering the train operation in the nodding and side sway environment, it is selected to analyze under the nodding and side sway conditions.

Fig. 2 Technical route

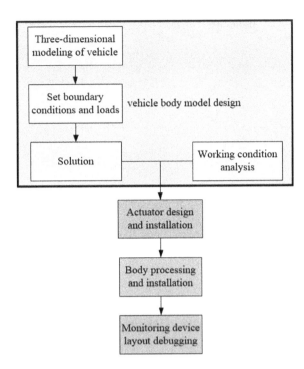

The finite element software is used to calculate the dynamics of the bogie frame. The main technical routes are as follows (Fig. 2).

5 Three-Dimensional Modeling of Bogie

5.1 Actuator Design

In order to realize the required working conditions such as nodding and side sway of the locomotive model, each actuator requires 1.5 ton of output with a weight of 1.2 tons of locomotive, stroke of ±80 mm, frequency of 30 Hz, speed of 350 mm/s, displacement accuracy ≤0.02 mm, and the amplitude is 1.0 mm when the frequency is 20 Hz. The actuator power is controlled within 50 kw. The actuator height is 600 mm. Its three-dimensional model is shown in Fig. 3.

Fig. 3 3D illustration of the actuator

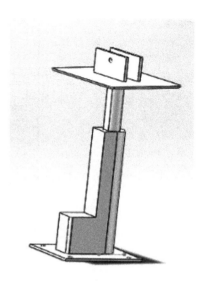

5.2 Architectural Design

According to the prototype parameters of the HXD$_2$ electric locomotive provided by Party A, the locomotive frame is first scaled down by 1:2. The overall length is 2300 mm and the width is 1413 mm. Figure 4 shows the construction of its 3D model with SolidWorks.

Fig. 4 3D diagram of the framework

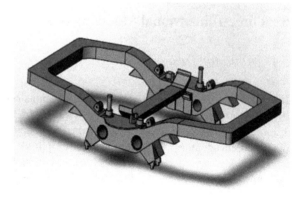

Fig. 5 Car body 3D

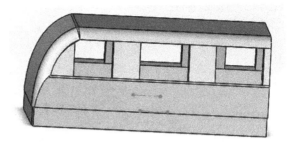

5.3 Car Body Design

Because the car body structure is not completely symmetrical, the whole vehicle model is generally used for modeling, and the semi-vehicle symmetry model is used for some working conditions [7]. After calculation, the size of the car body was 1512 mm wide, 3250 mm long, and 780 mm high. Its three-dimensional map is shown in Fig. 5.

5.4 Partial Assembly Drawing

After the initial completion of the frame, the wheel set and the suspension device, the assembly is carried out, wherein the bogie assembly drawing is shown in Fig. 6, and the test stand is shown in Fig. 7.

6 Condition Monitoring System Design

The test bench entity of the HXD_2 electric locomotive is shown in Fig. 8. The

Fig. 6 3D view of bogie assembly

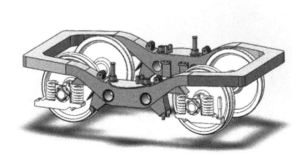

Fig. 7 3D diagram of the test bench assembly

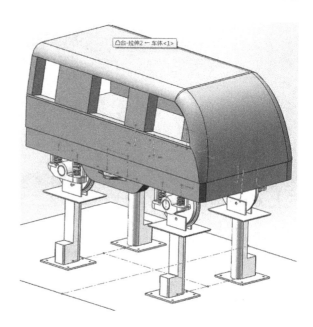

Fig. 8 Test bench entity diagram

test bench consists of actuators, bogies, suspensions, and car bodies. The actuators provide vibration acceleration in the vertical direction of the HXD_2 electric locomotive model. Through the coordination of the four actuators, the requirements of the rolling conditions, nodding, floating and shaking, and shaking head of the model car are realized.

6.1 Gearbox Housing Sensor Arrangement

Since the test wheel pair and the actuator are connected by bolts (as shown in Fig. 9); the gear box case is not vibrated by the bogie and the shock absorber, so it is subjected to vibration excitation and actuator. The excitation provided is almost the same, and the actuator contains an acceleration sensor (as shown in Fig. 10), so it can be used directly to collect the vibration signal of the gearbox of the HXD_2 model car. Therefore, the vibration signal of the gearbox housing is collected using acceleration sensors on four actuators.

Fig. 9 Schematic diagram of wheel set and actuator connection

Fig. 10 Schematic diagram of the actuator acceleration sensor

Fig. 11 Bogie acceleration sensor arrangement

6.2 Bogie Sensor Arrangement

The vibration excitation generated by the actuator is transmitted to the bogie through the wheel pair, a series of springs and the shock absorber. At the same time, the bogie is the main support mechanism of the vehicle body, so the vibration signal collection of the bogie needs to reflect the nodding and rolling of the vehicle, shaking heads, ups and downs, etc. Therefore, sensors (shown in Fig. 11) are placed at the four corners of the bogie for collecting the vibration signals of the bogie.

6.3 Body Sensor Arrangement

After being excited through the bogie, it is transmitted to the vehicle body through the second-stage spring and shock absorber. In order to record the vibration modes of the car body, such as nodding, rolling, shaking, and floating, a two-axis acceleration sensor is used. In the stability index test, according to the principle of Sperling [8], for a real train, the sensor should be installed on the floor surface of the body 1000 mm from the center of the bogie in the lateral direction, thus here it is installed scaled down 250 mm away from the center of the bogie to measure the rollover of the car body. In the longitudinal direction, it is installed at the front and rear ends of the center of the vehicle body to detect the nodding of the vehicle (as shown in Fig. 12).

6.4 Data Collector Selection

The test bench data acquisition system uses a 16-channel MDR data acquisition system. The acquisition system uses voltage, current, strain, force, acceleration, and

Fig. 12 Schematic diagram of the box acceleration sensor arrangement

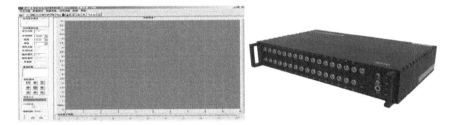

Fig. 13 MDR-80 16-channel data acquisition instrument

other sensors to capture the signal under the vibration state of the locomotive. The signal obtained by the acquisition is extracted and analyzed by the post-diagnosis method to obtain the dynamic response of the locomotive. The exterior of the acquisition instrument is shown in Fig. 13.

6.5 Test Verification

After the sensor monitoring system is set up, the reliability of the vibration damping test bench is verified by experiments. By operating the vibration damping controller to provide different degrees of control current for the damper, the damping of the damper is changed, and the collected real-time data is downloaded and saved by category through the data analysis system, and then the data is analyzed by the MATLAB software. And complete the calculation of the time domain feature parameters. The results under one working condition were selected as experimental results, as shown in Table 2, and the waveshapes were compared in Fig. 14.

It can be clearly seen from the time-domain diagram (Fig. 14) that when the control magnetorheological controller adjusts the current from 0 to 0.8 A, the vibration amplitude has a very significant drop, indicating that the vertical vibration of the bogie

Table 2 Experimental parameters under one working condition

Name	Parameter	Parameter
Frequency (Hz)	3	3
Amplitude (mm)	4	4
Channel	19	19
Electric current (A)	0	0.8

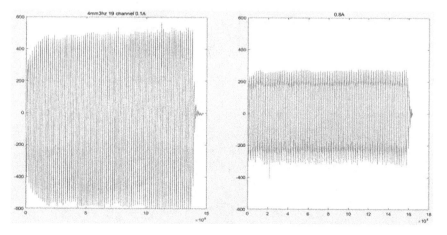

Fig. 14 3 Hz 4 mm 19 channel from 0 to 0.8 A time-domain diagram

can be applied by applying damping. Play a certain degree of vibration reduction (Table 3).

After processed by MATLAB software which is used to calculate the characteristic parameters of the time-domain signal, the data can also be analyzed and found that the damping effect of the bogie damper has a clear indication with the increase of damping.

Table 3 Changes of various time-domain parameters of the 19th channel signal in 3 Hz 4 mm working condition

Name	Electric current/0 A	Electric current/0.8 A
RMS	339.5907	200.2355
Peak index	1.664671	1.399352
Clearance factor	2.19889	1.55627
Pulse index	1.933859	1.481246
Waveform index	1.161707	1.058523

7 Conclusion

In this design, through the three-dimensional modeling of various parts of the test bench, relying on finite element analysis, the HXD2 electric locomotive bogie vibration test bench designed by dynamics calculation can effectively simulate the different degrees of the vehicle under different road conditions. The vertical vibration changes, and the vibration monitoring signal can be effectively monitored through the design of the sensor monitoring system. Finally, after the experimental data collection and analysis, it is verified that increasing the damping can reduce the vertical vibration of the bogie to a certain extent. The design of the test bench has greatly promoted the further development of the HXD2 electric locomotive bogie vertical damping system.

Acknowledgements This contribution is supported by national key R&D plan (No. 2018YFB1201702-06).

References

1. Lv S, Feng Q, Liu H, Ma C, Zhihe Z (2016) Wheel set drive system of the bogie for HXD2 electric locomotive. Electr Drive Locomotives 06:6–9
2. Nam JS, Park YJ, Chang HS (2016) Dynamic life prediction of pitch and yaw bearings for wind turbine. J Mech Sci Technol 30(1):249–256
3. Guo T, Peng Y, Yang J (2008) Study of frame structure and strength for HXD2 locomotive. Electr Drive Locomotives 06:9–12
4. Jianzhong Sun (2015) HXD2 locomotive gear box processing technology. Metal Process (Cold Process) 24:36–39
5. Luo J, Shang Y, Wang H, Xue H (2009) Finite element analysis of suspension spring of HXD2 electric locomotive. Sci Vol J Yangtze Univ (Natural Science Edition) 6(03): 259–260+354
6. Technical product documentation-Springs-Part 2: Presentation of data for cylindrical helical compression springs. (ISO 2162-2:1993)
7. Guo N, Yang J, Li X (2008) Strength calculation and experimental study of locomotive body of HXD2. Mech Manag Dev 03:3–6
8. Bao F (2010) Study on riding comfortability evaluating methods and the application to selection of line parameters of high speed railway. Jiaotong University, Beijing

Evaluation of Energy Consumption Level of Urban Rail Train Operation Plans: A Case Study

Jiao Zhang, Yujie Li and Hao Xie

Abstract Energy-efficient operation is an effective way to reduce the energy consumption of urban rail trains. Based on the train operation plans for rail dispatching, this paper studies the evaluation of the energy consumption level of urban rail transit operation. To effectively describe the capability and potentials of the time schedule, an optimization-based evaluation framework is presented to calculate the possible speed curves of planned trains. By comparing the optimized curves with the referencing results, the energy-saving capability of the trains under the time constraints can be computed. The presented framework is tested with a case study using field data. The results show that the utilization of the optimized solution can reduce the energy consumption by applying the presented optimization method in determining the speed profiles.

Keywords Energy consumption · Urban transit · Optimization · Time schedule

1 Introduction

Metro systems play an important role in urban economic and social development for metropolis due to the high loading capacity, punctuality, and low pollution [1]. More and more cities in China are establishing the urban rail lines to realize a higher transportation service capacity, and the energy consumption of the urban rail system has been a common concern of the developers and operators. The majority of studies on energy-saving operation of urban rail trains mainly focus on optimization of the driving control solutions for driver advisory and automatic operation [2, 3]. However, it should be noticed that the train operation plan according to the timetable also determines the capability of the energy consumption level and will be a precondition for the field activities. Energy conversation from a macro-perspective is becoming

---

J. Zhang (✉) · Y. Li
Beijing Mass Transit Railway Operation Corporation LTD, Beijing 100044, China
e-mail: zhangjiao2099@163.com

J. Zhang · H. Xie
Beijing Subway Operation Technology Centre, Beijing 102208, China

a hot topic for both the urban rail dispatching and rail signal control systems. It can be found that a lot of energy conversation algorithms have been proposed to optimize the driving strategy of rail trains. The growing R&D results in the numerical optimization area give more opportunity to solve the problem of reducing the energy consumption and emissions [4, 5]. It should not be ignored that the precondition to derive the optimization results is the time schedule of the train operations. The description of the inherent capability of the timetable will further enhance the efforts in practical operation stage. In this paper, we present a framework to evaluate the energy consumption level of urban rail train operation plans, which makes it possible of assessing the characteristics of the plans and promote the update for optimizing the train dispatching activities. A case study in Beijing Metro is provided to show the features of the presented solution.

2 The Bacterial Foraging Optimization Algorithm

In order to obtain the possible energy reduction results of the train operation plans during the real operation stage, the simulation of the energy-saving motion states of the trains according to the time schedule has to be carried out in advance. Different from the basic simulation strategy only using the traction calculation models, an optimized algorithm is adopted in this research to enhance the energy consumption characters by generating the optimal speed curves within the restriction of the departure and arrival time.

The bacterial foraging optimization algorithm (BFO) [6] is a bionic optimization algorithm that was proposed to simulate the foraging behavior of Escherichia coli. Different from exiting optimization methods, like particle swarm optimization (PSO) and bee colony optimization (BCO), the foraging process of bacteria is mainly divided into four operations: chemotaxis, swarming, reproduction, and migration. Its mathematical model simulates the chemotaxis, reproduction, and migration operations through three different levels of cycles as follows.

(1) Chemotaxis

The foraging behavior of an Escherichia coli (use *E. coli* for short) can be divided into two basic movements: swimming and tumbling, which depend on the direction of flagellum rotation. When the surroundings are good, the flagella rotate counterclockwise, and the bacteria swim forward quickly. On the contrary, when the surroundings are poor, the flagella rotate clockwise, causing the bacteria to roll frequently, so as to find the direction of high food concentration.

This process is realized by calculating and comparing the fitness of the two locations to determine bacterial swimming or tumbling. The movement of the bacterium can be presented by

$$\theta^i(j+1,k,l) = \theta^i(j,k,l) + C(i)\frac{\Delta(i)}{\sqrt{\Delta^T(i)\Delta(i)}} \quad (1)$$

where $\theta^i(j, k, l)$ represents the position of the ith bacterium in the jth chemotaxis, kth reproduction and lth migration step, $C(i)$ represents the step length in the chemotaxis step, and Δ represents a random direction vector in the range of $[-1, 1]$.

(2) Swarming

There is a group behavior among bacterial species such as *E. coli*. In the process of foraging, there are attractions and repulsions between individual bacteria; that is, individuals can communicate with each other through signals, so as to improve the speed of the foraging behavior. The forces between individuals may be presented by function as follows.

$$J_{cc}(\theta, P(j, k, l)) = \sum_{i=1}^{S} J_{cc}\left(\theta, \theta^i(j, k, l)\right)$$

$$= \sum_{i=1}^{S}\left[-d_{attractant}\exp\left(-w_{attractant}\sum_{m=1}^{p}\left(\theta_m - \theta_m^i\right)^2\right)\right]$$

$$+ \sum_{i=1}^{S}\left[h_{repellant}\exp\left(-w_{repellant}\sum_{m=1}^{p}\left(\theta_m - \theta_m^i\right)^2\right)\right] \quad (2)$$

where $J_{cc}(\theta, P(j, k, l))$ represents the interaction force between the bacteria which may affect the fitness value, $d_{attractant}$, $w_{attractant}$, $h_{repellant}$, and $w_{repellant}$ are four parameters, S indicates the group number, P is the spatial dimension, and θ_m^i is the mth component of the ith bacterium.

(3) Reproduction

After a period, a part of the group does not find enough food and they will become extinct, while others with the strong foraging ability will reproduce. The new individuals will replace the extinct individuals and maintain the size of the population. Half of the individuals with lower fitness will be removed selectively, while the other half with higher fitness will reproduce. This process is presented in the following function.

$$J(i, j, k, l) = J(i, j, k, l) + J_{cc}\left(\theta^i(j, k, l), P(j, k, l)\right) \quad (3)$$

where $J(i, j, k, l)$ is the fitness of the ith bacterium in the jth chemotaxis step, the kth reproduction step and the lth migration step, and $J_{cc}(\theta, P(j, k, l))$ represents the interaction force.

(4) Migration

When the environment of bacterial colonies changes, it will have a great impact on their foraging behavior. For example, when local temperature rises or food is scarce, bacterial colonies will migrate to new areas for foraging. Migratory operations enable bacterial populations to have a better ability to find food, and their stochastic performance enables them to jump out of local optimum and approach global optimum better.

According to the above-mentioned operations, the procedures of the bacterial foraging optimization algorithm can be summarized as follows.

- Step 1: Initialize the required parameters:

  - j: the index of the chemotactic step,
  - k: the index of the reproduction step,
  - l: the index of the migration step,
  - p: the dimension of the search space,
  - S: the size of the population,
  - Nc: the number of chemotactic steps,
  - Ns: the maximum number chemotactic steps,
  - Nre: the number of reproduction steps,
  - Ned: the number of migration steps,
  - Ped: migration probability,
  - $C(i)$: the length of the step taken in the random direction.

- Step 2: migration loop: $l = l+1$
- Step 3: reproduction loop: $k = k+1$
- Step 4: chemotactic loop: $j = j+1$.
- Take a chemotactic step for the ith bacterium
- Compute the fitness function $J(i, j, k, l)$.

Let $J_{\text{last}} = J(i, j, k, l)$ to preserve fitness, and constantly update the value of J_{last} so that it always stores the optimal value.

The $p \times 1$ random vector $\Delta(i)$ representing the direction is generated to tumble the bacterium.

The bacteria swim in the direction of rollover according to $C(i)$, and we can calculate the bacterial position $\theta^i(j+1, k, l)$.

Recalculate the fitness $J(i, j+1, k, l)$.
The swimming stop condition is determined as follows.
Initialize $M = 0$, while $m < Ns$
$m = m+1$;
If $J(i, j+1, k, l) < J_{last}$, the new position is better than the previous position, makes $J_{last} = J(i, j+1, k, l)$, and recalculates the next position and fitness;
Else, let $m = Ns$ and stop the swimming.
Continue to treat the next bacteria with $i = i+1$.
Step 5: If $j < Nc$, go back to step 4 and continue the chemotactic operation.
Step 6: Calculate the cost function of each bacterium

$$J_{\text{health}}^i = \sum_{j=1}^{N_c+1} J(i, j, k, l)$$

The bacteria are sorted according to the cost function, and half of the them with the higher cost function are destroyed and the remaining will be reproduced to produce new bacterial colonies.

- Step 7: If k <Nre, go back to step 3.
- Step 8: Migratory operation of each bacterium is performed according to Ped.
- if l <Ned
-     Go back to step 2;
- else
-     End the whole process.
- end

3 Energy Consumption Level Evaluation Framework

Based on the above-mentioned BFO algorithm, we can simulate the possible online operation of the trains according to the time schedule in the operation plan. A reference trajectory is also computed to evaluate the capability of the planned trip to reduce the energy consumption as much as possible. There are mainly four steps in an energy consumption level evaluation framework.

(1) Prepare the original data set

Before the evaluation is carried out, the train operation plan data has to be extracted from the time schedule from the operation management department of the urban rail transit. The train number, the stations in the whole line, the departure, and arrival time will be obtained. In addition, the fundamental data of the trains and rail tracks have to be determined to perform force analysis during the trip within the line according to the temporal constraints.

(2) Establish the energy consumption model

In order to realize the quantitative analysis and comparison of the energy consumption of the trains planned in the time schedule, a model is required to describe the energy consumption of the traction condition. Usually, the energy is estimated based on the motion state and the traction force of the train at a specific location and speed level, which can be described as follows.

$$Es(v_t) = \int_0^{S_t} \frac{\sigma F(v_t)}{\lambda(v_t)} ds + P_a t \qquad (4)$$

where $Es(v_t)$ is the energy consumption at instant t, σ represents the mechanical power coefficient of the train, $F(v_t)$ is the traction force associated with the running speed v_t, $\lambda(v_t)$ is the transformation efficiency, and P_a denotes the auxiliary power.

(3) Calculate the speed curve

Based on the basic traction control method and the BFO-based optimization method, two different speed curves can be calculated to describe the relationship between the running distance and speed of the train. The departure and arriving time information from the train operation plans provide prior restrictions to the traction control calculation and the optimization, and thus the derived curves can reflect the possible operations when the plan is performed in the future.

The flowchart of the optimization using the BFO algorithm is as follows (Fig. 1).

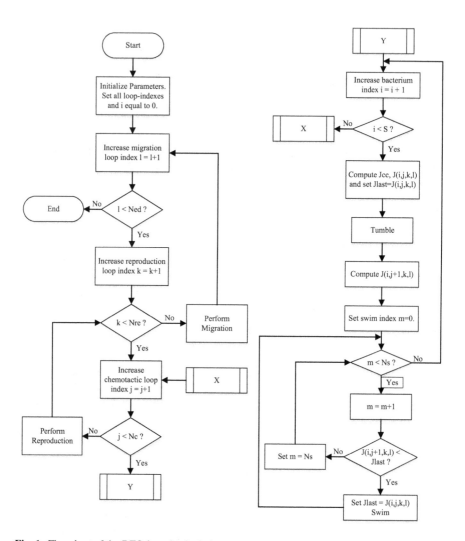

Fig. 1 Flowchart of the BFO-based calculation

(4) Assessment of energy-saving capability

Results of the derived curves can be summarized for the whole rail line. By comparing the actual energy consumption of different curves, the achievable energy-saving rate can be evaluated, and the similar operation can be carried out for different train plans to realize a large-scale evaluation.

4 Results of Case Study

The presented method is tested through a case study in Batong Line of Beijing Metro. The time schedule of the line is extracted from the operation plan map. The practical parameters of the trains operating in this line, including the weight, traction characters, and braking characters, are introduced in the evaluation. In addition, the parameters of the rail tracks are also provided to calculate the speed curves for evaluation. The time plan of a train (No. 182046) from Sihui Station to Tuqiao Station is utilized to carry out the comparison of the energy consumption levels of traction control. Figure 2 shows the derived speed–distance curves from the traction calculation and the energy-saving optimization, where the speed limit in the track sections is 80 km/h. Figure 3 describes the speed–time curves from the two evaluation methods accordingly.

In order to show more details about the difference between the optimized and reference curve, two typical track sections are selected to give more detailed information of the results. Figure 4 shows the high-resolution speed–distance curves in a track section between Sihui Station and Sihui East Station. The detailed results for the track section between the Guoyuan Station and Jiukeshu Station are given in Fig. 5. It is obvious that these curves from different methods are with the similar mode but different in details, especially the distribution of operation conditions and the conversion points.

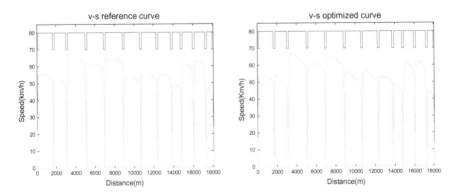

Fig. 2 Speed–distance curve for energy consumption calculation (left: reference curve; right: optimized curve)

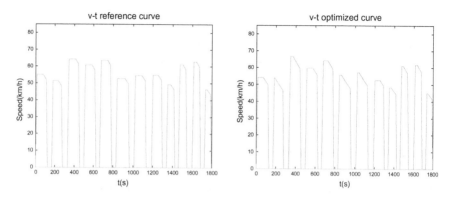

Fig. 3 Speed–time curve for energy consumption calculation (left: reference curve; right: optimized curve)

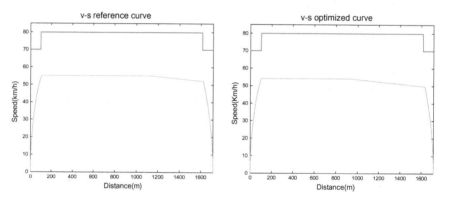

Fig. 4 Details of speed–distance curves between Sihui Station and Sihui East Station (left: reference curve; right: optimized curve)

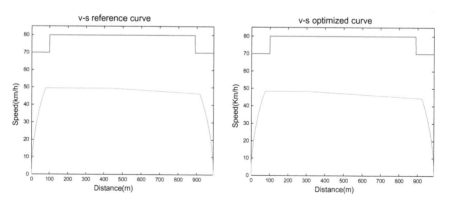

Fig. 5 Details of speed–distance curves between Guoyuan Station and Jiukeshu Station (left: reference curve; right: optimized curve)

Table 1 Statistical results of the reference curve and the optimized curve

Section No.	1	2	3	4	5	6	7	8	9	10	11	12
E_reference (kwh)	15.56	13.23	21.32	19.04	20.78	14.49	15.28	15.06	11.68	18.09	19.07	10.07
E_BFO (kwh)	14.39	12.36	20.19	17.86	19.9	13.51	14.26	13.81	10.77	17.32	17.9	9.25
Saving rate (%)	7.50	6.58	5.30	6.2	4.2	6.72	6.69	8.32	7.75	4.27	6.16	8.14

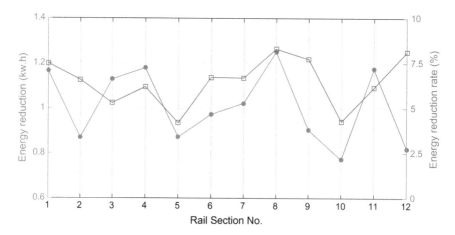

Fig. 6 Energy reduction and the reduction rates in all the track sections

To summarize the evaluation results, Table 1 lists the detailed reduced energy by the optimization for motion prediction against the reference curve through traction calculation, and the reduction rates for all the twelve sections are also computed. The results can be found in Fig. 6. From the results, it can be found that an average energy reduction rate of 6.48% can be achieved by using the BFO-based optimization in evaluation, which reveals the capability of the operation plan and time schedule in potential applications. In filed operation, a suitable driving strategy by the drivers or the automatic train operation (ATO) system will be a decisive factor to explore the capability of the fixed time schedule in realizing a green and environment-friendly transportation mode.

5 Conclusions

The evaluation of energy consumption level of urban transit trains based on the time schedule is significant to update and optimize the train operation plans in a pre-trip stage. By simulating the possible operation mode of the trains under the constraints of time plans, the potential energy-saving capability through the fixed operation plans can be assessed with the comparison of the referencing strategy. From the case study, it can be found that the optimized speed profiles result in an enhanced energy-saving capability, and the consumption is decreased by 6.8% for a whole trip in the test line. The presented optimization method provides an effective way to describe the possible operations of the trains in practical operation, and that is also meaningful to realize the improved driving operation by the drivers and onboard control systems.

Acknowledgements This work was financially supported by the National Key R&D Program of China (2017YFB1201105).

References

1. Yang S, Wu J, Yang X, Sun H, Gao Z (2018) Energy-efficient timetable and speed profile optimization with multi-phase speed limits: Theoretical analysis and application. Appl Math Model 56:32–50
2. Shoichiro W, Takafumi K, Eisuke I (2017) Evaluation of automatic train operation design for energy saving based on the measured efficiency of a linear-motor train. IEEJ Trans Ind Appl 137(6):460–468
3. Sheu J, Lin W (2012) Energy-saving automatic train regulation using dual heuristic programming. IEEE Trans Veh Technol 61(4):1503–1514
4. Lu Q, Feng X, Wang Q (2012) Energy-saving optimal control of following trains based on genetic algorithm. J Southwest Jiaotong Univ 47(2):265–270
5. Zheng W, Chen X, Huang H, Zhang Y (2017) Genetic algorithm based energy-saving ATO control algorithm for CBTC. Comput Syst Sci Eng 32(5):353–367
6. Hou Y (2015) Particle swarm optimization research based on bacterial foraging algorithm. J Softw Eng 9(4):838–847

Study on Dynamic Energy-Saving Adjustment Strategy of Metro Vehicle Air-Conditioning and Ventilation System

Yujie Li, Xuyang Wang, Jiao Zhang, Lu Wang and Liang Ma

Abstract Energy saving and emission reduction is an important work of metro operation management; the energy consumption of ventilation and air-conditioning system in subway account for about 25% of the total energy consumption in subway operation, To study the dynamic energy-saving adjustment strategy of metro vehicle air-conditioning and ventilation system is great significance for subway operation. This paper established a kind of dynamic energy-saving adjustment strategy of metro vehicle air-conditioning and ventilation system based on traffic monitoring data, including the VAV control system, temperature regulating system and a new design method of air volume control system. It has some guiding significance for similar projects.

Keywords Energy saving · Metro air-conditioning system · Dynamic energy-saving adjustment strategy · Air volume control system

1 Introduction

With the continuous increase of urban subway operating mileage in China, the importance of energy conservation and emission reduction is becoming increasingly prominent. According to statistics, the electricity cost of domestic subway operation accounts for about 50% of the total operation cost of subway [1–3]. The energy consumption of metro system consists of traction network energy consumption and power consumption.

Y. Li · J. Zhang · L. Wang · L. Ma
Beijing Mass Transit Railway Operation Corporation LTD, Beijing 100044, China

X. Wang
Beijing Key Lab of Urban Intelligent Traffic Control Technology, North China University of Technology, Beijing 100144, China

J. Zhang (✉)
Beijing Subway Operation Technology Centre, Beijing 102208, China
e-mail: zhangjiao2099@163.com

The energy consumption of ventilation and air-conditioning system in subway account for about 25% of the total energy consumption [4–6]. Therefore, the ventilation and air-conditioning system of subway vehicles should not only meet the needs of passengers for thermal comfort, but also take into account its comprehensive energy efficiency. By means of spatial layout adjustment, frequency conversion technology and dynamic load adjustment, the maximum energy saving can be achieved. This paper focuses on the dynamic energy-saving adjustment strategy of an air-conditioning and ventilation system for metro vehicles.

2 Principle of Dynamic Adjustment

In order to regulate the air parameters and air quality in train cars, the air-conditioning system must deliver cold (or hot) air with fresh air to the air-conditioned room. In summer, in order to make the cold air of low temperature drying, need to provide the cold source (frozen water or refrigerant) with refrigeration equipment; in winter, air heaters or heat pumps must be used to prepare to warm air. If humidity is required in the carriage, an air humidification system should be provided. To sum-up, the energy input (such as electricity, steam, hot water) of the relevant equipment and machine constitutes the energy consumption in operation when the air conditioner is in operation.

The operating energy consumption of air-conditioning system is related to the size of air-conditioning equipment, cold and heat source equipment and power equipment. Usually, the load of the train air-conditioning includes the external surface of the car, the heat transfer through the windows, as well as the cold and hot load and wet load of lighting, human body and indoor mechanical equipment. Therefore, the operating energy consumption of the train air-conditioning system is directly related to the temperature and humidity conditions of the line section during operation, the set value of indoor temperature and humidity, the power configuration and use status of indoor lighting and mechanical equipment, the number of train passengers and their activities and the amount of fresh air intake in the system. In addition, the efficiency and running time of all kinds of equipment are also directly related to the running energy consumption.

So, want to save energy consumption, should design small air-conditioning load above all. Consider train ventilation system parameters such as the structure, mechanical and electrical equipment operation time are basic fixed, and train operation section parameters such as temperature, humidity and local meteorological conditions is are difficult to control, in addition to choose high efficiency (should have higher efficiency in part load) of the equipment, mainly should consider adjusting the passenger flow density on the train air-conditioning system with the load impact on energy consumption.

3 Dynamic Adjustment Strategy

The energy-saving operation strategy of the train air-conditioning system based on passenger flow parameters mainly includes three links: passenger flow detection, air volume control and temperature CO_2 parameter feedback. The specific strategy model is shown in Fig. 1. Passenger flow detection is an input result, which mainly relies on train weighing detection, video detection, infrared laser detection and other methods to obtain the train density, feedback to the control system, guide the train VAV air-conditioning control system, temperature regulation system and new air volume control system.

(1) VAV air-conditioning control system

VAV air-conditioning system is a kind of air-conditioning system which can adjust and control the temperature of an air-conditioning area by changing the air supply volume. Different from the regulation mode of the traditional air-conditioning system with fixed air volume, the variable air volume air-conditioning system controls the change of sensible heat load of the subway car by adjusting the air volume into the subway car, so that the controlled temperature and the air quality in the subway car can meet the required standards. Variable air volume air-conditioning system meets the requirement of sensible heat load in subway cars by keeping the air supply temperature constant and changing the air supply to adjust the temperature in subway cars. Compared with the fixed air volume system, the variable air volume air-conditioning system can not only reduce the energy consumption of the fan, but also avoid the double energy consumption caused by heat and cold offset, thus achieving obvious energy-saving effect. According to the results of practical engineering calculation, the energy saving of the VAV system is up to 35–75% compared with that of the VAV system [7].

The most basic requirement of the design of the air-conditioning system is to be able to send a sufficient amount of treated air into the subway car, which is used to absorb the residual heat and moisture in the subway car, so as to achieve the required temperature and humidity in the subway car. The basic calculation formula is as follows:

$$Q \approx 1.2L(t_n - t_s) \qquad (1)$$

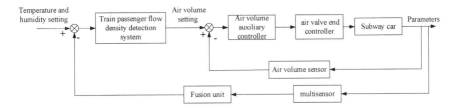

Fig. 1 Energy-saving operation strategy of train air-conditioning system

In the formula,

- L air supply volume, m³/h;
- t_n air temperature (or return air temperature) in the carriage;
- t_s air supply temperature;
- Q sensible heat load to be absorbed by air-conditioning supply, W.

Since the total air volume of the air-conditioning system of subway cars and the air volume sent to each compartment area will change with the change of the heat load of the air-conditioning, only by effectively controlling the VAV air-conditioning system can the normal, energy-saving and comfortable operation of the VAV air-conditioning system be realized.

(2) Temperature regulating system

The main loop controlled by this system is a set value system for temperature, that is, a follow-up system that can control the temperature in the subway car within a certain value, and the secondary loop is the air volume, as shown in Fig. 2. The temperature controller is the main controller, and the air volume controller is the secondary controller. The temperature regulator will change with the change of measured load and set conditions, so that the air volume regulator is constantly changing its given value, so that the air volume regulator changes with the change of its given value. In other words, cascade control system mainly relies on its secondary loop and has certain adaptive energy. The subway car air-conditioning controller is used to detect the actual temperature of the subway car, comparing with humidity values again, according to its comparative worth out of the set value of air volume required, then adjusted according to the value of wind valve, so that the total air volume changes, to keep the subway car in setting temperature range.

(3) New air volume control system

As the most important study on the air-conditioning system of subway cars, the effective control of fresh air volume is able to automatically adjust the temperature [8], relative humidity and air quality in the subway cars, so as to meet the ventilation needs of subway cars. For the consideration of air quality and energy saving in the carriage, the CO_2 concentration was selected as an important index in the monitoring of the control of fresh air volume. On the basis of CO_2 concentration representing the degree of air pollution caused by CO_2 itself as a pollutant in the carriage, the requirement of fresh air for the personnel in the carriage is determined by monitoring the condition of the personnel in the carriage, that is, the number of personnel and

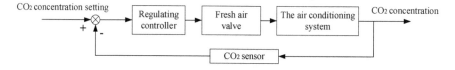

Fig. 2 Temperature control system of subway cars

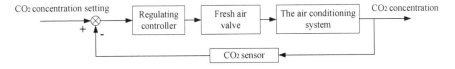

Fig. 3 Block diagram of new air volume control in subway cars

the activity of people. This control method has certain energy-saving effect. Since CO_2 is closely related to human metabolism, the data show that the daily intake of air per person is 10 m³, of which oxygen is 2 m³. When people exhale, CO_2 accounts for 5–6% and oxygen for 15–16%. Therefore, a fresh air volume of 30 m³ must be guaranteed in the subway car to achieve the goal of limiting CO_2 concentration. In the closed and more strict subway cars, the shield opening system is used to separate the subway cars from the tunnel, which increases the enclosed space greatly. On this basis, fresh air volume control can play a greater role in the air-conditioning of subway cars [9]. The content of CO_2 in the air is 0.035%, and once the content in the air exceeds 0.54%, it will exceed the safe limit of CO_2 and aggravate the greenhouse effect. The calculation of fresh air volume is based on mass balance, and the specific calculation method is as follows:

$$Q = \frac{G}{(C_i - C_o)E_v} \text{l/S} \quad (2)$$

In the formula,

Q fresh air volume,m³/h;
G total expansion rate,mg/s;
C_i concentration value,mg/l;
C_o outdoor air concentration,mg/l;
E_v ventilation efficiency.

Figure 3 shows the new air volume control chart of the air-conditioning system of subway cars. In the diagram, by comparing the measured carbon dioxide concentrations and setting density within subway car, according to the comparison of difference to control air valve opening, by the way can dilute the concentration of carbon dioxide in the subway car, so as to ensure the purpose of cabin air pure and fresh degree, ultimately to achieve the requirement of the new air volume control, sending way not only can ensure the comfort of subway car, but also can save energy.

4 Conclusion

The characteristics of the subway car widely for the space, large area, regional personnel fluidity is big, so the subway car air-conditioning control process is a big inertia, pure time-delay, time-varying and nonlinear complex system; this article is based on testing data to establish a kind of subway vehicle passenger flow dynamic energy-saving air-conditioning and ventilation system adjustment strategy, including the VAV control system, temperature regulating system and a new design method of air volume control system, which has certain guiding significance on construction of similar projects.

Acknowledgements This work was financially supported by National Key R&D Program of China (2017YFB1201105).

References

1. Liu W, Zhou H, He Q (2008) Modeling pedestrians flow on stairways in shanghai metro transfer station. In: 2008 international conference on intelligent computation technology and automation (ICICTA), Hunan, pp 263–267
2. Li G-Q, You S-J (2011) A new energy saving ventilation and air conditioning system for city rail transit. In: 2011 international conference on business management and electronic information, Guangzhou, pp 808–811
3. Amri H, Hofstädter RN, Kozek M (2011) Energy efficient design and simulation of a demand controlled heating and ventilation unit in a metro vehicle. In: 2011 IEEE forum on integrated and sustainable transportation systems, Vienna, pp 7–12
4. Wang Z, Han N, Wang Y (2011) Studies on neural network modeling for air conditioning system by using data mining with association analysis. In: 2011 international conference on internet computing and information services, Hong Kong, pp 423–427
5. Wang Y, Feng H, Xi X (2014) Sense, model and identify the load signatures of HVAC systems in metro stations. In: 2014 IEEE PES general meeting conference & exposition, National Harbor, MD, pp 1–5
6. Li Y, Zhang Y, Lin X, Xiao F, Li J (2014) Study on ventilation and smoke control system of subway transfer stations. In: 2014 7th international conference on intelligent computation technology and automation, Changsha, pp 746–749
7. Yanping A, Ping Q (2015) Air-conditioning energy-saving control strategy at subway station based on MAS evolutionary algorithm. In: 2015 8th international conference on intelligent computation technology and automation (ICICTA), Nanchang, pp 122–125
8. Li L, Jia D (2017) Research on air conditioning system of subway station based on fuzzy PID control. In: 2017 4th international conference on information science and control engineering (ICISCE), Changsha, pp 1131–1134
9. Pudikov VV, Litvinova NB, Grushkovsky PA (2018) Automatic control system of ventilation and air conditioning systems based on the organization of energy-saving air treatment modes. In: 2018 international multi-conference on industrial engineering and modern technologies (FarEastCon), Vladivostok, pp 1–3

Study on Energy Efficiency Performance of Rail Timetable Considering Track Condition

Jiao Zhang, Yujie Li, Lu Wang and Liang Ma

Abstract Energy saving is a significant issue in enhancing the capabilities of urban rail operations. To estimate the general performance of a specific timetable, the energy consumption model and the probable speed curves of the trains are involved to derive the evaluation result. In this paper, we present an energy efficiency performance evaluation method. A numerical optimization algorithm is adopted to derive the energy-efficient speed profiles and calculate the energy-saving performance. By utilizing the rail track parameters, the influence of track condition to the evaluation is concerned. The presented method is tested using the real track condition parameters and timetable data. The results demonstrate that the presented method is capable of calculating the energy-saving result based on the specific train plans.

Keywords Energy efficiency · Urban rail transit · Time schedule · Track condition

1 Introduction

Rail transportation has been a crucial transport branch in modern society due to its competitiveness of national or regional economy. Multiple advanced techniques have been applied in the current and the next generation rail systems to achieve the enhanced advantages and business benefits [1, 2]. In urban rail systems, the energy efficiency is significantly considered to optimize the trains' operations. The driving strategy of the trains is a decisive factor to reduce the unnecessary energy consumption over the conventional operation mode. There have been many results in generating the energy efficiency deriving strategies using specific algorithms and methods [3–5]. In the energy-efficient design of the train operation control system, it should be noticed that the trains' operation is constrained by the time schedule according to the characteristics of the urban rail system. In the time schedule design

---

J. Zhang · Y. Li · L. Wang · L. Ma
Beijing Mass Transit Railway Operation Corporation LTD, Beijing 100044, China

J. Zhang (✉)
Beijing Subway Operation Technology Centre, Beijing 102208, China
e-mail: zhangjiao2099@163.com

phase, the energy efficiency performance of a specific timetable will be a decisive step to optimize the possible filed operation.

In most studies on the energy-efficient operation topic, the numerical optimization using swarm intelligence strategies has been adopted to predict the optimized speed profiles [6]. For the operation plan of the trains, the energy efficiency performance can be estimated to evaluate the general level of the timetable. In addition, the track condition like the track gradient parameters is also an important factor to enhance the accuracy of prediction. In this paper, a method of energy efficiency performance prediction is presented, which makes it possible of evaluating the probable performance of a specific timetable. The real track condition is involved in the evaluation to enhance the evaluation. Practical timetable data from Beijing Metro is adopted to demonstrate the features of the presented method.

2 Theoretical Foundations of Bacterial Foraging Optimization Algorithm

Bacterial foraging optimization algorithm is a new evolutionary algorithm and a global random search scheme. It mainly includes three operations: chemotaxis, reproduction, and elimination and dispersal. The iterative computation is used to solve the optimization problem. The three operations and their processes are described as follows.

(1) Chemotaxis

There are two basic operations performed by an Escherichia coli in the process of foraging: swimming and tumbling. These two operations are achieved by the flagellums spreading all over the bacterial surface swinging in the same direction at a speed of 100–200 r/s. When all flagellums rotate counter-clockwise, E. coli swims forward at a speed of 10–20 um/s, and the average swimming time is about (0.86 + 1.18) s. On the contrary, when all flagellums rotate clockwise, E. coli tumbles in situ and randomly chooses one direction for the next swimming operation, and the average time of rotation is about (0.14 + 0.19) s. Generally, the bacteria tumble more frequently in areas with poor environment and swim more in areas with a good environment condition.

The whole life cycle of E. coli is a transformation between the two basic operations, swimming and tumbling, and flagellums almost never stop rotating. The purpose of swimming and tumbling is to find the food and avoid toxic substances.

(2) Reproduction

The law of biological evolution is survival of the fittest. Some bacteria with weak ability to find food will be eliminated naturally after a period of food searching. In order to maintain the population size, the remaining bacteria will reproduce. In the original BFO algorithm, the population size of the algorithm remains the same after the reproduction operation. Firstly, bacteria are sorted according to their fitness.

After that, a half of the bacteria with lower fitness eventually die while each of the rest bacteria asexually split into two bacteria, which are placed in the same location.

(3) Elimination and dispersal

Local areas where bacteria live in may suddenly change (e.g., sudden temperature rise) or gradually change (e.g., food consumption), which may lead to collective death of bacterial population living in this area or collective migration to a new area.

Migration operations occur with a certain probability which is named by Ped. If one bacterium in the population satisfies the probability of migration, the bacterium will die out and randomly generate a new one at any position in the solution space. The new bacterium may have different positions compared with the extinct bacterium, which means different foraging abilities. The new individual generated randomly by the migration operation may be closer to the global optimal solution, which is more conducive to the chemotaxis operation to jump out of the local optimal solution and find the global optimal solution.

3 Energy Efficiency Performance Evaluation with Force Analysis

The whole operating process of metro trains running at a section can be described as follows.

The train starts at the departure station at the initial speed $v_0 = 0$, and it is required to arrive at the terminal station within a given time interval, where the running distance is S and the final running speed is $v_n = 0$. The force diagram of the train during operation is shown in Fig. 1.

If the traction force is F_q, the running resistance is F_f, and the braking force is B_z. The combined force of the train is C. Furthermore, the combined force of the train in the process of accelerating is $C_a = F_q - F_f$, the combined force in the process of braking is $C_b = -B_z - F_f$, and the combined force in the process of coasting is $C_c = -F_f$.

Assuming that the acceleration of the train under the combined force is a, the recursive calculation process of train operation is as follows.

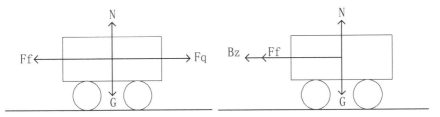

Fig. 1 Forces analysis of a train during operations (left: traction phase, right: braking phase)

$$a = \frac{C}{M} \tag{1}$$

$$v_i = \sqrt{v_{i-1}^2 + 2a \cdot \Delta s} \tag{2}$$

$$t_i = t_{i-1} + \frac{v_i - v_{i-1}}{a} \tag{3}$$

$$E_i = E_{i-1} + \Delta E \tag{4}$$

The parameters and variables in the above-given equations can be described as follows.

- M Mass of trains and passengers, kg.
- v_{i-1} The running speed of the train at step $i - 1$, m/s.
- v_i The running speed of the train at step i, m/s.
- t_{i-1} The running time of the train at step $i - 1$, s.
- t_i The running time of the train at step i, s.
- E_{i-1} Energy consumption of the train at step $i - 1$, kW h.
- E_i Energy consumption of the train at step i, kW h.

The running time and energy consumption of the metro trains running within a section depend on the effective allocation of four working conditions, i.e., the traction, cruising, coasting, and braking phases. It corresponds to the formation of the time-saving operation controls strategy with the minimum running time, the energy-saving operation control strategy with the minimum running energy consumption, and the time- and energy-saving operation control strategy with the optimal combination of running time and running energy consumption.

Two typical train operation processes can be found as shown in Fig. 2. The energy consumption of the train operation is mainly produced in three processes: the traction, cruising, and braking phases. Therefore, the energy efficiency optimization of the metro train is mainly achieved by choosing the appropriate initial and termination speed of coasting phase and increasing the coasting time.

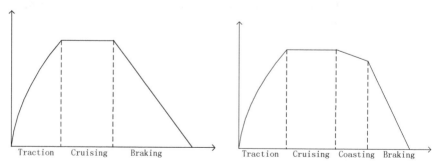

Fig. 2 Operation processes of trains (left: time-saving operation of trains, right: energy-saving operation of trains)

Based on the analysis of train operation processes, the energy consumption of the possible train operation according to the timetable can be calculated based on the energy consumption model for the traction condition. The procedures of energy efficiency performance evaluation for the timetable can be summarized as follows.

1. Data extraction: As the initial step, the schedule of multiple planned trains is extracted from the timetable. Some parameters are set before the calculation, including the train, railway track, and BFO algorithm.
2. Speed curve calculation: The speed curves of the planned trains are calculated and compared by utilizing the track gradient parameters.
3. Energy consumption evaluation: By using the specific energy consumption model, the energy-saving rate can be evaluated to all the planned trains and the timetable.

4 Results of Energy Consumption with Timetable Data

In order to test the energy efficiency performance evaluation method presented in this paper, the real urban rail timetable data and the track parameters are involved to perform the comparison. The timetable information is obtained from Batong Line of Beijing Metro. A scheduled train (no. 052099) is planned to move from the SiHui station to TuQiao station during a fixed time interval (13:40:30, 14:10:20). A fixed passenger capacity is assumed to carry out the evaluation, with which totally 600 passengers are loaded in the target train during the whole trip. The track gradient information of the whole line covering 13 stations is adopted to discuss the influence of track conditions to the energy efficiency calculation using the presented optimization strategy. The time schedule of the target train is listed in Table 1.

By calculating the candidate referencing curves and the optimized curves of the train according to the time schedule, the energy-saving performance driven by the optimized operation strategy can be summarized as the following tables, where Table 2 summarizes the statistical result of energy consumption in different track sections without considering the track condition. With the same evaluation method, Table 3 summarizes the calculation results based on the real track gradient data. All the track sections are calculated separately to depict the differences resulted by the track conditions.

To indicate the detailed results of the energy efficiency performance derived by the optimization solution, the speed curves from the referencing calculation strategy and the optimized solution are compared in Figs. 3 and 4. The speed–distance curves are shown in Fig. 3 with the speed limits during the whole trip, where the track gradient parameters are not involved in the calculation. It can be found that there are obvious coasting driving conditions in all the track sections, which enable the reduction of energy consumption level under the ideal 2D operating process. To further enhance the consistency between the evaluation and the probable field scenario, the gradient

Table 1 Timetable of the target train (No. 052099)

Station No.	Station name	Arriving time	Departure time
1	SiHui	–	13:40:30
2	SiHui Dong	13:42:50	13:43:20
3	GaoBeiDian	13:45:15	13:45:55
4	ChuanMeiDaXue	13:48:10	13:48:45
5	ShuangQiao	13:51:00	13:51:30
6	GuanZhuang	13:53:50	13:54:20
7	BaLiQiao	13:56:50	13:57:20
8	TongZhouBeiYuan	13:59:40	14:00:15
9	GuoYuan	14:02:05	14:02:45
10	JiuKeShu	14:04:05	14:04:50
11	LiYuan	14:06:30	14:07:00
12	LinHeLi	14:08:40	14:09:05
13	TuQiao	14:10:20	–

Table 2 Energy efficiency results without considering the track condition

Section No.	Distance (m)	E_ref (kW h)	E_opt (kW h)	ΔE_opt (kW h)	ΔE_opt (%)
1	1715	15.9502	14.7538	1.1964	7.5008
2	1375	14.6993	13.8271	0.8722	5.9336
3	2002	23.3814	22.1740	1.2074	5.1638
4	1894	20.8879	19.6323	1.2556	6.0110
5	1912	19.8973	19.0362	0.8611	4.3278
6	1763	14.8503	13.8498	1.0005	6.7371
7	1700	15.6677	14.6172	1.0506	6.7052
8	1465	18.2056	16.8022	1.4034	7.7086
9	990	15.0518	14.0020	1.0497	6.9742
10	1225	15.1437	14.4336	0.7101	4.6890
11	1257	15.9655	14.8843	1.0812	6.7723
12	776	10.3275	9.4866	0.8410	8.1429

parameters of all the track sections are introduced into the calculation with the same timetable data, and Fig. 4 shows the derived speed–distance curves with the gradient curve.

By comparing the results in Fig. 3, the ratio of costing condition has been reduced due to the differences in track condition. The force analysis of the target train is significantly influenced by the slope of the longitudinal section. To properly utilize the practical track condition, the profiles of the probable driving conditions have to be adjusted according to the strict limitation of the timetable. The involvement of

Table 3 Energy efficiency results considering the track condition

Section No.	Distance (m)	E_ref (kW h)	E_opt (kW h)	ΔE_opt (kW h)	ΔE_opt (%)
1	1715	15.7013	14.8721	0.8292	5.2811
2	1375	19.0542	18.2742	0.7800	4.0935
3	2002	24.3449	23.3083	1.0365	4.2577
4	1894	24.4067	23.4469	0.9598	3.9325
5	1912	25.1150	24.2072	0.9078	3.6148
6	1763	19.6377	18.3575	1.2802	6.5191
7	1700	22.7459	21.7945	0.9514	4.1827
8	1465	27.0477	26.1037	0.9440	3.4902
9	990	16.1837	15.2569	0.9268	5.7268
10	1225	15.0710	13.9785	1.0925	7.2493
11	1257	17.0534	15.9003	1.1531	6.7618
12	776	13.8324	12.9536	0.8788	6.3535

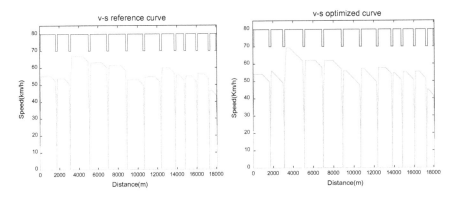

Fig. 3 Speed–distance curves without track condition data (left: reference curve; right: optimized curve)

the track conditions makes it possible of predicting the probable driving profiles and the energy consumption characteristics. The average energy reduction level of all the twelve track sections under the gradient-free evaluation reaches 6.39%, while the gradient involved calculation has been reduced to 5.12%. The utilization of the track parameters reflects more detailed characters of the track conditions and provides the exact prediction result of the energy-saving capability from the given timetable. The case study using only one target train in the timetable indicates the capability of this solution to calculate the energy-saving performance. More evaluations can be carried out to a specific timetable with multiple train plans. The derived speed curve profiles can provide an effective guidance to the drivers or the automatic train operation system.

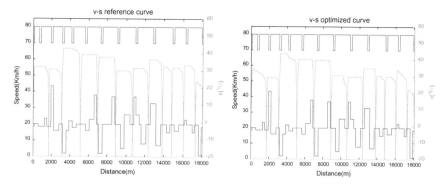

Fig. 4 Speed–distance curves using real track condition data (left: reference curve; right: optimized curve)

5 Conclusions

The precise calculation of the energy efficiency characteristics is a significant issue in optimizing the urban rail train operations. Different from the driving strategy optimization for a specific train, in this paper, we consider the evaluation of the energy consumption performance of the whole plan. The procedures of the calculation method including three steps are described. By using the field data of the timetable and the track gradient parameters, a case study is given to show the calculation result of the energy-saving performance. The results from the calculation using the real track parameters give more reliable evaluation of the time schedule for the target train. The presented method makes good use of the train plans and practical track conditions. It is capable of providing an effective guidance to the update or the optimization of the operation plans of the urban rail system.

Acknowledgements This work was financially supported by National Key R&D Program of China (2017YFB1201105).

References

1. Luan X, Wang Y, Schutter B, Meng L, Lodewijks G, Corman F (2018) Integration of real-time traffic management and train control for rail networks—Part 2: extensions towards energy-efficient train operations. Transp Res Part B 115:72–94
2. Albrecht A, Howlett P, Pudney P, Zhou P (2015) Energy-efficient train control: the two-train separation problem on level track. J Rail Transp Plan Manag 5:163–182
3. Wang P, Goverde R (2019) Multi-train trajectory optimization for energy-efficient timetabling. Eur J Oper Res 272:621–635
4. Liu W, Wang D, Li Q, Cui M (2016) A novel time-approaching search algorithm for energy-saving optimization of urban rail train. J Southwest Jiaotong Univ 51(5):918–924

5. Shoichiro W, Takafumi K, Eisuke I (2017) Evaluation of automatic train operation design for energy saving based on the measured efficiency of a linear-motor train. IEEJ Trans Ind Appl 137(6):460–468
6. Hu H, Fu Y, Hu C (2010) PSO-based optimal operation strategy of energy saving control for train. In: Proceedings 2010 IEEE 17th international conference on industrial engineering and engineering management, pp 1560–1563

Coordinated Control Method of Passenger Flow Based on Anylogic

Xuyang Wang, Huijuan Zhou, Yu Zhao and Ying Liu

Abstract Based on the analysis of the characteristics of subway passenger flow, the coordinated control method of subway passenger flow is studied. The simulation and modeling of 12 stations of Beijing Subway Line 5 were carried out by using Anylogic. The results of simulation were analyzed by using interval transport capacity utilization, number of people stranded at the station and average station retention rate. The research shows that the coordinated control of passenger flow optimizes the indicators of each system to make the utilization rate of the train interval transportation more balanced, and the average retention rate of the passenger flow of the whole line is more uniform and gentler.

Keywords Anylogic · Subway · Passenger flow characteristics · Simulation · Passenger flow coordinated control

1 Introduction

Under the influence of increasing passenger demand, the operation difficulty of rail transit is increasing, especially in the aspects of passenger flow control and safety operation. The transportation capacity cannot fully cope with the peak passenger

---

X. Wang · H. Zhou (✉) · Y. Liu
Beijing Key Lab
of Urban Intelligent Traffic Control Technology, North China University of Technology, Beijing 100144, China
e-mail: zhouhuijuan@ncut.edu.cn

X. Wang
e-mail: wangxy950114@163.com

Y. Liu
e-mail: 2017309030123@mail.ncut.edu.cn

Y. Zhao
Standard and Metrology Research Institute, China Academy of Railway Sciences Co. Ltd., Beijing 100081, China
e-mail: up2u@sina.com

flow, and the management and control method of passenger flow is single. Under this condition, reasonable and scientific coordinated control of the entire line or even the entire network of passenger flow is needed to improve the operational efficiency of the entire rail transit system [1]. Many people take the Beijing Subway as an example to analyze the law of passenger flow in Beijing and summarize the problems of insufficient coverage of the network and uneven distribution of passengers' time and space [2–4]. In the aspect of passenger flow control of rail transit, the linear flow planning method is used to describe the passenger flow and the objective function and constraint conditions are selected to control the passenger flow [5–9].

This paper uses Beijing Subway Line 5 as a model for modeling and analysis, which can effectively verify the validity and practicability of various passenger flow coordinated control methods.

2 Modeling Ideas and Evaluation Indicators

Using Anylogic simulation software, select Tiantongyuan North Station of Beijing Subway Line 5 to Lama Temple Station, a total of 12 stations for modeling, editing the Pedestrian Behavior Logic Module, Train Operation Logic Module, Line Physical Environment Module Construction and 3D Display Module Construction.

In this model simulation, the real-time data of interval transport capacity utilization, and number of people stranded at the station and average station retention rate are used as evaluation indicators.

3 Modeling Process

3.1 Logic Module Construction

Pedestrian Behavior Logic Module. In this model, the passenger flow mainly includes three streamlines, namely the inbound passenger flow, the passenger flow and the outbound passenger flow. Since it is a single-line system, the transfer passenger flow is not considered. The pedestrian logic module mainly simulates the passenger's inbound and outbound behaviors.

The generation of pedestrians is implemented by using the Ped Source module in the pedestrian library. Pedestrians need to perform card swipe after entering the station. The Ped Service module is used to define the card swiping behavior. Pedestrians enter the waiting area on the platform through the credit card machine. The passengers choose the waiting area by Ped Select. The Output module defines that the probability of selecting each waiting area is the same, so that the waiting passenger flow is evenly distributed in the waiting area. After the train arrives, follow the principle of "first go up and then go up," and the passengers who get off the bus are realized by the Ped Source module. The elevators that are used during the outbound process are implemented using the Ped Escalator module. The passengers travel using the Ped Go To module. When the passengers arrive at the Target Line, the passengers are considered to be out of the station. The passenger waiting for the bus enters the vehicle after the passenger gets off the train. The boarding action is implemented by the Go To Train module. When the passenger gets on the train, the train is considered to be over and the train leaves the station.

Figure 1 shows the pedestrian boarding logic module, and Fig. 2 shows the Go To Train agent module.

Train Operation Logic Module. The train operation logic module includes the running behaviors such as the appearance, running and disappearance of the train. The running behavior of the train is realized by the modules in the track library.

The train is generated by the Train Source module in the track library. The initial position of the train is placed in the track field. The running process of the train is realized by the Train Move To module. The stop position is marked with the Position on track function. The train getting on and off is realized by two Delay modules. The first Delay realizes the passenger getting off the vehicle, and the second Delay realizes the passenger getting on the train. After the action of the platform is completed, the train still uses the Train Move To module to set the trajectory of the train so that the train leaves the platform and reaches the next station. The logic diagram of the station is shown in Fig. 3.

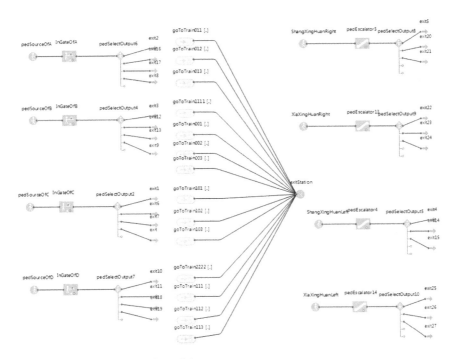

Fig. 1 Pedestrian boarding logic module

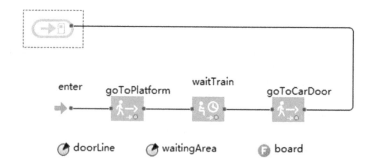

Fig. 2 Go to train agent module

Fig. 3 Single station train operation logic diagram

3.2 Line Physical Environment Module Construction

The construction of the physical environment is the basis of modeling. The modules that need to be built in this simulation have the physical environment of 12 stations and the laying of rails. The environment of the station is set to the side platform, the floor of the station is set to two floors, the lower floor is the two-way transfer channel of the platform and the upper floor is A, B, C, D four import and export and gates. At the same time, the pillars, fences, elevators and other facilities in the platform will be built. Figure 4 shows the physical environment of a station.

3.3 3D Display Module Construction

Using the 3D module in Anylogic to transform the model into a body space helps to more intuitively observe the behavior of pedestrians and trains. The 3D display module created this time includes the train view 3D display module and the station view 3D display module. Figure 5 shows the train view 3D display module.

4 Passenger Flow Coordination Control Model for Subway Lines

In this paper, the simulations of passenger flow coordinated control and passenger flow coordinated control are carried out separately, and the results are compared. The passenger flow coordination control model [1] used is:

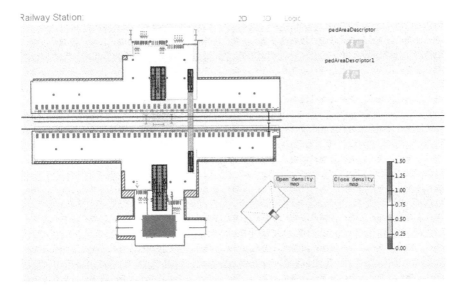

Fig. 4 Physical environment of a single station

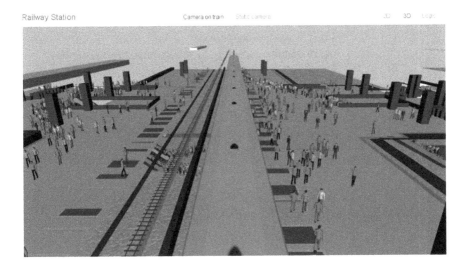

Fig. 5 Train view 3D display module

$$\min J = \sum_{n=1}^{N} \sum_{k=1}^{K} v_n^k \cdot \text{DET}_n^k \cdot \text{TS}_n^k \qquad (1)$$

$$\text{s.t} \begin{cases} \sum_{n=1}^{N}\sum_{k=1}^{K} PA_n^k \leq N \cdot K \cdot C \cdot \theta \\ PI_n^k + PO_n^k + v_n^{k-1} \cdot DET_n^{k-1} \leq Z_{n,\max} \\ \sum_{n=1}^{j-1} PA_n^k \cdot \sum_{m=i+1}^{N} r(n,m) \leq (j-i) \cdot K \cdot C \cdot \theta \\ \sum_{k=1}^{K}\sum_{n=i}^{j} PA_n^k \leq \sum_{k=1}^{K}\sum_{n=i}^{j} RC_n^k \end{cases} \quad (2)$$

In the formula:

$n = 1, 2, 3, \ldots, N$	the nth station on the line;
$k = 1, 2, 3, \ldots, K$	the kth train from the departure station;
PA	number of passengers successfully boarding on the platform during one sampling period;
PO	number of passengers who successfully got off the train during a sampling period;
PI	number of passengers waiting to enter the station on a platform during one sampling period;
DET	a stranded passenger who cannot get on the train after a train stops in a sampling period;
RC	the remaining capacity of a train stopped at the station;
TS	length of one sampling period;
C	the number of passengers rated for a train;
θ	maximum full load factor of a train;
v	retention coefficient;
r	destination station selection rate;
J	objective function.

5 Model Simulation and Result Analysis

According to the passenger flow coordinated control model, the demand for the passenger flow in the first few stations during the early peak period is strong, and the number of passengers on the train needs to be controlled. Station 9 is a transfer station, and there are more passengers entering the line, and the number of passengers on the train is also required. Control, from the overall observation, the generated passenger flow control scheme is closer to the passenger flow control station of Line 5 in actual life. Figure 6 shows the station average current limit rate.

Figure 7 shows the utilization rate of the interval transportation capacity. After the coordinated passenger flow control, the passenger flow pressure is evenly distributed

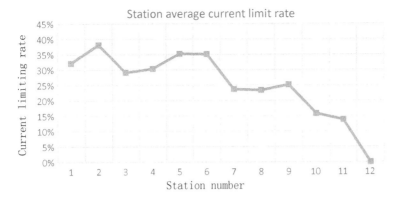

Fig. 6 Station average current limit rate

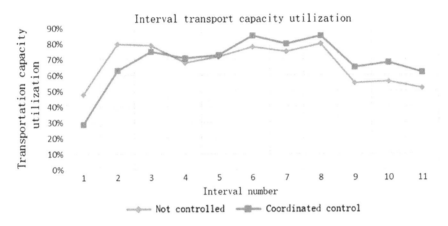

Fig. 7 Interval transport capacity utilization

in each station and not concentrated in a certain number of stations, and the utilization rate of the section transportation capacity is more balanced.

Figure 8 shows the number of people stranded at the station and the average station retention rate. When the passenger flow is not controlled at the stations on the line, the stranded passengers are mainly concentrated in the downstream stations because the upstream stations occupy the transport capacity resources of the trains, and the passenger retention rate is generally high. Through the coordinated control of passenger flow, the large passenger flow on board can be controlled to reserve transport capacity for downstream stations. The number of stranded passengers in downstream stations decreases, making the number of boarding passengers more uniform and the average retention rate of passenger flow in the whole line more uniform and gentler.

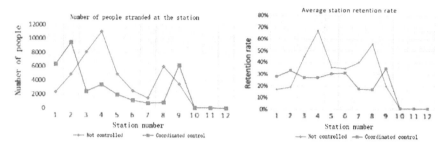

Fig. 8 Number of people stranded at the station and average station retention rate

6 Conclusion

Based on the analysis of the characteristics of subway passenger flow, the modeling and simulation of Beijing Subway Line 5 were carried out by using Anylogic. The three indicators of station capacity utilization capacity, station detention number and station average retention rate were adopted. The results of the simulation are analyzed. The results show that through the coordinated control of passenger flow, the indicators of each system are optimized, and the utilization rate of the train interval is more balanced. The average retention rate of the passenger flow of the whole line is more uniform and gentler. Combined with the physical environment of the station and other specific means, the specific means of passenger flow control has a greater guiding significance for the actual operation of rail transit.

In the follow-up work, the model can be further refined through more comprehensive data analysis to make the control results more effective and accurate.

Acknowledgements The authors would like to acknowledge the support of National Key R&D Program of China (No. 2016YFB1200402).

References

1. Zhao Y (2017) Coordinated control method for passenger flow on subway lines based on hierarchical hierarchy. North China University of Technology (in Chinese)
2. Wang J, Liu J, Ma Y et al (2013) Temporal and spatial distribution characteristics of passenger flow in Beijing rail transit stations. City Traffic 6:18–27 (in Chinese)
3. Liu J (2012) Traffic assignment modeling and empirical study of Urban rail transit network based on transfer. Beijing Jiaotong University (in Chinese)
4. Zhang C (2006) Analysis of passenger flow characteristics of urban rail transit. Southwest Jiaotong University (in Chinese)
5. Wang L, Yan X, Zhang Y (2015) Coordinative passenger flow control during peak hours in Urban rail stations. In: Cota international conference of transportation professionals
6. Palma AD, Kilani M, Proost S (2015) Discomfort in mass transit and its implication for scheduling and pricing. Transp Res Part B Methodol 71:1–18

7. Zhu W, Xu R, Jiang YP (2014) Calibration of rail transit assignment model with automated fare collection data and a parallel genetic algorithm. Transp Res Rec J Transp Res Board 2442(2442):20–28
8. Castelain E, Mesghouni K (2002) Regulation of a public transport network with consideration of the passenger flow: modeling of the system with high-level Petri nets. 6
9. Heilporn G, Giovanni LD, Labbé M (2008) Optimization models for the single delay management problem in public transportation. Eur J Oper Res 189(3):762–774

Train Operational Plan Optimization of Rail Transit Connecting to Airport Ground Transport Center

Yamin Li, Xiaohan Xu, Rui Xu and Ailing Huang

Abstract With the increasing demand for air transport, the problem of traffic connection between airport and city is becoming more and more severe. Rail transit with the characteristics of speed, punctuality, and larger capacity plays an increasingly significant role in connecting traffic between airport and urban center. From the perspectives of both the passengers and train operators, this paper establishes a multi-objective optimization model of rail transit train operation plan for airport ground transport center (GTC), constructs three objective functions, including passenger waiting time, passenger comfort, and operator income, and develops a genetic algorithm to solve the fuzzy optimal solution of the model. Taking Beijing Rail Transit Airport Line as the research object, this paper validates the proposed model and evaluates the scheme by using typical workday passenger flow data. The results show that our model can provide decision-making basis for the optimization of train dispatch.

Keywords Airport GTC · Rail transit dispatching · Vehicle scheduling · Multiple objectives · Headway

1 Introduction

With a growing demand for air transportation, the influx and outflow of passengers at an airport grow correspondingly. The difficulties of the connection between the airport and the city are becoming more and more severe; at times, such difficulties even restrict the development of air transportation. Rail transit is characterized by its speed, punctuality, reliability, and large capacity, which puts it on an increasingly important role in connecting airports and cities. Rail transit will become a dominant

Y. Li
China Academy of Transportation Sciences, Beijing 100029, China

X. Xu · R. Xu · A. Huang (✉)
Beijing Jiaotong University, Beijing 100044, China
e-mail: alhuang@bjtu.edu.cn

mode of transportation connecting large airports to urban areas. With the improvement of the urban rail transit network and the expansion of airport scale, the main means of transport connecting the airport to the urban area in the future will be rail transit [1].

In view of the high level of affluency of air passengers as compared to the general population in China, from this demographic's perspective, airport train services are required to provide a higher level of service (LOS), such as reducing intervals hence shorter passenger *waiting* time, improving passenger waiting environment, improving journey experience, so that passengers can reach their destination safely and quickly with better LOS. However, the train operator's objective is to maximize its operational efficiency while minimizing its operational cost. This means that the departure interval should be extended as much as possible while meeting the demand of passenger flow to reduce the number of train runs, thus minimizing operating costs [2].

At present, the research on rail transit scheduling at home and abroad mainly focuses on the research of passenger transfer time [3–5] and transport capacity [6–8] and considers less on the economic benefits of train operators. That is to say, the optimization objectives in existing research are usually centered on (1) minimizing total system cost [9, 10] of interchange hub or (2) minimizing passenger transfer time. Coordinated train dispatching models are hence constructed based on these optimization objectives. In these studies, solving the model could determine dispatch intervals and their corresponding number of rolling stock. For example, Chen et al. considered the passenger comfort and incorporated it into the passenger's travel cost and built a rail transit dispatch optimization model with the objective of minimizing passenger and operating cost [11]. Wang et al. divided the passenger flow into a large and a small traffic demand scenario according to the starting point and constructed an optimization model of the train operation plan to minimize the passenger travel cost and the minimum operating cost. When the model is solved, the model is transformed into the ideal point method. The single-objective problem finally analyzes the impact of the re-entry station on operating costs [12].

Based on the above background, aiming at maximizing the interests of both passengers and train operators, this paper establishes a multi-objective optimization model and solution method for rail transit traffic planning. Combining with the passenger flow data of Beijing airport line on typical working days, an empirical analysis of the proposed optimization model for traffic planning is carried out, and the validity of the model method is verified.

2 Multi-objective Optimization Model of Train Operational Plan

2.1 Model Assumptions and Explanations

This paper builds multi-objective functions on the driving interval from two aspects including passengers and train operators. Since the waiting time and comfort are two typical indicators to represent the LOS of rail transit, on the passenger level, we mainly focus on passenger waiting time and passenger crowdedness in the train carriage. These two indicators can be obtained from the number of people in the train and the waiting time of the passengers. On the train operator level, the operating income is considered. A multitude of other variables may also interfere with model calibration. To simplify and generalize, the model proposed in this study is based on the following assumptions, which are observed in real-world operations:

(1) Passenger demand is not affected by the frequency of departure and the operating conditions of the adjacent lines.
(2) Passengers are "first come, first served." That is, passengers arriving at the station first will board the train first.
(3) Train departure intervals remain uniform throughout a specified period of time.
(4) The running speed of the vehicle in the inter-station interval is constant, and no special events occur on the way.
(5) Passengers' acceptable waiting time is a fixed value, beyond which passengers will be dissatisfied.

Variables and symbols are described as follows:

n	The total number of departures during the period;
m	Number of total line stations;
$\lambda_i(t)$	Passenger boarding rate (person/min) at t time of station i;
$\beta_i(t)$	The passenger disembarkation rate (person/min) at time t of station i;
u_{ij}	Passenger waiting time (min);
V_{ij}	Number of passengers on the jth bus in the interval from station i to station $i+1$;
B_{ij}	Number of passengers boarding the jth bus at station i;
h	Driving time interval (min);
C_{max}	Maximum passenger capacity of trains;
L	Maximum departure interval (min);
s_{ij}	The number of spare seats on the jth train in the interval from station i to station $i+1$;
r_0	Passengers in the train do not feel the crowded standing number threshold;
ρ	The fare per person (yuan/person).

2.2 Objective Function Based on Passenger Waiting Time and Ride Comfort

On the passenger level, passenger waiting time and number of passengers in the train are considered at the same time, and two objective functions of passenger waiting time and passenger crowdedness are obtained. For the passenger waiting time, the objective function of establishing the minimum waiting time for passengers is as follows:

$$\min z_1 = \sum_{j=1}^{n} \sum_{i}^{m-1} u_{ij} \tag{1}$$

$$\text{s.t } h \leq L \tag{2}$$

The passenger comfort of the passenger is mainly affected by the density of passengers inside the vehicle. When the passenger has a seat, the passenger is satisfied. When a passenger is standing, if the number of standees does not exceed the threshold number r_0 of the number of standings that makes the passenger feel crowded, the satisfaction of the standing passenger is k_0; if the passengers in the train feel crowded, the satisfaction of the standing passenger is 0. Satisfaction subordinating degree function of passengers standing in the bus is as follows:

$$\mu(V_{ij} - s_{ij}) = \begin{cases} k_0, & 0 < V_{ij} - s_{ij} < r_0 \\ 0, & \text{else} \end{cases} \tag{3}$$

In the formula, k_0—a satisfactory subordinating degree value for standees when there is no feeling of congestion in the bus.

When the passenger has a seat, the passenger's satisfactory degree of membership is 1. The total comfort value of passengers in the jth train from i station to $i+1$ station is:

$$v_{ij} = \begin{cases} V_{ij}, & V_{ij} - s_{ij} \leq 0 \\ s_{ij} + (V_{ij} - s_{ij}) * \mu(V_{ij} - s_{ij}), & \text{else} \end{cases} \tag{4}$$

The objective function of the passenger comfort is thus constructed as follows:

$$\max z_2 = \frac{\sum_{j=1}^{n} \sum_{i=1}^{m-1} v_{ij}}{\sum_{j=1}^{n} \sum_{i=1}^{m-1} V_{ij}} \tag{5}$$

$$\text{s.t } h \leq L \tag{6}$$

Among them, v_{ij} is the passenger comfort value, which is composed of the satisfactory subordinating degree value of the standees and the satisfactory subordinating degree value of the seated passengers.

2.3 Objective Function Based on Business Operating Income

Under the premise of a certain cost, the operating company expects the higher the operating income, the better the objective function of building the enterprise income is as follows:

$$\max z_3 = \frac{\sum_{j=1}^{n} \sum_{i=1}^{m-1} B_{ij} \cdot \rho}{n} \tag{7}$$

$$\text{s.t } h \leq L \tag{8}$$

2.4 Fuzzy Optimal Solution for Multi-objective Programming

This paper establishes the model from the perspective of passengers and enterprises. There are three objective functions in total, but since it is difficult to make multiple objective functions reach the optimal solution, this paper adopts the fuzzy optimal solution. The solution is as follows:

Firstly, the optimal value $z_i^*(i = 1, 2, 3)$ $(i = 1, 2, 3)$ of each objective function is obtained under the constraint condition, and then the scaling factor $d_i (d_i \geq 0)$ is obtained, and the corresponding fuzzy target set G_i, whose subordinating degree function [11, 13] is defined as:

$$G_i(x) = \begin{cases} 0 & z_i < z_i^* \\ 1 - (z_i^* - z_i)/d_i & z_i^* - d_i \leq z_i < z_i^* \\ 1 & z_i > z_i^* \end{cases} \tag{9}$$

Note that $G = \bigcap_{i=1}^{3} G_i(x)$, A is a feasible solution set, find x^* to make $G(x^*) = \max_{x \in A}\{G(x)\}$, and then x^* is the optimal solution of $G(x)$ in the feasible solution set A. The above process is equivalent to solving the following problem:

$$\text{Max } Z = \lambda \tag{10}$$

$$\text{s.t } \begin{cases} 1 - (z_i^* - z_i)/d_i \geq \lambda \\ S_i \leq 0 \\ \lambda \geq 0 \end{cases} \tag{11}$$

In the formula, z_i—the value of each objective function of the original problem; $S_i \leq 0$—the original problem constraint.

2.5 Model Solving Algorithms

Genetic algorithm imitates the principle of "survival of the fittest" in nature. Since Holland put forward in the 1960s, it has become a common intelligent heuristic algorithm. Since the model constructed in this paper has three objective functions, the genetic algorithm is used to obtain the optimal solution of a single target for each objective function, and these three solutions are substituted into each objective function to obtain the range of each objective function, thereby obtaining each target, and the scaling factor is finally obtained by the fuzzy optimal solution.

3 Model Validation and Evaluation

The paper takes the Beijing Rail Transit Airport Line as the research object. Based on the typical working day airport line passenger flow data, the paper conducts a case analysis to validate the proposed driving plan optimization model. By applying real data into the proposed model, train intervals and dispatched rolling stock numbers at different time periods can be obtained. A full-day train service plan can therefore be determined. The plan is then evaluated from both the passengers' and train operator's perspective.

3.1 Processing Rail Transit Passenger Flow Data

(1) Line and passenger comfort parameters

Beijing rail transit airport line is operated on a Y-shaped circular loop through Beijing city and Beijing Capital International Airport. The circular service starts from Dongzhimen Station, calls at Sanyuanqiao, Terminal 3, Terminal 2, Sanyuanqiao and terminates at Dongzhimen Station. The operation time is from 6:00 to 23:30, and the spacing of each station is shown in Table 1.

The airport line uses L-shaped four-carriage rolling stock, and the number of seats s is 224. In order to obtain the parameters of passenger comfort, this paper designs questionnaires for data collection, and a total of 1017 questionnaires are collected. The survey results show that the crowdedness threshold value r_0 is 40, meaning that

Table 1 Airport line distance table

Road section	Station distance (km)
Dongzhimen—Sanyuanqiao	3.02
Sanyuanqiao—Terminal 3	18.32
Terminal 3—Terminal 2	7.24
Terminal 2—Sanyuanqiao	20.74

Table 2 Passenger boarding rate (person/min)

Time interval	Dongzhimen	Sanyuanqiao	Terminal 3	Terminal 2
6:00–7:00	0.50	0.34	0.65	0.49
7:00–8:00	3.95	2.69	5.08	3.85
8:00–9:00	6.52	4.44	8.38	6.36
9:00–10:00	5.21	3.55	6.70	5.08
10:00–11:00	3.59	2.45	4.62	3.51
11:00–12:00	2.44	1.66	3.14	2.38
12:00–13:00	4.62	3.15	5.95	4.51
13:00–14:00	4.57	3.11	5.87	4.45
14:00–15:00	4.87	3.32	6.27	4.76
15:00–16:00	3.19	2.17	4.09	3.11
16:00–17:00	2.39	1.63	3.07	2.33
17:00–18:00	4.78	3.26	6.15	4.67
18:00–19:00	3.22	2.20	4.15	3.15
19:00–20:00	4.78	3.26	6.15	4.67
20:00–21:00	4.10	2.79	5.28	4.01
21:00–22:00	3.67	2.50	4.72	3.58
22:00–23:00	2.40	1.63	3.08	2.34
23:00–23:30	0.54	0.36	0.00	0.52

passengers start to feel crowded when the number of standees exceeds 40. The level of satisfaction of the standees k_0 is set at 0.7 when the train is not crowded. Therefore, the maximum passenger capacity C_{max} of the whole train is 336 people, consisting of 224 seats and 112 standee capacity. As for other parameters, the maximum departure interval L is 10 min, and the fare ρ per person is 25 yuan/person.

(2) Data processing of passenger boarding rate and alighting rate

This paper collected IC card swipe data of Beijing Airport Line on Thursday, November 3, 2016, and extracted OD data of passenger flow. On this basis, passenger boarding rate and alighting rate are calculated. For the purpose of this study, only boarding and alighting rate on Dongzhimen to Airport direction are used. The results are shown in Tables 2 and 3.

3.2 Model Solution

The solution of the model is based on the idea of fuzzy optimal solution. The genetic algorithm is used to solve the optimal solution of the three objective functions of passenger waiting time, ride comfort, and enterprise income by programming in

Table 3 Passenger alighting rate (person/min)

Time interval	Dongzhimen	Sanyuanqiao	Terminal 3	Terminal 2
6:00–7:00	0.76	0.52	0.97	0.74
7:00–8:00	3.95	2.69	5.08	3.85
8:00–9:00	4.35	2.96	5.59	4.24
9:00–10:00	4.44	3.02	5.70	4.33
10:00–11:00	4.76	3.24	6.12	4.65
11:00–12:00	4.34	2.96	5.59	4.24
12:00–13:00	3.49	2.38	4.48	3.40
13:00–14:00	4.95	3.37	6.36	4.83
14:00–15:00	4.87	3.32	6.27	4.76
15:00–16:00	3.74	2.55	4.81	3.65
16:00–17:00	3.30	2.25	4.24	3.22
17:00–18:00	5.62	3.82	7.22	5.48
18:00–19:00	5.26	3.58	6.76	5.13
19:00–20:00	4.42	3.01	5.68	4.31
20:00–21:00	4.27	2.91	5.49	4.17
21:00–22:00	2.16	1.47	2.77	2.10
22:00–23:00	2.49	1.70	3.21	2.43
23:00–23:30	0.54	0.36	0.00	0.52

MATLAB. On this basis, the scaling factors of each objective are obtained, and finally, the fuzzy optimal solution is obtained. The parameters of the genetic algorithm are set as shown in Table 4.

Since the passenger boarding and alighting rate are different at different times throughout the day, the obtained driving interval will be different. The solution process of the model will be described by taking the time range of 8:00–9:00 as an example.

(1) After the airport line passenger flow data is substituted into the model, it was found that the passenger's ride comfort value is close to 100%, which was not worthwhile for this optimization objective. Therefore, in solving for service intervals, riding comfort z_2 does not have an optimal solution.

Table 4 Values of algorithm parameters

Parameter	Value
Maximum iteration number	100
Chromosome length	10
Initial population size P_0	100
Crossing probability P_c	0.6
Mutation probability P_m	0.05

(2) In order to solve the problem easily, the corresponding negative value of passenger waiting time is taken, the passenger waiting time is taken as the fitness of the genetic algorithm, and the fitness curve is shown in Fig. 1. The optimal solution for obtaining the passenger waiting time objective function z_1 is $h_1 = 3.1$; at this time, $z_1 = -2317$, $z_3 = 1255$.

(3) Taking the train operator's income per train as the fitness of the genetic algorithm, the fitness curve is shown in Fig. 2. The optimal solution of the train

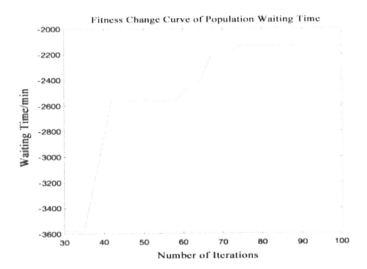

Fig. 1 Curve of waiting time fitness change

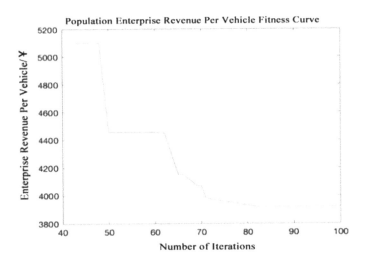

Fig. 2 Curve of enterprise revenue per vehicle fitness

operator's income per train objective function z_3 is $h_3 = 8.6$; at this time, $z_1 = -5034$, $z_3 = 3920$.

Therefore, $z_1 \in [-5034, -2317]$, $z_3 \in [1255, 3920]$, thereby obtaining the expansion factor $d_1 = -2317 - (-5034) = 2717$, $d_3 = 3920 - 1255 = 2665$. By substituting Formula 10, the original problem is transformed into the following planning problem:

$$\max Z = \lambda \tag{12}$$

$$\text{s.t} \begin{cases} z_1 - 2717\lambda \geq -5034 \\ z_3 - 2665\lambda \geq 1255 \\ h \leq 10 \\ \lambda \geq 0 \end{cases} \tag{13}$$

The result shows that the fuzzy optimal solution is $h = 3.5$, where $z_1 = 2406$ and $z_3 = 1513$.

Similarly, the corresponding departure interval at different times throughout a day can be obtained, which becomes the operation plan for the Dongzhimen–Airport direction. The details are shown in Table 5.

Table 5 Airport line operation plan

Time interval	Departure train pairs	Departure interval
6:00–7:00	7	8 min
7:00–8:00	13	4.6 min
8:00–9:00	18	3.5 min
9:00–10:00	14	4.3 min
10:00–11:00	12	5.5 min
11:00–12:00	10	6.0 min
12:00–13:00	11	5.2 min
13:00–14:00	13	4.7 min
14:00–15:00	10	6.0 min
15:00–16:00	9	6.2 min
16:00–17:00	13	4.8 min
17:00–18:00	16	3.9 min
18:00–19:00	13	4.6 min
19:00–20:00	14	4.4 min
20:00–21:00	9	6.3 min
21:00–22:00	9	6.6 min
22:00–23:00	8	7.4 min
23:00–23:30	3	9.6 min

Table 6 Comparison of evaluation indicators before and after optimization

Evaluating indicator	Before optimization	After optimization
Total passenger waiting time (min)	76,402	53,940
Maximum sectional load factor	74.52%	59.31%
Train pairs	126	202
Operator's revenue per train (yuan)	6,235	3,889

3.3 Evaluation of the Operation Plan

In order to better evaluate the optimized driving plan, the Beijing airport line original driving plan and the optimized driving plan are compared and analyzed, and the effectiveness of the model is verified by comparison under different evaluation indicators. The evaluation indicators encompass passenger total waiting time, maximum section full load rate, and enterprise revenue per vehicle.

As shown in Table 6, the total waiting time of passengers decreased by 24,646 min, a decrease of 29.4%, indicating that the optimized scheme has a very significant effect on reducing passenger waiting time; the maximum sectional load factor was reduced by 15.21 percentage points, that is a 20.4% reduction, indicating that the optimized scheme has a significant effect on reducing the maximum sectional load factor, i.e., the crowdedness.

However, the number of trains increased by 76 pairs (+60.3%), and the number of trains running in the optimized scheme increased greatly compared with that before optimization; the revenue per vehicle dropped by 2346 yuan, a decrease of 37.6%. This shows that the optimized plan is to reduce the waiting time of passengers by nearly doubling the number of trains. Therefore, although the optimized scheme satisfies the passenger's demand for reducing the time cost, considering the significant increase in the number of trains, the operating cost and organizational difficulty of the operator will also be greatly improved, and the operating income of the enterprise will also drop significantly.

4 Conclusion

The airport line full-day operation plan solved by the model satisfies passenger demand, greatly reduces passenger waiting time, significantly reduces crowdedness, and ensures passengers' ride comfort. At the same time, the number of trains in the optimized scheme is much higher than the original train schedule of the airport line. This means that if the train is operated directly according to the solution, the operating cost of the company will rise quite a lot. Therefore, the train operation plan directly solved by the model needs to be optimized and adjusted in combination with the actual situation. As far as the Beijing Airport Line is concerned, the expansion and reconstruction of the depot in November 2017 have begun. It is reported that the departure interval of the airport line is expected to be shortened to 4 min after the completion of the project, indicating that the relevant operational management departments attach great importance to the need for passengers to reduce waiting time for waiting. Therefore, the model solution results still have important practical significance.

Acknowledgements This work is supported by National Key R&D Program of China (Grant No. 2018YFB1601200).

References

1. Wang A (2014) Study on passenger volume forecasting of airport rail transit. Chang'an University (in Chinese)
2. Hanhui W (2017) Passenger flow short-time prediction and operation organization of urban rail transit. Chang'an University (in Chinese)
3. Wang Q, Li F (2003) On the coordination of switching non-rail transport modes to rail transport modes in urban passenger traffic system. J Xi'an Univ Arch Sci Technol (Nat Sci Ed) 35(2):136–139 (in Chinese)
4. Sigurd L, Pesenti R, Ukovieh W (2004) Scheduling multimodal transportation systems. Eur J Oper Res 155:603–615
5. Yin H, Wong SC, Xu JN, Wong CK (2002) Urban traffic flow prediction using a fuzzy-neural Approach. Transp Res Part C 10:95–98
6. Fang Z (2003) Optimization of transit network to minimize transfers. Environmental Engineering college of Florida International University
7. Cai H (2012) Study on the optimization of urban rail transit train operation organization under networking conditions. Beijing Jiaotong University (in Chinese)
8. Zhan K, Shi Y, Wang X (2013) Research on optimization of train operation organization efficiency of Guangzhou metro. Railw Transp Economy 6:102–106 (in Chinese)
9. Huang W (2004) Research on coordination of rail transit and routine public exchange. Chang'an University (in Chinese)
10. Prabhat SOM (2006) A model for development of optimized feeder routes and coordinated schedules—a genetic algorithms approach. Transp Porous Media 13:413–425
11. Chen S, He S, He B (2013) Train ervice planning for urban rail Transit with passenger fluctuation. Urban Rail Transit Res 16(10):53–58 (in Chinese)
12. Wang Y, Ni S (2013) Optimization of train operation plan for urban rail transit large-sized traffic mode. J China Railw Soc 35(07):1–8 (in Chinese)
13. Yan B (2006) Optimization study on operation organization of urban rail transiy. Southeast University, Nanjing (in Chinese)

Defect Detection for Catenary Sling Based on Image Processing and Deep Learning Method

Jing Cui, Yunpeng Wu, Yong Qin and Rigen Hou

Abstract With the development of high-speed railway, catenary sling defects have become a big hidden danger that affects the driving safety. Therefore, it is extremely important to detect and troubleshoot the sling defects. A new visual inspection method for catenary sling defects is proposed in this paper, focusing on two key points: foreign body, hard bending detection and no-stress detection. With regard to the first aspect, a light YOLOv3 network is presented to extract sling regions, and then Faster R-CNN is employed to inspect foreign body and hard bending. This can acquire an attractive inspection result, due to removal of redundant areas and defect amplification in a catenary image. With regard to the second, a new method based on traditional visual called self-defined linear difference (SLDD) is proposed to inspect no-stress after extracting catenary sling area using the light YOLOv3. This solves the problem of low detection rate caused by less train sample for deep learning. The experiment shows that Faster R-CNN has a good effect on foreign body and hard bending inspection after sling areas extraction, and the precision rate is more than 90%. And further experiment also shows that the light YOLOv3 combined with SLDD can achieve a better detection result for sling no-stress.

Keywords Catenary sling · Defect detection · Deep learning · Image processing

---

J. Cui · Y. Wu · Y. Qin (✉)
State Key Laboratory of Rail Traffic Control and Safety, Beijing Jiaotong University, No. 3 Shangyuancun, Haidian District, Beijing, People's Republic of China
e-mail: 16271244@bjtu.edu.cn; yqin@bjtu.edu.cn

Y. Wu
School of Traffic and Transportation, Beijing Jiaotong University, No. 3 Shangyuancun, Haidian District, Beijing, People's Republic of China

R. Hou
Beijing-Shanghai High Speed Railway Company Limited, Beijing, China

1 Introduction

The catenary sling is an important part of the high-speed railway facilities, which guarantees the safety of the contact line and the bow net. And the defect of the catenary sling will cause serious accidents to railway safety, which is reported in Ref. [1]. At the present stage, the defect of catenary sling mainly relies on the data collected by rail inspection vehicle, followed by artificial detection. Due to the large cost of manpower and low efficiency, a better solution must be sought. In Ref. [2], at present, there is no mature intelligent detection system to apply for the defect detection of catenary sling, so it is bound to become a new trend to solve the existing detection problems by using image processing, deep learning and other relevant new computer automatic detection technologies.

This paper combines traditional image processing with deep learning technology, focus on the study of three major defects in the catenary sling area: foreign body (fracture), hard bending and no-stress. The process can be mainly divided into three steps: preprocessing of catenary sling images, sling area positioning and defect detection. The catenary sling preprocessing mainly enhances the image data for improving the visual effect and highlight the sling area. The positioning of the catenary sling area is mainly calibrated and cut out by the improved deep learning network YOLOv3 for reducing the useless areas in an image. In our experiment, the Faster R-CNN network is used to detect three kinds of defects in the sling area. The test results showed that the results of foreign body (fracture) and hard bending were good, with the precision rate of more than 90%, while the effect no-stress was only 40%. Therefore, the method based on traditional image processing called self-defined linear difference (SLDD) is further adopted to detect the no-stress defect for acquiring a better result.

The rest of the paper is organized as follows: the methodology of this paper is presented in Sect. 2. The experimental results are elaborated in Sect. 3. The conclusion is described in Sect. 4.

2 Methodology

In this paper, a new method is proposed for catenary sling. The main idea is to calibrate and cut out the sling areas with a light YOLOv3 network, and then Faster R-CNN algorithm is used to detect the foreign body (fracture) and hard bending of slings. In addition, the SLDD used to detect no-stress defect. The whole algorithm flow chart is shown in Fig. 1.

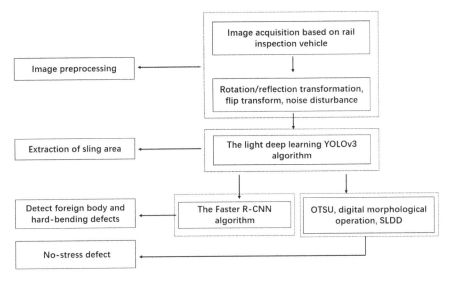

Fig. 1 Algorithm flow chart

2.1 Image Preprocessing

At the present stage, the defect of catenary sling mainly relies on the data collected by track inspection vehicle, followed by manual inspection. Similarly, our data is also from a track inspection vehicle running on a section of Beijing-Shanghai Line. As the pictures were taken at night, the quality was poor. It is necessary to preprocess the images in order to achieve a better result for subsequent detection.

In Refs. [3, 4], rotate theta around the origin, the formula is given by:

$$x' = x\cos\theta + y\sin\theta, \ y' = -x\sin\theta + y\cos\theta \tag{1}$$

Global histogram equalization, the formula is given by:

$$h(v) = \text{round}\left(\frac{\text{cdf}(v) - \text{cdf}_{\min}}{(M \times N) - \text{cdf}_{\min}} \times (L-1)\right) \tag{2}$$

Before image processing and analysis, image preprocessing is usually carried out to obtain images more suitable for post-processing.

2.2 Sling Area Extraction Using the Light YOLOv3 Network

Because the picture has large unused areas, we need to use the deep learning algorithm to identify and segment the sling area for accurate detection and less computation.

Fig. 2 Picture before and after the positioning of the string area

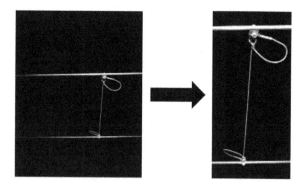

We chose the deep learning network YOLOv3 algorithm and removed the YOLO layer (52 * 52) which is used for detecting the small scale in the model. In Ref. [5], the network is mainly composed of a series of convolutional layers of 1 * 1 and 3 * 3.

Since YOLOv3 algorithm is more suitable for detecting small objects, it has the advantages of fast speed, simple pip-line, low background error detection rate and strong universality. The YOLO layer (52 * 52) used for detecting the small scale in the model is removed, making the model algorithm more suitable for detecting the catenary sling area in the picture. Figure 2 is the picture before and after the positioning of the string area.

2.3 Sling Foreign Body and Hard-Bending Detection Using Faster R-CNN

Through deep learning algorithm, we detect three types of defects: foreign body (fracture), hard bending and no-stress. In order to ensure the accuracy of detection effect, we use the Faster R-CNN algorithm to inspect the defects.

In Ref. [6], as a CNN network target detection method, Faster R-CNN first extracts feature maps of input image using a set of basic conv + relu + pooling layers. RPN network is mainly used to generate region proposals. Firstly, a bunch of anchors boxes are generated, which are filtered, and softmax judges that anchors belong to foreground or background.

2.4 Sling No-Stress Detection Using the Light YOLOv3 Combined with SLDD

For sling no-stress inspection, we use traditional image processing technologies which contain OTSU, digital morphological operation. In particular, a method

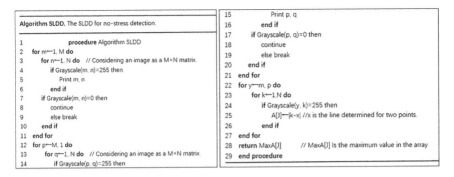

Fig. 3 Algorithm SLDD

called self-defined linear difference detection (SLDD) is proposed for sling no-stress inspection after extracting catenary sling area using the light YOLOv3.

In Refs. [7, 8], OTSU algorithm is an efficient binary image algorithm. A threshold value is used to divide the data in the image into two categories. Morphological operation is an image processing method developed according to the set theory method of mathematical morphology of binary images. The formula of expand is given by:

$$des(x, y) = \max_{(x',y'):\, element(x',y') \neq 0} src(x + x', y + y') \qquad (3)$$

The formula of erosion is given by:

$$dst(x, y) = \min_{(x',y'):\, element(x',y') \neq 0} src(x + x', y + y') \qquad (4)$$

The authors of Ref. [9] proposed image edge detection greatly reduces the amount of data, eliminates the information that can be considered irrelevant, and preserves the important structural properties of the image. As shown in Fig. 3, the followed algorithm SLDD is used to pick up the maximum and defect the no-stress.

3 Experimental Results

3.1 Experiment Setup

Data Source

The data was collected from the track inspection vehicles running on a section of the Beijing-Shanghai high-speed railway. As shown in Fig. 4, the data is captured by a HD camera installed on the top of a railway inspection train as the train is running on the tracks.

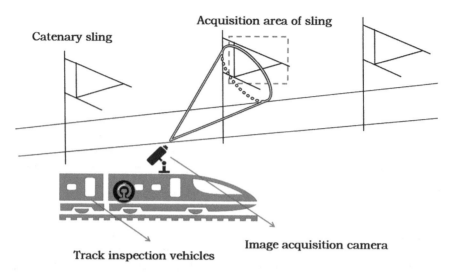

Fig. 4 Data acquisition process

Evaluation Criterion

The precision (*P*) is respectively given by:

$$P = \text{TP}/(\text{TP} + \text{FP}) \tag{5}$$

where TP is the number of defects that were inspected correctly, FP is the number of wrongly inspected defects.

Data Processing

The number of enhanced images is adequate for foreign body and hard-bending defects; each of them has one thousand pictures. But the number of no-stress is poor, subsequent experiments showed that the effect of deep learning algorithm on enhanced images was not good. Thus, the paper proposes a method based on traditional image processing to detect no-stress. And its initial image size is 5120 * 3840. The image size of subsequent experiments is about 2500 * 1200. Figure 5 is the data enhancement process.

Fig. 5 Data enhancement process

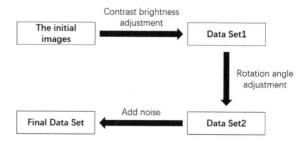

3.2 Sling Foreign Body and Hard Bending Inspection

According to the methodology in Sect. 2, the paper first extracts the string region using the light YOLOv3 network. And then Faster R-CNN is used to detect defects in the extracted regions. It can be seen from the training data that the method Faster R-CNN based has a good effect on foreign body (fracture) and hard bending, with the precision rate over 90%, while the detection effect no-stress is poor.

For comparison, YOLOv3 also be used for the detection, which has a precision rate about 60%. Figure 6 shows the detection effect of three kinds of defects. Table 1 is the comparison results of different algorithms.

Therefore, the next step is to adopt the traditional computer vision method for the defect no-stress, and to separate out the sling region of the current-carrying ring by the light YOLOv3 algorithm [10].

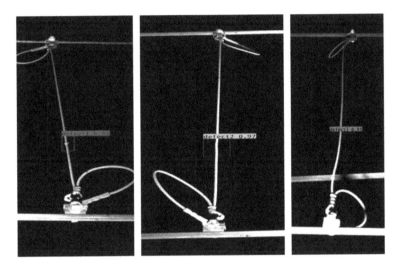

Fig. 6 Effect diagram of detection of three kinds of defects

Table 1 Comparison results of different algorithms

Precision rate	Foreign body (%)	Hard bending (%)	No-stress (%)
YOLOv3	65	60	30
Faster R-CNN	91	90	39

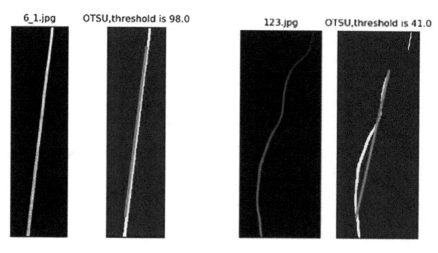

Fig. 7 Normal lifting chord and non-stressed lifting chord

Table 2 Comparison results of no-stress

No-stress defect	YOLOv3	Faster R-CNN	SLDD
Precision rate	30%	39%	80%

3.3 Sling No-Stress Inspection

After extracting catenary sling area using the light YOLOv3, OTSU algorithm is used to binarize the sling area to distinguish the string from the background area, and then digital morphological operation (expansion and corrosion) is performed on the binarized image. Since the sling of the catenary is placed vertically, the pixel points are traversed from the upper and lower ends of the image. Therefore, the proposed self-defined linear difference (SLDD) can be used to determine whether the slings are no-stress, as shown in Fig. 7. The SLDD obtains a precision rate with 80%, which achieves favorable performance against state-of-the-art methods such as YOLOv3 and Faster R-CNN, as shown in Table 2.

4 Conclusion

This paper is based on image processing technology and deep learning technology to carry out three main defect detection of catenary sling area, which has made some achievements. Among them, the effect of foreign body (fracture) and hard bending was better, and the precision rate was more than 90%. Traditional image processing technology is adopted to detect the no-stress, and OTSU is used for threshold segmentation, digital morphological operation and self-defined linear difference detection (SLDD) are used to achieve a better result.

Although this paper on the defect detection of catenary sling has made certain progress, it still has a long way to go. Many practical problems remain unresolved, such as complex tunnel conditions, cable interference complexity and many other problems. Breakthroughs will be made in the follow-up research.

Acknowledgements The study is sponsored by National Natural Science Foundation of China (No. 91738301).

References

1. Dong GD, Qin F, Chen LM (2019) Dynamic stress analysis of catenary string of high-speed railway. In: The twenty-fifth annual academic conference of beijing mechanics association, BSTAM, Beijing, pp 1186–1187 (in Chinese)
2. Xun YB (2018) Application of image processing in defect detection of catenary sling. Southwest Jiao Tong University, pp 1–106 (in Chinese)
3. Hu YF (2009) Research on digital watermarking in image transform domain. Zhejiang University, pp 1–139 (in Chinese)
4. Gao YP (2005) Research and implementation of image enhancement. Shandong University of Science and Technology, pp 1–69 (in Chinese)
5. Cui JW, Zhang JP, Sun GL, Zheng BW (2019) Extraction and research of crop feature points based on computer vision. Sensors
6. Ma J (2019) Image recognition of UAV based on convolutional neural network. Beijing University of Posts and Telecommunications, pp 1–73 (in Chinese)
7. Hui D, Chen XB, Xi JT (2019) An improved background segmentation algorithm for fringe projection profilometry based on OTSU method. Shanghai Jiao Tong University
8. Lu ZQ (2006) Fast algorithm of mathematical morphology corrosion expansion operation. In: The 13th national academic conference on image and graphics, NCIG, Nanjing, Jiangsu, pp 319–324
9. Hun ZB (2019) An improved edge detection algorithm based on binary wavelet and morphological fusion. Harbin University of Science and Technology, pp 1–58 (in Chinese)
10. Tan J (2018) Research on an improved YOLOv3 target recognition algorithm. Huazhong University of Science and Technology, pp 1–66 (in Chinese)

Rail Fastener Defect Inspection Based on UAV Images: A Comparative Study

Ping Chen, Yunpeng Wu, Yong Qin, Huaizhi Yang and Yonghui Huang

Abstract Rail fastener seriously affects the safety of railway system. The main approaches for detection of rail fastener defect are human inspection and rail vehicle inspection, which have many drawbacks such as low efficiency, high cost and so on. The paper presents a novel UAV-based visual inspection mean for fastener defect and focuses on two kinds of approaches based on UAV rail images: methods using traditional visual and using deep learning. With regards to the first aspect, a new traditional detection method using SVM and HOG features is proposed, which acquires a low mAP, due to heterogeneous background, various illumination, small target and so on. With regards to the second, some approaches based on deep learning including Fastener RCNN, YOLOv3, improved YOLOv3 and FPN are compared and applied for fastener defect detection, which can achieve a satisfied result. Finally, a quantitative comparison experiment shows FPN acquires a mAP of 95.78%, which achieves favorable performance against state-of-the-art methods. Therefore, it is possible to carry out fastener defect inspection with UAV images.

Keywords Rail fastener · UAV image · SVM · HOG · Deep learning

1 Introduction

Rail transportation plays a significant role in the development of economic and industrial growth. More and more catastrophic accidents were caused by the failures of railway facilities [1]. Therefore, with the development of high-speed and high-load

P. Chen · Y. Wu · Y. Qin (✉)
State Key Laboratory of Rail Traffic Control and Safety, Beijing Jiaotong University, No. 3 Shangyuancun, Haidian District, Beijing, People's Republic of China
e-mail: yqin@bjtu.edu.cn; 16221285@bjtu.edu.cn

Y. Wu
School of Traffic and Transportation, Beijing Jiaotong University, No. 3 Shangyuancun, Haidian District, Beijing, People's Republic of China

H. Yang · Y. Huang
Beijing-Shanghai High Speed Railway Company Limited, Beijing, China

rail transit, detection of fastener defects in high-speed railway is especially important. Railway track fastener is used to fix the railway track on the sleeper, which maintains the distance between tracks and prevents the longitudinal and lateral movement of the track relative to the sleeper [2, 3]. If the rail fastener is loose or missing, it will lose the function of stabilizing the railway track, thereby causing serious accidents.

At present, many countries have developed their own comprehensive inspection trains. Image and video acquisition equipment are installed on the inspection trains. Some systems use machine vision technology to detect and warn track components and line environment. Zhang et al. [4] introduced Hough line detection algorithm into fastener positioning process, and they used neural network algorithm based on image moment invariant to detect fastener defects. Beijing Jiaotong University has done research on rail defect detection, road fragmentation detection and fastener status detection, classification and recognition [5–7]. Zhang et al. [8] introduced structured light into the process of fastener defect detection. Firstly, the fastener was illuminated by structured light, and then the fastener status was classified and identified by linear detection technology.

Currently, for fastener defect detection using traditional means based on inspection trains, a high-definition camera is fixed on an inspection train to capture rail images as train is running on tracks. Subsequently, a detection system with a customized algorithm is used for defect detection. However, there are also some unavoidable disadvantages such as limited running time, high cost and poor flexibility.

At present, unmanned aerial vehicles (UAVs) have become a research hotspot in many fields such as crop measurements, bridge crack and power facilities inspection and so on. Compared with rail inspection vehicles, UAVs have incomparable advantages in cost, flexibility and so on [9].

To summarize, a detection method for rail fastener defect based on UAVs combined with visual inspection is presented in this paper. For rail fastener defect inspection, the paper lists traditional visual methods using SVM and HOG features and deep learning methods including Fastener RCNN, YOLOv3, improved YOLOv3 and FPN. All these models are trained and compared in the interest of getting the best detection for fasteners. Finally, experiment shows that FPN achieves favorable performance against others, and the detection performance of improved YOLOv3 is upgraded greatly.

Other sections of the paper are organized as follows: The methodology of this paper is presented in Sect. 2. The experimental results are elaborated in Sect. 3. The conclusion is described in Sect. 4.

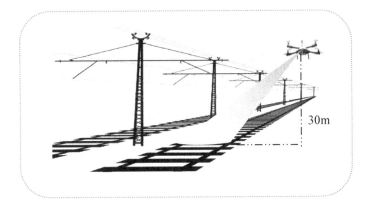

Fig. 1 Pilot flight diagram of UAV

2 Methodology

2.1 UAV Image Capture Program

In this paper, the UAV images capture program is adapted to capture rail fastener images. The railway track images can be acquired efficiently through a high-definition camera installed on a UAV. After finishing a certain flight mission along railway tracks, the collected images will be processed and analyzed in customized software with our algorithm. Figure 1 illustrates that the UAV flies under a height of 30 m on one side of the railway track, in case of some unexpected problems influencing train running. All images used in this study are captured by HD cameras fixed on a UAV.

2.2 Fastener Defect Detection Based on HOG Features and SVM

As a fastener can be photographed repeatedly, shadow only has a small part to influence the train process and the result.

Shape feature extraction and description are some of the important research topics in content-based image retrieval. It is applied to image retrieval and classification extensively, as it does not change with illumination and image color. The edge gradient can descript the topographic shape and spatial position of fasteners effetely. Due to little effect of color to image, grayscale is equivalent to this study, which is conducive to further image processing.

HOG feature with SVM classifiers has been widely used in image recognition and it is also adopted in this study to compare with the deep learning algorithms.

In addition, in order to decrease the dimension of features greatly and reduce the complexity of computation, the principal components analysis (PCA) method is used. And, particle swarm optimization (PSO) is used to optimize parameters C and gamma, at the same time. The initial model is gotten by training with previous parameters. Before detection with sliding window and non-maximum suppression (NMS) border regression, we apply hard negative mining to optimize the model. And, the result of its object detection will be given in Sect. 3.

2.3 Fastener Defect Detection Based on Deep Learning Method

Since deep learning brings breakthrough in image recognition, the speed and accuracy of detection get great improvement after image preprocessing by traditional visual methods. In this section, the paper lists deep learning method for fastener defect detection including Faster RCNN, YOLOv3, improved YOLOv3 and FPN.

Defect detection based on YOLOv3

You Only Look Once (YOLO) is a series of detection method raised in 2016 first. YOLO integrates target region prediction and target category prediction in a single neural network model and improves the real-time performance of target detection and recognition [10]. The neural network uses the whole image information to predict the boundary frame of the target (bounding box) and to predict the type of the target. YOLO is the first one to divide images into grid cells.

Comparing to the previous version, YOLOv3 has made great improvements in classification method and network structure, since it consists of a new network structure called Darknet-53 [11]. A series of 3×3 convolution and 1×1 convolution is included in 53 layers of convolution. It further optimizes network performance by residual block.

Defect detection based on improved YOLOv3

YOLOv3 applied a multi-scale prediction mechanism. It follows the anchor's mechanism in YOLOv2, and nine anchors of different sizes from YOLOv3 are clustered by k-means method. What is more, in order to make full use of anchors, YOLOv3 refines the mesh and distributes anchors to three scales according to size. However, as the fasteners are small targets, the feature map with output tensor $13 \times 13 \times [3 \times (5 + \text{class number})]$ is only suitable for large target detection. We modify the branch of YOLOv3 network and delete the feature map mentioned above. The ability of detection for big target is worse; it enhances the feature map of small target and rises the accuracy of detection for fastener. The modified network structure is shown in Fig. 2.

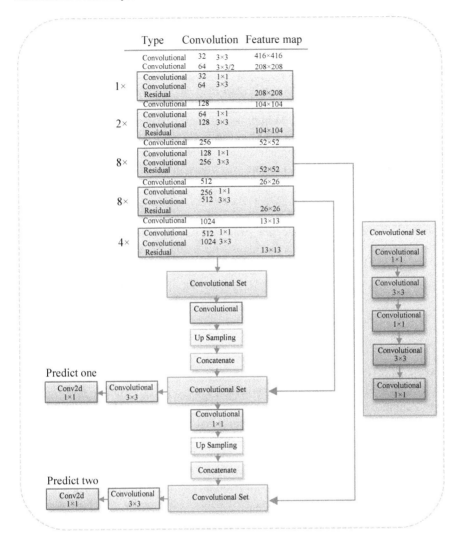

Fig. 2 The structure of modified YOLOv3 network

Defect detection based on Faster RCNN

As an object detection method for CNN network, Faster RCNN extracts feature maps of images using a set of basic conv + relu + pooling layers, which are shared to subsequent RPN layers and full connection layers. RPN network is used to judge anchors belong to foreground or background, and Roi Pooling sends proposal feature maps extracted by multiple integrating feature maps and proposals to full connection layers to judged objects. And, bounding box regression is used to modify anchors and to obtain accurate proposals. What is more, proposal feature maps are used to calculate the category of proposal, and bounding box regression is used to get the final exact location of the detection box [12].

Defect detection based on FPN

Most object detection algorithms only use top-level features for prediction. It is well known that the semantic information of low-level features is little, but the accuracy of target position is high. While high-level features have more abundant semantic information, the position of target is not as accurate as the former. Feature Pyramid Network (FPN) adopted high resolution in low-level features and rich semantic information at the same time. It makes predictions in each converged feature, which is different to other networks.

Experiment Result

In order to achieve a better detection result and demonstrate the advantage of UAV-based inspection method, this paper presents comparative experiments based on Faster RCNN, YOLOv3, modified YOLOv3 and FPN networks in this section.

3 Experiment Setup

UAV parameters

As shown in Fig. 3a, the rail images were captured by DJI Matrice 600 equipped with Zenmuse Z30 (DJI-Innovations, Shenzhen, China). The Matrice 600 is a six-rotor flying platform designed for professional aerial photography and industrial applications. The aircraft can effectively extend the time of flight through equipped six Intelligent Flight Batteries. It has built-in API Control feature and expandable center frame, and its maximum takeoff weight is 15.1 kg, which makes Matrice 600 ideal for connecting other devices to meet the specific needs of different applications. The Zenmuse Z30 enables non-contact distance detection by a high-performance camera system with a zoom lens, which offers 30 × optical zoom, 6 × digital zoom

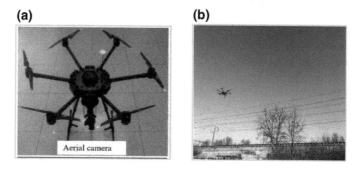

Fig. 3 a Physical map of UAV; b activity drawing of UAV

and HD 1080P video. The UAV adopts an industrial level Zenmuse platform with a precision of 0.01 degrees, so the problem of image blur caused by jitter is effectively solved.

The UAV using an aerial camera was sent to capture images or video of rails. It flies at one side of the railway as Fig. 3b shows.

What is more, the UAV collects related POS information to images, which contains the coordinates of the aircraft at that moment. Therefore, it is easy to find the location of rail defects based on these coordinates.

Evaluation Criterion

The mean average precision (mAP) is calculated according to the relationship $P(R)$ of precision (P) and recall (R) as:

$$\text{Precision} = \frac{TP}{TP + FP} \times 100\% \tag{1}$$

$$\text{Recall} = \frac{TP}{TP + FN} \times 100\% \tag{2}$$

$$\text{mAP} = \int_0^1 P(R) dR \tag{3}$$

where TP, FP and FN are true positive, false positive and false negative, respectively.

Fastener Defect Category

The practice database was built with 517 images captured by UAV from a section of Beijing–Shanghai high-speed railway line.

There are many kinds of deduction fracture cases, and missing of fastener is not negligible. In addition, we set three classes objects to train model, including normal fasteners, fractured fasteners and missing fasteners.

3.1 Performance Analysis

In order to get a better result, Faster RCNN, YOLOv3, modified YOLOv3 and FPN networks are adopted for defect detection.

Qualitative comparison of different models

As we set the same steps to 30,000 and similar training and test sets, the mAP comparison of model is persuasive in the specific detection of fastener. Figure 4 shows the comparison of average precision among these networks. The result of HOG feature and SVM gets an unsatisfactory result, due to heterogeneous background, various illumination, small target and so on. As the cut of 13×13 feature map, the mAP of modified YOLOv3 achieves a better performance than original YOLOv3 in fastener

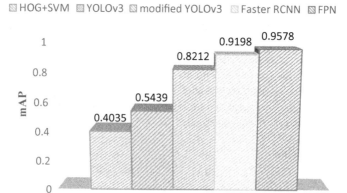

Fig. 4 The detection results of different network

Table 1 Detection time comparison of the models

	SVM	YOLOv3	Improved YOLOv3	Faster RCNN	FPN
cost	2.21 s	0.01 s	0.01 s	0.23 s	0.28 s

detection. It is worth to know that the YOLOv3 has an outstanding performance in detection time than other methods, as shown in Table 1, which means YOLOv3 has its incomparable dominance in real-time detection. FPN and Faster RCNN are also adopted with getting excellent results. In addition, since FPN acquires more features of small target, it gets an outstanding performance in detection of fastener.

The accuracy and speed of detection are different among these models, since, in addition to the core of the algorithm, there are many other factors, such as the threshold of NMS, evaluation of IOU, number of proposals and so on. As YOLOv3 is one-stage method, it shows an absolute advantage in speed regions of interest (ROI) pooling is applied to receive fixed-size feature maps, but YOLOv3 inputs multi-size feature maps. What is more, VGG16 is adopted in Faster RCNN for feature extraction, and neural nets for feature extraction in YOLOv3 use Darknet-53. FPN is combined with CNN to generate stronger feature maps.

The detection result using FPN

The detection result of original image is shown in Fig. 5. As we can see, the result of target detection is ideal. The fasteners are detected accurately without deviation. The approach presented in this paper is effective and feasible.

Fig. 5 Detection result of the fasteners

4 Conclusion

In this paper, a novel approach of detection based on UAV images is presented. Traditional visual and deep learning methods are applied to the detection of fastener. After processing the database of fastener, the training of different networks is made. And, the average precisions of these models are calculated to illustrate the suitability for fastener detecting of different models. Faster RCNN acquires a mAP of 91.98%, which achieves favorable performance against state-of-the-art methods. Finally, the location corresponding to fasteners is positioned. Therefore, it is feasible to carry out fastener defect inspection with UAV images.

In future, two aspects will be studied: firstly, we will explore better models for rail defect classification and health based on UAV images; secondly, with the development of high UAV photography, faster detection models will be developed in complex environments to increase detection efficiency.

Acknowledgements This research are supported by National Natural Science Foundation of China (No. 91738301), High-level Talents Training Program of Ministry of Transportation of China (No. I18I00010), Major Program of National Natural Science Foundation of China (No. 61833002): Fundamental theory and methods in operational risk assessment and control of high speed train, Research project of Beijing Shanghai High Speed Railway company: Research on key technologies of UAV inspection for Beijing Shanghai high speed railway infrastructure (No.I20D00010).

References

1. Arivazhagan S, Shebiah RN, Magdalene JS, Sushmitha G (2015) Railway track derailment inspection system using segmentation-based fractal texture analysis. ICTACT J Image Video Process 6:1060–1065
2. Mazzeo PL, Nitti M, Stella E, Distante A (2004) Visual recognition of fastening bolts for railroad maintenance. Pattern Recognit Lett 25(6):669–677

3. Stella E, Mazzeo PL, Nitti M, Cicirelli G, Distante A, Orazio TD (2002) Visual recognition of missing fastening elements for railroad maintenance. In: IEEE 5th international conference on intelligent transportation system, IEEE Intelligent Transportation Systems Council, National University of Singapore, Singapore, pp 94–99
4. Zhang Y (2008) Research on key technologies of video processing system for patrol vehicle. Master's degree thesis of Beijing Jiaotong University, Beijing Jiaotong University, Beijing (in Chinese)
5. Chang J (2009) Design of video detection system for patrol vehicle and realization of key technologies. Master's degree thesis of Beijing Jiaotong University, Beijing Jiaotong University, Beijing (in Chinese)
6. Shi ZH (2009) Research on optical detection algorithms of rail surface scratch. Master's degree thesis of Beijing Jiaotong University, Beijing Jiaotong University, Beijing (in Chinese)
7. Li RF (2011) Research on light source feedback and control technology in rail surface scratch detection system. Master's degree thesis of Beijing Jiaotong University, Beijing Jiaotong University, Beijing (in Chinese)
8. Zhang HB, Yang JF, Tao W (2011) Vision method of inspecting missing fastening components in high-speed railway. Opt Soc Am 50(20):3658–3665
9. Wu YP, Qin Y, Wang ZP, Jia LM (2018) A UAV-based visual inspection method for rail surface defects. 8(7):1028
10. Redmon J, Divvala S, Girshick R, Farhadi A (2016) You only look once: unified, real-time object detection. In: IEEE conference on computer vision and pattern recognition, 2016 (CVPR), Seattle, WA, pp 779–788
11. Redmon J, Farhadi A (2018) Yolov3: an incremental improvement
12. Ren SQ, He KM, Girshick R, Sun J (2017) Faster R-CNN: towards real-time object detection with region proposal networks. IEEE Trans Pattern Anal Mach Intell 39(6):1137–1149

Defect Detection for Bird-Preventing and Fasteners on the Catenary Support Device Using Improved Faster R-CNN

Jiahao Liu, Yunpeng Wu, Yong Qin, Hong Xu and Zhenglu Zhao

Abstract With the rapid speed-up of electrified railway and the construction of high-speed electrified railway, automatic defect detection on the catenary support device is of crucial importance for operation and cost reduction. The paper presents an innovative and effective method based on image processing technologies and deep learning networks for bird-preventing and fastener defect. Aim at the problem of low detection accuracy caused by the small defect target on catenary support device in images, an improved Faster R-CNN network added a top-down-top feature pyramid fusion structure is proposed in this paper. The improved faster R-CNN gets 81.2 mAP which achieves a 5.7% improvement over faster R-CNN in railway dataset and 80.1 mAP which achieves a 5.5% improvement over faster R-CNN in VOC2007 dataset. Furthermore, the method can achieve favorable performance against state-of-the-art methods.

Keywords Deep learning · Faster R-CNN · Feature pyramid fusion · Catenary support device

1 Instruction

In the last two decades, as one of the most important modes of transportation, railway plays an important role in the development of economic and industry growth. Due to the excitation and vibration in repeated and long-term operation, the bird-preventing tends to loosen, leading to skewing or even falling off. Fasteners serving as the

---

J. Liu · Y. Wu · Y. Qin (✉)
State Key Laboratory of Rail Traffic Control and Safety, Beijing Jiaotong University, No. 3 Shangyuancun, Haidian District, Beijing, People's Republic of China
e-mail: yqin@bjtu.edu.cn; 18120833@bjtu.edu.cn

J. Liu · Y. Wu
School of Traffic and Transportation, Beijing Jiaotong University, No.3 Shangyuancun, Haidian District, Beijing, People's Republic of China

H. Xu · Z. Zhao
JH High Speed Railway Operation and Maintenance Company CEEB Co Ltd, Beijing, China

connection of the cantilevers on the catenary support devices may loosen, damage, or are even missing. Therefore, it is critically important to detect the loosen and missing of the bird-preventing and fasteners on the catenary support device. At present, the detection means are personnel manually to read a large volume of data from captured images' offline. However, with the massive construction of high-speed electrified railway, the high-speed rail operation mileage exceeded 29,000 km. The number of artificially detected photographs is huge. Cameras usually shoot at night at different angles and the defects to be detected in the images are very small and even inconspicuous. Such a method of detection can produce a large number of blurred photographs, visual fatigue is prone to occur in manual inspection, and some defects will be missed accordingly.

In the last decades, many institutions and researchers have devoted their efforts to the development of automatic inspection methods for railway. In Ref. [1], multi-task learning is used railway crosstie and fastener detection by combining multiple SVM detectors. With the development of computing resources on parallel computing systems (such as GPU clusters), the deep learning algorithms have becoming popular in many computer vision fields; in particular, the deep convolutional neural networks (CNNs) have achieved the state-of-the-art performance in image detection tasks. Deep learning-based object detection frameworks can be primarily divided into two families: (i) two-stage detectors, such as region-based CNN (R-CNN) [2] and its variants fast R-CNN [3], faster R-CNN [4] and (ii) one-stage detectors, such as YOLO [5] and its variants YOLOv2 [6], YOLOv3 [7]. Two-stage detectors generate region proposal in the first stage and classify and location fine-tuning in the second stage. Compared with two-stage detectors, one-stage detectors are significantly more time-efficient, but two-stage detectors achieve better detection performance. In Ref. [8], faster R-CNN is used for fastener defect detection to advance detection rate and efficiency. However, the objections in this paper are quite large; the detection effect of faster R-CNN based on VGG16 for small target is not good. In Ref. [9], three DCNN-based detections include two detectors to localize the cantilever joints and their fasteners and a classifier to diagnose the fasteners' defects. However, this paper does not report the occupation of computing resources.

This paper presents a new based on pyramid fusion architecture for images' object detections. This architecture has great improvement for small objection in image. The rest sections of the paper are shown as follows: The related work is shown in Sect. 2. Top-down-top images pyramid fusion architecture and improved faster R-CNN in Sect. 3. Experimental results are presented in Sect. 5. The conclusion is described at the end of this paper.

2 Related Work

Before the era of deep learning, traditional defection method is usually composed of two parts, namely feature extraction and classifier training. The most widely used feature extractions are the SIFT feature [10] and HOG feature [11]. SVM [12] and its derivative algorithms are often trained as classifiers to predict possible locations of targets by sliding windows over the entire image. Recently, various object detection algorithms based on deep convolutional neural networks have become ubiquitous and achieved good results in the vision benchmark [13]. Based on region proposal, Girshick [4] and Girshick [2] proposed a region convolution neural network (R-CNN) and fast R-CNN. Faster R-CNN was proposed to unify the region proposal generation and the object classification by using region proposal network (RPN). RPN is a fully convolutional network which includes a 3×3 convolutional kernel size and two parallel 1×1 convolutional sizes which output channels corresponding 2 k and 4 k (k is the amount of anchors). The output of RPN is run through two 1×1 kernel convolutional layers to produce background/foreground class scores and probabilities for classification and corresponding bounding box regression coefficients for box regression. With the attention mechanisms, the RPN can tell the rest of network where to look. After RPN, we can extract region proposal accordingly for foreground. These region proposals can be divided into fixed quantities through ROI pooling layer. After feature extraction, feature vectors were fed into a sequence of fully connected layers before two parallel output layers: classification layer and regression layer. Classification layer was responsible for generating the probabilities of object classification and regression layer to refine the region proposal to fit the ground truth correspondingly.

As one-stage network, YOLO can realize faster detection speed than R-CNN. YOLO considered object detection as a regression problem and spatially partition of fixed grids on the whole image. Each grid performs detection based on whether the target center falls in the lattice or not. Based on a carefully designed lightweight architecture, YOLO could make prediction at 40 FPS and reach 150 FPS with a more simplified backbone. However, YOLO is prone to migration in coordinate regression due to lack of the location of region proposal and low accuracy for small target detection.

Automatic defect defection of the bird-preventing and fasteners on the catenary support device has not been achieved to the best knowledge of the authors.

3 Methodology

Because the railway maintenance is in the night skylight period, the shooting image needs to be supplemented with light, which causes some images to be seriously exposed and some images to have very dark targets. Although the catenary support devices are usually orderly arranged and fixed on the rail, the cameras mounted on the top of the train are not uniform. Due to different objects of the lens, the camera parameters and shooting angle are different. As a result, the size of the bird-preventing and fasteners of catenary support device may vary greatly in different images as shown in Fig. 1a. Fasteners of catenary support usually very small, and fasteners usually account for only a few dozen pixels in an image, such as the open

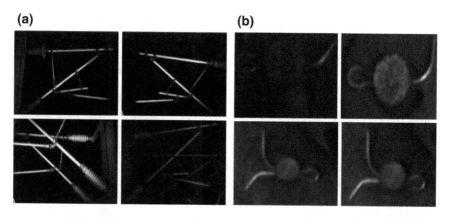

Fig. 1 Four examples of defect detection. **a** Captured catenary device images. **b** The open pins of joints from cantilevers

pins, as shown in Fig. 1b. This makes detection very difficult. In this paper, we will propose a new network architecture which has significant improvement for small object detection.

3.1 Network Architecture

The paper proposes a new top-down-top feature pyramid fusion architecture as shown in Fig. 2a. Typical object detection network based on two stages such as faster R-CNN only used a single layer's feature map to detect objects as shown in Fig. 2b. This will make it difficult to detect small objects in images, because there is little information left in the final feature map layer by convolution and downsampling by pooling operation of small objects with tens of pixels in the images through the

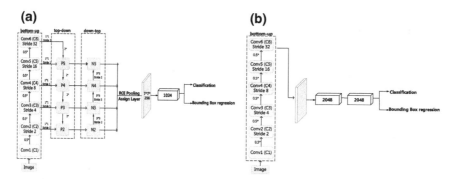

Fig. 2 DCNN architecture. **a** Architecture in this paper. **b** Faster R-CNN architecture based on ResNet101

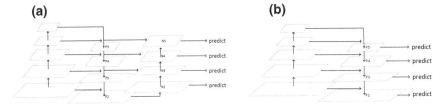

Fig. 3 **a** Using top-down-top pyramid levels fusion. **b** FPN structure

network. Deep convolutional networks learn hierarchical features in different layers which capture different scale feature. In particular, abundant spatial feature information and detailed feature information for small object detection concentrates on shallow feature layer, while deeper feature layer generally expresses strong semantic information for object classification. When detecting fasteners on the catenary support device, high-resolution spatial feature information is required and these objects information may not available in the deeper convolution layers. When detecting the larger objects such as the bird-preventing, deep semantic information can guide classification tasks better. Singh et al. [14] conducted comprehensive experiments on small object detection. They argued that learning a single-scale detector to handle all scale objects was much more difficult than learning multi-scale detectors with image pyramids. Ref. [15] proposed a feature pyramid network (FPN) as shown in Fig. 3 [1] which include a top-down architecture with lateral connections for building high-level semantic feature maps at all scales. However, the deeper convolution layers of this structure have less spatial detail feature information. This paper proposed to add a down-top architecture fusing more spatial details features to enhance the locations of these features based on FPN.

3.2 Pyramid Structure Based on Extracted Feature Maps

The bottom-up pathway is feed-forward computation of the backbone ConvNet, which adopted the residual network [16] for extracting different scales of feature maps. The ResNets are divided into four stages which have different residual blocks. We choose the output of the last layer of each stage as these layers have larger receptive field and more abundant feature information for itself stage. We denote {C2, C3, C4, C5} as pyramid layers which relatively have strides of {4,8,16,32} pixels with respect to the input image before entering the network. Due to C1 layer occupying large computation memory, we did not include C1 for pyramid network.

3.3 Top-down-top Pathway and Lateral Connections

Figure 3a shows the building block that constructs our top-down-top pyramid feature network. This structure is divided into top-down pathway and down-top pathway. The top-down pathway only enhances the semantic features by upsampling spatially coarser feature maps from higher pyramid levels. However, features in low levels which are helpful for accurately localized are ignored. The down-top pathway makes low-layer information easier to propagate. We use {P2, P3, P4, P5} to denote feature levels generated by FPN. From P2 to P5, we upsample the spatial resolution by using bilinear interpolation and then merged with the corresponding C2 to C5 layers by 1 × 1 convolutional layer to keep dimension consistency. We use {N2, N3, N4, N5} to denote newly feature map by down-top pathway corresponding to {P2, P3, P4, P5}. As shown in Fig. 3b, each feature map N_i first down-samples through a 3 × 3 convolutional layer with stride 2 to keep the same size of feature map P_{i+1} and then lateral fuse with P_{i+1} by 1 × 1 convolutional layer. The channels of each feature map fixed 256.

4 Experimental Results

In the section, various comparative experiments with two-stage networks and one-stage networks are conduct on railway dataset taken in real railway scenes in Test-I and VOC2007 dataset in Test-II.

4.1 Experiment Setup

Experiment Capturing Railway Images and Experimental Environment The dataset used in the experiments consists of the catenary support device images captured from a section of Beijing–Shanghai high-speed railway. The images are collected by the imaging inspection vehicle during the night. To build the training set, we manually draw bounding boxes and assign the labels including bird-preventing loosening and broken pin 1000 images, in which 800 images are in the training set and 200 images are in the test set. The defection classes are divided into eight categories including transverse cantilever bird-preventing loosening, slant cantilever bird-preventing loosening, joints, diagonals, pin a, pin b, clevis, and normal.

Evaluation Standard The true positive (TP), false negative, and false positive (FP) are counted to compute the indicator precision and recall. The average precision (AP) of each class and mean average precision (mAP) are the significant evaluation index of object detection. The value of AP is the area enclosed by the precision/recall curve. The mAP is the average of APs.

$$\text{precision} = \frac{TP}{TP+FP} \times 100\% \tag{1}$$

$$\text{recall} = \frac{TP}{TP+FN} \times 100\% \tag{2}$$

$$AP = \int_0^1 P(R) dR \tag{3}$$

4.2 Implementation Details

In dataset of railway images, the 6560 × 4384 pixels HD raw images are first resized to 1500 × 1000 in order to alleviate the memory footprint of the model. Then, image augmentation is made including rotation, mirror, add noise, image deformation to prevent overfitting. The VOC2007 dataset has not changed. For each image, we sample 256 region of interests (RoIs) with positive-to-negative ratio 1:3. Weight decay is 0.001, and momentum is set to 0.9. For each anchor in different scale levels, we set 5 scales are 32^2, 64^2, 128^2, 256^2, 512^2 and four aspect ratios are 1:1, 1:1.5, 1:2, 2:1, respectively. Figure 4a, b shows several visualized defection examples.

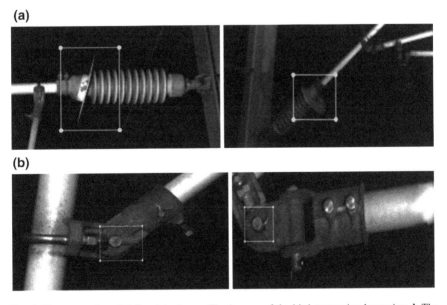

Fig. 4 Four examples of defect detection. **a** The images of the bird-preventing loosening. **b** The images of broken open pins

4.3 Comparison with Other Models

We compared our methods with HOG + SVM traditional detector and recently fast detectors like faster R-CNN, FPN-based faster R-CNN of two-stage networks, and YOLOv3 of one-stage networks. From the following results, we can see that our model achieves best results.

Evaluation on Test-I (railway dataset): the steps of train are 20,000. Figure 5 shows the visualized defection examples and results by our model in railway dataset. Table 1 shows the defection results of railway dataset.

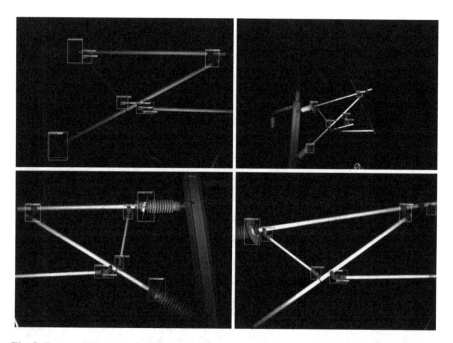

Fig. 5 Four examples of defect detection results which labeled with different color rectangles, the threshold of showing is 0.6

Table 1 Comparison of the defection results

Model	Backbone	mAP
HOG + SVM	–	40.3
YOLOv3	Darknet	75.2
Faster R-CNN	VGG16	65.7
Faster R-CNN	ResNet50	70.3
Faster R-CNN	ResNet101	74.5
FPN (faster R-CNN)	ResNet101	79.2
Our model	ResNet101	81.2

Table 2 Comparison in VOC2007 dataset

Model	Backbone	mAP
YOLOv3	Darknet	55.6
Faster R-CNN	VGG16	63.7
Faster R-CNN	ResNet50	73.1
Faster R-CNN	ResNet101	74.6
FPN (faster R-CNN)	ResNet101	76.2
Our model	ResNet101	80.1

Evaluation on VOC2007(railway dataset): the steps of train are 90,000, and Table 2 shows the defection results of VOC2007 dataset.

5 Conclusion

In this work, we apply improved faster R-CNN to a new application of defect detection of bird-preventing and fasteners on the catenary support device. This paper presents a new structure for object detection. We design several simple and yet effective components to enhance object location information. For the size of detection, objects vary greatly, and our model has impressive results.

Acknowledgements This research are supported by National Natural Science Foundation of China (No. 91738301), High-level Talents Training Program of Ministry of Transportation of China (No. I18I00010), Major Program of National Natural Science Foundation of China (No. 61833002): Fundamental theory and methods in operational risk assessment and control of high speed train, Research project of Beijing Shanghai High Speed Railway company: Research on key technologies of UAV inspection for Beijing Shanghai high speed railway infrastructure (No.I20D00010).

References

1. Gibert X, Patel VM, Chellappa R (2016) Deep multitask learning for railway track inspection. IEEE Trans Intell Transp Syst 18(1):153–164
2. Girshick R et al (2014) Rich feature hierarchies for accurate object detection and semantic segmentation. In: Proceedings of the IEEE conference on computer vision and pattern recognition
3. Ren S et al (2015) Faster r-cnn: towards real-time object detection with region proposal networks. In: Advances in neural information processing systems
4. Girshick RJCS (2015) Fast R-CNN
5. Redmon J et al (2016) You only look once: unified, real-time object detection. In Proceedings of the IEEE conference on computer vision and pattern recognition
6. Redmon J, Farhadi A (2017) YOLO9000: better, faster, stronger. In: IEEE conference on computer vision & pattern recognition
7. Redmon J, Farhadi A (2018) YOLOv3: an incremental improvement

8. Wei X et al (2019) Railway track fastener defect detection based on image processing and deep learning techniques: a comparative study. Eng Appl Artif Intell 80:66–81
9. Hao F et al (2014) Automatic fastener classification and defect detection in vision-based railway inspection systems. IEEE Trans Instrum Meas 63(4):877–888
10. Rublee E et al (2011) ORB: an efficient alternative to SIFT or SURF. In: ICCV, Citeseer
11. Wang X, Han TX, Yan S (2009) An HOG-LBP human detector with partial occlusion handling. In: 2009 IEEE 12th international conference on computer vision
12. Joachims T (1998) Making large-scale SVM learning practical. Technical report, SFB 475: Komplexitätsreduktion in Multivariaten
13. Everingham M et al (2007) The PASCAL visual object classes challenge 2007 (VOC2007) results
14. Singh B, Davis LS (2018) An analysis of scale invariance in object detection snip. In: Proceedings of the IEEE conference on computer vision and pattern recognition
15. Lin T-Y et al (2017) Feature pyramid networks for object detection. In Proceedings of the IEEE conference on computer vision and pattern recognition
16. He K et al (2016) Deep residual learning for image recognition. In: Proceedings of the IEEE conference on computer vision and pattern recognition

Research on the Flexible Operation and Maintenance Management of Urban Rail Vehicle

Xiaoqing Cheng, Ruohong Lan, Zhiwen Liao and Yong Qin

Abstract The safety of urban rail vehicles is a crucial part in the operation of urban rail transit. Combining the concept of flexible management and the characteristics of urban rail vehicle operation and maintenance, this paper proposed the definition and process model of flexible operation and maintenance management of urban rail vehicles. Taking the vehicle door as a case, the activity-based classification is used to classify the vehicle equipment, and the decision tree method is used to determine the maintenance mode. The research work of this paper has theoretical significance and practical value for reducing the operation and maintenance cost of urban rail vehicles, improving the efficiency of maintenance and ensuring the safety of urban rail system.

Keywords Urban rail vehicle · Flexible management · Operation and maintenance · Maintenance mode

1 Introduction

As the carrier of passenger transportation, urban rail vehicle is the core part of the whole transportation system. Vehicle safety is a crucial part in the operation of urban rail transit. The operation and maintenance process of urban rail vehicle is complex and expensive, accounting for a large part of the equipment maintenance cost in urban rail system [1]. Reasonable and efficient arrangement of vehicle operation and maintenance is significant to reduce the operation cost of urban rail transit.

To improve the efficiency of urban rail vehicle maintenance, many scholars are actively exploring the maintenance decision with a variety of methods. Xu [2] classified the vehicle system and built a decision model of vehicle maintenance. Song [3] used the failure tree method to determine the risk points affecting the safety. Wang [4] carried out FMECA analysis on the frequent failures of the EMU. Kim and Jeong

---

X. Cheng · R. Lan · Z. Liao (✉) · Y. Qin
State Key Laboratory of Rail Traffic Control and Safety, Beijing Jiaotong University, No. 3 Shangyuan Village, Haidian District, Beijing, China
e-mail: 19120832@bjtu.edu.cn

[5] analyzed the brake system of railroad vehicle through an evaluation of the failures consequences. Cheng [6] adopts an analytic network process (ANP) technique for the rolling stock maintenance strategy evaluation.

At present, most urban rail maintenance mode in China still needs to be further optimized because of unreasonable design cycle and resource waste. The flexible management is applied in various areas, such as finance industry [7], power grid investment [8] and 5G network [9]. In this paper, the flexible management is applied to the operation and maintenance of vehicles in order to reduce the operation and maintenance cost and improve the efficiency of vehicle maintenance.

2 Overview of Flexible Operation and Maintenance Management of Urban Rail Vehicles

Flexible management is an optimal and stable combination. Based on the concept of flexible management, the paper defines the flexible operation and maintenance management of urban rail vehicles as: The flexible operation and maintenance management of urban rail vehicles is a dynamic vehicle maintenance strategy generation problem. This management classifies the equipment of urban rail vehicle starting from the safety and reliability and three dimensions of flexibility: scope, time and cost, then quickly generates a scientific and optimal operation and maintenance strategy according to the characteristics of different equipment, in order to achieve the highest maintenance efficiency, the smallest workload and the lowest maintenance cost (Fig. 1).

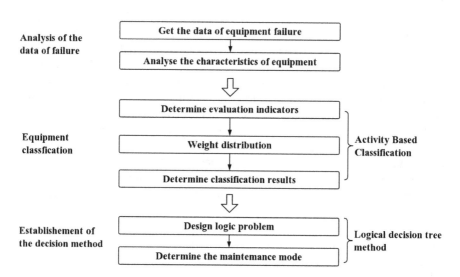

Fig. 1 Flexible operation and maintenance management flowchart

Table 1 Statistics on vehicle failures of a subway line in 2017 (partial)

Vehicle subsystem	Failure times	Proportion
Car door	475	0.13
Passenger information system	407	0.11
Air supply system	194	0.05
Auxiliary system	97	0.03

The paper selected subsystem failure data (Table 1) of the urban rail vehicle for analysis. The systems with the most failures are: traction/electric braking system, air-conditioning system, door system and passenger information system. The failure probability of the door system is high, and once the failure occurs, it causes a great impact on the operation, so chose the door system as the research object.

3 Urban Rail Vehicle Equipment Classification

In order to reduce the operation and maintenance cost of vehicle effectively and improve efficiency, the activity-based classification is used to classify the components of urban rail vehicles. The steps are: (1) analyze the influencing factors of classification, (2) determine the evaluation indicators and their weights, (3) determine the classification evaluation calculation method, (4) collect the data for evaluation calculation, (5) determine the classification criteria, classify the equipment.

3.1 Analysis of the Factors Affecting Classification

(1) Safety: the severity of the failure consequence. The severity of failure consequence is the primary factor in the classification of the equipment. For equipment with serious consequences of failure, it should be overhauled to minimize the impact of such failures on the normal operation of the train.
(2) Failure frequency: the probability of failure. Equipment with a high probability of failure should be valued in the maintenance work, and the probability of failure of such equipment can be reduced by increasing the frequency of inspections.
(3) Economy: purchase price. Maintenance costs can be measured in economy. The paper considers the purchase price of equipment as an important analytical factor.
(4) Maintenance time. Components with long maintenance time should be listed as the key object. Through routine maintenance, the probability of failure and the probability of long-term maintenance after a failure can be reduced.

Table 2 Results of the weight of indicators

Indicator	The severity of the failure consequence	Failure frequency	Purchase price	Maintenance time
Weight	0.61	0.22	0.10	0.07

In summary, the indicators for equipment classification evaluation are as follows: the severity of the failure consequence, the probability of failure, the purchase price and the maintenance time.

3.2 Determination of the Weight of Classification Indicators

The weight of the equipment classification indicators is determined by the analytic hierarchy process, and the urban rail staff and experts are invited to evaluate the importance of the four indicators. The result is given in Table 2.

3.3 Classification Score Calculation

3.3.1 The Severity of the Failure Consequence

The failure mode impact, effects and criticality analysis (FMECA) are used to analyze the severity of the failure consequence and the probability of failure. According to the maintenance regulations of urban rail company, the classification score for the impact of urban rail vehicle failures on safety is given in Table 3.

There is much equipment in the door system, and 15 representative parts with high failure rate are selected for analysis. Firstly, the failure categories of key components are filled out, and the severity of the failure consequence is analyzed. Finally, scored the components according to Table 3. Table 4 is the partial result.

Table 3 Equipment failure impact rating reference

Raking	Severity level	Description of the impact of the severity of the failure	Score
1	Mild	General failure	1, 2, 3
2	Medium	Late	4, 5, 6
3	Major	Remove passengers	7, 8
4	Disaster	Rescue	9, 10

Table 4 Door system components FMECA table (partial)

Equipment components	Failure category	Impact on the system	Impact on train operation	Maintenance measures	Level	Score
Buzzer alarm	Electrical components	No alarm when closing the door	General failure	Replace components	1	2
	other	No alarm when closing the door	General failure	Adjustment installation	1	
LDCU	Electrical components	The door cannot be opened	May cause delay	Replace components	2	3
	Electronic board	The vehicle display status is incorrect	May cause delay	Replace components	3	
...	...	...	...	...	...	...

3.3.2 The Probability of Failure

To rate the failure mode occurrence probability ranking (OPR), first calculate the criticality of the failure mode with CA quantitative analysis method; the failure mode criticality C_{mj} and the system total criticality C_r are calculated as shown in Eqs. (1) and (2).

$$C_{mj} = \alpha_j \beta_j \lambda_p t \tag{1}$$

$$C_r = \sum_{j=1}^{M} C_{mj} \tag{2}$$

where j is 1, 2, 3, ..., M, and M is the total number of failure modes of the component. α_j is the ratio of the number of occurrences of the type j failure mode to the number of occurrences of all failure modes. λ_p is the failure rate of the equipment in the statistical interval. The average is taken as the ratio of the total number of failures occurring during working hours to the working hours. t is the working time of the equipment in a certain task stage; this article is 266,084 h. The probability of failure β_j is determined by the analyst based on experience; it is the probability that the product will be invalid due to the type j failure. Take the values according to the following Table 5.

The classification and scoring of the failure mode occurrence probability ranking (OPR) is based on the standards specified in GJB/Z 1391–2006 as shown in Table 6.

Calculated the criticality of the door system and obtained the score of the probability of failure according to the criteria in Table 6. Table 7 shows some results.

Table 5 Failure impact probability value

Description	Failure impact probability β_j
Be bound to cause an accident	1
Very likely to cause an accident	0.75
Likely lead to accidents	0.5
May cause an accident	0.25
Rarely causes an accident	0.1
Unlikely to cause an accident	0

Table 6 Equipment criticality ranking standards

Ranking	Definition	Qualitative description rating	Score
1	Very high	More than 20% of the total system criticality	9, 10
2	Higher	More than 10% of the total system criticality, less than 20%	7, 8
3	General	More than 1% of the total system criticality, less than 10%	4, 5, 6
4	Lower	More than 0.1% of the total system criticality, less than 1%	2, 3
5	Very low	Less than 0.1% of total system criticality	1

Table 7 Door system CA table (partial)

Component	Failure category	C_{mj}	α_j	λ_ρ	β_j	C_r	Proportion (%)	Score
Buzzer alarm	Electronic component	1.46	0.5	0.11	1	2.56	7.98	6
	Other	1.1	0.5	0.11	0.75			
LDCU	Electronic board	0.35	0.33	0.04	0.04	1.8	5.61	5
	Electronic component	0.27	0.25	0.04	1			
...	...	...	...	...	...	...	...	...

3.3.3 Purchase Price

According to the equipment procurement situation of the urban rail company and the evaluation of experts, the scoring standards are shown in Table 8.

According to the above standards, the score of purchase price of the door component can be obtained as shown in Table 9 (partial).

Table 8 Component purchase price scoring standards

Ranking	Definition	Purchase price	Score
1	Very expensive	More than 100,000 yuan	9
2	Generally expensive	More than 10,000 yuan, less than 100,000 yuan	7
3	Generally cheap	More than 100 yuan, less than 10,000 yuan	5
4	Very cheap	Less than 100 yuan	2

Table 9 Door System components purchase price scoring (partial)

Components	Score
Buzzer alarm	7
LDCU	7
MDCU	7
…	…

3.3.4 Maintenance Time

The maintenance time refers to the time when the staff detects and repairs the failure. Classified the maintenance time according to the urban rail company maintenance record and asked the staff to rate it. The results are shown in Table 10.

Calculated the average repair time of the components and scored according to the standards (Table 10). The results are shown in Table 11.

Table 10 Component maintenance time scoring standards

Ranking	Definition	Maintenance time	Score
1	Very long	More than 1 h	9
2	Generally long	More than 0.5 h, less than 1 h	7
3	Short	More than 10 min, less than 0.5 h	5
4	Very short	Less than 10 min	2

Table 11 Score of component maintenance time

Components	Score
Buzzer alarm	5
LDCU	2
MDCU	2
…	…

3.3.5 Comprehensive Score Calculation

Through the above analysis and calculation, the equipment comprehensive score was calculated according to Eq. (3).

$$H = \omega_1 S + \omega_2 O + \omega_3 C + \omega_4 T \tag{3}$$

where H is the comprehensive score of component. ω_1, ω_2, ω_3, and ω_4 is the weight of each indicator. S, O, C and T are the indicators' score.

Taking the door system as the research object, the calculation results are as shown in Table 12.

3.4 Components Classification

According to the classification criteria of the activity-based classification, the number of products in class A is about 5–10% of the total, and the comprehensive score is the highest; the number of products in class B is about 20–30% of the total, and the products in class C is about the total 60–75%; the score is the lowest [10]. The classification results of the component of the urban rail vehicle door system are provided in Table 13.

On the basic of the actual situation of urban rail vehicle maintenance, define the vehicle equipment of Class A, B and C (Table 14).

Table 12 Classification score of the door components

Components	S	O	C	T	H
Buzzer alarm	2	7	7	5	3.3
LDCU	5	6	7	2	4.88
MDCU	4	7	7	2	4.99
…	…	…	…	…	…

Table 13 Classification of door components

Level	Number	Proportion (%)	Components
A	1	7	Door motor
B	3	27	Door S1 value, LDCU, MDCU
C	11	73	Buzzer alarm, door limit switch, insert block, communication line, lower stop pin, door, handle, diode, lower rail, door position sensor, lock tongue, roller rocker

Table 14 Definition of vehicle equipment activity-based classification

Class	Rank	The severity of the failure consequence	Failure consequences	The possibility of failure	Purchase price	Maintenance time
A	Crucial	Very seriously	May cause passengers' personal property threats or have a great impact on the normal operation of the train	Very high	Very expensive	Very long
B	Key	Seriously	May cause less damage to passenger property safety or cause less impact on the normal operation of the train	High	Expensive	Long
C	General	Mild	Almost no impact	Low	Cheap	Short

4 Urban Rail Vehicle Flexible Maintenance Decision

4.1 Maintenance Mode Decision Model

Under the premise of ensuring the operation safety, the goal of maintenance mode optimization is highest efficiency and lowest cost. This paper uses the logic decision tree method to analyze the requirements and conditions of decision objects, design logical problems, complete the decision process by judging logical problems and determine the appropriate maintenance mode [11] (Fig. 2).

4.1.1 Analysis of Factors Affecting Maintenance Mode Selection

The decision-making method of equipment maintenance in urban rail vehicle is affected by many factors, such as the maintainability, economy and effectiveness of maintenance modes. This paper proposed maintenance decision-making method for different class equipment.

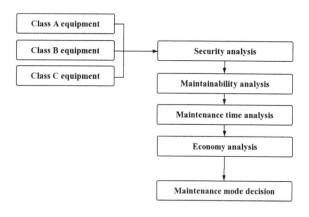

Fig. 2 Equipment maintenance mode decision flowchart

Table 15 Maintenance modes corresponding to the severity of the failure

The severity of failure consequence	Maintenance mode
General failure	BM and PM mainly, supplemented by CBM
Late	CBM and PM mainly, supplemented by BM
Remove passengers	CBM and PM
Rescue	CBM and PM

(1) Security. The severity of vehicle failure is the primary consideration in the maintenance mode decision. For equipment with serious failure consequences, planned maintenance (PM) and condition-based maintenance (CBM) shall be adopted. If severity of the failure is relatively mild, breakdown maintenance (BM) and planned maintenance can be adopted. The corresponding relationship between the severity of failure consequence and maintenance mode is shown in Table 15.

(2) Failure manifestation. Failure manifestation refers to the manifestation of system equipment failure, that is, potential failure and functional failure. Potential failure refers to the existing but undiscovered failure [12]. Functional failure refers to the state in which the product fails to fulfill the specified functions, which can be distinguished into hidden functional failure and obvious functional failure according to its performance. The relationship between maintenance modes and failure type is shown in Table 16.

(3) Maintainability. When the vehicle equipment is repaired according to the prescribed procedures and modes, the failure can be eliminated and the equipment can be maintained or restored to the prescribed state.

(4) Maintenance time. When a variety of maintenance modes is available, the modes that can reduce the overall maintenance time can be selected on the premise of ensuring the safe operation of the train.

Table 16 Maintenance modes corresponding to the failure type

Failure type	Maintenance mode
Potential failure	CBM and PM
Obvious functional failure	CBM and PM mainly, supplemented by BM
Hidden function failure	CBM and PM

(5) Economy. On the premise of ensuring the normal and safe operation of trains, the most economical maintenance mode is usually chosen.

4.1.2 Establishment of Maintenance Mode Decision Model

Class A and Class B are mainly equipment with high severity of fault consequences and have high failure rate, high purchase price and long maintenance time. They are key maintenance parts. The Class C equipment consequences of the failure are less serious. The maintenance decision flowchart for the three types of equipment is shown in Fig. 3.

For equipment with serious consequences, the maintenance decision process is as follows: First, judge whether the PM and CBM are applicable and valid. If such modes are applicable and effective, the maintenance mode is adopted. If neither of them is applicable nor invalid, the maintenance mode needs to be improved.

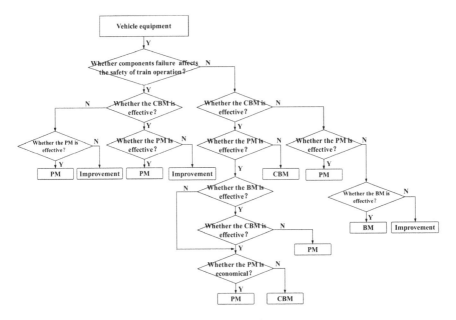

Fig. 3 Equipment maintenance mode decision tree mode

In the Class C, the equipment with less serious consequences has the following decision-making process: Firstly, whether the BM, PM and CBM are applicable to the equipment. If only one mode is applicable and effective, then the maintenance mode is selected. If not only one mode is applicable and effective, the next step is to judge whether the maintenance time is long. If there is an option of BM, remove the BM. If there is only one mode of maintenance, the mode is the final option. If there is more than one mode of maintenance, then make economic decisions and choose the most economical mode.

4.2 Decision Analysis of Maintenance Modes for Door System

According to the established decision model of the maintenance mode, taking the door system as a case, the door motor in the Class A equipment is selected for decision verification.

(1) Judgment the safety of the door motor: Failure of the door motor may cause the door to fail to open, and the passengers cannot get on and off normally, which may cause the train to be late, but will not affect the safe operation of the train and the safety of the passengers. Choose N for this question.
(2) Judgment of applicability and validity: A real-time monitoring system is installed on the door, and its operating condition can be monitored in real time through the train network, so the CBM is also applicable to the door motor. In addition, PM and BM are applicable and effective for the door motor, so answer is Y.
(3) Judgment of maintenance time. The door motor failure repair time is long, so the BM is excluded.
(4) Judgment of economic. There are two main types of failures of the door motor. One is the failure of the overall function, and the other is the failure of the internal parts. When the overall function of the door motor fails, the entire door motor needs to be replaced, and the purchase price of the door motor is expensive. Therefore, CBM is selected. For the internal parts failure of the door motor, the small parts have lower procurement costs; replaced part can be adopted. The CBM is more economical. Therefore, the maintenance mode of the door motor is CBM. The decision-making process for the maintenance method of the door motor is shown in Fig. 4.

The same method is used to analyze the maintenance decision of other components of the door system, and the maintenance mode of the door system equipment is shown in Table 17.

Fig. 4 Flowchart of the door motor maintenance mode decision

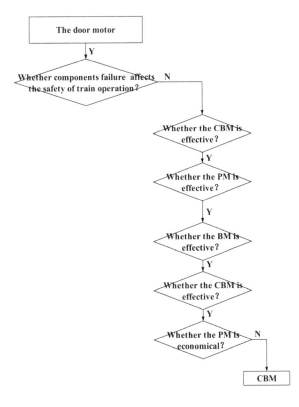

Table 17 Maintenance modes of door system equipment (partial)

Component	Failure type	Maintenance mode	Maintenance measure
Buzzer alarm	Electronic components	PM	Regular inspection
	Other	PM	Regular inspection
LDCU	Electronic board	CBM	State detection
	Electronic component	CBM	State detection
MDCU	Electronic board	CBM	State detection
	Electronic component	CBM	State detection
…	…	…	…

5 Conclusion

The paper proposed the definition and the basic process of the flexible operation and maintenance management of urban rail vehicles. Taking the door system as a case, established the maintenance mode decision method of urban rail vehicle based on activity-based equipment classification.

The determination of the maintenance strategy of urban rail vehicles involves many factors. There are many uncertain factors in actual production, which will also change the maintenance mode. The main shortcomings of this paper are: The evaluation indicators of equipment classification could not fully consider the factors affecting the maintenance strategy in the actual production process. The results of the expert grading method are not objective enough. Subsequent research would carry out more accurate quantitative calculation of the maintenance cycle, establish a software system for intelligent equipment maintenance strategy optimization and provide technical support for the maintenance strategies.

Acknowledgements The authors gratefully acknowledge the support provided by National Key R&D Program of China (No. 2017YFB1201201).

References

1. Zhang S (2015) Optimization of maintenance strategy for urban rail vehicle doors based on reliability and maintenance cost analysis. Southwest Jiaotong University (in Chinese)
2. Xu S (2015) Statistical analysis of subway vehicle fault information and optimization of maintenance strategy. Southwest Jiaotong University (in Chinese)
3. Song Y (2012) Research on service and maintenance strategy of passenger car bogie based on fault data analysis. Tsinghua University (in Chinese)
4. Wang C (2012) Research on the damage law and maintenance period of main equipment of high-speed EMU. Beijing Jiaotong University (in Chinese)
5. Kim J, Jeong HY (2013) Evaluation of the adequacy of maintenance tasks using the failure consequences of railroad vehicles. Reliab Eng Syst Saf 117(117):30–39
6. Cheng YH, Tsao HL (2010) Rolling stock maintenance strategy selection, spares parts' estimation, and replacements' interval calculation. Int J Prod Econ 128(1):404–412
7. Wang Z, Wang T, Yang Y (2015) Research on the flexible strategy of enterprise risk management. J Inn Mongia Agric Univ (Soc Sci Ed) 17(1):20–25 (in Chinese)
8. Liu Y (2015) Exploration of the application of flexible decision technology in power grid investment management under the new economic normal (23):70–70 (in Chinese)
9. Zabala A, Rojas E, Roldan JM et al (2018) Towards per-user flexible management in 5G
10. Lan J, Xu Y, Huo L et al (2006) Research on the weight of fuzzy analytic hierarchy process. Syst Eng Theory Pract 26(9):107–112 (in Chinese)
11. Liu P (2014) Reliability analysis of urban rail vehicle plug door based on fault tree and fuzzy Petri net. Nanjing University of Science and Technology (in Chinese)
12. Feng T, Tang J, Yang J et al (2008) Net P-F interval judgment and maintenance strategy for potential faults in RCM. China Equip Eng 8:21–23 (in Chinese)

Structural Health Monitoring of Cracks on Bogie Frame Using Lamb Waves

Kexin Liang, Guoqiang Cai and Ye Zhang

Abstract Structural health monitoring (SHM) is a technology that can change the maintenance mode of structure from passive to active. This paper introduces Lamb waves into the monitoring of cracks on bogie frame, which is beneficial to monitor the damage of blind areas in traditional inspection that are unable or inconvenient to detect. At the same time, this method is conducive to realize the life-cycle management. In addition, the detection method based on vibro-acoustic modulation theory is very sensitive to the closed micro cracks. Therefore, it can recognize the damage in initial stage, thus reduce the loss. Experiments show that this method can effectively identify the fatigue cracks on bogie frame with no need for healthy baseline data.

Keywords SHM · Lamb waves · Cracks · Bogie frame · Life-cycle management · VAM

1 Introduction

As the "skeleton" of Bogie, the frame, which welded with supports, hangers and other structural components, bears the random impact loads of complex frequencies and different periods. Especially under the terrible operating situation, the irregular geometry of the frame and the welded joints will lead to local dynamic stress concentration, resulting in damage and even fracture [1].

However, the current "periodic inspection" methods make it difficult to detect structural damages, which cannot be observed by human eyes in time, and even more impossible to monitor damage on line in real time. As a result, the train may continue to run normally in case of unknown damages, which has buried potential safety hazards. At the same time, the existing non-destructive testing technologies (such as magnetic particle inspection and penetration inspection) need to dismantle

---

K. Liang (✉) · G. Cai
State Key Lab of Rail Traffic Control and Safety, Beijing Jiaotong University, Beijing, China
e-mail: 18120762@bjtu.edu.cn

Y. Zhang
Beijing Key Lab of Traffic Engineering, Beijing University of Technology, Beijing, China

the bogie frame and remove the coating from surface firstly, which can be really complicated, time-consuming and unfriendly to environment.

Lamb wave is widely used for real-time non-destructive testing of complex structures due to its fast propagation speed, wide detection range, and small energy attenuation. According to the difference of characteristic parameters, damage detection methods can be divided into two parts: one is based on the peak value of Lamb wave signal, the other is based on the scattering signal caused by damages. However, it is difficult to be applied in practice because of its excessive dependence on healthy baseline signal. By contrast, time reversal algorithm overcomes the above defects. The received signal is reversed and reloaded so that multiple Lamb wave signals related to the damage source will converge to the site of damage at the same time. It has been applied in the field of NDT of large plate structures and pipes for a long time [2, 3]. However, the time-reversal method requires the tested structure to be completely symmetrical. Thus, its application scope is greatly limited.

To overcome the shortcoming that the efficiency of ultrasonic inspection is hard to be improved for its "point-by-point" property, weaken the heavy dependence of current Lamb wave-based methods on non-destructive baseline signal, and release the restriction of time reversal for the strict symmetry of the structure to be tested, this paper proposals a baseline-free damage identification method based on the theory of vibration–sound modulation, which does not need healthy signal and is suitable for complex structures. By analyzing the monitoring Lamb wave signal properly, it can be realized to identify tiny cracks on the bogie frame and thus provide strong guarantee for the running safety of rail traffic.

2 Theoretical Fundamentals

Lamb wave is defined as the propagation of an elastic disturbance in a free-boundary plate. It is a stress wave in a structure with two parallel surfaces, in which the shear and longitudinal waves are coupled to each other. There are two modes, symmetric mode and anti-symmetric mode, in Lamb waves [4]:

$$\frac{\tan(qh)}{\tan(ph)} = -\frac{4k^2qp}{\left(k^2 - q^2\right)^2} \quad \text{(Symmetric mode)} \tag{2.1}$$

$$\frac{\tan(qh)}{\tan(ph)} = -\frac{\left(k^2 - q^2\right)^2}{4k^2qp} \quad \text{(Anti - symmetric mode)} \tag{2.2}$$

in which $q^2 = \omega^2/c_T^2 - k^2$, $k = 2\pi/\lambda_{\text{wave}}$.

Where k is wave number; ω is angular frequency; λ_{wave} is wavelength; c_L and c_T are longitudinal and transverse/shear velocities (Fig. 1).

Fig. 1 Symmetric mode and anti-symmetric mode of lamb wave

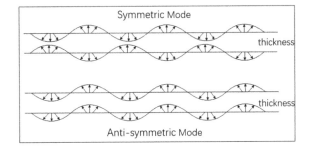

Different from the traditional ultrasonic "spot-by-spot" inspection, Lamb waves have the characteristics of high speed as well as high efficiency. Therefore, it is very suitable for large-area non-destructive inspection of plate-shaped structures [5].

Meanwhile, vibro-acoustic modulation (VAM) is a nonlinear acoustic detection method, which is very sensitive to structural damages like cracks [6]. A set of low-frequency vibration signals and a set of high-frequency ultrasonic signals are excited to the structure at the same time. When the structure is intact, the spectrum of received signal consists of only two main clusters of frequencies with relatively large amplitude. However, if the structure has any tiny crack, the contact interface of the structure will open and close repeatedly under the action of vibration. The two input signals will interact and generate modulation effect, thus other components such as modulated side frequency and high-order harmonic will occur in the received signal. The modulated side frequency of the modulation effect is [7]:

$$f_{sn} = f_0 \pm nf_1 \qquad (2\text{-}3)$$

where f_{sn} is the n-order step modulated side frequency ($n = 1, 2, \ldots, n$); f_0 is the high frequency of the input ultrasonic signal; f_1 is the low frequency of the input vibration signal.

By observing the spectrum component of the received signal, we can determine whether there are defects in the structure [8–10]. The detection principle is shown in Fig. 2 [11].

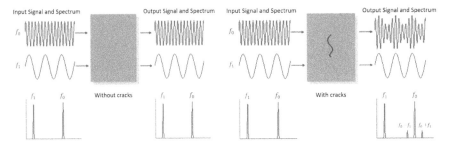

Fig. 2 Detection principle based on vibration–sound modulation

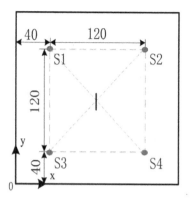 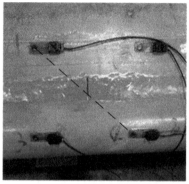

Fig. 3 Layout of crack and sensors

3 Specimen Information and Experimental Setup

3.1 Specimen Information

In this paper, the bogie made of Q345 steel was tested, of which the material density is 7850 kg/m^3, elastic modulus is 2.06×10^5 MPa, and Poisson's ratio is 0.28. The main test area is the side beam of bogie frame [12].

3.2 Experimental Setup

By scanning, 162 Hz was chosen among the characteristic frequencies of the tested bogie for vibration. Moreover, the 3.5-peak Lamb wave signal was modulated by Hanning window with a central frequency of 280 KHz was taken as ultrasonic signal.

Firstly, the high-frequency Lamb wave signal was motivated by PZT sensor, and then collected by other three PZT sensors for comparison. After that, the exciter generated vibration signal with large amplitude and low frequency. While the vibration state was steady, we added Lamb waves to the specimen and collected signals as above (Fig. 3).

4 Results

Based on previous researches, we used EMD and HHT to analysis the received signal and converted it to frequency domain. Since the amplitude of the modulation side frequency is very small compared with the carrier amplitude in the early stage of the damage, for the convenience of observation, the vertical axis was taken as the logarithmic form of the amplitude and then normalized.

As can be seen from Fig. 4, there are modulated side-frequency harmonics around 280 KHz and details are shown as magnification.

When high-frequency Lamb wave was excited without low-frequency vibration, the modulated side-frequency harmonics disappeared with the removal of low-frequency vibration (see Fig. 5).

Therefore, it can be proved that the modulated side-frequency harmonics can be used as a gist of damage determination.

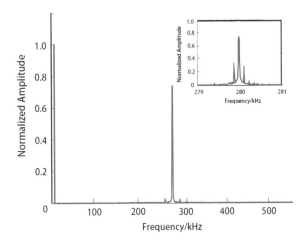

Fig. 4 Spectrum of the received signal under vibration

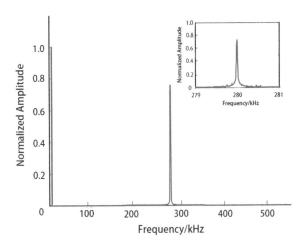

Fig. 5 Spectrum of the received signal without vibration

5 Discussion

The above test was repeated on a non-damaged aluminum plate and frequencies of input signals were changed correspondingly. It was found that no side-frequency harmonics appeared. Therefore, it can be proved that it is the crack damage that causes the signal to undergo the above modulation.

6 Conclusions

Through the interaction of high-frequency ultrasound wave and low-frequency vibration wave, structural cracks will generate the modulated side frequency, which is an obvious characterization for detection. Under certain conditions, the self vibration of bogie during running period can also be used as low-frequency excitation, which provides a method for crack monitoring without baseline data.

The nonlinearity introduced by material itself and test system will influence the results. How to minimize this influence and how to quantitative analysis cracks are worth further study and research.

Acknowledgements The paper is partial support by National Key R&D Program of China (2018YFB1201601-07) and Beijing Municipal Commission of Education Social Science Foundation (SM201810005002).

References

1. Guo ZQ et al (2015) Structural strength and modal analysis of a certain type of heavy vehicle frame. Appl Mech Mater 757:69–73 (in Chinese)
2. Ge L, Wang X, Jin C (2014) Numerical modeling of PZT-induced lamb wave-based crack detection in plate-like structures. Wave Motion 51(6):867–885
3. Tua PS, Quek ST, Wang Q (2005) Detection of cracks in cylindrical pipes and plates using piezo-actuated Lamb waves. Smart Mater Struct 14(6):1325
4. Rose Joseph L (2000) Ultrasonic waves in solid media, pp 1807–1808
5. Cawley P, Alleyne D (1996) The use of lamb waves for the long range inspection of large structures. Ultrasonics 34(2):287–290
6. Duffour P, Morbidini M, Cawley P (2006) Comparison between a type of vibro-acoustic modulation and damping measurement as NDT techniques. NDT & E Int 39(2):123–131
7. Singh AK et al (2017) A theoretical and numerical study on the mechanics of vibro-acoustic modulation. J Acoust Soc Am 141(4):2821–2831
8. Duffour P, Morbidini M, Cawley P (2006) A study of the vibro-acoustic modulation technique for the detection of cracks in metals. J Acoust Soc Am 199(3):1463
9. Pieczonka L et al (2015) Damage imaging in nonlinear vibro-acoustic modulation tests. Health Monit Struct Biol Syst
10. Parsons Z, Staszewski WJ (2006) Nonlinear acoustics with low-profile piezoceramic excitation for crack detection in metallic structures. Smart Mater Struct 15(4):1110–1118

11. Liu B, Luo Z, Gang T (2018) Influence of low-frequency parameter changes on nonlinear vibro-acoustic wave modulations used for crack detection. Struct Health Monit 17(2):218–226
12. Zhang Q (2009) Discussion on code for design of steel structures (GB50017—2003). Ind Constr 39(6):36–38

Research on Emergency Capacity Training System for Rail Transit Dispatchers

Qingmei Hu, Nan Lv, Lan Zhang and Yukun Gu

Abstract With the rapid development of rail transit, in order to solve the increasingly severe contradiction between supply and demand of rail transit traffic dispatchers and to improve the emergency response capability of traffic dispatchers, it is urgent to establish a complete train system for emergency response capability of rail transit traffic dispatchers. Based on the current situation and demand of Beijing subway dispatching command, this paper designed a training simulation system composed of two subsystems: scenario playback system and simulation training system, analyzed the functions of dispatcher operation environment simulation and instructor system simulation, and designed simulation software system architecture. As a comprehensive simulation system integrating multiple dispatching links, the system can simulate various emergencies and train dispatcher's emergency handling ability in an all-round way. It will give full play to the incomparable advantages of traditional methods in traffic dispatcher training and emergency drills.

Keywords Rail transit · Traffic scheduling · Simulation training system · Emergency response capability

1 Introduction

With the unconventional development of rail transit, the operation lines are continuously opened and the operation situation is becoming increasingly complicated. On the one hand, the passenger flow has increased dramatically, the interval between trains is getting shorter and shorter, the requirements of passengers are getting higher and higher, and it is more and more difficult for traffic controllers to deal with emergencies. On the other hand, the increasing number of new subway employees and their low safety quality and business skills pose potential risks to subway operation safety belts. As the maker and main implementer of the emergency response plan, the emergency response capability of traffic controllers will directly affect the

Q. Hu (✉) · N. Lv · L. Zhang · Y. Gu
Beijing Mass Transit Railway Operation Corp. Ltd., Beijing, China
e-mail: 13944206@qq.com

operation safety of the whole line, affect the speed of restoring normal operation order, and the magnitude of adverse impact on passengers. Therefore, according to the job characteristics of traffic controllers, combined with the emergency handling methods of rail transit emergencies, designing the corresponding rail transit traffic controller simulation training system is an important way to solve the current demand contradiction.

In terms of training subway employees, foreign countries and rail transit developed areas are ahead of domestic ones. The subway operation departments in Europe, America, Japan, South Korea, Taiwan, and Hong Kong are equipped with advanced computer simulation training systems to provide practical training required by professional training for subway employees such as drivers [1], but the training system for dispatching personnel is still under study.

2 Traffic Controller Work Analysis

The duty of the traffic controller is to be responsible for daily dispatching and emergency command in case of emergency, and to cooperate with the power dispatching and environmental dispatching to jointly complete the operation organization work, ensure the safety of the team operation, command and coordinate the operation of each post in the traffic, organize and implement various traffic work plans and construction operations, and ensure the normal operation of the traffic system. The specific work includes:

- Monitor the train operation, supervise and maintain the order of the main train operation, and ensure the safe and punctual operation of the train.
- To convey the instructions of the superior about the operation work, timely issue scheduling orders, layout, inspection and implementation of the driving work plan, and to ensure the smooth progress of the driving work.
- Organize the operation of commissioning vehicles and engineering vehicles, reasonably arrange construction operations, and supervise the construction operations and personnel safety.
- Reasonable handling of emergencies in operation, recording of the handling process, analysis of the degree of impact, notification of failures and delays, timely adjustment of train services, and resumption of normal operation as soon as possible.
- Monitor the operation of all kinds of driving equipment, and make fault records and announcements.

From the job duties of traffic controllers, it can be seen that one of the job duties of traffic controllers is to deal with emergencies in operation, adopt effective train adjustment strategies, and resume normal operation as soon as possible. However, the quality of emergency handling is closely related to the dispatcher's emergency handling ability, which is an important problem that needs to be solved by the dispatcher's emergency handling ability training system.

3 Functional Analysis and Design of Training Simulation System

3.1 Functional Module Design of Training Simulation System

Dispatching simulation training system is the core training system for traffic dispatchers in Beijing Metro Dispatching Command Center. It undertakes the tasks of operation training, emergency drills, case explanation, and business training for traffic dispatchers in the center. The simulation training system for dispatching consists of two subsystems: scenario playback system and simulation training system.

(1) *The scenario playback system*: It has synchronous playback functions such as historical operation information, dispatching operation commands, alarm events, and dispatching telephone recording. Through the real-scene playback function, meet the needs of emergency disposal analysis, through the annotation, sketch and parameter save function, realize the emergency disposal case analysis courseware and make dispatcher emergency disposal case library needs.

(2) *The simulation training system*: It uses the data of typical lines (e.g., Beijing Line 1 is selected as the typical line) to realize the simulation training function. This system consists of two subsystems: teacher machine and student machine. Through the teacher workstation, faults of vehicles and equipment injected into the site can be simulated. Students can take corresponding measures according to ATS and large-screen real-life display, so as to train students' emergency handling ability in case of faults. At the same time, the system can save the disposal recording and ATS playback, and through the real-scene playback system playback, realize the training effect analysis, and evaluation.

3.2 Dispatcher Operation Environment Simulation

3.2.1 Simulation of Train Dispatching System

As a platform for operation and command of train dispatchers, subway train dispatching system is the management center of subway operation and production activities, and is the command center for organizing train operation, ensuring operation, and carrying out real-time monitoring and adjustment.

According to the characteristics of the ATC system [2, 3], the contents of the train dispatching simulation training system should include the following ATS simulation functions: drawing of the operation diagram, recording of fault status, automatic routing, train speed adjustment, etc. At the same time, the system also needs to realize ATO and ATP function simulation of the car. The automatic driving function of the train is realized, and the protection function under the conditions of overspeed,

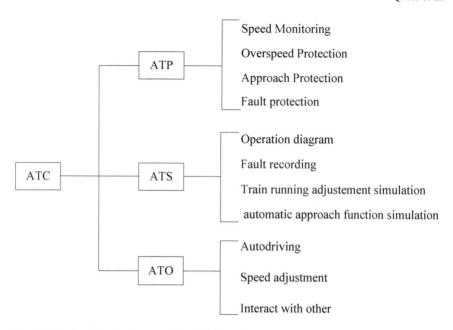

Fig. 1 Main functions implemented by ATC simulation system

approach, fault, and the like is realized. The main functional block diagram of ATC simulation system is as follows (Fig. 1):

3.2.2 Simulation of Traffic Communication System

In addition to realizing ATC function simulation, the simulation training system also needs to establish a communication simulation system for traffic controllers to simulate the communication between traffic controllers and drivers and station controllers, so as to simulate a realistic working environment for traffic controllers.

3.3 Instructor System Functions

As a simulation training system, besides completing the dispatcher's operation environment simulation, the system will also realize the management, monitoring, fault and emergency setting, training data storage, etc., of the simulation training process. Most of these tasks are completed by the teacher system responsible for managing the simulation system. The work accomplished by the faculty system includes:

(1) *Setting or modifying training courses*: setting simulation running time, running initial conditions, simulation training period, and the role of trainees; Specify the conditions for various failures or emergencies;
(2) *Prompting or guiding trainees*: informing trainees of matters needing attention in the training process through the communication system;
(3) *Observe the operation of the trainees*: observe and analyze the operation of the trainees through the monitoring of station lines, actual operation diagrams, trainees' screens, etc.
(4) *Intervening the operation process of the trainees*: testing the trainees' ability to handle faults and emergency incidents by setting or canceling various faults or emergencies in real time;
(5) *Data maintenance and management*: complete the data update of lines, signals, switches, etc., and the revision of course management, etc. Complete statistics and management of training records and data; User information management and authority allocation; Run log data management;
(6) *State detection and maintenance of the system*: detect whether the state and function of each part of the system are normal, record the operation log, and obtain relevant technical support in time through remote maintenance function (Fig. 2).

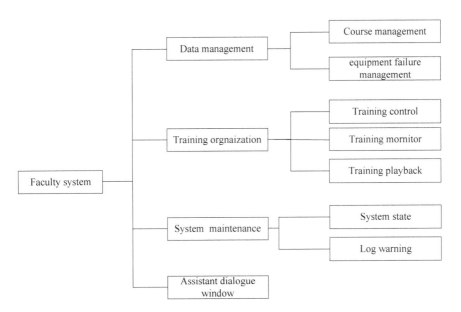

Fig. 2 Functional diagram of instructor system

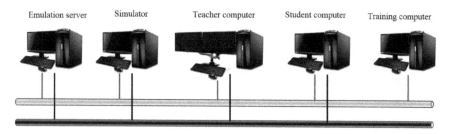

Fig. 3 Structural diagram of simulation training system

3.4 Software Design of Training Simulation System

The simulation training system for dispatching is a distributed system, which can adopt C/S structure and TCP/IP protocol, and the application software adopts modular structure. The system is to be deployed in Xiaoying control center, including a set of simulation servers, a set of simulation simulators, a teacher machine and several student machines. The following diagram shows the overall system structure of the system (Fig. 3).

(1) *Simulation server software*: The simulation server software completes the related functions of the center server and station server in the actual operation equipment, and is responsible for the whole line function processing, and simulates and realizes the remote automatic control of signal equipment and trains.
(2) *Simulation simulator software*: Simulate the train operation on the virtual line of the simulator and the corresponding interlocking, ATC and external interface system data to generate a simulated ATS operation environment.
(3) *Cadet machine software*: Cadet machine software realizes the relevant representation and control functions of the central traffic controller. The ATS system can be used to train the trainees on the trainees' computers. Students can also take corresponding measures according to ATS real-life display on the trainees' aircrafts, so as to train the trainees' emergency handling ability in case of failure.
(4) *Teacher machine software*: The teacher machine software has all the functions of the trainee machine, and has the operation function of the station local workstation in the station control state, the fault injection function, the fault pre-configuration function (pre-setting the fault, executing the preset fault injection at the time point), the temporary control function in the case of the fault, and the automatic auxiliary function of simulation training.
(5) *Scenario playback software*: The scenario playback software realizes synchronous playback of historical operation information, dispatching operation commands, alarm events and dispatching telephone recording, and also has the function of sketching and annotating.

4 Conclusion

At present, there are relatively a few researches on the training system of urban rail transit dispatching in China, and most of them are only carried out for a certain link. There is no complete simulation training system including all relevant subsystems of dispatching. As a comprehensive simulation system integrating multiple dispatching links, the emergency capability training system for rail transit dispatchers will play an incomparable advantage over traditional methods in dispatcher training, preview of various schemes and other uses.

The simulation training system can create a realistic working environment for trainers. Through the corresponding curriculum setting, various operating conditions can be reproduced, and the trainees can repeatedly operate and practice, so that the working skills of new employees can be improved more intuitively, effectively, comprehensively, and rapidly, and at the same time, standardized retraining can be carried out on the operation of old employees. The subway dispatching simulation training system can not only simulate various dispatching operations under normal conditions to achieve the purpose of standardized training, but also set various faults in the operation process, thus strengthening the emergency response training of dispatchers. In the simulation environment, special operations and emergency situations that are not easy to encounter can also be simulated according to needs, and the emergency response capability of dispatchers in various environments can be strengthened in a targeted manner.

In short, the design of a complete train system for rail traffic dispatchers' emergency response capability can play an important role in alleviating the increasingly severe contradiction between supply and demand of rail traffic dispatchers, greatly improving the emergency response capability of traffic dispatchers, minimizing the impact of emergencies on operations, and reducing operational safety risks.

References

1. Shuguang L, Lei Q, Jun Y (2009) Research on urban rail transit dispatching and command simulation system. Urban Rail Transit 22(5):50–53 (in Chinese)
2. Hua S, Junqi D (2007) Component design and implementation of automatic train monitoring simulation system for urban rail transit. Urban Rail Transit Res 10(2):41–43 (in Chinese)
3. Xu M, Zhibin J, Ruihua X (2004) Research on simulation system of train delay in rail transit. Res Urban Rail Transit 7(6):35–38 (in Chinese)

A Study on Detection Technology of Rail Transit Vehicle Wheel Web Based on Lamb Wave

Ge Wang, Guoqiang Cai and Xunhe Yin

Abstract In recent years, the rapid development of rail transit and train speed has improved the economic development, but also brought safety problems. Wheel safety is the foundation of the safety of rail transit vehicles. If there is a problem with the wheels, the consequences are unimaginable. However, there is little research on the wheel web detection method of the vehicle. Through investigating the wheel web crack fault, the occurrence of spokes crack of the vehicle may cause the wheels to crack and cause the derailment of the train, with extremely serious consequences. At present, there are many researches on the detection methods of wheel tread and wheel rim. There are few methods to detect wheel spokes, and they mainly rely on manual observation and sound listening, which is easy to produce missing detection. Therefore, in this paper, a damage detection method based on Lamb wave was proposed to detect, identify and locate the damage to verify the effectiveness of the method.

Keywords Rail transit · Wheel web · Lamb wave · Damage detection

1 Introduction

In recent years, the high-speed trains of harmony and fuxing have been gradually put on line, and the railway trains have been speeded up continuously. Many cities are competing for the development of urban rail transit. The rise of rail transit will facilitate people's life and bring huge social and economic benefits. In 1842, British railway engineer Rankine observed that there was a danger of stress concentration on the axle, which was the first research on the safety of rail transit vehicles. For nearly two hundred years, domestic and foreign experts and scholars have carried out extensive research on the structural safety of each component of railway vehicle. In

G. Wang (✉)
School of Electronic and Information Engineering, Beijing Jiaotong University, Beijing, China
e-mail: 18125131@bjtu.edu.cn

G. Cai · X. Yin
Beijing Key Lab of Traffic Engineering, Beijing University of Technology, Beijing, China

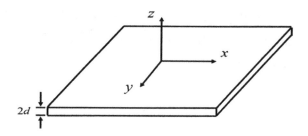

Fig. 1 Infinite plate

train operation, wheel safety is the basis of rail transit vehicle safety, and its running state may be directly related to vehicle speed and train safety. Wheel pairs include treads, rims, rims, hubs, spokes and other parts. Common faults of wheel pairs include wheel tread circumferential wear, thin rims, rim wear, tread abrasion, peeling and partial depression, wheel rims or tread Hazards of defects, cracks in the holes of the spokes exceeding the limit. These failures of the wheels will cause vehicle safety accidents, bring huge losses to railway transportation, and affect people's property and life safe. Common wheel failure mainly includes wheel tread abrasion, stripping and excessive circumferential abrasion; wheel rim damage, wheel rim crack, brake disk crack, etc., as shown in Fig. 1, are the common damage diagram of each part of the wheel of a rail vehicle with spoke hole. After the acceleration of the train, the damage and faults of various parts of the wheels gradually increase, which brings great safety risks to the driving [1].

At present, the main methods of wheel flaw detection are original observation and magnetic particle flaw detection. The traditional method for detecting wheel faults is to observe with the naked eye and tap the wheel with a hammer to hear the sound. This method is simple to operate and mainly targets the more obvious wheel faults. The difficulty of magnetic particle flaw detection lies in surface derusting. The cleaning degree of the surface of flaw detection site affects the quality of flaw detection. However, too low quality of flaw detection is not high, and too high quality will cause waste [2, 3].

In view of the shortcomings of the current wheel flaw detection methods, in order to strengthen the detection of spokes damage of rail vehicles, a structural crack damage identification method based on Lamb wave is proposed in this paper to detect the crack damage of spokes of rail vehicles.

2 Background

2.1 Structural Health Monitoring Technology Based on Lamb Wave

Damage identification technology based on Lamb wave is one of the emerging means of structural health monitoring technology. The characteristics of Lamb wave are its long propagation distance, small attenuation and sensitivity to minor damage, attract aviation, rail transit and other fields to study its feasibility in flaw detection [4]. Structural health monitoring technology is regarded as the key technology to ensure material reliability and reduce maintenance cost. It collects information by means of embedded or surface-pasted sensors and realizes online monitoring of the "healthy" state of the structure by analyzing signals to perceive internal defects and damage of the structure [5]. Damage detection based on Lamb wave is a research subject with extensive engineering application background. At present, the application of Lamb wave flaw detection technology excited by the piezoelectric probe is the main nondestructive flaw detection method for metal sheet [6].

2.2 Advantages of Lamb Wave Flaw Detection

Since Worlton first applied Lamb wave to practical detection, Lamb wave technology-based nondestructive detection methods for plate structures have received more and more attention [7]. Compared with traditional stress wave, Lamb wave has the following advantages in flaw detection:

(1) Multi-mode features
   Lamb wave signal contains much more information than traditional stress wave and can generate a large number of data points in a limited frequency range.
(2) Long propagation distance
   The propagation distance of Lamb wave is up to several hundred millimeters, with less attenuation, which can be used for damage detection in a large area.
(3) Sensitive to injury
   In the process of Lamb wave propagation, the energy damage is less, and it is very sensitive to the slight damage of the structure, which can effectively measure the damage [7].
(4) Simple operation
   When using Lamb wave for wheel flaw detection, it is convenient and fast to operate, and there is no need to remove paint, rust and other operations on the structure of the wheel, so as to protect the structure of the wheel and avoid secondary damage.

3 Propagation Characteristics and Excitation Optimization of Lamb Wave

3.1 Dispersion Characteristics of Lamb Wave

By arranging a proper number of piezoelectric sensors on the surface of wheel spokes, the Lamb wave damage detection method is used to collect and process the response signals of the tested structure, so as to realize the structural state monitoring. When the Lamb wave is excited in the spoke plate, complex vibration will be generated by particles in the plate, and its mode is related to the thickness of the plate and the excitation frequency. Therefore, Lamb wave will exhibit multi-mode characteristics in the propagation process of the tested structure, which is called the dispersion phenomenon [8, 9].

The propagation mode equation of Lamb wave in the plate is solved. As shown in Fig. 1, if the thickness of the thin plate is 2D and the width of the thin plate is infinite [10], there are boundary conditions: $z = \pm d$, $\delta ZX = 0$, $\delta ZY = 0$. The frequency equation of Rayleigh–Lamb plates with isotropic plate structure can be obtained [11], as shown in Eq. (1):

$$\frac{\tan(qd)}{\tan(pd)} + \left[\frac{4k^2 pq}{(q^2 - k^2)^2}\right]^{\pm 1} = 0 \tag{1}$$

where p and q can be obtained from Eq. (2):

$$p^2 = \frac{\omega^2}{c_L^2} - k^2; \quad q^2 = \frac{\omega^2}{c_T^2} - k^2 \tag{2}$$

where

c_L p-wave velocity
c_T s-wave velocity

In the above equation, if the power exponent is $+1$, then Eq. (1) is the symmetric mode equation of Lamb wave. If the power exponent is -1, then Eq. (1) is the antisymmetric mode equation of Lamb wave.

3.2 Incentive Optimization

Compared with traditional longitudinal wave and shear wave, Lamb wave is faster and more efficient in detecting thin plate and has a better effect in nondestructive testing of large area plate structure. However, the Lamb wave itself has frequency dispersion and multi-mode phenomenon, which is relatively complex [12]. The thickness of the

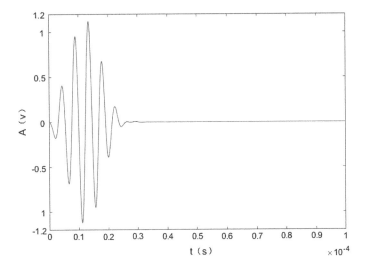

Fig. 2 Excitation signal

tested plate structure and the selected frequency of the excitation signal affect the mode of Lamb wave and its propagation speed, and the propagation mode of Lamb wave has a certain influence on the sensitivity of damage detection. Therefore, it is necessary to weaken the influence of dispersion and multi-mode characteristics in order to make better use of Lamb wave to realize NDT.

In the experiment, sinusoidal wave modulated by Hanning window was selected as the excitation signal of Lamb wave narrowband [13]. The expression of the excitation waveform is shown in Eq. (3):

$$y = \frac{1}{2}\left[1 - \cos(\frac{2\pi ft}{n})\right]\sin(2\pi ft) \quad (3)$$

Its waveform is shown in Fig. 2.

4 Wheel Spoke Crack Damage Detection Test

4.1 Test Device

In the process of detecting spokes damage, the main test equipment used includes computer system, signal collector, piezoelectric sensor, sensor connection box, connection wire, super glue, etc.

Table 1 Material parameters of wheel spoke plate

E (GPa)	M	ρ (kg/m3)	h (mm)	c_L (m/s)	c_T (m/s)
210	0.3	7800	19	6020	3218

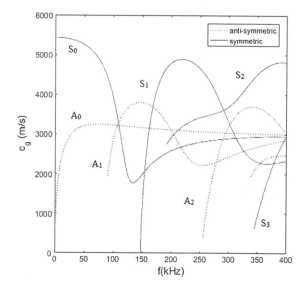

Fig. 3 Material dispersion curve of wheel spokes

4.2 Excitation Frequency

As shown in Table 1, the thickness of wheel spoke plate is about 19 mm. Therefore, the scatter curve diagram of frequency propagation of Lamb wave in the wheel spoke plate can be obtained, as shown in Fig. 3.

As shown in Fig. 3, when the excitation frequency is less than 90 kHz, the anti-symmetric mode is only A_0. It is found through experiments that the signal excited at this frequency will lose a part of the signal at the receiving end, so the excitation frequency needs to be increased. In this paper, 200, 220, 240, 260, 280, and 300 kHz are selected for signal excitation. The excitation signals at these frequencies are shown in Fig. 4.

As can be seen from Fig. 4, 260 kHz is selected as the excitation frequency to facilitate subsequent signal analysis.

4.3 Wheel Spokes Damage Detection Experiment

In the experiment, a crack damage detection system was used to detect the wheel with crack damage. Wheel damage and sensor layout are shown in Fig. 5.

The excitation frequency of 260 kHz was adopted, and the path 2–4 (2 sending and 4 receiving) was measured at the same time. The received signal is shown in Fig. 6.

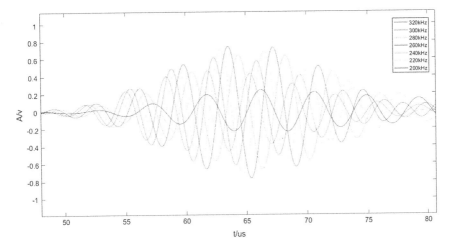

Fig. 4 Direct wave signals at different frequencies

Fig. 5 Sensor layout of wheel spokes

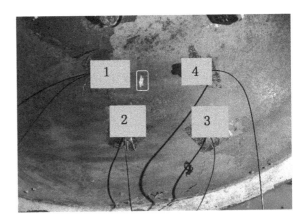

According to the layout diagram of the sensor, the distance between the excitation sensor 2 and the receiving sensor 4 is 144 mm. The A_0 mode wave propagation speed is about 2947.4 m/s through calculation, and the distance between the two sensors and the damage position is about 193 mm, which is 3 mm different from the actual distance.

5 Conclusion

This paper mainly studies the feasibility of the damage detection method based on Lamb wave. Based on the experimental results, it can be concluded that the method is effective. The propagation characteristics of Lamb wave are studied for the spokes

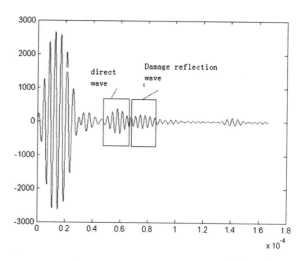

Fig. 6 Signal received by 4 in path 2–4

plate of the car. On this basis, Lamb wave is applied to the damage detection of the spokes plate of the car. This method is mainly concerned with the selection of excitation frequency and sensor layout in the test device and the feature extraction of signals to complete damage identification.

Argument: It is feasible to apply Lamb wave in rail transit damage detection.

Prove: Damage defect can change wave propagation in spoke plate, and there is obvious wave packet reaction damage detection.

Verification: The propagation of Lamb wave signal in wheel spokes was obtained through the test, and the damage wave packet signal was obtained.

References

1. yongjiang C (2015) Application of ultrasonic array flaw detection technology in online detection of locomotive wheels. Railw Locomot Bullet Train (10) (in Chinese)
2. Yuande S, Yongwei H, Mingzhao R et al (2012) Pulsed electromagnetic detection of cracks in spokes of railway locomotive and rolling stock. Railw Tech Superv 40(6):15–17 (in Chinese)
3. Liu X (2009) Cracking and flaw detection method of rolling steel wheel of locomotive. Locomot Electr Transm (1):66–69 (in Chinese)
4. Qiu J, Li F, Wang J (2017) Damage detection for high-speed train axle based on the propagation characteristics of guided waves. Struct Control Health Monit 24(3)
5. Wang Y, Xiaoqing D (2011) Track and test of wheel wear and vibration performance of CRH3 emu (1): 2010 annual track and test phase report Beijing: China academy of railway sciences, 11 (in Chinese)
6. Frirestone FA, Lying DS (1954) Methods and means for generating and utilizing vibration waves in plates, Patent, USA
7. Worlton DC (1957) Ultrasonic testing with lamb waves. Nondestruct Test 15(4):218–222
8. Shan S (2015) Research on damage monitoring method of boom of concrete pump truck based on ultrasonic guided wave (in Chinese)

9. Golub MV, Eremin AA, Shpak AN, Lammering R (2019) Lamb wave booms, (2). Bonded rectangular block. Appl Acoust 155
10. Rose Joseph L, He C et al (2004) Ultrasonic waves in solids. Science press, pp 82–92 (in Chinese)
11. Okabe Y, Fujibayashi K, Shimazaki M et al (2010) The delamination detection in composite laminates using dispersion change -based on mode conversion of lamb waves. Journal of J Smart Mater Struct 19(11):115013
12. Liu Z, Zhong X, Dong T, He C, Wu B (2017) The delamination detection in composite plates by synthesizing time—reversed Lamb waves and a modified damage imaging algorithm-based on RAPID. Struct Control Health Monit 24(5)
13. Qin W, Ying L, Ziping W et al (2011) Selection of the best excitation waveform for lamb wave driver. Piezoelectric Acoustrooptic (06):863–866 (in Chinese)

CPSIA information can be obtained
at www.ICGtesting.com
Printed in the USA
LVHW080454060421
683530LV00001B/2